T0319248

Genomic and Precision Medicine

Genomic and Precision Medicine

Infectious and Inflammatory Disease

Series Editors

Geoffrey S. Ginsburg
Duke Center for Applied Genomics & Precision Medicine,
Duke University School of Medicine, Durham, NC, United States

Huntington F. Willard
Director & Principal, Geisinger National Precision Health,
North Bethesda, Maryland, United States

Volume Editors

Ephraim L. Tsalik
Duke Center for Applied Genomics & Precision Medicine,
Duke University School of Medicine;
Emergency Department Service,
Durham VA Health Care System, United States

Christopher W. Woods
Professor of Medicine and Global Health,
Duke Center for Applied Genomics & Precision
Medicine, Duke University School of Medicine;
Chief, Infectious Disease Division,
Durham VA Health Care System, United States

Academic Press is an imprint of Elsevier
125 London Wall, London EC2Y 5AS, United Kingdom
525 B Street, Suite 1650, San Diego, CA 92101, United States
50 Hampshire Street, 5th Floor, Cambridge, MA 02139, United States
The Boulevard, Langford Lane, Kidlington, Oxford OX5 1GB, United Kingdom

Notices
Knowledge and best practice in this field are constantly changing. As new research and experience broaden our understanding, changes in research methods, professional practices, or medical treatment may become necessary.

Practitioners and researchers must always rely on their own experience and knowledge in evaluating and using any information, methods, compounds, or experiments described herein. In using such information or methods they should be mindful of their own safety and the safety of others, including parties for whom they have a professional responsibility.

To the fullest extent of the law, neither the Publisher nor the authors, contributors, or editors, assume any liability for any injury and/or damage to persons or property as a matter of products liability, negligence or otherwise, or from any use or operation of any methods, products, instructions, or ideas contained in the material herein.

Library of Congress Cataloging-in-Publication Data
A catalog record for this book is available from the Library of Congress

British Library Cataloguing-in-Publication Data
A catalogue record for this book is available from the British Library

ISBN 978-0-12-801496-7

For information on all Academic Press publications
visit our website at https://www.elsevier.com/books-and-journals

Publisher: Andre Gerhard Wolff
Acquisition Editor: Peter Linsley
Editorial Project Manager: Sam Young
Production Project Manager: Kiruthika Govindaraju
Cover Designer: Matthew Limbert

Typeset by SPi Global, India

Contents

5. Host-pathogen interactions

Scott D. Kobayashi and Frank R. DeLeo

6. Sepsis genomics and precision medicine

Hector R. Wong

7. Sepsis: Metabolomics and proteomics

Yong B. Tan and Raymond J. Langley

14. Genomics and precision medicine for malaria: A dream come true?

Desiree Williams and Karine G. Le Roch

15. The role of disease surveillance in precision public health

David L. Blazes and Scott F. Dowell

16. Multiple sclerosis

Athanasios Ploumakis, Howard L. Weiner and Nikolaos A. Patsopoulos

17. Systemic sclerosis

Sevdalina Lambova and Ulf Müller-Ladner

18. Leveraging genomics to uncover the genetic, environmental and age-related factors leading to asthma

Brian D. Modena, Ali Doroudchi, Parth Patel and Varshini Sathish

Contributors

Numbers in paraentheses indicate the pages on which the authors' contributions begin.

Sai Krishna Athuluri-Divakar (413), Liver Tumor Translational Research Program, Simmons Comprehensive Cancer Center, Division of Digestive and Liver Diseases, Department of Internal Medicine, University of Texas Southwestern Medical Center, Dallas, TX, United States

Peter J Barnes (383), National Heart and Lung Institute, Imperial College, London, United Kingdom

David L. Blazes (257), Surveillance and Epidemiology, Bill and Melinda Gates Foundation, Seattle, WA, United States

Eileen Tsai Chambers (401), Department of Pediatrics, Duke University, Durham, NC, United States

Danielle V. Clark (105), Austere environments Consortium for Enhanced Sepsis Outcomes (ACESO), The Henry M Jackson Foundation, Bethesda, MD, United States

Daisy D. Colón-López (141), Diagnostic Systems Division, United States Army Medical Research Institute of Infectious Diseases, Fort Detrick, Frederick, MD, United States

Frank R. DeLeo (61), Laboratory of Bacteriology, Rocky Mountain Laboratories, National Institute of Allergy and Infectious Diseases, National Institutes of Health, Hamilton, MT, United States

Ali Doroudchi (331), Division of Allergy, National Jewish Health, Denver, CO, United States

Scott F. Dowell (257), Surveillance and Epidemiology, Bill and Melinda Gates Foundation, Seattle, WA, United States

Ann Regina Falsey (117), University of Rochester; Rochester General Hospital, Rochester, NY, United States

Laura Filkins (35), Department of Pathology, University of Texas Southwestern, Dallas, TX, United States

Yujin Hoshida (413), Liver Tumor Translational Research Program, Simmons Comprehensive Cancer Center, Division of Digestive and Liver Diseases, Department of Internal Medicine, University of Texas Southwestern Medical Center, Dallas, TX, United States

Ming-Feng Hsueh (429), Duke Molecular Physiology Institute, Duke University School of Medicine, Durham, NC 27701, United States

Letitia D. Jones (167), Duke Human Vaccine Institute and Department of Pediatrics, Duke University Medical Center, Durham, NC, United States

Purvesh Khatri (25), Institute for Immunity, Transplantation, and Infection, Department of Medicine; Department of Medicine, Division of Biomedical Informatics Research, Stanford University, Stanford, CA, United States

Scott D. Kobayashi (61), Laboratory of Bacteriology, Rocky Mountain Laboratories, National Institute of Allergy and Infectious Diseases, National Institutes of Health, Hamilton, MT, United States

Jeffrey W. Koehler (141), Diagnostic Systems Division, United States Army Medical Research Institute of Infectious Diseases, Fort Detrick, Frederick, MD, United States

Virginia Byers Kraus (429), Department of Medicine; Duke Molecular Physiology Institute, Duke University School of Medicine, Durham, NC 27701, United States

Sevdalina Lambova (291), Department of Propaedeutics of Internal Diseases, Medical University, Plovdiv, Bulgaria

Raymond J. Langley (95), Department of Pharmacology, University of South Alabama College of Medicine, Mobile, AL, United States

Karine G. Le Roch (223), Department of Molecular, Cell and Systems Biology, University of California Riverside, Riverside, CA, United States

Sophie Limou (185), Institut de Transplantation Urologie-Néphrologie, CHU de Nantes, CRTI UMR1064, Inserm, Université de Nantes; Ecole Centrale de Nantes, Nantes, France; Frederick National Laboratory for Cancer Research, Frederick, MD, United States

Brian D. Modena (331), Division of Allergy, National Jewish Health, Denver, CO, United States

M. Anthony Moody (167), Duke Human Vaccine Institute and Department of Pediatrics, Duke University Medical Center, Durham, NC, United States

Ulf Müller-Ladner (291), Department of Internal Medicine and Rheumatology, Justus-Liebig University Giessen, Giessen; Department of Rheumatology and Clinical Immunology, Campus Kerckhoff, Bad Nauheim, Germany

Keyur Patel (155), University Health Network, University of Toronto, Toronto, ON, Canada

Parth Patel (331), Division of Allergy, National Jewish Health, Denver, CO, United States

Nikolaos A. Patsopoulos (267), Department of Neurology, Brigham and Women's Hospital, Ann Romney Center for Neurological Diseases, Harvard Medical School, Boston, MA, United States

Athanasios Ploumakis (267), Department of Neurology, Brigham and Women's Hospital, Ann Romney Center for Neurological Diseases, Harvard Medical School, Boston, MA, United States

Schlaberg Robert (35), Department of Pathology, University of Utah, Salt Lake City; ARUP Laboratories, Salt Lake City, UT; IDbyDNA Inc., San Francisco, CA, United States

Varshini Sathish (331), Division of Allergy, National Jewish Health, Denver, CO, United States

Kevin L. Schully (105), Austere environments Consortium for Enhanced Sepsis Outcomes (ACESO), The Henry M Jackson Foundation, Bethesda, MD, United States

Nicholas A Shackel (155), University of New South Wales, Sydney; Ingham Institute, Liverpool; Liverpool Hospital, Liverpool, NSW, Australia

Brian I. Shaw (401), Department of Surgery, Duke University, Durham, NC, United States

Christopher P. Stefan (141), Diagnostic Systems Division, United States Army Medical Research Institute of Infectious Diseases, Fort Detrick, Frederick, MD, United States

Neeraj K. Surana (445), Departments of Pediatrics, Molecular Genetics and Microbiology, and Immunology, Duke University, Durham, NC, United States

Timothy E. Sweeney (25), Inflammatix, Inc., Burlingame, CA, United States

Yong B. Tan (95), Department of Surgery, University of South Alabama College of Medicine, Mobile, AL, United States

Amelia B. Thompson (167), Duke Human Vaccine Institute and Department of Pediatrics, Duke University Medical Center, Durham, NC, United States

Ephraim L. Tsalik (1), Duke Center for Applied Genomics & Precision Medicine, Duke University School of Medicine; Emergency Department Service, Durham VA Health Care System, United States

Howard L. Weiner (267), Department of Neurology, Brigham and Women's Hospital, Ann Romney Center for Neurological Diseases, Harvard Medical School, Boston, MA, United States

C. William Wester (185), Vanderbilt University Medical Center (VUMC), Department of Medicine, Division of Infectious Diseases; Vanderbilt Institute for Global Health (VIGH), Nashville, TN; Harvard School of Public Health, Department of Immunology and Infectious Diseases, Boston, MA, United States

Huntington F. Willard (9), Director & Principal, Geisinger National Precision Health, North Bethesda, Maryland, United States

Desiree Williams (223), Department of Molecular, Cell and Systems Biology, University of California Riverside, Riverside, CA, United States

Cheryl A. Winkler (185), Frederick National Laboratory for Cancer Research, Frederick, MD, United States

Hector R. Wong (83), Division of Critical Care Medicine, Cincinnati Children's Hospital Medical Center and Cincinnati Children's Research Foundation; Department of Pediatrics, University of Cincinnati College of Medicine, Cincinnati, OH, United States

Christopher W. Woods (1), Professor of Medicine and Global Health, Duke Center for Applied Genomics & Precision Medicine, Duke University School of Medicine; Chief, Infectious Disease Division,Durham VA Health Care System, United States

Acknowledgments

We wish to express our appreciation and gratitude to our many colleagues, especially in the Duke scientific community, who have shared their knowledge and ideas about *Genomic and Precision Medicine* and who, by doing so, have continued to provide inspiration for this project. We particularly thank our editors at Academic Press/Elsevier, Peter Linsley and Sam Young, who have encouraged us to develop and evolve the concepts of *Genomic and Precision Medicine* as it pertains to infectious and inflammatory disease. We acknowledge and especially thank the approximately 40 authors of the chapters that comprise this volume. Through the years spent cultivating their expertise and the time invested in writing these chapters, these authors have provided fascinating insights into the current and future applications of genomic and precision medicine to a wide array of diseases.

We also thank our families for their patience and understanding for the many hours we spent creating this volume of *Genomic and Precision Medicine*, focused specifically on Infectious and Inflammatory Disease.

Chapter 1

Overview: Genomic and precision medicine for infectious and inflammatory disease

Christopher W. Woods[a,b], Ephraim L. Tsalik[c,d]
[a]Professor of Medicine and Global Health, Duke Center for Applied Genomics & Precision Medicine, Duke University School of Medicine, [b]Chief, Infectious Disease Division, Durham VA Health Care System, United States, [c]Emergency Department Service, Durham VA Health Care System, Durham, NC, United States, [d]Duke Center for Applied Genomics & Precision Medicine, Duke University School of Medicine, dEmergency Department Service,Durham VA Health Care System, United States

The era of systems biology has leveraged the emergence of genomics, transcriptomics, proteomics, metabolomics, lipidomics, and even microbiomics, opening up new possibilities for understanding the host response to disease, identifying therapeutic targets, and developing diagnostic tools that do not fit the traditional biomarker paradigm. Access to large electronic health databases with granular clinical metadata complements these biological data. To best use these complex, rich datasets, biostatisticians and computational biologists have revolutionized approaches by developing methods to reduce data dimensionality, match phenotype to molecular changes, compare groups and make predictions. Furthermore, the discovery of rare and common genetic variants that may contribute to disease susceptibility and response to therapy have encouraged a precision approach to diagnosis, treatment, and understanding of prognosis for individual patients.

In particular, this burgeoning of information has led to the hope of improved diagnostics with the ability to distinguish among diseases with similar clinical phenotypes representing heterogeneous etiologies or pathophysiological processes. Similarly, these data may hold the key to improved prognostics to tailor treatment and assign clinical resources more appropriately. This is especially important for infectious and inflammatory disease states. Various statistical algorithms have been utilized to construct these disease "classifiers," including sparse factor modeling, Bayesian constructions of the elastic net, sparse principal component analysis, and the molecular distance to health [1]. When derived on an appropriate patient population and then tested in new cohorts, these classifiers demonstrate improved performance when compared to the single analyte biomarkers that are available. In particular, unbiased approaches that utilize machine

Genomic and Precision Medicine. https://doi.org/10.1016/B978-0-12-801496-7.00001-0

learning to identify candidate biomarkers have improved classifier sen- sitivity and specificity. Unbiased approaches that utilize machine learning to identify candidate biomarkers have improved classifier sensitivity and specificity.

Particular attention should be paid to study design for the development of classifiers in both infectious and inflammatory diseases. Previously, many classifiers have been generated relative to healthy controls. Furthermore, patient cohorts for biomarker development often only include patients with phenotypic extremes, such as confirmed microbial etiology and clearly uninfected patients or classical presentations of inflammatory disease states such as lupus nephritis. These clearly defined populations facilitate biomarker development, but bias analysis by only capturing two ends of the clinical disease spectrum. Selection of patients on this basis is also constrained to patient populations in which pathogen-based testing is successful and may exclude some patient populations. Also, the use of "omics" has the potential to generate a large number of biomarkers from a small patient cohort, which may not extend to a broader population. While being scientifically useful, these signatures will need to be refined and tested in broader, more relevant patient populations to be clinically applied.

Defining clinical challenges for infectious diseases

Precision medicine approaches are relevant for most infectious diseases. A classic pharmacogenomic example is the use of the HLA-B*5701 genotype associated with hypersensitivity to abacavir [2]. However, many of the more vexing issues of infectious diseases require acute disease management decisions. This impacts the ultimate design of precision medicine tools. Some relevant examples are listed below.

Sepsis

Severe infections can lead to the physiological state known as sepsis, which despite advanced supportive care measures, still carries a mortality of 17–26% [3]. The management of sepsis remains challenging due to its heterogeneous nature, including factors such as the causative pathogen, site of infection, clinical management, and many other identified and unidentified variables interacting to define clinical severity and outcomes. Furthermore, clinical trials of sepsis have been plagued by failures, with diversity in host response likely playing a key role. This highly variable mix of patient, pathogen, and clinical factors means that sepsis is not a single disease. Yet, sepsis treatment is largely uniform and far from personalized. This is likely the major reason that sepsis clinical trials research is considered the "graveyard" for pharmaceutical and biotechnology industries. However, the use of omics-derived classifiers to allow early definition of pathogen class, host factors, and risk groups might begin to define sepsis sub-categories [4,5]. Doing so can aid in new sepsis clinical trials research and can also help re-evaluate failed studies that could have been effective for certain sepsis sub-types. The ultimate goal is to personalize sepsis diagnosis and treatment, first in clinical research and then in clinical care. A similar approach to sepsis prognosis may lead to an early awareness of risk to progression to severe sepsis and death which could prioritize patients to a more aggressive treatment and monitoring arm [6].

Antimicrobial resistance and stewardship

With the increase in antibacterial resistance and a decreasing antibacterial pipeline, there is a need for coordinated efforts to promote appropriate use of antibacterial agents. Such "stewardship" encourages the appropriate use of antimicrobials by promoting the selection of the optimal drug regimen. In particular, a reliable bacterial versus viral host response diagnostic could prevent millions of unnecessary prescriptions for antibiotics. Precision medicine can help solve the crisis of antimicrobial resistance by changing the way antibacterial agents are prescribed and developed. Ultimately, important tools for combatting resistance are rapid, inexpensive, accurate, point of care diagnostics to guide antimicrobial decision making. Such early disease diagnostics not only can help reduce antibiotic overuse, but also encourage clinicians to identify other causes for the inflammatory response when infection is absent.

Clinical trial efficiency

An additional benefit of precision medicine for bacterial infection is that improved diagnostics will help drive new drug development by facilitating clinical trials for new antimicrobials, particularly agents to be reserved for resistant pathogens or agents that are intended for use only against a limited range of pathogens. Without the benefit of rapid, point of care diagnostics, clinical trials for bacterial infections generally enroll patients with relevant clinical symptoms, but only a small percentage of those enrolled patients are infected with the target pathogen (e.g., a particular species or multi-drug resistant pathogens). Targeted therapy will support new business models for antibacterial agents, focusing treatments for the right patients at the right time. For some clinical trials, rapid diagnostics can decrease the cost of a clinical trial, which in turn helps industry maintain research and development for new agents. Precision medicine could also harmonize regulatory guidelines by increasing the comparability of patient populations.

Pre-symptomatic disease detection

The genomic revolution has also provided an opportunity to harness the host response to assess the earliest responses to pathogen exposure. Through the use of novel study designs such as human challenge studies with unattenuated live virus, clinical and molecular data can be mined to derive signatures of early signs of infection, even before the onset of symptoms [7,8]. These signatures can be validated in real life models of disease transmission through the use of index:cluster studies. Specifically, viral-exposed subjects can be monitored for emergent disease as demonstrated by the presence of pre-symptomatic signatures which would allow for early isolation and treatment when indicated. One can envision the screening of concentrated groups of people to exonerate them from illness before high-value deployment such as among military personnel.

Vaccines

Vaccine development is limited by reliance on animal models that may not always predict human responses. Overall, modeling age-specific human immunity outside the body, coupled with big data approaches, could help developers formulate vaccines for distinct populations. Research teams leading Precision Vaccines Programs are learning how vaccines work in different populations and working to develop new versions that optimally target the most vulnerable. To speed vaccine development, model systems have been developed using human monocytes, which are important contributors to innate immune responses to vaccines antigens or adjuvants. Adjuvant-type vaccines have been shown to turn up or down a different set of biological pathways at different times of life. These differential responses provide clues to how effective a vaccine may be in preventing infections in specific age groups. Further, investigators have shown that proteins expressed by adult or newborn cell types could function as potential predictors of adverse effects such as swelling or fever—leading to new ways to create smarter, more tolerable formulations.

Precision population health

Although not directly related to genomic medicine, the emergence of big data techniques combined with geographic information systems allows for refined approaches to identify vulnerable or at-risk populations. These precision public health approaches allow for more efficient and equitable distribution of resources, particularly in resource-limited settings.

Defining challenges for inflammatory diseases

Historically, inflammatory diseases were synonymous with rheumatic diseases. These syndromes were highly variable, even among patients who seemingly had the same condition such as systemic lupus erythematosus. Over time, diagnostic criteria were developed to allow patients to be assigned to specific disease states. For example, systemic lupus erythematosus has gone through a series of diagnostic criteria, being updated periodically to reflect new tests and new understanding. The most recently issued diagnostic criteria were offered in 2012 by the Systemic Lupus International Collaborating Clinics (SLICC) [9]. Those criteria require the presence of at least 4 of 17 components, including at least 1 of 11 clinical criteria and 1 of 6 immunologic criteria, or that the patient has biopsy-proven SLE nephritis in conjunction with antinuclear antibodies or anti-double-stranded DNA antibodies. This approach allows for the standardization of clinical treatment as well as research in an otherwise heterogeneous disease. However, this standardization also neutralizes the differences between individuals and can hinder personalized care for patients with lupus. Two patients with SLE may not have any overlapping signs, symptoms, or laboratory abnormalities yet are treated as having the same disease, particularly in the context of research.

Similar heterogeneity is evident across the spectrum of rheumatic diseases. For example, juvenile idiopathic arthritis encompasses a multitude of conditions. A scheme offered by the International League of Associations for Rheumatology includes systemic arthritis, polyarthritis, oligoarthritis (persistent and extended), enthesitis-related arthritis, and psoriatic arthritis [10]. However, children may meet criteria for more than one subclass, may change from one class to another over time, and can be further stratified by the presence of other conditions such as fibromyalgia [11].

Beyond these diseases, which classically fall under the rheumatology umbrella, we are learning that many other disease states have an inflammatory component as an intrinsic aspect of their pathology. For example, coronary artery disease, cancer, osteoarthritis, and others have been shown to have inflammatory mediators that are central to pathogenesis [12–14]. This has given rise to a number of immunomodulatory strategies with variable results so far.

Our recognition of the importance of inflammation in a large diversity of diseases creates both challenges and opportunities. The challenges inherent in this scenario stem largely from the heterogeneity among individuals and even within individuals over time. This complicates efforts to diagnose, treat, and predict outcomes. This heterogeneity, which is not adequately accounted for in current algorithms, runs counter to the goals of precision medicine. However, exciting efforts are underway to embrace this heterogeneity, allowing for more personalized understanding of disease with direct bearing on diagnosis, treatment, and prognosis. Models enabling precision medicine in rheumatic disease may incorporate demographic information (gender, race, ethnicity, age), molecular data (genetics, transcriptomics, functional immunology), and disease specific variables (duration, severity, activity, historical response to treatment). Once defined, these models can be applied to disease prediction, prevention, treatment personalization, and improved patient participation.

The tools needed to enable these advances in precision medicine are becoming more prevalent and robust. They stem in part from large patient data repositories such as the Childhood Arthritis and Rheumatology Research Alliance (CARRA). CARRA has greater than 50 enrollment sites throughout North America and Puerto Rico that enroll and gather information about children with JIA, SLE, dermatomyositis with additional rheumatic diseases coming. Similar registries have been created for lupus and rheumatoid arthritis (Corrona) among others. Whereas these registries initially focused on patient-reported and routine clinical measures, there is a nascent movement toward more comprehensive biological phenotyping such as genomics, transcriptomics, proteomics, and immune profiling. One innovative example was an effort by DxTerity Diagnostics, which used social media to reach thousands of patients with SLE in the LIFT study. Upon completing an online consent form, subjects were mailed a kit to collect a blood sample, which was then used to generate transcriptomic data. Bypassing the need for enrollment in clinical environments substantially reduced costs and time. Results from this study may help identify SLE subtypes that cluster together based on gene expression, which may itself define distinct disease states, link biomarkers to therapy, and improve prognosis of disease progression.

These new directions that aggregate clinical data from thousands of patients combined with systems biology measurements create hope and promise for a new era in precision medicine. They come at a time when the number of immunomodulatory agents, particularly biologics, continues to escalate. Yet clinicians have very little to guide their decisions regarding what drug to start, when to switch, and how long to treat.

Summary

The chapters presented in this volume will expand on these themes. The authors will explore various infectious and rheumatic diseases as well as conditions that are increasingly recognized as inflammatory in nature. Challenges inherent to the diagnosis and management of these diseases will be reviewed as well as strategies to address shortcomings with a vision for how more precise care may be delivered to the individual, but also for improvement of population health.

References

[1] Yang WE, Woods CW, Tsalik EL. Host-based diagnostics for detection and prognosis of infectious diseases. In: Sails A, Tang Y-W, editors. Methods in microbiology. vol. 42. Elsevier Ltd; 2015. p. 465–500.

[2] Mallal S, Phillips E, Carosi G, et al. HLA-B*5701 screening for hypersensitivity to abacavir. N Engl J Med 2008;358(6):568–79.

[3] Fleischmann C, Scherag A, Adhikari NK, et al. Assessment of global incidence and mortality of hospital-treated sepsis. Current estimates and limitations. Am J Respir Crit Care Med 2016;193(3):259–72.

[4] Tsalik EL, Henao R, Nichols M, et al. Host gene expression classifiers diagnose acute respiratory illness etiology. Sci Transl Med 2016;8(322):322ra311.

[5] Sweeney TE, Azad TD, Donato M, et al. Unsupervised analysis of transcriptomics in bacterial sepsis across multiple datasets reveals three robust clusters. Crit Care Med 2018;46:915–25.

[6] Sweeney TE, Perumal TM, Henao R, et al. A community approach to mortality prediction in sepsis via gene expression analysis. Nat Commun 2018;9(1):694.

[7] McClain MT, Nicholson BP, Park LP, et al. A genomic signature of influenza infection shows potential for presymptomatic detection, guiding early therapy, and monitoring clinical responses. Open Forum Infect Dis 2016;3(1):ofw007.

[8] Woods CW, McClain MT, Chen M, et al. A host transcriptional signature for presymptomatic detection of infection in humans exposed to influenza H1N1 or H3N2. PLoS One 2013;8(1):e52198.

[9] Petri M, Orbai AM, Alarcon GS, et al. Derivation and validation of the systemic lupus international collaborating clinics classification criteria for systemic lupus erythematosus. Arthritis Rheum 2012;64(8):2677–86.

[10] Petty RE, Southwood TR, Baum J, et al. Revision of the proposed classification criteria for juvenile idiopathic arthritis: Durban, 1997. J Rheumatol 1998;25(10):1991–4.

[11] Nordal E, Zak M, Aalto K, et al. Ongoing disease activity and changing categories in a long-term nordic cohort study of juvenile idiopathic arthritis. Arthritis Rheum 2011;63(9):2809–18.

[12] Libby P. Inflammation in atherosclerosis. Nature 2002;420(6917):868–74.

[13] Kroemer G, Senovilla L, Galluzzi L, Andre F, Zitvogel L. Natural and therapy-induced immunosurveillance in breast cancer. Nat Med 2015;21(10):1128–38.

[14] Liu-Bryan R, Terkeltaub R. Emerging regulators of the inflammatory process in osteoarthritis. Nat Rev Rheumatol 2015;11(1):35–44.

Chapter 2

The human genome as a foundation for genomic and precision health

Huntington F. Willard
Director & Principal, Geisinger National Precision Health, North Bethesda, Maryland, United States

Introduction

That genetic variation can influence health and disease has been a central, if not yet fully practiced, principle of medicine for over a hundred years. What has limited application of this principle until recently has been the special nature and presumed rarity of clinical circumstances or conditions to which genetic variation was relevant. Now, however—with the availability of a reference sequence of the human genome and a rapidly growing number of genome sequences from both asymptomatic and symptomatic individuals, with emerging appreciation of the extent of genome variation among different individuals and different populations worldwide, and with a growing understanding of the role of common as well as rare variation in disease—we are increasingly able to exploit the impact of that variation on human health on a broad scale, in the context of genomic medicine and precision health [1–3].

Variation in the human genome has long been the cornerstone of the field of human genetics (Box 1), and its study led to the establishment of the medical specialty of medical genetics some four decades ago. The general nature and frequency of gene variants in the human genome became apparent with work over 50 years ago on the incidence of polymorphic protein variants in populations of healthy individuals, work that is the conceptual forerunner to the much larger and detailed efforts that mark modern human genetics and genomics. Such data underlie the conclusion that virtually every individual has his or her own unique constitution of gene products, the implications of which provide a foundation for what today we call personalized medicine or precision health as a modern application of what the British physician Archibald Garrod called "chemical individuality" in the very early years of the last century [2].

Genomic and Precision Medicine. https://doi.org/10.1016/B978-0-12-801496-7.00002-2

Box 1 Genetics and genomics in precision medicine and health

Throughout this and the many other chapters in this volume, the terms "genetics" and "genomics" are used repeatedly, both as nouns and in their adjectival forms. While these terms seem similar, they in fact describe quite distinct (though frequently overlapping) approaches in biology and in medicine. Having said that, there are inconsistencies in the way the terms are used, even by those who work in the field.

Here, we provide operational definitions to distinguish the various terms and the subfields of medicine to which they contribute.

The field of *genetics* is the scientific study of heredity and of the genes that provide the physical, biological, and conceptual bases for heredity and inheritance. To say that something—a trait, a disease, a code, or an information—is "genetic" refers to its basis in genes and in DNA.

Heredity refers to the familial phenomenon whereby traits (including clinical traits) are transmitted from generation to generation, due to the transmission of genes from parent to child. A disease that is said to be inherited or hereditary is certainly genetic; however, not all genetic diseases are hereditary (witness cancer, which is always a genetic disease, but is only occasionally an inherited disease).

Genomics is the scientific study of a genome or genomes. A *genome* is the complete DNA sequence, referring to the entire genetic information of a gamete, an individual, a population, or a species. As such, it is a subfield of genetics when describing an approach taken to study genes. The word "genome" originated as an analogy with the century-old term "chromosome," referring to the physical entities (visible under the microscope) that carry genes from one cell to its daughter cells or from one generation to the next. "Genomics" gave birth to a series of other "-omics" that refer to the comprehensive study of the full complement of genome products—for example, proteins (hence, *proteomics*), transcripts (*transcriptomics*), or metabolites (*metabolomics*). The essential feature of the "-omes" is that they refer to the complete collection of genes or their derivative proteins, transcripts, or metabolites, not just to the study of individual entities. The distinguishing characteristics of genomics and the other "omics" are their comprehensiveness and scale, their integration with and dependence on technology development, an emphasis on rapid data release and availability, and an awareness of the policy and ethical implications of such work in research, in the practice of medicine, and increasingly in the social arena [2, 3].

By analogy with genetics and genomics, *epigenetics* and *epigenomics* refer to the study of factors that affect gene (or, more globally, genome) function, but without an accompanying change in genes or the genome. The *epigenome* is the comprehensive set of epigenetic changes in a given individual, tissue, tumor, or population. It is the paired combination of the genome and the epigenome that appear to best characterize and determine one's phenotype.

Medical Genetics is the application of genetics to medicine with a particular emphasis on inherited disease. Medical genetics is a broad and varied field, encompassing many different subfields, including clinical genetics, biochemical genetics, cytogenetics, molecular genetics, the genetics of common diseases,

Box 1 Genetics and genomics in precision medicine and health—cont'd

and genetic counseling. Medical Genetics and Genomics is one of 24 medical specialties recognized by the American Board of Medical Specialties, the medical organization overseeing physician certification in the United States.

Genetic Medicine is a term used to refer to the application of genetic principles to the practice of medicine and thus overlaps medical genetics. However, genetic medicine is somewhat broader, as it is not limited to the specialty of Medical Genetics and Genomics, but is relevant to health professionals in many, if not all, specialties and subspecialties. Both medical genetics and genetic medicine approach clinical care largely through consideration of individual genes and their effects on patients and their families.

By contrast, *Genomic Medicine* refers to the use of large-scale genomic information and to consideration of the full extent of an individual's genome and other "omes" in the practice of medicine and medical decision making. The principles and approaches of genomic medicine are relevant well beyond the traditional purview of medical genetics and include, as examples, gene expression profiling to characterize tumors or to define prognosis in cancer, genotyping variants in the set of genes involved in drug metabolism or action to determine an individual's correct therapeutic dosage, scanning the entire genome for millions of variants that influence one's susceptibility to disease, or analyzing multiple protein or RNA biomarkers to detect exposure to potential pathogens, to monitor therapy and to provide predictive information in presymptomatic individuals.

Finally, *Precision Medicine* (or, increasingly, *Precision Health*) refers to the rapidly advancing field of health care that is informed by each person's unique clinical, genetic, genomic, and environmental information [1]. The goals of precision health/medicine are to take advantage of a molecular understanding of disease—and a vast array of data from sources that range from genomes to Electronic Medical Records to mobile health devices—to optimize preventive health care strategies and drug therapies while people are still well or at the earliest stages of disease. Because these factors are different for every person, the nature of disease, its onset, its course, and how it might respond to drug or other interventions are as individual as the people who have them and the communities in which they live. In order for precision medicine/health to be used by health care providers and their patients, these findings, analyzed with significant input from the data sciences, must be translated into precision diagnostic tests and targeted therapies, using implementation tools that are compatible with the modern health care environment [4]. Since the overarching goal is to optimize medical care and outcomes for each individual, treatments, medication types and dosages, and/or prevention strategies may differ from person to person—resulting in customization and precise targeting of patient care.

In this introductory chapter, the organization, variation, and expression of the human genome are presented as a foundation for the chapters to follow on approaches in translational genomics and on the principles of genomic and precision health as applied to inflammatory and infectious disease.

The human genome

The typical human genome consists of approximately 3 billion (3×10^9) base pairs of DNA, divided among the 24 types of nuclear chromosomes (22 autosomes, plus the sex chromosomes, X and Y) and the much smaller mitochondrial chromosome) (Table 1).

Individual chromosomes can best be visualized and studied at metaphase in dividing cells, and karyotyping of patient chromosomes has been a valuable and routine clinical laboratory procedure for a half century, albeit at levels of resolution that fall well short of most pathogenic DNA variants (Fig. 1). The ultimate resolution, of course, comes from direct DNA sequence analysis, and an increasing number of new technologies have facilitated comparisons of individual genomes with the reference human genome sequence, enabling clinical sequencing of patient samples to search for novel variants or mutations that might be of clinical importance, whether in a diagnostic or screening context [5].

Genes in the human genome

While the human genome contains an estimated 20,000 protein-coding genes, the coding segments of those genes—the exons—comprise only about 1–2% of the genome; most of the genome consists of DNA that lies between genes, far from genes, or in vast areas spanning several million base pairs (Mb) that appear to contain no genes at all. Despite much progress in gene identification and genome annotation, it is nearly certain that there are some genes, including clinically relevant genes, that are currently undetected or that display characteristics that we do not currently recognize as being associated with genes.

TABLE 1 Characteristics of the reference human genome[a]

Length of the human genome (base pairs)	3,609,003,417
Number of known protein-coding genes	20,418
Average gene density (number of genes/Mb)[b]	5.66[c]
Number of ncRNA genes	22,107
Number of known short sequence variants[d]	665,695,433
Number of known structural variants[e]	6,013,111

[a] *From Ensembl, database GRCh38.p12 (version 95.38, accessed February 2019).*
[b] *Mb, megabase pairs.*
[c] *Protein-coding genes only.*
[d] *For example, SNPs, substitutions, in/dels.*
[e] *For example, CNVs, inversions.*

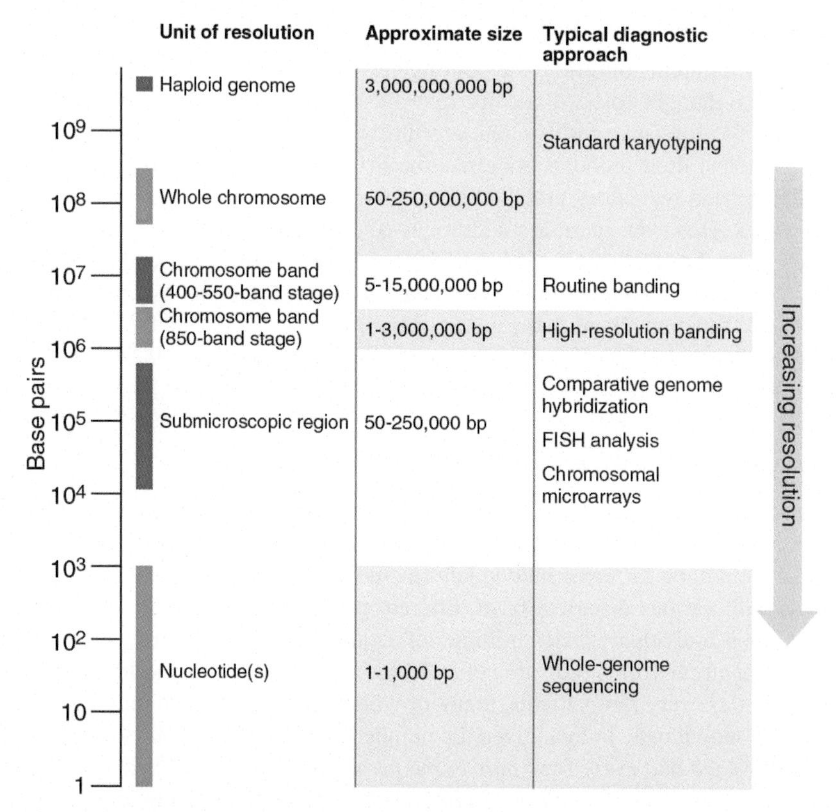

FIG. 1 Spectrum of resolution in chromosome and genome analysis. The typical resolution and range of effectiveness are given for various diagnostic approaches used routinely in clinical and research practice. FISH, fluorescence *in situ* hybridization. Source*: From Nussbaum RL, McInnes RR, Willard HF. Genetics in medicine. 8th ed. Philadelphia, PA: W.B. Saunders, Co.; 2016. 546 pp, with permission.*

Nonetheless, the statement that the vast majority of the genome consists of spans of DNA that are non-genic, of no known function, and of uncertain clinical relevance remains true and remains a challenge for interpreting the clinical relevance of genetic variants across the genome [6].

In addition to being relatively sparse in the genome, genes are distributed quite nonrandomly along the different human chromosomes. Some chromosomes are relatively gene-rich, while others are quite gene-poor, ranging from approximately 3 genes/Mb of DNA to >20 genes/Mb (excluding the Y chromosome and the tiny mitochondrial chromosome). And even within a chromosome, genes tend to cluster in certain regions and not in others, a point of clear clinical significance when evaluating genome integrity, dosage, or arrangement in different patient samples.

Coding and noncoding genes

There are a number of different types of gene in the human genome. Most genes known or thought to be clinically relevant are protein-coding and are transcribed into messenger RNAs that are ultimately translated into their respective proteins; their products comprise the list of enzymes, structural proteins, receptors, and regulatory proteins that are found in various human tissues and cell types. However, there are additional genes whose functional product appears to be the RNA itself. At least some of these so-called noncoding RNAs (ncRNAs) have a range of functions in the cell, but most do not as yet have any identified function. Overall, the genes whose transcripts make up the collection of ncRNAs could represent as many as a half of the total of ~40,000 identified human genes (Table 1).

Variation in the human genome

With completion of the initial reference human genome sequence some 20 years ago, attention has turned to the discovery and cataloguing of variation in that sequence among different individuals (including both healthy individuals and those with various diseases) from different populations worldwide [4, 7–11]. Any given individual carries millions of sequence variants that are known to exist in multiple forms (i.e., are polymorphic) in our species. In addition, there are countless very rare variants, many of which probably exist in only a single or a few individuals. In fact, given the number of individuals in our species, *essentially each and every base pair in the human genome is expected to vary in someone somewhere around the globe*. It is for this reason that the original human genome sequence is considered only a "reference" sequence, derived as a consensus of the limited number of genomes whose sequencing was part of the Human Genome Project, but likely identical to no individual's genome.

Types of variation

Early estimates were that any two randomly selected individuals have sequences that are 99.9% identical or, viewed another way, that an individual genome would be heterozygous at approximately 3–5 million positions, with different bases at the maternally and paternally inherited copies of that particular sequence position. The majority of these differences involve simply a single unit in the DNA code and are referred to as single-nucleotide variants (SNVs) or polymorphisms (SNPs) (Table 1). The remaining variation consists of insertions or deletions (in/dels) of (usually) short sequence stretches, variation in the number of copies of repeated elements or inversions in the order of sequences at a particular locus in the genome (Fig. 2). Any and all of these types of variation can influence disease and thus must be accounted for in any attempt to understand the contribution of genetics to clinical medicine and to precision health (Table 2).

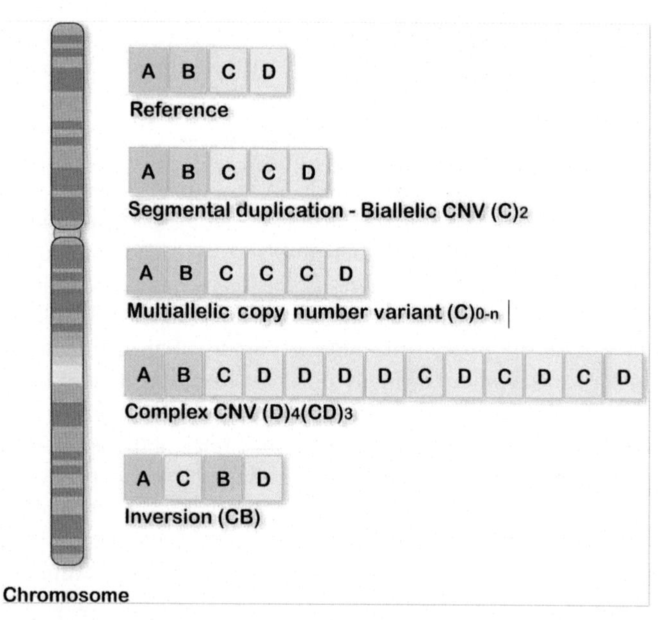

FIG. 2 Schematic representation of different types of structural polymorphism in the human genome, leading to deletions, duplications, inversions, and CNV changes relative to the reference arrangement. Source*: From Estivill X, Armengol L. Copy number variants and common disorders: filling the gaps and exploring complexity in genome-wide association studies. PLoS Genet 2007;3:e190, with permission.*

TABLE 2 Common variation in the human genome

Type of variation	Size range (approx.)[a]	Effect(s) in biology and medicine
Single-nucleotide polymorphisms/ variants	1 bp	Nonsynonymous → functional change in encoded protein? Others → potential regulatory variants? Most → no effect? (neutral)
Copy number variants (CNVs)	10 kb to 1 Mb	Gene dosage variation → functional consequences? Most → no effect or uncertain effect
Insertion/deletion polymorphisms (in/dels)	1 bp to 1 Mb	In coding sequence: frameshift mutation? → functional change Most → uncertain effects
Inversions	Few bp to 100 kb	? break in gene sequence ? long-range effect on gene expression ? indirect effects on reproductive fitness Most → no effect? (neutral)
Segmental duplications	10 kb to >1 Mb	Hotspots for recombination → polymorphism (CNVs)

[a] *bp, base pair; kb, kilobase pair; Mb, megabase pair.*

Copy number variation

Over the past decade, increasing attention has focused on the prevalence of structural variants in the genome, which, in any given genome, collectively account for far more variation in genome sequence (expressed in terms of the amount of genomic DNA affected) than do SNVs [12]. The most common type of structural variation involves changes in the local copy number of sequences (including genes) in the genome. This variation is based on blocks of different sequences that are present in multiple copies, often with extraordinarily high sequence conservation, in many different locations around the genome. Rearrangements between such duplicated segments are a source of significant variation between individuals in the number of copies of these DNA sequences and these are generally referred to as copy number variants (CNVs) (Fig. 2). When the duplicated regions contain genes, genomic rearrangements can result in the deletion of the region (and the genes) between the copies and thus give rise to disease. It is of considerable ongoing interest to evaluate the role of CNVs and other structural variants in the etiology of a range of clinical conditions.

De novo mutations

While much emphasis is placed on inherited genome variation, each such variant had to originate as a *de novo* or new change occurring in germ cells at some point in time—whether 9 months ago (in the case of a newborn infant's genome) or up to 100,000 years ago (in the case of ancient polymorphisms). At whatever point a *de novo* change occurred, such a variant would be extremely rare in the population (occurring just once), and its ultimate frequency in the population over time depends on chance and on the principles of Mendelian inheritance and population genetics. The ability to sequence genomes directly provides a robust method for measuring mutation rates genome-wide, by, for example, comparing the sequence of an offspring's genome (or a portion of that genome) with that of his or her parents [13].

Such studies have shown that every individual carries an estimated 30–70 new mutations per genome that were not present in the genomes of her or his parents. This rate, however, is known to vary from gene to gene around the genome and from individual to individual and is dependent on the age of the parents [14, 15]. Overall, the mutation rate, combined with considerations of population growth and dynamics, predicts that there must be an enormous number of relatively new (and thus rare) mutations in the current worldwide population of 7 billion individuals [12, 13, 16].

Conceptually similar studies have explored *de novo* mutations in CNVs, where the generation of a new length variant depends on recombination, rather than on errors in DNA synthesis to generate a new base pair. Indeed, the measured rate of formation of new CNVs is orders of magnitude higher than that of base substitutions [17, 18].

Variation in individual genomes

The most extensive current inventory of the amount and type of variation to be expected in any given genome comes from the direct analysis of individual diploid human genomes. Any given human genome typically carries about 5 million SNVs, many of which are previously unknown. This suggests that the number of SNVs described for our species is still incomplete, although presumably the fraction of such novel SNVs will decrease as more and more genomes from more individuals and from more populations are sampled.

Typically, each genome carries thousands of nonsynonymous SNVs—variants that encode a different amino acid in thousands of protein-coding genes around the genome. These measurements underscore the potential impact of gene and genome variation on human biology and on medicine. A comprehensive and influential international study compared several thousand genome sequences from the 1000 Genomes Project [19]. This study documented that each genome carries 100 or more likely loss-of-function mutations, about 10,000 nonsynonymous changes, and some 500,000 variants that overlap known gene regulatory regions. Ongoing studies continuingly extend and refine these data in the context of individual populations [10, 11].

These and other findings also indicate—perhaps surprisingly—that thousands of genes in the human genome are highly tolerant to many mutations that appear likely to result in a loss of function [7, 9, 11, 20, 21]. Within the clinical setting, this awareness has important implications for the interpretation of data from sequencing of patient material, particularly when predicting the impact of mutations in genes of currently unknown function and whether they increase or decrease risk relative to the population as a whole [22, 23].

Notwithstanding the remarkable amount of information on genomes and genome variation over the past decade and its readiness for at least some evidence-driven applications in clinical care, it is clear that we are overall still in a mode of discovery; no doubt many millions of additional SNVs and other variants remain to be uncovered, as does the degree to which any of them might impact an individual's phenotype in the context of wellness and health care. The broad question of "what is normal?"—an essential concept in human biology and in clinical medicine—remains an open question when it comes to the human genome [24].

Variation in populations

Leveraging key technological developments (including whole-exome and whole-genome sequencing) that have greatly increased the throughput of genotyping on a genome-wide scale, a growing number of large-scale projects have gathered genotypic information on millions of SNVs and structural variants in up to many hundreds of thousands of individuals from hundreds of populations worldwide [9, 12, 19, 25]. Two major conclusions emerge: First, some 85–90% of the common variation found in our species is shared among different population groups; a relative minority of common variants are specific to or highly enriched/depleted in genomes from a particular population. And second, most variable sites in the genome are rare, not common, and are private to specific populations or even families rather than ancient and shared among populations [16].

These findings reflect an explosion of population growth from an ancestral population of likely fewer than 10,000 individuals, with Eurasians diverging from an ancestral African population an estimated 38–64 thousand years ago. It is now possible to trace or reconstruct the history and the genetic/geographic origins of many population groups around the world. These findings are of innate interest to specific groups, but also have profound implications for health care delivery to different groups of individuals worldwide characterized by different DNA variants and thus different susceptibility to different medical conditions, notably evident in the global diversity of inflammatory and infectious diseases, as explored in this volume.

Expression of the human genome

A key question in exploring the function of the human genome is to understand how proper expression of our 20,000–40,000 genes is determined, how it can be influenced by either genetic variation or by environmental exposures or inputs, and by what mechanisms such alterations in gene expression can lead to pathology evident in the practice of clinical medicine. The control of gene activity—in development, in different tissues and cell lineages, during the cell cycle, and during the lifetime of an individual, both in sickness and in health—is determined by a complex interplay of genetic, epigenetic, and environmentally-influenced features [2].

By "genetic" features, we here refer to those found in the genome sequence (see Box 1), which plays a role, of course, in determining the identity of each gene, its particular form (termed "alleles"), its level of expression (requiring a consideration of various regulatory elements), and its particular genomic landscape (three-dimensional configuration, base composition, chromatin composition). By "epigenetic" features, here we mean packaging of the DNA into chromatin, in which it is complexed with a variety of histones as well as innumerable non-histone proteins that influence the accessibility and activity of genes and other genomic sequences. The structure of chromatin—unlike the genome sequence itself—is highly dynamic and underlies the control of gene expression that shapes in a profound way both cellular and organismal function.

It has been appreciated for decades that there is high variability in gene expression levels among individuals. Much of this is due to differences among the genes themselves, a result consistent with local sequence variation influencing the expression of such genes. It is likely that the ongoing discovery of regulatory variants will correlate with variation in these patterns of gene expression [20], with an anticipated but as-yet-unknown degree of impact on human disease.

Genes, genomes, and disease

In the context of genomic medicine and precision health, an overriding question is to what extent variation in the sequence and/or expression of one's genome influences the likelihood of disease onset, determines or signals the natural history of disease, and/or provides clues relevant to the management of wellness or disease. As just discussed, variation in one's constitutional genome can have a number of different direct or indirect effects on gene expression, thus contributing to the likelihood of disease.

Comprehensive catalogues of genomic and other "omic" data, from sequence to functional elements encoded in the genome, to interacting networks of RNAs and proteins, and to metabolites, carbohydrates and small molecules in a variety of cell and tissue types, are emerging (Table 3) [26–29]. The integrative nature of physiology and medicine, aided in major part by advances in the data sciences, lends itself well to "omic" approaches that seek to gather comprehensive datasets that can be queried informatically to gain insights into patterns that promise to reveal distinctive insights about health or disease.

TABLE 3 Personalized "Omic" signatures of health or disease.

Dataset	"Omic" approach	Technology platform or approach
Human genome sequence	Genomics	SNPs; SNVs; other genome variants; whole exome sequencing; complete personal genome sequences
Gene expression profiles	Transcriptomics	Sequencing of RNA products (RNA-seq)
Protein abundance	Proteomics	Protein arrays of specific protein products
Metabolites	Metabolomics	Analysis of hundreds to thousands of metabolites
Chromatin	Epigenomics	Array- or sequence-based assessment of chromatin modification
Gene networks, interactions	Systems biology	Large-scale interactions among genes or proteins
Carbohydrates, glycomedicine	Glycomics	Comprehensive assessment of carbohydrates and protein modification
Microbiomes	Metagenomics	Analysis of viral, fungal, and bacterial communities in human specimens
Genomic and clinical data	Informatics	Integrated databases of "omic" data and electronic health records

Clinical sequencing in search of causal variants

The development of new sequencing technologies [30] has extended earlier efforts at genome-wide genotyping of common SNPs to uncover rare or novel variants responsible for (or at least statistically associated with) a disease or phenotype of interest [7, 9–11]. At least four conceptually related sequencing-based strategies have emerged to relate or implicate particular variants in a range of clinical conditions or phenotypes, providing individualized information in a screening or diagnostic mode that provides information of demonstrable value and impact for patients and families. Importantly, each of these differs in terms of its coverage of the genome and the impact of knowing variants detected by each, in the context of diagnostic testing or population-based screening.

Genotyping for polygenic scores

Genotyping of patient material to scan variants across the genome (hence, genome-wide) is frequently misunderstood or overinterpreted, as—despite its being included among the range of other sequence-based approaches—it is limited to analysis of a relatively small fraction of the genome, seeking variants that, in aggregate, are associated with (and thus may predict risk of) a clinical condition. Such genome scans (as they are sometimes called) are limited to a few million potentially variant positions across the genome—in reality, a tiny glimpse of the totality of genetic variation that any individual's genome harbors.

That said, detailed analysis of such genotypic data, in combination with data from electronic health records and other clinical datasets, have created what have become known as “polygenic risk scores” (PRS) or simply “polygenic scores” (PGS)—effectively low-density scans of the genome, without prior indication of what genes (hence, polygenic) might be relevant to the condition(s) of interest. This is a highly promising avenue of research, both for finding segments of the genome (and thus candidate genes) that might be relevant to disease risk, as well as for potentially triaging populations into clinical groups that are, at least statistically, at much elevated or reduced risk relative to the population as a whole [31, 32].

Targeted resequencing

In the targeted sequencing approach, original efforts focused on one or several genes that were believed to be strong candidates for the phenotype under study. However, as reduced costs have made it more cost-effective to sequence more extensive gene panels (i.e., a hundred or so genes known or thought to be relevant to the condition under study) or whole exomes, targeted resequencing, with its inherent and self-limiting bias toward candidate genes that “seem plausible,” has fallen out of favor in most research studies.

Whole-exome sequencing

The approach of isolating and sequencing entire exomes from patient samples has increasingly become a favored approach, both for research and for clinical care, in both diagnostic and screening modalities. Many hundreds of thousands of patient samples have been examined in searches for potentially causative variants, either targeted to particular phenotypes or broadened in large population-based cohorts to any clinical finding or laboratory value that covaries with a genome variant [7, 10, 11].

Exome sequencing does, however, present challenges, especially when needing to distinguish between legitimate disease associations and population stratification. A potentially overwhelming number of rare or novel variants need to be evaluated for potential functional importance before one can conclude with confidence that a particular variant is causal. This task appears tractable for Mendelian disorders within families, but is significantly more challenging for complex diseases and traits at the moment [32, 33]. Nonetheless, clinical whole-exome sequencing is now being offered as a diagnostic service at an increasing number of academic medical centers, and there are growing efforts to integrate population-based genomic screening into clinical care on a routine basis [5, 34–38].

Whole-genome sequencing

These challenges are even more pressing for whole-genome sequencing. Issues of disease penetrance, possible allele–allele interactions around the genome, coding and noncoding variants, and protective versus predisposing alleles all have to be taken into consideration [7, 39]. At this point, as the lines between research and clinical care "blur" [40], it seems fair to say that the major challenges confronting whole-genome sequencing over time will fall in the categories of informatics and clinical utility, not technical capacity or even cost [41].

Microbiomes and microbiota

The human genome is not the only genome relevant to the practice of medicine. Both in states of health and disease, our own genome is vastly outnumbered by the genomes of a host of microorganisms, many living peacefully and continuously on various body surfaces as commensal or symbiotic microbiomes, others wreaking havoc as adventitious viral, bacterial, or fungal pathogens [42].

The genomes of many thousands of microorganisms have been determined and are being utilized to provide rapid diagnostic tests in clinical settings, to predict antibiotic or antifungal efficacy, to identify the source of airborne, water, or soil contaminants, to monitor hospital or community environments, and to better understand the contribution of communities of microbes—the microbiota—and various environmental exposures to diverse human phenotypes, especially across the range of inflammatory and infectious diseases discussed in subsequent chapters [43].

The field of metagenomics explores this heterogeneous ecosystem by comprehensive sequence analysis of the collected genomes from biological specimens (such as stool, urine, sputum, water sources, and air), followed by both taxonomic and bioinformatic analysis to deconvolute the many genomes contained in such specimens and to define the different organisms, their genes, and genome variants. The metagenomic approach is particularly informative for characterizing organisms that cannot be cultured in standard microbiology laboratories. The dynamic balance among the many species that make up the microbiome points to cooperative networks of communities of microbiota, evolving over time and in response to different environments [44].

Undoubtedly, the states of health and disease are determined in part by the balance of genomes both within us and external to us. The full complement of genomic information from both of these sources of genomes will both provide insights into defining the states of health and disease and contribute to a basis for precision in both the prevention of disease and its treatment.

From genomes to precision health

Of all the promises of the current scientific, social, and medical revolution stemming from advances in our understanding of the human genome and its variation, genomic medicine and precision health may be the most eagerly awaited. The prospect of examining an individual's entire genome (or at least a significant fraction of it) in order to make individualized risk predictions and treatment decisions is an attractive, albeit challenging, one that is just in its early stages of clinical implementation [3, 4, 38, 45, 46].

Having access to the reference human genome sequence has been transformative for the fields of human genetics and genome biology [8], but by itself is an insufficient prerequisite for genomic medicine. Equally important are the growing emergence of large-scale phenotypic and clinical data sets that allow discovery of genomic/clinical associations [10, 11, 32] and various advances in the data sciences to reliably capture and assess information on individual genomes, their epigenetic modification, and the transcriptome, proteome, microbiome, and metabolome for health and disease status (Table 3). Each of these technologies provides information that, in combination with phenotypic data and evaluation of environmental triggers, will contribute to assessment of individual risks and guide both clinical management and decision making.

References

[1] Collins FS, Varmus H. A new initiative on precision medicine. N Engl J Med 2015;372:793–5.
[2] Nussbaum RL, McInnes RR, Willard HF. Genetics in medicine. 8th ed. Philadelphia, PA: W.B. Saunders, Co; 2016. p. 546.

[3] Green ED, Guyer MS. National Human Genome Research Institute, charting a course for genomic medicine from base pairs to bedside. Nature 2011;470:204–13.

[4] Aronson SJ, Rehm HL. Building the foundation for genomics in precision medicine. Nature 2015;526:336–42.

[5] Adams DR, Eng CM. Next-generation sequencing to diagnose suspected genetic disorders. N Engl J Med 2018;379:1353–62.

[6] Salzberg SL. Open questions: how many genes do we have? BMC Biol 2018;16:94.

[7] Gudbjartsson DF, Helgason H, Gudjonsson SA, Zink F, Oddson A. Large-scale whole-genome sequencing of the Icelandic population. Nat Genet 2015;47:435–44.

[8] Lander ES. Initial impact of the sequencing of the human genome. Nature 2011;470:187–97.

[9] The UK10K Consortium. The UK10K project identifies rare variants in health and disease. Nature 2015;526:82–90.

[10] Bycroft C, Freeman C, Petkova D, Band G, Elliott LT. The UK Biobank resource with deep phenotyping and genomic data. Nature 2018;562:203–9.

[11] Dewey FE, Murray MF, Overton JD, Habegger L, Leader JB. Distribution and clinical impact of functional variants in 50,726 whole exome sequences from the DiscovEHR study. Science 2016;354:aaf6814.

[12] Sudmant PH, Mallick S, Nelson BJ, Hormozdiari F, Krumm N. Global diversity, population stratification, and selection of human copy number variation. Science 2015;349:aab3761.

[13] Shendure J, Akey JM. The origins, determinants, and consequences of human mutations. Science 2015;349:1478–83.

[14] Rahbari R, Wuster A, Lindsay SJ, Hardwick RJ, Alexandrov LB. Timing, rates, and spectra of human germline mutation. Nat Genet 2016;48:126–33.

[15] Conrad DF, Keebler JEM, DePristo MA, Lindsay SJ, Zhang Y. Variation in genome-wide mutation rates within and between human families. Nat Genet 2011;43:712–4.

[16] Olson MV. Human genetic individuality. Annu Rev Genomics Hum Genet 2012;13:1–27.

[17] Estivill X, Armengol L. Copy number variants and common disorders: filling the gaps and exploring complexity in genome-wide association studies. PLoS Genet 2007;3:e190.

[18] Itsara A, Wu H, Smith JD, Nickerson DA, Romieu I. De novo rates and selection of large copy number variation. Genome Res 2010;20(11):1469–81.

[19] The 1000 Genomes Project Consortium. A global reference for human genetic variation. Nature 2015;526:68–74.

[20] Lek M, Karczweski KJ, Minikel EV, Samocha KE, Banks E. Analysis of protein-coding genetic variation in 60,706 humans. Nature 2016;536:285–91.

[21] Narasimhan VM, Hunt KA, Mason D, Baker CL, Karczewski KJ. Health and population effects of rare gene knockouts in adult humans with related parents. Science 2016;352:474–7.

[22] Biesecker LG, Nussbaum RL, Rehm HL. Distinguishing variant pathogenicity from genetic diagnosis: how to know whether a variant causes a condition. JAMA 2018;320:1929–30.

[23] Emdin CA, Khera AV, Chaffin M, Klarin D, Natarajan P. Analysis of predicted loss-of-function variants in UK Biobank identifies variants protective for disease. Nat Commun 2018;9:1613.

[24] Manrai AK, Patel CJ, Ionnidis JPA. In the era of precision medicine and big data, who is normal? JAMA 2018;320:1496–7.

[25] Sudmant PH, Rausch T, Gardner EJ, Handsaker RE, Abyzov A. An integrated map of structural variation in 2,504 human genomes. Nature 2015;526:75–80.

[26] Huttlin EL, Ting L, Bruckner RJ, Gebreab F, Gygi MP. The BioPlex network: a systematic exploration of the human interactome. Cell 2015;162:425–40.

[27] Rolland T, Taşan M, Charloteaux B, Pevzner SJ, Zhong Q. A proteome-scale map of the human interactome network. Cell 2014;159:1212–26.

[28] Vidal M, Cusick ME, Barabasi A-L. Interactome networks and human disease. Cell 2011;144:986–98.

[29] Gamazon ER, Segre AV, van de Bunt M, Wen X, Xi HS. Using an atlas of gene regulation across 44 human tissues to inform complex disease- and trait-associated variation. Nat Genet 2018;50:956–67.

[30] Shendure J, Balasubramanian S, Church GM, Gilbert W, Rogers J. DNA sequencing at 40: past, present, and future. Nature 2017;550:345–53.

[31] Khera AV, Chaffin M, Aragam KG, Haas ME, Roselli C. Genome-wide polygenic scores for common diseases identify individuals with risk equivalent to monogenic mutations. Nat Genet 2018;50:1219–24.

[32] Bastarache L, Hughey JJ, Hebbring S, Marlo J, Zhao W. Phenotype risk scores identify patients with unrecognized Mendelian disease patterns. Science 2018;359:1233–9.

[33] Blair DR, Lyttle CS, Mortensen JM, Bearden CF, Jensen AB. A nondegenerate code of deleterious variants in Mendelian loci contributes to complex disease risk. Cell 2013;155:70–80.

[34] Rehm HL, Berg JS, Brooks LD, Bustamante CD, Evans JP. ClinGen—the clinical genome resource. N Engl J Med 2015;372:2235–42.

[35] Landrum MJ, Kattman BL. ClinVar at five years: delivering on the promise. Hum Mutat 2018;39:1623–30.

[36] Yang Y, Muzny DM, Xia F, Niu Z, Person R. Molecular findings among patients referred for clinical whole-exome sequencing. JAMA 2014;312:1870–9.

[37] Willard HF, Feinberg DT, Ledbetter DH. How Geisinger is using gene screening to prevent disease. Harv Bus Rev 2018;. March 14.

[38] Schwartz MLB, McCormick CZ, Lazzeri AL, Lindbuchler DM, Hallquist MLG. A model for genome-first care: returning secondary genomic findings to participants and their healthcare providers in a large research cohort. Am J Hum Genet 2018;103:328–37.

[39] Gloyn AL, McCarthy MI. Variation across the allele frequency spectrum. Nat Genet 2010;42:648–50.

[40] Angrist M, Jamal L. Living laboratory: whole-genome sequencing as a learning healthcare enterprise. Clin Genet 2014;87:311–8.

[41] Burke W. Making sense of the genome remains a work in progress. JAMA 2018;320:1247–8.

[42] Lloyd-Price J, Mahurkar A, Rahnavard G, Crabtree J, Orvis J. Strains, functions, and dynamics in the expanded Human Microbiome Project. Nature 2017;550:61–6.

[43] Gilbert JA, Quinn RA, Debelius J, Xu ZZ, Morton J. Microbiome-wide association studies link dynamic microbial consortia to disease. Nature 2016;535:94–103.

[44] Rakoff-Nahoum S, Foster KR, Comstock LE. The evolution of cooperation within gut microbiota. Nature 2016;533:255–9.

[45] Manolio TA, Abramowicz M, Al-Mulla F, Anderson W, Balling R. Global implementation of genomic medicine: we are not alone. Sci Transl Med 2015;7.

[46] Williams MS, Buchanan AH, Davis FD, Faucett WA, Hallquist MLG. Patient-centered precision health in a learning health care system: Geisinger's genomic medicine experience. Health Aff 2018;37:757–64.

Chapter 3

Data analytics for precision medicine

Timothy E. Sweeney*, Purvesh Khatri[†,‡]
**Inflammatix, Inc., Burlingame, CA, United States, [†]Institute for Immunity, Transplantation, and Infection, Department of Medicine, Stanford University, Stanford, CA, United States, [‡]Department of Medicine, Division of Biomedical Informatics Research, Stanford University, Stanford, CA, United States*

What do we see as "precision medicine"?

The term "precision medicine" often conjures, at least in the public consciousness, a future system wherein advanced diagnostics are able to accurately identify and subtype patients with complex diseases, such that treatment decisions can be made more accurate by tailoring to the specific patient. Between today and that promise of tomorrow is a chasm of data gathering, analysis, and validation. We will here focus specifically on pragmatic guidelines for discovery and validation of new precision medicine tools. Many novel diagnostics focus on interpreting "biomarkers," with varying degrees of assistance from machine learning algorithms. Such biomarkers can be hereditary (e.g., DNA mutations, conferring risk of an event), can be acutely expressed during the disease state (such as transcriptomics or proteomics), or can be phenotypic of the patient response (such as vital signs tracked by wearable devices). While all of these have advantages and disadvantages, we will focus on biomarkers of expression, since these may be of greatest interest in acute inflammatory or infectious conditions.

Precision medicine, by its name, seems to suggest that ever-finer patient subtypings may lead to improved treatments. We argue that in many fields, rather than more "personalized" solutions, the current need is often for generalizable solution that are broadly applicable. By this we mean that there is still routine diagnostic uncertainty for many infectious and inflammatory disease states, and that physician needs (and market forces) are likely to push novel "precision medicine" solutions toward getting these big general questions right before focusing on subtyping [1]. In oncology, for instance, one still usually has to get the tissue of origin right before applying an advanced subtyping panel. Similarly, a test for subtyping a disease of interest may not be effective if there is substantial diagnostic equipoise in who has the disease (for example, Crohn's disease

Genomic and Precision Medicine. https://doi.org/10.1016/B978-0-12-801496-7.00003-4

vs. ulcerative colitis, or bacterial vs. viral infections). Second, if the same question is being asked (e.g., whether antibiotics are needed) in different populations (e.g., children and adults), a single test, if properly calibrated, will lead to greater utility. Tests indicated for increasingly fragmented populations will fail to overcome barriers to market entry, and those that do make it may have overly specific indications for use, leaving many patients without help. Finally, since off-label uses of tests and therapies are common, if biomarkers fail to deliver similar performance in seemingly similar conditions, patients will be harmed. Thus, in the era of precision medicine, there is an unmet need to develop broad, generalizable, disease-defining diagnostic and prognostic solutions.

Data analytics for precision medicine

A data analyst practicing precision medicine is typically faced with two questions: (1) what the desired output is, and (2) what data and tools will lead to that desired output. The desired output is usually a tool for diagnosis, prognosis, specific treatment-response, or disease subtyping. Each of these (except disease subtyping) require labels (e.g., whether the patient has the disease, or had a poor outcome, or responded to treatment), is called supervised machine learning (ML). Disease subtyping, wherein we start without labels and an algorithm picks robust groups based on some structure in the data, is called unsupervised ML. A key point for the data analyst is evaluating whether their results are correct in that the biomarker continues to be associated with the desired output (diagnosis, prognosis or subtyping) in independent cohorts. For supervised-learning outcomes, it is relatively simple to test accuracy vs. true labels. For unsupervised learning, there are two methods of validation; one is to re-perform the same clustering/subtyping analysis in a new cohort and show that the same phenotypes hold; the other is to derive a subtype classifier in independent data, and then show that the classifier can reproduce its intended phenotypes in independent data. We favor the latter, as it provides a clinically actionable classifier, but both methods are appropriate in practice.

Traditionally, most groups dictate splitting a dataset into a training and testing portion to examine the accuracy of the diagnostic/prognostic tool. However, such a test is not representative of real-world challenges of application in a novel patient population. Thus, we favor fully independent testing in a cohort gathered under different circumstances. In any case, a model should not be considered to have been "validated" unless fixed coefficients are applied to novel patients that are fully independent from the training data. When possible, the STARD guidelines are an excellent checklist for publishing a new precision medicine tool [2].

Pros and cons of public data

The obvious followup question to the exhortation to test in independent data is where to find it. While there are many schools of thought on this point, we are strong proponents of careful use of public data for training and testing of robust precision medicine tools. Perhaps the largest repository is the NIH's Gene Expression Omnibus (https://www.ncbi.nlm.nih.gov/geo/), which as of the writing of the article has >100,000 experiments comprised of >2.6 million samples (usually microarrays or sequencing data). Many other repositories for data exist; in particular, *Nucleic Acids Research* publishes a yearly "database issue" which explains updates to many global bioinformatics databases (https://academic.oup.com/nar/issue/46/D1).

The commonly cited pitfalls of working with public data are: (1) no control over experimental condition, (2) possible mislabeling or errors in normalization by the original authors, (3) dealing with technical heterogeneity when datasets used different measurement platforms, and (4) variance in clinical trial criteria even for the same diseases. On the other hand, the benefits of public data are: (1) they may cover a broad range of clinical heterogeneity for a condition of interest, (2) the increased amount of data can substantially improve statistical power, especially in discovery work, (3) data re-use may prevent unnecessary human subject experimentation, and (4) they are readily available without the need for IRB approval [3].

A key paradigm in traditional biomedical research is the controlled experiment that aims to reduce heterogeneity by ensuring that all samples come from a carefully phenotyped population that differs only with respect to the disease of interest and that are profiled using the same technology. Though it reduces confounding, this paradigm is flawed in that it does not account for real-world biological and technical heterogeneity. Thus, the results of controlled experiments (especially in the "omics" space) are often overfit and not generalizable once a more heterogenous (more real-world) population is tested (Fig. 1).

We have shown that data from a large number of clinical diseases of interest, assayed from most tissues in the body, are available for reuse. In particular, work integrating gene expression profiles from hundreds of public experiments showed that both known and unexpected insights between different disease states can be found from public data, indicating a vast potential across a wide variety of conditions [4]. Across a broad range of diseases, our group has shown that an analysis using multiple independent heterogeneous public data sets is highly likely to yield more generalizable results than even the largest single-site study [5]. In fact, we have shown that, if a researcher were to fix the number of clinical subjects, more generalizable results would be found from the analysis of several moderately-powered studies rather than one very large study [5,6]. This is because multiple studies capture a broader picture of the clinical heterogeneity of a disease. As a result, heterogeneous public data (a collection of multiple controlled experiments) are useful because despite the high technical heterogeneity, most diseases have even more clinical heterogeneity. The heterogeneity in public data can thus be a blessing in disguise for translational medicine disguise and present a unique opportunity for accelerating translational medicine. The flip side, of course, is that precision medicine "at the frontier," where no one

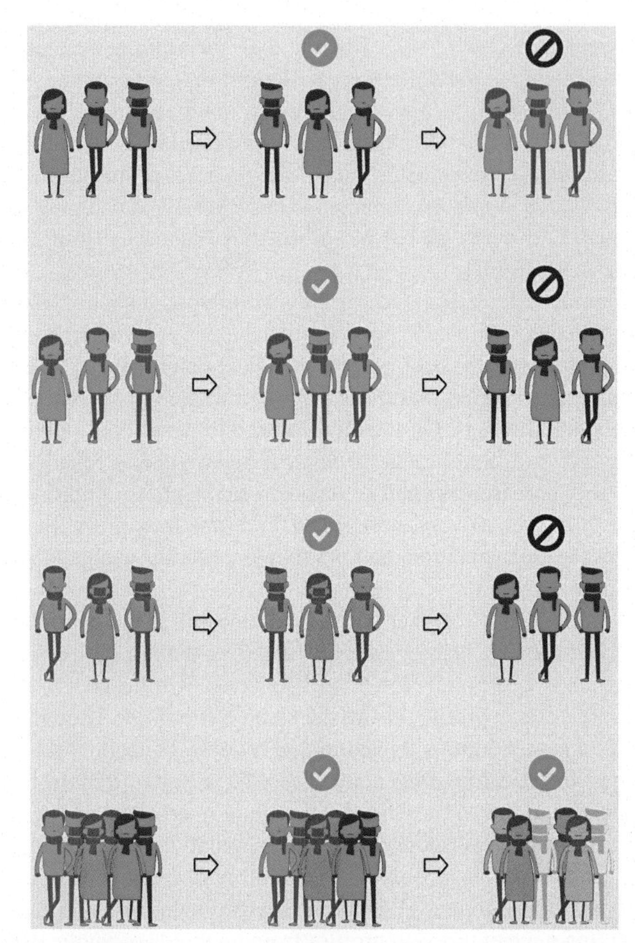

FIG. 1 Single-cohort studies do not capture the clinical heterogeneity of a disease. Validation in a held-out test set (which is not dissimilar from the training set) usually looks good, but is overfit. Training across clinical heterogeneity leads to robust validation performance in truly independent patients.

else has looked, will require gathering novel cohorts. Clinical trialists and those that ask novel questions in enrolling new cohorts are the roots that allow the tree of precision medicine to flourish.

Our group has published open-source software (MetaIntegrator on CRAN for R) to help bring together multiple cohorts into a single integrated analysis workflow [7] (as have several others) [8–10]. In general, multi-cohort analysis begins with a meta-analytic search for differentially expressed biomarkers of interest (we have typically used transcriptomics but any other expressed biomarker would also work). We then move to some method of integrating biomarkers into, for instance, a prognostic score (such as difference of geometric means, discussed below). The basic pieces of this pipeline have been demonstrated in organ transplant rejection, sepsis, acute infections, tuberculosis, cancer, and autoimmunity by our group [11–16]. Others have also demonstrated excellent work in a wide range of areas such as viral infection prediction [17] and cancer diagnosis [18]. The general outcome is that signatures discovered in heterogeneous cohorts and that do not rely on highly advanced ML models prone to overfitting are likely to be generalizable.

Sepsis: A case-in-point example

As a case-in-point argument for generalizable biomarkers, consider the field of biomarkers for infections and sepsis [19]. Sepsis is a syndrome defined as a dysregulated immune response to any infection [20]. It is not simply found in those with a bloodstream infection; in fact, most patients with sepsis do not have bloodstream infections. Sepsis is thus difficult to identify, since even non-microbiologically-proven infections can cause sepsis. Partly due to this high clinical heterogeneity and uncertainty over who has sepsis, over the last 50 years, >100 interventional trials have been completed in sepsis, with no drugs successfully brought to market [21,22]. Thus, the primary unmet need in sepsis is a test for whether a patient in fact has "sepsis." By first reducing heterogeneity (identifying the presence and type of infection that has led to a septic response) and then risk-stratifying those likely to have a worse outcome (prognostic enrichment), we can vastly improve all sepsis interventions. In the era of precision medicine, only a generalizable sepsis biomarker will be successful in defining a syndromic disease across a heterogeneous population [1]. Once this has been established, the final piece of precision medicine, which is disease subtyping or "predictive enrichment" for treatment response, can be successful to identify right patients that should be treated at the right time.

Supervised learning in sepsis

There are two particular needs in sepsis that groups around the world have tried to solve. The first is a diagnostic for the presence and type of an infection; the other is a prognostic tool for risk stratification [23–30]. This has proven to be an incredibly difficult challenge for a number of reasons; our view is that sepsis is too heterogeneous for single-cohort studies to find generalizable signatures. As a result, we have focused our work in sepsis on multi-cohort analyses [13,15,26]. We began each sepsis analysis by first conducting a systematic search for all datasets of possible interest, and then working with domain experts to remove datasets not of clinical relevance. For the remaining datasets, we grouped the datasets into either training or test data (but a single dataset is never used as both). We used the MetaIntegrator tool outlined above to identify differentially expressed genes with high statistical confidence, and then combined those genes using a non-parametric tool (difference of geometric means) to produce a single diagnostic score. This method was shown to be superior to other single-cohort discovery methods in identifying a panel of genes that remained diagnostic across multiple independent datasets [31].

Such a method works well for two-class data, but for multiclass data (such as in acute infections, which may be bacterial, viral, or neither), a multiclass ML algorithm is necessary [15,25,30]. The promise of such a multiclass ML algorithm is that complex relationships between biomarkers of interest may be learnable and so produce greater diagnostic accuracy than a non-parametric algorithm. A disadvantage of ML algorithms is that they are not easily ported across technical platforms. Thus, to our knowledge, no group has yet shown the external validity of a fixed ML algorithm for diagnosing sepsis. In fact, in a recent large academic collaboration that sought new prognostic tools for sepsis (trained on 30-day mortality instead of the presence of an infection), three groups compared the generalizability of both ML and non-parametric models in independent data, and found that the fixed ML models often performed poorly in independent data on different technical platforms [26]. Ultimately, a stable technical platform is of course necessary to translate a test through regulatory clearance and to market; it may prove that by removing technical heterogeneity, such a platform enables large-scale ML for a new sepsis biomarker. Across the work described above, it is clear that precision medicine tools are coming which will eventually be able to determine the presence, type, and severity of acute infections and sepsis Fig. 2.

Unsupervised learning in sepsis

We (and others) have thus begun to look into the "other" precision medicine focus, namely, subtyping of sepsis patients through unsupervised learning [32–38]. Sepsis is a fertile ground for subtyping because, as a loosely defined syndrome, it is likely to contain many different pathophysiological mechanisms. Here again, we see results from groups doing discovery work within single datasets and across multiple datasets. No group has established which subgroup is

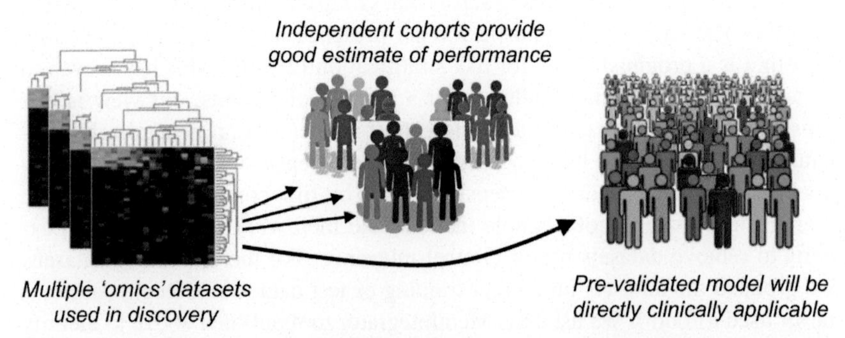

FIG. 2 A novel biomarker signature should be fixed in multiple discovery cohorts and applied externally using a fixed model in independent data prior to pivotal clinical studies.

"right" (in fact, several may be "right"); but we have seen both techniques of subtype validation used (training a subtype-classifier in discovery and using it to recapitulate phenotypes, and re-running a clustering algorithm on independent test data). Perhaps best known, Wong et al. have validated that their subtypes can identify which patients can benefit (and which suffer harm) from steroid treatment; this is the true promise of predictive enrichment in subtyping (to personalize treatment decisions) [32,36,37,39]. We also note that other groups have successfully validated non-transcriptomic subtypes in other critical illnesses as well, using either other molecular data, or even EMR-based phenotypic data [40,41]. While we argue that a molecular test that can define "sepsis" is a prerequisite to "sepsis" subtyping, the field is clearly benefitting for a number of novel discovery efforts being applied across domains.

Promise of the future

We have focused herein on some of the methods and broad categories of data analytics for precision medicine. In our view, two things are clear. First, clinical trialists who are willing to test novel hypotheses in new disease states are pushing the bounds on what is possible by applying new technologies (e.g., epigenomics) and analytics to single cohorts. Second, we find that such discovery work is often best understood as part of a broader whole, wherein multiple datasets can be combined to discover and validate precision medicine tools for disease diagnosis, prognosis, and predictive enrichment. Still, while hundreds of academic works have demonstrated potential tools, almost none have made it to commercial application. A new test must have a market impact that incentivizes commercial entities to spend resources on analytical validation, clinical validation, regulatory submission, and commercialization. Only fixed models, demonstrated to be calibrated in independent data, and generalizable enough to command a substantial market share, are likely to make a major clinical impact in any field. Still, the growing wealth of clinical data, decreasing costs of "omics" technologies, and rapid development of robust analytical pipelines for analysis increase the likelihood of success of precision medicine tools in the coming years.

References

[1] Sweeney TE, Khatri P. Generalizable biomarkers in critical care: toward precision medicine. Crit Care Med 2017;45(6):934–9.

[2] Bossuyt PM, Reitsma JB, Bruns DE, et al. STARD 2015: an updated list of essential items for reporting diagnostic accuracy studies. BMJ 2015;351:h5527.

[3] Bauchner H, Golub RM, Fontanarosa PB. Data sharing: an ethical and scientific imperative. JAMA 2016;315(12):1237–9.

[4] Haynes W, Vashisht R, Vallania F, et al. Integrated molecular, clinical, and ontological analysis identifies overlooked disease relationships. bioRxiv 2018;. 214833.

[5] Sweeney TE, Haynes WA, Vallania F, Ioannidis JP, Khatri P. Methods to increase reproducibility in differential gene expression via meta-analysis. Nucleic Acids Res 2016;45:e1.

[6] Inthout J, Ioannidis JP, Borm GF. Obtaining evidence by a single well-powered trial or several modestly powered trials. Stat Methods Med Res 2012;25:538–52.
[7] Haynes WA, Vallania F, Liu C, et al. Empowering multi-cohort gene expression analysis to increase reproducibility. Pac Symp Biocomput 2017;22:144–53.
[8] Sangaralingam A, Dayem Ullah AZ, Marzec J, et al. Multi-omic' data analysis using O-miner. Brief Bioinform 2017;20:130–43.
[9] Toro-Domínguez D, Martorell-Marugán J, López-Dominguez R, et al. ImaGEO: integrative gene expression meta-analysis from GEO database. Bioinformatics 2018;35:880–2.
[10] Sharov AA, Schlessinger D, Ko MS. ExAtlas: an interactive online tool for meta-analysis of gene expression data. J Bioinform Comput Biol 2015;13:1550019.
[11] Khatri P, Roedder S, Kimura N, et al. A common rejection module (CRM) for acute rejection across multiple organs identifies novel therapeutics for organ transplantation. J Exp Med 2013;210(11):2205–21.
[12] Mazur PK, Reynoird N, Khatri P, et al. SMYD3 links lysine methylation of MAP3K2 to Ras-driven cancer. Nature 2014;510(7504):283–7.
[13] Sweeney TE, Shidham A, Wong HR, Khatri P. A comprehensive time-course-based multicohort analysis of sepsis and sterile inflammation reveals a robust diagnostic gene set. Sci Transl Med 2015;7(287):287ra271.
[14] Sweeney TE, Braviak L, Tato CM, Khatri P. Genome-wide expression for diagnosis of pulmonary tuberculosis: a multicohort analysis. Lancet Respir Med 2016;4(3):213–24.
[15] Sweeney TE, Wong HR, Khatri P. Robust classification of bacterial and viral infections via integrated host gene expression diagnostics. Sci Transl Med 2016;8(346):346ra391.
[16] Lofgren S, Hinchcliff M, Carns M, et al. Integrated, multicohort analysis of systemic sclerosis identifies robust transcriptional signature of disease severity. JCI Insight 2016;1(21):e89073.
[17] Fourati S, Talla A, Mahmoudian M, et al. A crowdsourced analysis to identify ab initio molecular signatures predictive of susceptibility to viral infection. Nat Commun 2018;9(1):4418. https://doi.org/10.1038/s41467-018-06735-8.
[18] Botling J, Edlund K, Lohr M, et al. Biomarker discovery in non-small cell lung cancer: integrating gene expression profiling, meta-analysis, and tissue microarray validation. Clin Cancer Res 2013;19(1):194–204.
[19] Lydon EC, Ko ER, Tsalik EL. The host response as a tool for infectious disease diagnosis and management. Expert Rev Mol Diagn 2018;18(8):723–38.
[20] Singer M, Deutschman CS, Seymour CW, et al. The third international consensus definitions for Sepsis and septic shock (Sepsis-3). JAMA 2016;315(8):801–10.
[21] Opal SM, Dellinger RP, Vincent JL, Masur H, Angus DC. The next generation of sepsis clinical trial designs: what is next after the demise of recombinant human activated protein C?*. Crit Care Med 2014;42(7):1714–21.
[22] Cohen J, Vincent JL, Adhikari NK, et al. Sepsis: a roadmap for future research. Lancet Infect Dis 2015;15(5):581–614.
[23] Langley RJ, Tsalik EL, van Velkinburgh JC, et al. An integrated clinico-metabolomic model improves prediction of death in sepsis. Sci Transl Med 2013;5(195):195ra195.
[24] Zaas AK, Burke T, Chen M, et al. A host-based rt-PCR gene expression signature to identify acute respiratory viral infection. Sci Transl Med 2013;5(203):203ra126.
[25] Tsalik EL, Henao R, Nichols M, et al. Host gene expression classifiers diagnose acute respiratory illness etiology. Sci Transl Med 2016;8(322):322ra311.
[26] Sweeney TE, Perumal TM, Henao R, et al. A community approach to mortality prediction in sepsis via gene expression analysis. Nat Commun 2018;9(1):694.

[27] Wong HR, Shanley TP, Sakthivel B, et al. Genome-level expression profiles in pediatric septic shock indicate a role for altered zinc homeostasis in poor outcome. Physiol Genomics 2007;30(2):146–55.

[28] Scicluna BP, Klein Klouwenberg PM, van Vught LA, et al. A molecular biomarker to diagnose community-acquired pneumonia on intensive care unit admission. Am J Respir Crit Care Med 2015;192(7):826–35.

[29] Ramilo O, Allman W, Chung W, et al. Gene expression patterns in blood leukocytes discriminate patients with acute infections. Blood 2007;109(5):2066–77.

[30] Mahajan P, Kuppermann N, Mejias A, et al. Association of RNA biosignatures with bacterial infections in febrile infants aged 60 days or younger. JAMA 2016;316(8):846–57.

[31] Sweeney TE, Khatri P. Benchmarking sepsis gene expression diagnostics using public data. Crit Care Med 2017;45(1):1–10.

[32] Wong HR, Cvijanovich N, Lin R, et al. Identification of pediatric septic shock subclasses based on genome-wide expression profiling. BMC Med 2009;7:34.

[33] Davenport EE, Burnham KL, Radhakrishnan J, et al. Genomic landscape of the individual host response and outcomes in sepsis: a prospective cohort study. Lancet Respir Med 2016;4: 259–71.

[34] Burnham KL, Davenport EE, Radhakrishnan J, et al. Shared and distinct aspects of the sepsis transcriptomic response to fecal peritonitis and pneumonia. Am J Respir Crit Care Med 2017;196:328–39.

[35] Sweeney TE, Azad TD, Donato M, et al. Unsupervised analysis of transcriptomics in bacterial sepsis across multiple datasets reveals three robust clusters. Crit Care Med 2018;46:915–25.

[36] Wong HR, Cvijanovich NZ, Allen GL, et al. Validation of a gene expression-based subclassification strategy for pediatric septic shock. Crit Care Med 2011;39(11):2511–7.

[37] Wong HR, Cvijanovich NZ, Anas N, et al. Developing a clinically feasible personalized medicine approach to pediatric septic shock. Am J Respir Crit Care Med 2015;191(3):309–15.

[38] Scicluna BP, van Vught LA, Zwinderman AH, et al. Classification of patients with sepsis according to blood genomic endotype: a prospective cohort study. Lancet Respir Med 2017;5(10):816–26.

[39] Wong HR, Sweeney TE, Hart KW, Khatri P, Lindsell CJ. Pediatric sepsis endotypes among adults with Sepsis. Crit Care Med 2017;45(12):e1289–91.

[40] Vranas KC, Jopling JK, Sweeney TE, et al. Identifying distinct subgroups of ICU patients: a machine learning approach. Crit Care Med 2017;45(10):1607–15.

[41] Mayhew MB, Petersen BK, Sales AP, Greene JD, Liu VX, Wasson TS. Flexible, cluster-based analysis of the electronic medical record of sepsis with composite mixture models. J Biomed Inform 2018;78:33–42.

Chapter 4

Clinical metagenomics for infection diagnosis

Schlaberg Robert[a,b,c], Laura Filkins[d]
[a]*Department of Pathology, University of Utah, Salt Lake City, UT, United States,* [b]*ARUP Laboratories, Salt Lake City, UT, United States,* [c]*IDbyDNA Inc., San Francisco, CA, United States,* [d]*Department of Pathology, University of Texas Southwestern, Dallas, TX, United States*

Applications of metagenomics in clinical microbiology and precision diagnostics

Advanced diagnostics

Traditional laboratory workup for suspected infectious diseases largely relies on culture and nucleic acid amplification tests (NAAT). Culture-based tests require viable microorganisms, adequate culture conditions, and time for sequential growth, identification, antimicrobial susceptibility testing, and epidemiologic typing, if needed. NAAT are specific to a limited number of targeted pathogens, need to be frequently updated to accommodate new, newly recognize, or genetically variant pathogens, and provide no or very limited genotyping information. Thus, the efficacy of conventional diagnostics varies by the type of infection, patient demographics, immune status, and the breadth of the test menu available at a given institution. While effective for the diagnosis of diseases caused by a single pathogen, conventional tests are often non-informative for infectious syndromes that can be caused by dozens or even hundreds of different pathogens that cause similar clinical phenotypes but require different treatment [1–6].

Clinical metagenomics enables universal pathogen detection and therefore provides advantages for testing of patients with infectious syndromes that can have many causes, such as bloodstream infections, bone and joint infections, meningitis/encephalitis, ocular infections, gastroenteritis, and respiratory tract infections (see Table 1). PCR amplification of marker genes that are present in all bacteria (e.g. 16S ribosomal RNA gene, 16S rRNA) and fungi (e.g. internal transcribed spacer sequences, ITS) has been widely used in research studies and demonstrated utility in select clinical settings. However, the absence of such universal marker genes for viruses, the greater accuracy and richer information from shotgun metagenomics, and decreasing costs for next-generation

Genomic and Precision Medicine. https://doi.org/10.1016/B978-0-12-801496-7.00004-6

TABLE 1 Overview and examples of metagenomics studies for syndromic testing

Organ system	Diagnosis (examples)	Pathogen (examples)	References
Bloodstream and systemic infections	Fever of unknown origin, sepsis	Parvovirus B19, dengue virus, chikungunya virus, JC polyomavirus, human hepegivirus-1, anelloviruses, HIV, *Enterobacter cloacae*, *Pseudomonas aeruginosa*, *C. burnetii*, *Mycobacterium immunogenum*, *P. falciparum*, *Orientia tsutsugamushi*	[7–17]
Bone and joints	Prosthetic joint infection	*Mycoplasma salivarium*, *Parvimonas micra*, *Morganella morganii*, *Citrobacter koseri*, *Staphylococcus aureus*, *Cutibacterium acnes*	[18–22]
Central nervous system	(Chronic) meningitis, meningoencephalitis, (progressive) encephalitis	*Leptospira borgpetersenii*, *Neisseria meningitidis*, *Brucella melitensis*, Echovirus, Hepatitis E virus, Cache Valley virus, West Nile virus, Variegated squirrel bornavirus 1, astrovirus, Japanese Encephalitis virus, *Balamuthia mandrillaris*	[23–34]
Eye	Uveitis, endophthalmitis, keratitis	Rubella virus, HSV-1, CMV, VZV, HHV6, *Capnocytophaga* sp., *Streptococcus agalactiae*, *Staphylococcus aureus*, *Mycobacterium abscessus*, *Fusarium* sp., *Cryptococcus neoformans*, *Toxoplasma gondii*, *Brugia malayi*	[35–40]
Gastrointestinal tract	Gastroenteritis	Anellovirus, adenoviridae, astroviridae, caliciviridae, picornaviridae, reoviridae, protoparvoviruses, *Clostridium difficile*, *E. coli*, *Salmonella enterica*, protozoa	[41–48]
Respiratory tract	Pneumonia, cystic fibrosis, upper respiratory tract infection	Various, gemykibivirus, *Chlamydophila psittaci*	[49–55]

sequencing (NGS) make shotgun sequencing increasingly the method of choice for clinical metagenomics. Because of its unbiased nature and ability to simultaneously detect pathogens and profile commensals, metagenomics can also provide a more complete picture of interactions between microorganisms and the state of the patient's microbiota. By directly sequencing microbial genomes, metagenomics can identify and genetically profile pathogens as well as detect genetic markers of antimicrobial drug susceptibility or resistance. Clinical utility and improved diagnostic yield have been demonstrated over the past several years and diagnostic laboratories are now starting to incorporate clinical metagenomics into routine workflows. This chapter describes principles, applications, and common workflows for precision diagnostics in the management of patients with infectious diseases.

Pathogen discovery

Diagnostic metagenomics was pioneered for pathogen discovery when conventional diagnostics failed to identify the cause of clusters of suspected infections. Early examples include the following discoveries: Bas-Congo virus (a rhabdovirus) and Lujo virus (an arenavirus), both novel hemorrhagic fever-associated viruses; a novel arenavirus associated with a cluster of fatal, transplant-associated infections; and FTLS virus from patients with severe fever, thrombocytopenia and leukopenia syndrome [56–59]. More common than the discovery of entirely new viruses is the detection of new variants of known pathogens some of which may escape detection by NAATs designed to detect known strains. For example, Xu et al. reported a new human parainfluenza virus 3 variant that was missed by NAAT due to an insertion in the target gene [60]. Because metagenomic sequencing requires no a priori knowledge of expected pathogens or their genome sequences, it is a powerful tool to detect and profile genetic variants as long as minimal sequence similarities with known reference genomes exist.

Antimicrobial resistance

NGS also enables detection of genes or genetic mutations that confer resistance to antimicrobial drugs in viruses, bacteria, fungi, or parasites. Whole genome sequencing of bacterial isolates to identify and track transmission events and to identify mechanisms of drug resistance is among the earliest NGS applications in clinical microbiology. Bacterial whole genome analysis can be performed based on metagenomics without prior isolation of bacteria. For example, metagenomics has been used to detect and identify *Mycobacterium tuberculosis* from sputum specimens and identify genetic mutations that predict antimycobacterial drug resistance [61–63]. While often more challenging, progress has also been made for a number of other bacterial pathogens [64]. Recently, proof of concept studies have demonstrated that analysis of bacterial genomes can be used to predict not just resistance but also minimal inhibitory concentrations (MIC) when large enough training sets are combined with machine learning approaches [65, 66]. NGS is also rapidly replacing Sanger sequencing for identification of resistance-conferring mutations in viruses (e.g. HIV, CMV) resulting in increased detection of low prevalence resistance-associated mutations and workflow advantages due to high-throughput testing [67–70].

Molecular typing

Because metagenomics can provide microbial whole genome sequences, this methodology also generates genotypic information that may be applied for molecular epidemiology studies. Sequencing directly from a clinical specimen further enables genotypic analysis at the time of diagnosis and without the need to first obtain isolates in pure culture. This allows for real-time monitoring of transmission patterns inside the hospital or in the community. By breaking down the traditional differentiation of diagnostic and surveillance efforts, clinical metagenomics is a powerful tool to accelerate identification and containment of infectious disease outbreaks [71].

Community profiling

NGS-based metagenomics enables high-resolution profiling of microbial communities, both human-associated ones and others. Associations between distinct intestinal and respiratory microbial profiles with health or disease have generated immense interest in the microbiota as a modifiable risk factor for a wide range of conditions [72–76]. Fecal microbiota transplants (FMT) are only one example of successful interventions that aim to alter microbial communities [77]. As these interventions become part of precision medicine, the need to monitor microbial communities will grow, generating the need to perform high-quality, reproducible, diagnostic-grade microbial community profiling in regulated diagnostic laboratories.

Scope and definitions

This chapter focuses on the use of metagenomics for pathogen detection directly from clinical specimens collected from the site of infection. Applications of the detection of cell-free microbial DNA that may have been shed from the site of infection and is circulating in plasma are not covered here [78]. The term “metagenomics” is often used to refer to different amplification-based and amplification-independent methods. In this chapter, we use metagenomics for unbiased (also known as “shotgun”) sequencing of libraries that were prepared without sequence-based enrichment (such as by PCR, also known as amplicon sequencing). Shotgun metagenomics can be performed using DNA or RNA (after reverse transcription to cDNA). The latter is also referred to as "metatranscriptomics." For simplicity and because diagnostic metagenomics generally requires DNA and RNA-based detection, we will refer to shotgun sequencing of both DNA and RNA as metagenomics.

Current state of clinical metagenomic testing

Syndromic testing currently relies on microscopy, culture (aerobic and anaerobic bacterial culture, fungal culture, mycobacterial culture), serologic testing, and multiplex NAAT panels for viruses [79]. Culture often lacks sufficient sensitivity (especially in patients pretreated with antimicrobials) and can take days to weeks. While NAAT testing is sensitive, specific, and rapid, it has an inherently limited scope (especially for bacteria and fungi), limited resolution regarding genetic markers for drug resistance, and often return negative results. Besides detecting unexpected or uncommon pathogens, clinical metagenomics can also decrease time-to-results for slow growing or hard to diagnose organisms, detect bacteria in patients who have been treated with antibiotics, and provide rapid, high-resolution organism identification and antimicrobial resistance information to support optimal treatment choices [80]. By answering multiple diagnostic questions with a single test, this has the potential to reduce costs.

Until recently, high sequencing costs, complex specimen processing, and time-consuming data analysis requiring expert interpretation have hindered broad adoption of clinical metagenomics tests in diagnostic microbiology laboratories. A first application was as a method of last resort in patients with negative results from conventional tests despite high levels of suspicion of an infectious etiology [23]. At the time of writing, a growing number of diagnostic laboratories are introducing clinical metagenomics testing. For the full potential to be realized and infectious etiologies to be identified (or excluded) faster, clinical metagenomics testing will need to be incorporated early in diagnostic testing algorithms. This will help to reduce costs by eliminating unnecessary laboratory and imaging studies and personalize patient cases with the goal of shortening intensive care or inpatient treatment. Rapid progress is being made toward these goals, but several barriers for widespread adoption by diagnostic laboratories still exist. While the power of metagenomics has been well documented [23–25, 49, 81], few systematic performance comparisons have been published. These studies are complicated by the lack of established reference methods (gold standard) to help interpret frequent detections of microorganisms by clinical metagenomics tests that are missed by conventional approaches. Lastly, there is limited guidance available for laboratories to navigate regulatory requirements for this novel technology. Approaches to overcome these challenges are summarized below.

Applications and published performance evaluation studies

Few performance evaluations of clinical metagenomics tests have been published. One such initial study compared detection of respiratory viruses by unbiased RNA sequencing to detection by a commercial respiratory virus PCR panel [50]. This study demonstrated 86% sensitivity compared to respiratory virus panel (RVP)-positive specimens, detection of 12 additional viruses not targeted by the RVP, and high reproducibility of sequencing results between replicate specimen preparation and sequencing assays. Another pilot study employed RNA and DNA sequencing to brain tissues. This study demonstrated the utility of both positive and negative pathogen detection in making a clinical diagnosis as well as targeting further work-ups [82]. In a report by Parize et al., untargeted next-generation sequencing was compared to conventional methods in a head-to-head comparison for the diagnosis of diverse infections in immunocompromised patients and resulted in about 17% increased detection of likely bacterial or viral pathogens compared to conventional methods. This trial also yielded a high negative predictive value (NPV) of 98% [83]. While these and similar studies provide very encouraging results, additional work is needed, including standardized approaches to assess sensitivity, specificity, and accuracy for the large number of relevant pathogens that can be detected by clinical metagenomics in a given specimen type. An overview of metagenomics testing in a range of infectious syndromes is provided in Table 1.

Challenges and potential solutions for diagnostic application

While clinical metagenomics tests provide many advantages, risks and limitations of any new technology need to be understood and mitigated as much as possible. Here, we describe the technical and clinical challenges of metagenomic analyses for infectious disease diagnosis and suggest approaches to improve test characteristics while minimizing sources of potential error.

Pre-analytical factors

As with any laboratory test, pre-analytic factors affect performance. Specimen collection, transport, and storage can influence results of metagenomics tests in many similar manners as well as unique ways that differ from culture or PCR-based tests. Pre-analytical steps should be designed with this in mind as different types of tests often need to be performed on the same specimen. For example, culture requires preservation of viable or infective microorganisms and freezing of specimens is usually not a suitable way to store specimens. For NAAT, preserving intact nucleic acid (DNA or RNA) is important. This can be done by freezing or the addition of stabilizing buffers but preserving the overall composition of specimens (e.g. arresting growth of commensal microorganisms) is less critical. For metagenomics tests, the overall specimen composition (ratio of pathogen to host and other microbial cells) can affect test performance and contamination of specimens during collection or by reagents can complicate result interpretation [84]. For example, a sputum specimen stored at room temperature for 48 h may [1] reduce viability of pathogens for recovery by culture, [2] result in over-growth of commensal reducing sensitivity of metagenomics-based pathogen detection, and [3] result in degradation of pathogen nucleic acids reducing sensitivity of NAAT. Adequate collection, transport, and storage conditions are required to support multimodal testing.

Specimen processing

Preparation of specimens for NGS includes at a minimum RNA and/or DNA extraction, library preparation, and sequencing. Workflows currently require numerous steps and are more complex than many other molecular tests commonly performed in clinical laboratories. As a result, ensuring consistent quality is both more difficult and essential.

Choice of target nucleic acid, RNA vs DNA

To enable detection of all pathogen classes, RNA and DNA need to be analyzed. RNA is required for detection of RNA viruses and can improve sensitivity for the detection of bacteria, fungi, and parasites. This is because ribosomal RNA (rRNA) is present in multiple copies per cell, increasing sensitivity for detection, and because of the more favorable ratio of pathogen-to-host RNA abundance (compared to DNA) facilitating detection of pathogen RNA. In addition, RNA-based detection can help in differentiating latent from active infections (e.g. for DNA viruses) by providing evidence for active replication [85]. Conversely, sequencing of pathogen DNA enables whole genome analyses, which can provide important information for molecular typing and identification of genes involved in antimicrobial drug resistance and pathogenicity.

Nucleic acid extraction

High-quality and sufficient quantity of nucleic acid are required. Methods need to be optimized to ensure efficient lysis of all relevant pathogens and maximize the ratio of pathogen-to-host (or commensal) nucleic acid as this will increase analytical sensitivity [84]. Although metagenomics tests can theoretically detect any pathogen, it may be useful to determine which pathogens are relevant for a given application or specimen type as optimal extraction methods may differ. Organisms with thick cell walls (e.g. molds, mycobacteria) frequently require mechanical lysis, such as bead beating methods. Despite improvements in sequencing library preparation methods, specimens with low nucleic acid extraction yields (i.e. acellular specimens) may be insufficient for direct input into the library preparation workflow. Such samples may require pre-amplification prior to library preparation. In high cellularity specimens, total nucleic acid yield is often sufficient, but the ratio of pathogen-to-host nucleic acid may be low resulting in decreased sensitivity of pathogen detection. Relative concentration of pathogen nucleic acid can be increased by steps to remove host cells ("de-hosting," e.g. by preferential lysis of host cells followed by removal of released host DNA). De-hosting methods are based on the concept that pathogen DNA is protected by more robust bacterial cell walls or viral capsids and can substantially increase analytical sensitivity in adequately preserved specimens [86, 87]. Finally, many commonly used reagents and consumables (e.g. enzymes, extraction reagents) contain low levels of contaminating microbial nucleic acid [88]. Selection of high-purity reagents can help reduce detection of contaminating microbial nucleic acid, simplify result interpretation, and reduce the likelihood of false-positive or false-negative results.

Library preparation and library quality control (QC)

Most laboratories rely on commercial library preparation kits that may require customization for a given application. Available methods have become faster, involve fewer steps, and are increasingly compatible with automation on liquid handling instruments. These are welcome developments for diagnostic laboratories considering clinical metagenomics tests. In addition, nucleic acid extraction and library preparation protocols for some sequencing platforms (e.g. nanopore sequencing) inherently involve fewer steps that those for sequencing-by-synthesis based or semiconductor chip-based sequencing methods (e.g. Illumina, Thermo Fisher Scientific). Factors influencing the selection of library preparation methods include: [1] requirement for and extent of sample barcoding (allowing combination of multiple specimens in a single sequencing run to reduce costs); [2] requirement for quantitative analyses and tolerance to read duplication (e.g. during PCR steps); unique molecular identifiers (or molecular barcodes) can be used when necessary [89]; [3] performance of host-depletion or target enrichment steps, when applicable (see below); [4] costs; and [5] sequencing platform to be used. Unique, dual-indexed adapters are generally preferred as part of clinical metagenomics tests.

The addition of dual-indexed (barcoded) sequencing adapters for sample barcoding substantially reduces the frequency with which sequencing reads are associated with the wrong specimen (index cross-talk, barcode hopping), which can result in false-positive results appearing like sample-to-sample contamination events [90]. Read duplication occurs when a single nucleic acid fragment is amplified during library preparation steps resulting in multiple identical sequencing reads. This can inflate read counts assigned to a given microorganism, thus affecting community profiles or abundance estimates [91]. Library preparation methods with minimal sequence duplication and maximum diversity typically yield higher-quality results.

Next-generation sequencing

The optimal sequencing chemistry and platform will depend on the given application and factors such as required data output (amount of sequencing data and/or number of sequencing reads generated), sequencing read length, sequencing error profile, time required per sequencing run, and reliability of reagents and instrumentation. Review of NGS technologies is beyond the scope of this chapter. Currently available sequencing platforms and their performance characteristics have been recently reviewed [92–94]. Both systematic and random sequencing errors can occur in all sequencing technologies, but overall error profiles differ substantially between available platforms. Acknowledging the inherent limitations of each sequencing technology, strategies to compensate or correct for sequencing errors have been developed [95–97]. The degree of robustness needed for a given data analysis strategy and susceptibility of the clinical application to different types of errors will guide the selection of a sequencing platform. For clinical metagenomics, the total number of reads (depth) generally correlates with the analytical sensitivity of the test [98]. Therefore, a larger number (millions or tens of millions) of shorter reads (usually <500 nucleotides) per specimen may be preferable. The possibility of tailoring sequencing depth to a specific specimen has been demonstrated with the Oxford Nanopore MinION platform. Because sequencing data from the MinION platform can be analyzed in real time, sequencing depths may be reduced (sequencing time shortened and costs reduced) for specimens with higher pathogen load and once a positive result is obtained [99]. While this provides opportunities for more efficient use of resources, decreased costs, and shorter turn-around times, implementation of this concept in standardized diagnostic workflows may be challenging. While current short read platforms provide greater sequencing depth, advantages of longer reads include the ability to link antimicrobial resistance genes to a given microorganism (by sequencing a resistance gene in the context of the genome it is contained in), to improve species or strain-level classification, to phase mutations (i.e. determine whether two or more mutations are present in the same or on different genomes), and to improve assembly of plasmids [100–104].

Data analysis

Slow speed and variable accuracy of metagenomic data analysis, and the need for expertise in result interpretation has slowed adoption of clinical metagenomic tests by diagnostic laboratories. The scale and complexity of data analysis, incompleteness of reference sequence databases, and taxonomic inconsistencies across different classes of pathogens pose additional challenges [105]. In research applications, multiple analysis approaches are often used in combination with customized databases, interpretive strategies tailored to the given question, and usually require extensive bioinformatics expertise. Results even of artificial control datasets differ widely depending on what analysis tool and database have been used [106]. In contrast, diagnostic testing requires strict adherence to pre-determined procedures, version-controlled software and databases, and validated interpretative criteria. In addition, analysis steps need to be optimized for rapid turn-around time to maximize clinical utility. A comprehensive review of challenges and potential solutions is beyond the scope of this chapter. The following section highlights some of these challenges and their potential impact on clinical metagenomics tests.

Metagenomics classification tools

Numerous metagenomics classification tools have been developed [106–111]. They are often developed for a specific use and identifying the most suitable solution can be challenging. A first step in many metagenomic classification tools is to assign each sequencing read to the most relevant (generally the most similar) reference sequence. The result of this step is a catalog of taxa to which one or more sequencing reads were assigned. While the resulting read counts for different taxa can be useful to infer microbial community structures, clinical metagenomics tests need to ultimately provide information on the presence or absence of relevant microorganisms. Thus, interpretive criteria need to be established and validated. To minimize analysis times, a number of strategies have been developed, including reducing the number of query sequences, size of reference databases, and speeding up classification algorithms [112]. However, it is important to consider the impact of these strategies on the accuracy of results. For example, assembly of raw sequencing reads into a much smaller number of longer contiguous sequences (contigs) reduces the number of query sequences. However, assembly may not be successful when a pathogen is present at low abundance thus limiting analytical sensitivity. Similarly, smaller reference databases will speed up queries but reducing their completeness will increase the risk of false-negative (pathogen-derived sequences do not match the correct reference closely enough to be identified) and false-positive (mis-assignment of sequences to the wrong reference) results. Alignment-free tools can substantially increase the speed of classification algorithms but they may not always provide sequence comparison metrics that users are familiar with.

Reference sequence databases and meta-databases

As briefly discussed above, reference databases must be carefully selected and thoroughly validated. Options include comprehensive public reference sequence databases (e.g. NCBI [113]), curated databases focusing on the most relevant human pathogens (e.g. FDA-ARGOS [114]), on different classes of pathogens (e.g. fluDB [115]), or on select applications (e.g. CARD, comprehensive antibiotic resistance database [116]), to name just a few. Alternatively, specific databases can be developed in-house. Comprehensive publicly-available databases are attractive options. However, they are plagued by extensive mis-annotation of sequences, variable completeness, and substantial overrepresentation of some taxa (see Fig. 1). These issues are further complicated by the need for frequent

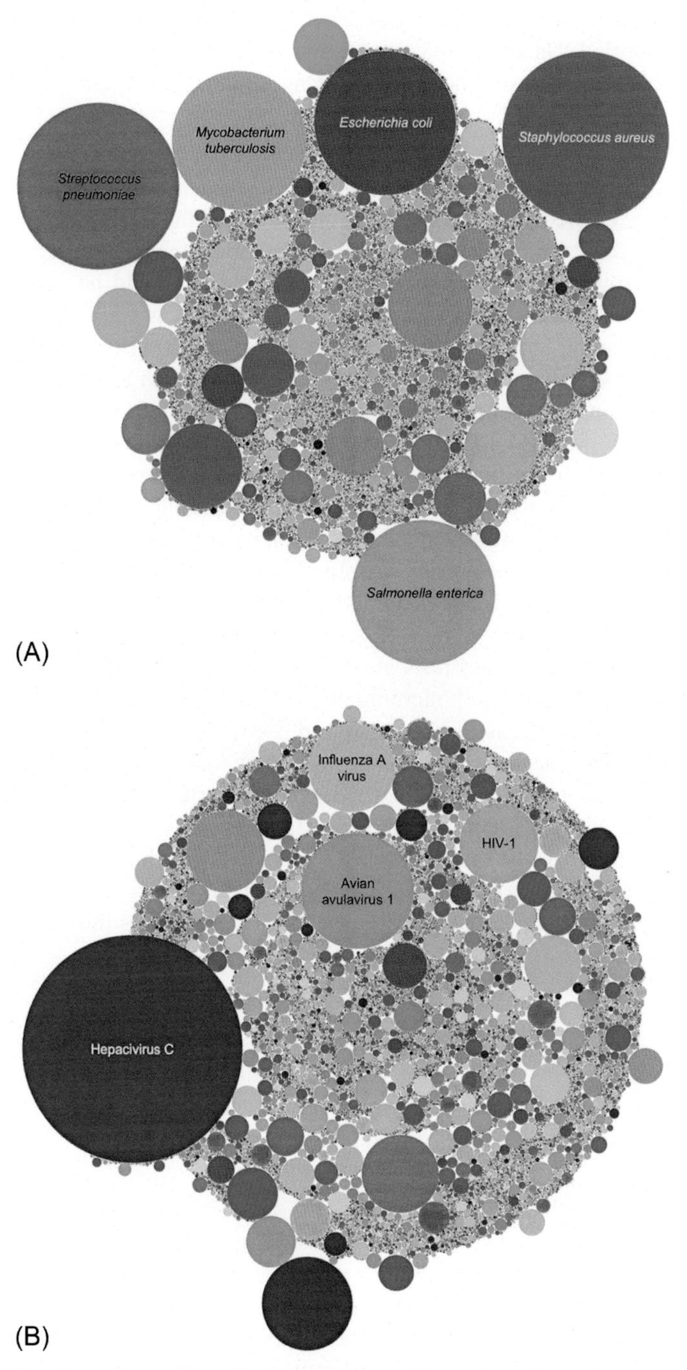

FIG. 1 Overrepresentation of commonly studied bacteria in the NCBI RefSeq genome database (A) and of viral reference sequences belonging to hepacivirus C (hepatitis C virus), HIV and influenza virus in the NCBI nt database (B). Less commonly studied organisms may be represented by a single reference sequence or completely absent in public databases.

database updates to ensure that newly sequenced or newly emerging pathogens can be identified. Database updates need to be validated to ensure consistent performance, which poses additional challenges to diagnostic laboratories.

Result reporting

The ability to detect and identify virtually any microorganism with a single test poses unique challenges to the reporting of results and the interpretation of clinical metagenomics tests. It is critical to balance completeness with actionability of results. How detailed and comprehensive a report should be may depend on the intended use of a test, the patient population (e.g. unusual pathogens may be relevant in immunocompromised patients), the experience of the ordering providers with the technology, and the degree to which additional consulting may be provided (some laboratories have implemented or are considering implementing "metagenomics boards" analogous to tumor boards to review and interpret results). To facilitate interpretation of diagnostic reports, it is important to prioritize results based on known pathogenicity and other available information regarding detected microorganisms. For specimens from non-sterile body sites (e.g. respiratory specimens), quantitative or semiquantitative results may facilitate interpretation. While generally assumed that highly abundant microorganisms are more likely to be relevant, this needs to (and can be) confirmed using clinical metagenomics tests. As part of these efforts, the expected abundance of commonly detected microorganisms in asymptomatic individuals (i.e. reference intervals) needs to be established. No single answer can accurately address all clinical scenarios, specimen types, patient populations, and test designs. Therefore, it is critical to clearly define what criteria are used for reporting, carefully validate their use, and encourage transparency of analysis and reporting criteria for the end users (e.g. clinicians). Mechanisms for revising reporting procedures and method re-validation enable continuous improvement as clinical metagenomics tests in response to clinical findings [84].

Quality control

Quality checks can be implemented at a variety of steps within the workflow. They are essential to ensure consistent quality and identify problematic runs or questionable results for individual specimens. Acceptability criteria need to be defined and questionable results need to be rejected or repeated.

Run-level quality control

Run-level quality is assessed by the inclusion of external positive and negative controls [84]. One challenge for determining QC pass/fail criteria is to ensure high quality results without setting overly conservative criteria for acceptability, which could lead to unnecessary specimen repeats, delay in result reporting, and extra costs. Positive controls ensure adequate performance of the entire workflow. Results can be interpreted qualitatively (expected pathogens are detected) and quantitatively (expected pathogens detected at the expected abundance). Negative controls can consist of matrix-matched specimens (reflecting the composition of the relevant specimen type) or blank specimens (maximizing the ability to detect contamination, e.g. from reagents) [84, 117].

Sequencing quality

Metrics to monitor sequencing quality will depend on the sequencing platform used. Some commonly used metrics include cluster density of the flow cell, amount of sequencing data generated in a given run, base call quality, the proportion of sequencing data that passes manufacturer-defined specifications, and the error rate as determined by inclusion of a control library of consistent quality [118].

Specimen adequacy

As discussed above, quality and quantity of extracted nucleic acid can affect quality of sequencing libraries and data. Criteria need to be defined and validated to reject or accept extracted nucleic acid and/or resulting sequencing libraries for next-generation sequencing analysis. Internal controls (whole organisms) spiked into patient specimens and associated external controls can be used as process controls and to ensure adequate analytical sensitivity was achieved in a given specimen [84]. Alternatively, specimens can be spiked with modified or synthetic nucleic acid molecules, but this approach does not control for extraction efficiency. Nucleic acid and/or library size distribution and library concentration need to be monitored, as does the quantity and quality of resulting sequences for each specimen.

Reagent, environmental, and sample-to-sample contamination

Enzymes, reagents, and nucleic acid extraction kits used in metagenomics workflows often contain variable amounts of contaminating microbial nucleic acid. These residual nucleic acids originate from numerous sources, including: recombinant manufacturing processes, raw materials, common reagents, plasticware, and the laboratory environment [88, 119]. These reagent contaminants can vary from lot to lot and can be identified by the adequate use of negative and/or blank controls. Because of the unbiased nature of library preparation steps, all nucleic acid molecules compete for inclusion in the sequencing library. Therefore, contaminants will be more apparent in specimens with low cellularity (less competition from the specimen) than in highly cellular specimens (more competition with the background contaminants) [84]. Internal controls can be used to correct and control for this effect as they are spiked into all specimens in constant amounts. Environmental contamination can also occur during specimen processing and multiple negative controls may be needed to ensure detection of those events. "Index hopping" can manifest itself as an apparent sample-to-sample contamination event [90, 120, 121]. Methods to reduce the risk and/or frequency of "index hopping" were discussed above. Identification of low-frequency contaminants, particularly when the organism is also a potential human pathogen, is especially challenging. One approach is the continuous performance of meta-analyses including prior specimens run in the lab to help identify infrequent organism detection that may be contamination rather than specimen derived [122].

Areas for ongoing and future improvements

Advances in specimen preparation workflows, sequencing technology, and data analysis have enabled the introduction of the first clinical metagenomics tests in reference and public health laboratories. Further progress is being made at an accelerating pace to continuously improve test performance, speed up laboratory and data analysis workflows, reduce costs, and expand applications for metagenomics tests. These improvements will help make testing more widely available and clinically useful.

Analytical sensitivity

An important feature of infectious disease tests is the ability to detect pathogens at low abundance. This ability is usually quantified by determining the test's limit of detection (LOD) or the lowest target quantity that can be consistently detected. While the LOD of PCR-based tests is relatively independent of the specimen matrix (i.e. the composition of the specimen), non-pathogen DNA or RNA in the specimen competes with pathogen detection in clinical metagenomics tests and can affect their LOD. Therefore, it is important to establish the LOD in a representative specimen matrix using known pathogen concentrations determined by reference methods such as quantitative PCR, quantitative culture, or plaque assays [84, 94]. Representative specimen matrices and quantified reference pathogen preparations that are suitable for metagenomics testing are not readily available and laboratories often prepare their own materials. Standardization and commercial availability of materials used to assess the LOD of clinical metagenomics tests will make it easier to compare performance of different tests and protocols, improve test performance, and communicate test results in a consistent and clear manner. The composition of individual specimens may differ from the representative matrix in which the LOD was determined. Therefore, it is important to identify specimens with highly unusual composition and communicate on the report that the LOD may differ from the established one. Internal controls that are spiked into specimens in constant quantities can be used to determine how many individual specimens may differ in their composition from the representative matrix and analytical sensitivity may be reduced.

Host depletion and pathogen enrichment

An intuitive way to improve the sensitivity of a clinical metagenomics test is to remove competing cells or nucleic acids (e.g. those from the patient, "host depletion") or enrich for pathogen cells or nucleic acids (pathogen enrichment) during specimen processing. Numerous protocols for depletion or enrichment have been published and some are commercially available [87, 123–126]. Unfortunately, host depletion methods often add to the complexity of workflows, increase turn-around times, are difficult to standardize, and may require highly-controlled specimen collection and storage that may not be achievable in routine diagnostic practice. Pathogen enrichment protocols can also introduce bias and preferential detection of some pathogens, reduce the scope of a test intended to be all inclusive, and limit pathogen genetic information generated to genomic regions that can be enriched. For these reasons, host depletion protocols may provide greater benefits than pathogen enrichment, but optimal solutions have not become available yet.

Specimen preparation workflows

Laboratory workflows need to be further improved to increase speed, automate specimen preparation steps, minimize the risk for sample-to-sample contamination by using closed systems similar to current real-time PCR workflows, and develop reagents and consumables that are free of contaminating microbial nucleic acid.

Next-generation sequencing technology

While turn-around times of days or even weeks may be tolerable for some NGS applications, clinical metagenomics tests need to produce results as quickly as possible to provide clinical utility [92]. Ideally, results should be available in 24 h or less. Sequencing times contribute 50% or more of the overall turn-around time of current clinical metagenomics tests. Clearly, faster sequencing technologies are needed to realize the full potential of clinical metagenomics tests. Reduced sequencing costs and flexible batch sizes (i.e. consistent sequencing depth and costs per specimen without the need for batching a large number of specimens) will further improve adoption and utility.

Data analysis and result interpretation

Improvements in data analysis and result interpretation steps are aimed at [1] more rapid processing, [2] generation and validation of comprehensive databases to support the different clinical needs (e.g. pathogen detection, quantification, prediction of antimicrobial susceptibility and resistance, molecular profiling of pathogens), and [3] increasingly automated interpretation and removal of expert interpretation steps to further improve consistency of results.

Data analysis speeds have dramatically increased with the use of alignment-free strategies [108, 109, 127, 128]. As discussed above, in addition to compute times, current data analysis solutions generally require a human (often expert) interpretation step. The time required for expert interpretation needs to be taken into consideration when comparing data analysis speeds, especially as this interpretive step is often not readily scalable. Efforts such as machine learning-assisted interpretation promise to further speed up and simplify data analysis in the near future. However, increasingly automated result interpretation will also increase the need for standardized approaches to evaluate and ensure adequate performance of these steps. Curated, regulatory-grade reference sequence databases, such as the FDA-ARGOS database [114], are an important part of these efforts. Standardized datasets with known results and generated with (or simulated based on) different sequencing platforms will also provide means to compare and ensure adequate performance. Finally, easily implementable computer hardware solutions and intuitive user interfaces for data analysis and result review are essential for clinical metagenomics tests to be adopted by diagnostic laboratories without requiring large investments in computing and bioinformatics infrastructure [122].

Clinical utility

While clinical utility has been demonstrated in individual case reports, case series, and first prospective trials, additional studies are needed to determine in what specific clinical scenarios metagenomics testing provides the greatest utility. As these studies are being conducted, it is important to keep in mind that specimen preparation, sequencing, and data analysis technologies are evolving rapidly and that results and conclusions will change over time.

As with many new technologies in medicine, the most common initial application of clinical metagenomics testing has been in critically-ill patients after standard testing approaches have been exhausted. The disadvantages of this approach are even longer times to diagnosis and accumulating costs for numerous individual tests (see [34] for an example of the battery of tests performed on a patient with an unusual cause of meningoencephalitis that was identified by clinical metagenomics testing). With decreasing turn-around times and costs for clinical metagenomics testing, implementation earlier in diagnostic algorithms may improve outcomes and be cost effective, especially is specific patient populations. While pathogen detection with conventional tests may often be successful in previously healthy patients, uncommon and/or opportunistic pathogens play an important role in immunocompromised patients and may not be detectable by conventional tests. Therefore, clinical metagenomics tests may be warranted earlier in the diagnostic work-up of transplant recipients, cancer patients, severely ill patients, infants, the elderly, and other vulnerable populations to cast a wide net and potentially reduce time to adequate treatment as well as cost.

Another clinical scenario in which metagenomics testing may provide substantial advantages early in the diagnostic workup is for testing of specimens that have very limited volume available. For example, intraocular infections can be caused by a large number of different pathogens, including bacteria, fungi, viruses, and parasites. However, available specimen volumes (often 100–200 μL or less) limit the number of individual tests that can be performed. In these situations, a single test with the ability to detect any relevant pathogen provides many advantages. Therefore, the need to generate data justifying use of this powerful but novel technology needs to be balanced with the immediate benefit that can be provided as part of the diagnostic workup in reducing diagnostic odysseys and shortening time to diagnosis and adequate therapy.

Host-based diagnostics

In situations where the etiologic agent remains elusive despite broad testing, interrogation of the host immune response has the potential to inform treatment decisions. When RNA sequencing is used as part of clinical metagenomics testing, host gene expression information is available as a "byproduct" and may provide actionable information in the future. For example, host gene expression profiles have been used to differentiate true sepsis from systemic inflammatory response syndrome (SIRS) [129]. Similarly, distinct host gene expression profiles in peripheral blood white blood cells have been shown to discriminate between bacterial, viral, and fungal causes of pneumonia and other acute illnesses [130–133]. Host-based testing can provide information to guide appropriate treatment and may help reduce unnecessary or unnecessarily broad antimicrobial therapy [134]. While initial studies demonstrated great promise, thorough evaluation of the diagnostic performance of host-based testing for infectious diseases will be needed, especially when results may lead to discontinuation of antimicrobial therapy.

Conclusions

Broad adoption of clinical metagenomics testing has the potential to revolutionize the infectious disease diagnostics and improve our understanding of the etiology of many infectious diseases. Published case reports and clinical studies demonstrate the power of this technology. To date, most studies have been performed in challenging patient populations and difficult clinical settings. They have established advantages for the diagnosis of unexpected or rare pathogens for which no conventional tests are available. Studies are now being performed to determine the utility of clinical metagenomics tests when included earlier in diagnostic algorithms, providing the opportunity to simplify test selection and shorten time to diagnosis and adequate treatment. The broad scope of clinical metagenomics tests enables detection of any relevant pathogen with a single test from, even from low-volume specimens. This is especially important for testing of precious specimens, such as intraocular fluid, and when recollection of an exhausted specimen would be invasive (e.g. CSF). When antibiotic therapy precludes culture-based detection of bacteria, clinical metagenomics can provide an alternative as PCR-based tests are not available for many important bacterial pathogens. Given the rich information provided by clinical metagenomics tests, medical microbiologists may need to play a more active role in result interpretation and communication than is the case with many conventional tests. By providing pathogen genome information at the time of diagnosis, metagenomics testing has the potential to also inform infection control (e.g. strain typing, mechanisms of drug resistance) and public health efforts (e.g. pathogen surveillance, identification of emerging pathogens) sooner and without additional costs [135].

References

[1] Glaser CA, Honarmand S, Anderson LJ, Schnurr DP, Forghani B, Cossen CK, et al. Beyond viruses: clinical profiles and etiologies associated with encephalitis. Clin Infect Dis 2006;43(12):1565–77.

[2] Choi SH, Hong SB, Ko GB, Lee Y, Park HJ, Park SY, et al. Viral infection in patients with severe pneumonia requiring intensive care unit admission. Am J Respir Crit Care Med 2012;186(4):325–32.

[3] Jain S, Self WH, Wunderink RG, Fakhran S, Balk R, Bramley AM, et al. Community-acquired pneumonia requiring hospitalization among U.S. adults. N Engl J Med 2015;373(5):415–27.

[4] Jain S, Williams DJ, Arnold SR, Ampofo K, Bramley AM, Reed C, et al. Community-acquired pneumonia requiring hospitalization among U.S. children. N Engl J Med 2015;372(9):835–45.

[5] Freifeld AG, Bow EJ, Sepkowitz KA, Boeckh MJ, Ito JI, Mullen CA, et al. Clinical practice guideline for the use of antimicrobial agents in neutropenic patients with cancer: 2010 update by the Infectious Diseases Society of America. Clin Infect Dis 2011;52(4):e56–93.

[6] Murdoch DR, Corey GR, Hoen B, Miro JM, Fowler Jr. VG, Bayer AS, et al. Clinical presentation, etiology, and outcome of infective endocarditis in the 21st century: the international collaboration on endocarditis-prospective cohort study. Arch Intern Med 2009;169(5):463–73.

[7] Grumaz S, Stevens P, Grumaz C, Decker SO, Weigand MA, Hofer S, et al. Next-generation sequencing diagnostics of bacteremia in septic patients. Genome Med 2016;8(1):73.

[8] Blauwkamp TA, Thair S, Rosen MJ, Blair L, Lindner MS, Vilfan ID, et al. Analytical and clinical validation of a microbial cell-free DNA sequencing test for infectious disease. Nat Microbiol 2019;4(4):663–74.

[9] Sanford E, Farnaes L, Batalov S, Bainbridge M, Laubach S, Worthen HM, et al. Concomitant diagnosis of immune deficiency and pseudomonas sepsis in a 19 month old with ecthyma gangrenosum by host whole-genome sequencing. Cold Spring Harb Mol Case Stud 2018;4(6):https://doi.org/10.1101/mcs.a003244.

[10] Dias M, Pattabiraman C, Siddappa S, Gowda M, Shet A, Smith D, et al. Complete assembly of a dengue virus type 3 genome from a recent genotype III clade by metagenomic sequencing of serum. Wellcome Open Res 2018;3:44.

[11] Kafetzopoulou LE, Efthymiadis K, Lewandowski K, Crook A, Carter D, Osborne J, et al. Assessment of metagenomic nanopore and Illumina sequencing for recovering whole genome sequences of chikungunya and dengue viruses directly from clinical samples. Euro Surveill 2018;23(50).

[12] Schreiber PW, Kufner V, Hubel K, Schmutz S, Zagordi O, Kaur A, et al. Metagenomic virome sequencing in living donor-recipient kidney transplant pairs revealed JC polyomavirus transmission. Clin Infect Dis 2018;https://doi.org/10.1093/cid/ciy1018.

[13] Bal A, Sarkozy C, Josset L, Cheynet V, Oriol G, Becker J, et al. Metagenomic next-generation sequencing reveals individual composition and dynamics of anelloviruses during autologous stem cell transplant recipient management. Viruses 2018;10(11):https://doi.org/10.3390/v10110633.

[14] Piantadosi A, Freije CA, Gosmann C, Ye S, Park D, Schaffner SF, et al. Metagenomic sequencing of HIV-1 in the blood and female genital tract reveals little quasispecies diversity during acute infection. J Virol 2019;93(2).

[15] Xu M, Yang Y, Zhou Y, Liu Z, Liu Y, He M. Metagenomics in pooled plasma, with identification of potential emerging infectious pathogens. Transfusion 2018;58(3):633–7.

[16] Kandathil AJ, Breitwieser FP, Sachithanandham J, Robinson M, Mehta SH, Timp W, et al. Presence of human hepegivirus-1 in a cohort of people who inject drugs. Ann Intern Med 2017;167(1):1–7.

[17] Ngoi CN, Siqueira J, Li L, Deng X, Mugo P, Graham SM, et al. The plasma virome of febrile adult kenyans shows frequent parvovirus B19 infections and a novel arbovirus (Kadipiro virus). J Gen Virol 2016;97(12):3359–67.

[18] Huang Z, Zhang C, Li W, Fang X, Wang Q, Xing L, et al. Metagenomic next-generation sequencing contribution in identifying prosthetic joint infection due to parvimonas micra: a case report. J Bone Jt Infect 2019;4(1):50–5.

[19] Thoendel MJ, Jeraldo PR, Greenwood-Quaintance KE, Yao JZ, Chia N, Hanssen AD, et al. Identification of prosthetic joint infection pathogens using a shotgun metagenomics approach. Clin Infect Dis 2018;67(9):1333–8.

[20] Sanderson ND, Street TL, Foster D, Swann J, Atkins BL, Brent AJ, et al. Real-time analysis of nanopore-based metagenomic sequencing from infected orthopaedic devices. BMC Genomics 2018;19(1):714.

[21] Thoendel M, Jeraldo P, Greenwood-Quaintance KE, Chia N, Abdel MP, Steckelberg JM, et al. A novel prosthetic joint infection pathogen, mycoplasma salivarium, identified by metagenomic shotgun sequencing. Clin Infect Dis 2017;65(2):332–5.

[22] Street TL, Sanderson ND, Atkins BL, Brent AJ, Cole K, Foster D, et al. Molecular diagnosis of orthopedic-device-related infection directly from sonication fluid by metagenomic sequencing. J Clin Microbiol 2017;55(8):2334–47.

[23] Mongkolrattanothai K, Naccache SN, Bender JM, Samayoa E, Pham E, Yu G, et al. Neurobrucellosis: unexpected answer from metagenomic next-generation sequencing. J Pediatric Infect Dis Soc 2017;6(4):393–8.

[24] Wilson MR, Naccache SN, Samayoa E, Biagtan M, Bashir H, Yu G, et al. Actionable diagnosis of neuroleptospirosis by next-generation sequencing. N Engl J Med 2014;370(25):2408–17.

[25] Wilson MR, Suan D, Duggins A, Schubert RD, Khan LM, Sample HA, et al. A novel cause of chronic viral meningoencephalitis: cache valley virus. Ann Neurol 2017;82(1):105–14.

[26] Piantadosi A, Mukerji SS, Chitneni P, Cho TA, Cosimi LA, Hung DT, et al. Metagenomic sequencing of an echovirus 30 genome from cerebrospinal fluid of a patient with aseptic meningitis and orchitis. Open Forum Infect Dis 2017;4(3):ofx138.

[27] Wilson MR, Zimmermann LL, Crawford ED, Sample HA, Soni PR, Baker AN, et al. Acute west nile virus meningoencephalitis diagnosed via metagenomic deep sequencing of cerebrospinal fluid in a renal transplant patient. Am J Transplant 2017;17(3):803–8.

[28] Bozio CH, Vuong J, Dokubo EK, Fallah MP, McNamara LA, Potts CC, et al. Outbreak of Neisseria meningitidis serogroup C outside the meningitis belt-Liberia, 2017: an epidemiological and laboratory investigation. Lancet Infect Dis 2018;18(12):1360–7.

[29] Hoffmann B, Tappe D, Hoper D, Herden C, Boldt A, Mawrin C, et al. A variegated squirrel bornavirus associated with fatal human encephalitis. N Engl J Med 2015;373(2):154–62.
[30] Wilson MR, Shanbhag NM, Reid MJ, Singhal NS, Gelfand JM, Sample HA, et al. Diagnosing balamuthia mandrillaris encephalitis with metagenomic deep sequencing. Ann Neurol 2015;78(5):722–30.
[31] Simner PJ, Miller S, Carroll KC. Understanding the promises and hurdles of metagenomic next-generation sequencing as a diagnostic tool for infectious diseases. Clin Infect Dis 2018;66(5):778–88.
[32] Mai NTH, Phu NH, Nhu LNT, Hong NTT, Hanh NHH, Nguyet LA, et al. Central nervous system infection diagnosis by next-generation sequencing: a glimpse into the future? Open Forum Infect Dis 2017;4(2):ofx046.
[33] Naccache SN, Peggs KS, Mattes FM, Phadke R, Garson JA, Grant P, et al. Diagnosis of neuroinvasive astrovirus infection in an immunocompromised adult with encephalitis by unbiased next-generation sequencing. Clin Infect Dis 2015;60(6):919–23.
[34] Murkey JA, Chew KW, Carlson M, Shannon CL, Sirohi D, Sample HA, et al. Hepatitis E virus-associated meningoencephalitis in a lung transplant recipient diagnosed by clinical metagenomic sequencing. Open Forum Infect Dis 2017;4(3):ofx121.
[35] Shigeyasu C, Yamada M, Aoki K, Ishii Y, Tateda K, Yaguchi T, et al. Metagenomic analysis for detecting *Fusarium solani* in a case of fungal keratitis. J Infect Chemother 2018;24(8):664–8.
[36] Seitzman GD, Thulasi P, Hinterwirth A, Chen C, Shantha J, Doan T. Capnocytophaga keratitis: clinical presentation and use of metagenomic deep sequencing for diagnosis. Cornea 2019;38(2):246–8.
[37] Li Z, Breitwieser FP, Lu J, Jun AS, Asnaghi L, Salzberg SL, et al. Identifying corneal infections in formalin-fixed specimens using next generation sequencing. Invest Ophthalmol Vis Sci 2018;59(1):280–8.
[38] Doan T, Acharya NR, Pinsky BA, Sahoo MK, Chow ED, Banaei N, et al. Metagenomic DNA sequencing for the diagnosis of intraocular infections. Ophthalmology 2017;124(8):1247–8.
[39] Doan T, Wilson MR, Crawford ED, Chow ED, Khan LM, Knopp KA, et al. Illuminating uveitis: metagenomic deep sequencing identifies common and rare pathogens. Genome Med 2016;8(1):90.
[40] Gao D, Yu Q, Wang G, Wang G, Xiong F. Diagnosis of a malayan filariasis case using a shotgun diagnostic metagenomics assay. Parasit Vectors 2016;9:86.
[41] Yinda CK, Vanhulle E, Conceicao-Neto N, Beller L, Deboutte W, Shi C, et al. Gut virome analysis of cameroonians reveals high diversity of enteric viruses, including potential interspecies transmitted viruses. mSphere 2019;4(1).
[42] Petronella N, Ronholm J, Suresh M, Harlow J, Mykytczuk O, Corneau N, et al. Genetic characterization of norovirus GII.4 variants circulating in Canada using a metagenomic technique. BMC Infect Dis 2018;18(1):521.
[43] Vaisanen E, Mohanraj U, Kinnunen PM, Jokelainen P, Al-Hello H, Barakat AM, et al. Global distribution of human protoparvoviruses. Emerg Infect Dis 2018;24(7):1292–9.
[44] Ward DV, Scholz M, Zolfo M, Taft DH, Schibler KR, Tett A, et al. Metagenomic sequencing with strain-level resolution implicates uropathogenic E. coli in necrotizing enterocolitis and mortality in preterm infants. Cell Rep 2016;14(12):2912–24.
[45] Huang AD, Luo C, Pena-Gonzalez A, Weigand MR, Tarr CL, Konstantinidis KT. Metagenomics of two severe foodborne outbreaks provides diagnostic signatures and signs of coinfection not attainable by traditional methods. Appl Environ Microbiol 2017;83(3).

[46] Zhou Y, Wylie KM, El Feghaly RE, Mihindukulasuriya KA, Elward A, Haslam DB, et al. Metagenomic approach for identification of the pathogens associated with diarrhea in stool specimens. J Clin Microbiol 2016;54(2):368–75.

[47] Lokmer A, Cian A, Froment A, Gantois N, Viscogliosi E, Chabe M, et al. Use of shotgun metagenomics for the identification of protozoa in the gut microbiota of healthy individuals from worldwide populations with various industrialization levels. PLoS One 2019;14(2):e0211139.

[48] Andersen SC, Hoorfar J. Surveillance of foodborne pathogens: towards diagnostic metagenomics of fecal samples. Genes (Basel) 2018;9(1).

[49] Paul L, Comstock J, Edes K, Schlaberg R. Gestational psittacosis resulting in neonatal death identified by next-generation RNA sequencing of postmortem, formalin-fixed lung tissue. Open Forum Infect Dis 2018;5(8):ofy172.

[50] Graf EH, Simmon KE, Tardif KD, Hymas W, Flygare S, Eilbeck K, et al. Unbiased detection of respiratory viruses by use of RNA sequencing-based metagenomics: a systematic comparison to a commercial PCR panel. J Clin Microbiol 2016;54(4):1000–7.

[51] Langelier C, Zinter MS, Kalantar K, Yanik GA, Christenson S, O'Donovan B, et al. Metagenomic sequencing detects respiratory pathogens in hematopoietic cellular transplant patients. Am J Respir Crit Care Med 2018;197(4):524–8.

[52] Wang J, Li Y, He X, Ma J, Hong W, Hu F, et al. Gemykibivirus genome in lower respiratory tract of elderly woman with unexplained acute respiratory distress syndrome. Clin Infect Dis 2019;.

[53] Kalantar KL, Moazed F, Christenson SC, Wilson J, Deiss T, Belzer A, et al. A metagenomic comparison of tracheal aspirate and mini-bronchial alveolar lavage for assessment of respiratory microbiota. Am J Physiol Lung Cell Mol Physiol 2019;316(3):L578–84.

[54] Feigelman R, Kahlert CR, Baty F, Rassouli F, Kleiner RL, Kohler P, et al. Sputum DNA sequencing in cystic fibrosis: non-invasive access to the lung microbiome and to pathogen details. Microbiome 2017;5(1):20.

[55] Schlaberg R, Queen K, Simmon K, Tardif K, Stockmann C, Flygare S, et al. Viral pathogen detection by metagenomics and pan-viral group polymerase chain reaction in children with pneumonia lacking identifiable etiology. J Infect Dis 2017;215(9):1407–15.

[56] Grard G, Fair JN, Lee D, Slikas E, Steffen I, Muyembe JJ, et al. A novel rhabdovirus associated with acute hemorrhagic fever in central Africa. PLoS Pathog 2012;8(9):e1002924.

[57] Palacios G, Druce J, Du L, Tran T, Birch C, Briese T, et al. A new arenavirus in a cluster of fatal transplant-associated diseases. N Engl J Med 2008;358(10):991–8.

[58] Xu B, Liu L, Huang X, Ma H, Zhang Y, Du Y, et al. Metagenomic analysis of fever, thrombocytopenia and leukopenia syndrome (FTLS) in Henan province, China: discovery of a new bunyavirus. PLoS Pathog 2011;7(11):e1002369.

[59] Briese T, Paweska JT, McMullan LK, Hutchison SK, Street C, Palacios G, et al. Genetic detection and characterization of Lujo virus, a new hemorrhagic fever-associated arenavirus from southern Africa. PLoS Pathog 2009;5(5):e1000455.

[60] Xu L, Zhu Y, Ren L, Xu B, Liu C, Xie Z, et al. Characterization of the nasopharyngeal viral microbiome from children with community-acquired pneumonia but negative for luminex xTAG respiratory viral panel assay detection. J Med Virol 2017;89(12):2098–107.

[61] Bachmann NL, Rockett RJ, Timms VJ, Sintchenko V. Advances in clinical sample preparation for identification and characterization of bacterial pathogens using metagenomics. Front Public Health 2018;6:363.

[62] Votintseva AA, Bradley P, Pankhurst L, Del Ojo Elias C, Loose M, Nilgiriwala K, et al. Same-day diagnostic and surveillance data for tuberculosis via whole-genome sequencing of direct respiratory samples. J Clin Microbiol 2017;55(5):1285–98.

[63] Kavvas ES, Catoiu E, Mih N, Yurkovich JT, Seif Y, Dillon N, et al. Machine learning and structural analysis of *Mycobacterium tuberculosis* pan-genome identifies genetic signatures of antibiotic resistance. Nat Commun 2018;9(1):4306.
[64] Dunne Jr. WM, Jaillard M, Rochas O, Van Belkum A. Microbial genomics and antimicrobial susceptibility testing. Expert Rev Mol Diagn 2017;17(3):257–69.
[65] Nguyen M, Long SW, McDermott PF, Olsen RJ, Olson R, Stevens RL, et al. Using machine learning to predict antimicrobial MICs and associated genomic features for nontyphoidal salmonella. J Clin Microbiol 2019;57(2):https://doi.org/10.1128/JCM.01260-18.
[66] Moradigaravand D, Palm M, Farewell A, Mustonen V, Warringer J, Parts L. Prediction of antibiotic resistance in *Escherichia coli* from large-scale pan-genome data. PLoS Comput Biol 2018;14(12):e1006258.
[67] Alidjinou EK, Deldalle J, Hallaert C, Robineau O, Ajana F, Choisy P, et al. RNA and DNA Sanger sequencing versus next-generation sequencing for HIV-1 drug resistance testing in treatment-naive patients. J Antimicrob Chemother 2017;72(10):2823–30.
[68] Noguera-Julian M, Edgil D, Harrigan PR, Sandstrom P, Godfrey C, Paredes R. Next-generation human immunodeficiency virus sequencing for patient management and drug resistance surveillance. J Infect Dis 2017;216(suppl. 9):S829–33.
[69] Hage E, Wilkie GS, Linnenweber-Held S, Dhingra A, Suarez NM, Schmidt JJ, et al. Characterization of human cytomegalovirus genome diversity in immunocompromised hosts by whole-genome sequencing directly from clinical specimens. J Infect Dis 2017;215(11):1673–83.
[70] Houldcroft CJ, Bryant JM, Depledge DP, Margetts BK, Simmonds J, Nicolaou S, et al. Detection of low frequency multi-drug resistance and novel putative maribavir resistance in immunocompromised pediatric patients with cytomegalovirus. Front Microbiol 2016;7:1317.
[71] Deurenberg RH, Bathoorn E, Chlebowicz MA, Couto N, Ferdous M, Garcia-Cobos S, et al. Application of next generation sequencing in clinical microbiology and infection prevention. J Biotechnol 2017;243:16–24.
[72] Duvallet C, Gibbons SM, Gurry T, Irizarry RA, Alm EJ. Meta-analysis of gut microbiome studies identifies disease-specific and shared responses. Nat Commun 2017;8(1):1784.
[73] Parekh PJ, Balart LA, Johnson DA. The influence of the Gut microbiome on obesity, metabolic syndrome and gastrointestinal disease. Clin Transl Gastroenterol 2015;6:e91.
[74] Cryan JF, Dinan TG. Mind-altering microorganisms: the impact of the gut microbiota on brain and behaviour. Nat Rev Neurosci 2012;13(10):701–12.
[75] Huang YJ, LiPuma JJ. The microbiome in cystic fibrosis. Clin Chest Med 2016;37(1):59–67.
[76] Zhao J, Schloss PD, Kalikin LM, Carmody LA, Foster BK, Petrosino JF, et al. Decade-long bacterial community dynamics in cystic fibrosis airways. Proc Natl Acad Sci U S A 2012;109(15):5809–14.
[77] Kao D, Roach B, Silva M, Beck P, Rioux K, Kaplan GG, et al. Effect of oral capsule- vs colonoscopy-delivered fecal microbiota transplantation on recurrent clostridium difficile infection: a randomized clinical trial. JAMA 2017;318(20):1985–93.
[78] Kowarsky M, Camunas-Soler J, Kertesz M, De Vlaminck I, Koh W, Pan W, et al. Numerous uncharacterized and highly divergent microbes which colonize humans are revealed by circulating cell-free DNA. Proc Natl Acad Sci U S A 2017;114(36):9623–8.
[79] Ramanan P, Bryson AL, Binnicker MJ, Pritt BS, Patel R. Syndromic panel-based testing in clinical microbiology. Clin Microbiol Rev 2018;31(1).
[80] Motro Y, Moran-Gilad J. Next-generation sequencing applications in clinical bacteriology. Biomol Detect Quantif 2017;14:1–6.

[81] Fischer N, Rohde H, Indenbirken D, Gunther T, Reumann K, Lutgehetmann M, et al. Rapid metagenomic diagnostics for suspected outbreak of severe pneumonia. Emerg Infect Dis 2014;20(6):1072–5.

[82] Salzberg SL, Breitwieser FP, Kumar A, Hao H, Burger P, Rodriguez FJ, et al. Next-generation sequencing in neuropathologic diagnosis of infections of the nervous system. Neurol Neuroimmunol Neuroinflamm 2016;3(4):e251.

[83] Parize P, Muth E, Richaud C, Gratigny M, Pilmis B, Lamamy A, et al. Untargeted next-generation sequencing-based first-line diagnosis of infection in immunocompromised adults: a multicentre, blinded, prospective study. Clin Microbiol Infect 2017;23(8):574 e1–e6.

[84] Schlaberg R, Chiu CY, Miller S, Procop GW, Weinstock G, Professional Practice C, et al. Validation of metagenomic next-generation sequencing tests for universal pathogen detection. Arch Pathol Lab Med 2017;141(6):776–86.

[85] Schlaberg R, Ampofo K, Tardif KD, Stockmann C, Simmon KE, Hymas W, et al. Human bocavirus capsid messenger rna detection in children with pneumonia. J Infect Dis 2017;216(6):688–96.

[86] Marotz CA, Sanders JG, Zuniga C, Zaramela LS, Knight R, Zengler K. Improving saliva shotgun metagenomics by chemical host DNA depletion. Microbiome 2018;6(1):42.

[87] Hasan MR, Rawat A, Tang P, Jithesh PV, Thomas E, Tan R, et al. Depletion of human DNA in spiked clinical specimens for improvement of sensitivity of pathogen detection by next-generation sequencing. J Clin Microbiol 2016;54(4):919–27.

[88] Salter SJ, Cox MJ, Turek EM, Calus ST, Cookson WO, Moffatt MF, et al. Reagent and laboratory contamination can critically impact sequence-based microbiome analyses. BMC Biol 2014;12:87.

[89] Peng Q, Vijaya Satya R, Lewis M, Randad P, Wang Y. Reducing amplification artifacts in high multiplex amplicon sequencing by using molecular barcodes. BMC Genomics 2015;16:589.

[90] MacConaill LE, Burns RT, Nag A, Coleman HA, Slevin MK, Giorda K, et al. Unique, dual-indexed sequencing adapters with UMIs effectively eliminate index cross-talk and significantly improve sensitivity of massively parallel sequencing. BMC Genomics 2018;19(1):30.

[91] Gomez-Alvarez V, Teal TK, Schmidt TM. Systematic artifacts in metagenomes from complex microbial communities. ISME J 2009;3(11):1314–7.

[92] Goldberg B, Sichtig H, Geyer C, Ledeboer N, Weinstock GM. Making the leap from research laboratory to clinic: challenges and opportunities for next-generation sequencing in infectious disease diagnostics. mBio 2015;6(6):e01888-15.

[93] Goodwin S, McPherson JD, McCombie WR. Coming of age: ten years of next-generation sequencing technologies. Nat Rev Genet 2016;17(6):333–51.

[94] Lefterova MI, Suarez CJ, Banaei N, Pinsky BA. Next-generation sequencing for infectious disease diagnosis and management: A report of the association for molecular pathology. J Mol Diagn 2015;17(6):623–34.

[95] Giordano F, Aigrain L, Quail MA, Coupland P, Bonfield JK, Davies RM, et al. De novo yeast genome assemblies from MinION, PacBio and MiSeq platforms. Sci Rep 2017;7(1):3935.

[96] Schirmer M, Ijaz UZ, D'Amore R, Hall N, Sloan WT, Quince C. Insight into biases and sequencing errors for amplicon sequencing with the illumina MiSeq platform. Nucleic Acids Res 2015;43(6):e37.

[97] Loman NJ, Misra RV, Dallman TJ, Constantinidou C, Gharbia SE, Wain J, et al. Performance comparison of benchtop high-throughput sequencing platforms. Nat Biotechnol 2012;30(5):434–9.

[98] Cheval J, Sauvage V, Frangeul L, Dacheux L, Guigon G, Dumey N, et al. Evaluation of high-throughput sequencing for identifying known and unknown viruses in biological samples. J Clin Microbiol 2011;49(9):3268–75.
[99] Greninger AL, Naccache SN, Federman S, Yu G, Mbala P, Bres V, et al. Rapid metagenomic identification of viral pathogens in clinical samples by real-time nanopore sequencing analysis. Genome Med 2015;7:99.
[100] Balazs Z, Tombacz D, Szucs A, Csabai Z, Megyeri K, Petrov AN, et al. Long-read sequencing of human cytomegalovirus transcriptome reveals RNA isoforms carrying distinct coding potentials. Sci Rep 2017;7(1):15989.
[101] Leggett RM, Alcon-Giner C, Heavens D, Caim S, Brook T, Kujawska M, et al. Rapid MinION metagenomic profiling of the preterm infant gut microbiota to aid in pathogen diagnostics. bioRxiv 2017;. https://www.biorxiv.org/content/10.1101/180406v1.
[102] Lam MMC, Wyres KL, Wick RR, Judd LM, Fostervold A, Holt KE, et al. Convergence of virulence and MDR in a single plasmid vector in MDR *Klebsiella pneumoniae* ST15. J Antimicrob Chemother 2019; https://doi.org/10.1093/jac/dkz028.
[103] Lemon JK, Khil PP, Frank KM, Dekker JP. Rapid nanopore sequencing of plasmids and resistance gene detection in clinical isolates. J Clin Microbiol 2017;55(12):3530–43.
[104] Tyler AD, Mataseje L, Urfano CJ, Schmidt L, Antonation KS, Mulvey MR, et al. Evaluation of oxford nanopore's MinION sequencing device for microbial whole genome sequencing applications. Sci Rep 2018;8(1):10931.
[105] Postler TS, Clawson AN, Amarasinghe GK, Basler CF, Bavari S, Benko M, et al. Possibility and challenges of conversion of current virus species names to Linnaean binomials. Syst Biol 2017;66(3):463–73.
[106] McIntyre ABR, Ounit R, Afshinnekoo E, Prill RJ, Henaff E, Alexander N, et al. Comprehensive benchmarking and ensemble approaches for metagenomic classifiers. Genome Biol 2017;18(1):182.
[107] Nooij S, Schmitz D, Vennema H, Kroneman A, Koopmans MPG. Overview of virus metagenomic classification methods and their biological applications. Front Microbiol 2018;9:749.
[108] Breitwieser FP, Lu J, Salzberg SL. A review of methods and databases for metagenomic classification and assembly. Brief Bioinform 2017; https://doi.org/10.1093/bib/bbx120.
[109] Lindgreen S, Adair KL, Gardner PP. An evaluation of the accuracy and speed of metagenome analysis tools. Sci Rep 2016;6:19233.
[110] Peabody MA, Van Rossum T, Lo R, Brinkman FS. Evaluation of shotgun metagenomics sequence classification methods using in silico and in vitro simulated communities. BMC Bioinf 2015;16:363.
[111] Sczyrba A, Hofmann P, Belmann P, Koslicki D, Janssen S, Droge J, et al. Critical assessment of metagenome interpretation—a benchmark of metagenomics software. Nat Methods 2017;14(11):1063–71.
[112] Ames SK, Hysom DA, Gardner SN, Lloyd GS, Gokhale MB, Allen JE. Scalable metagenomic taxonomy classification using a reference genome database. Bioinformatics 2013;29(18):2253–60.
[113] NCBI. Available from, https://www.ncbi.nlm.nih.gov/nucleotide/.
[114] Sichtig H, Minogue T, Yan Y, Stefan C, Hall A, Tallon L, et al. FDA-ARGOS: a public quality-controlled genome database resource for infectious disease sequencing diagnostics and regulatory science research. bioRxiv 2018;482059. https://www.biorxiv.org/content/10.1101/482059v1.full.

[115] Zhang Y, Aevermann BD, Anderson TK, Burke DF, Dauphin G, Gu Z, et al. Influenza research database: an integrated bioinformatics resource for influenza virus research. Nucleic Acids Res 2017;45(D1):D466–74.
[116] Jia B, Raphenya AR, Alcock B, Waglechner N, Guo P, Tsang KK, et al. CARD 2017: expansion and model-centric curation of the comprehensive antibiotic resistance database. Nucleic Acids Res 2017;45(D1):D566–73.
[117] Endrullat C, Glokler J, Franke P, Frohme M. Standardization and quality management in next-generation sequencing. Appl Transl Genom 2016;10:2–9.
[118] (CAP) CoAP. CAP accreditation checklists—2018 edition; 2018.
[119] Lusk RW. Diverse and widespread contamination evident in the unmapped depths of high throughput sequencing data. PLoS One 2014;9(10):e110808.
[120] D'Amore R, Ijaz UZ, Schirmer M, Kenny JG, Gregory R, Darby AC, et al. A comprehensive benchmarking study of protocols and sequencing platforms for 16S rRNA community profiling. BMC Genomics 2016;17:55.
[121] Nelson MC, Morrison HG, Benjamino J, Grim SL, Graf J. Analysis, optimization and verification of illumina-generated 16S rRNA gene amplicon surveys. PLoS One 2014;9(4):e94249.
[122] Allcock RJN, Jennison AV, Warrilow D. Towards a universal molecular microbiological test. J Clin Microbiol 2017;55(11):3175–82.
[123] Briese T, Kapoor A, Mishra N, Jain K, Kumar A, Jabado OJ, et al. Virome capture sequencing enables sensitive viral diagnosis and comprehensive virome analysis. mBio 2015;6(5):e01491-15.
[124] Allander T, Emerson SU, Engle RE, Purcell RH, Bukh J. A virus discovery method incorporating DNase treatment and its application to the identification of two bovine parvovirus species. Proc Natl Acad Sci U S A 2001;98(20):11609–14.
[125] Somasekar S, Lee D, Rule J, Naccache SN, Stone M, Busch MP, et al. Viral surveillance in serum samples from patients with acute liver failure by metagenomic next-generation sequencing. Clin Infect Dis 2017;65(9):1477–85.
[126] Gu W, Crawford ED, O'Donovan BD, Wilson MR, Chow ED, Retallack H, et al. Depletion of abundant sequences by hybridization (DASH): using Cas9 to remove unwanted high-abundance species in sequencing libraries and molecular counting applications. Genome Biol 2016;17:41.
[127] Flygare S, Simmon K, Miller C, Qiao Y, Kennedy B, Di Sera T, et al. Taxonomer: an interactive metagenomics analysis portal for universal pathogen detection and host mRNA expression profiling. Genome Biol 2016;17(1):111.
[128] Wood DE, Salzberg SL. Kraken: ultrafast metagenomic sequence classification using exact alignments. Genome Biol 2014;15(3):R46.
[129] Sweeney TE, Shidham A, Wong HR, Khatri P. A comprehensive time-course-based multicohort analysis of sepsis and sterile inflammation reveals a robust diagnostic gene set. Sci Transl Med 2015;7(287):287ra71.
[130] Holcomb ZE, Tsalik EL, Woods CW, McClain MT. Host-based peripheral blood gene expression analysis for diagnosis of infectious diseases. J Clin Microbiol 2017;55(2):360–8.
[131] Ramilo O, Allman W, Chung W, Mejias A, Ardura M, Glaser C, et al. Gene expression patterns in blood leukocytes discriminate patients with acute infections. Blood 2007;109(5):2066–77.
[132] Woods CW, McClain MT, Chen M, Zaas AK, Nicholson BP, Varkey J, et al. A host transcriptional signature for presymptomatic detection of infection in humans exposed to influenza H1N1 or H3N2. PLoS One 2013;8(1):e52198.

[133] Langelier C, Kalantar KL, Moazed F, Wilson MR, Crawford ED, Deiss T, et al. Integrating host response and unbiased microbe detection for lower respiratory tract infection diagnosis in critically ill adults. Proc Natl Acad Sci U S A 2018;115(52):E12353–62.
[134] Hudson LL, Woods CW, Ginsburg GS. A novel diagnostic approach may reduce inappropriate antibiotic use for acute respiratory infections. Expert Rev Anti Infect Ther 2014;12(3):279–82.
[135] Gardy JL, Loman NJ. Towards a genomics-informed, real-time, global pathogen surveillance system. Nat Rev Genet 2018;19(1):9–20.

Chapter 5

Host-pathogen interactions

Scott D. Kobayashi, Frank R. DeLeo
Laboratory of Bacteriology, Rocky Mountain Laboratories, National Institute of Allergy and Infectious Diseases, National Institutes of Health, Hamilton, MT, United States

Introduction

The innate immune system is essential for host protection against invading microorganisms and is indispensable for overall maintenance of human health. Polymorphonuclear leukocytes (PMNs or neutrophils) are key components of the innate immune system and are the most abundant cellular component of the host immune system. The importance of PMNs in protection against pathogenic microorganisms is exemplified by a heightened susceptibility to infection in individuals with neutrophil deficiencies [1]. One of the primary functions of neutrophils is to eliminate invading microorganisms. Neutrophil recognition of invading microorganisms initiates a complex cascade of signal transduction pathways that ultimately direct ingestion and killing of the pathogen. Notwithstanding, the complexity of neutrophil biology extends far beyond microbicidal activity and host defense. Advances in genomics technologies have provided a diversity of tools that have been instrumental in dissecting the molecular mechanisms of PMN-pathogen interactions. Moreover, the ability to query the neutrophil transcriptome has provided detailed insight into essential processes such as PMN development [2], and post-phagocytosis sequelae including apoptosis and resolution of inflammation [3], and the role in tissue repair [4]. Perhaps more importantly, transcriptome studies helped pave the way toward increased understanding of the pathophysiology of several neutrophil disorders [5, 6].

Neutrophils are a primary reason that the human host is highly adept at defending itself against invading bacteria. As such, the vast majority of bacterial infections are minor and may go unnoticed, and are resolved readily. On the other hand, some pathogenic microorganisms have evolved means to circumvent the innate and/or adaptive immune responses and thereby cause disease. Although much of the focus in our efforts to understand host-pathogen interactions has been placed on responses of the host or host cells, genomics and related

Genomic and Precision Medicine. https://doi.org/10.1016/B978-0-12-801496-7.00005-8

technologies permit comprehensive molecular analyses of bacterial pathogens during this process. We are currently in an era that allows routine interrogation of pathogens at the whole-genome, transcriptome, and proteome levels to better understand infection processes and disease transmission. These approaches have been made possible by high-throughput DNA sequencing [7–14].

Here we review our understanding of the neutrophil transcriptome in the context of the inflammatory process and interaction with bacteria. In the second half of the chapter, we provide a brief overview of bacterial responses to the host or host cells. This text is not meant to provide a comprehensive review of the literature; rather, we have used selected examples to indicate advances in the area of host-pathogen interactions.

Host response to bacterial pathogens

Neutrophil phagocytosis and host defense

Binding and uptake of microbes by neutrophils is mediated by phagocytosis (Fig. 1). Bacteria produce conserved molecules that bind pattern recognition receptors on the neutrophil surface. Such receptors include *Toll*-like receptors and the formyl peptide receptor [15]. Binding of these receptors activates signaling pathways that contribute to innate host defense. Neutrophil phagocytosis is enhanced significantly by opsonization of microbes with host antibody and serum complement. Specific antibody binds to the surface of microbes and facilitates the deposition of complement components. Ligation of PMN opsonin receptors initiates cytoskeletal changes that facilitate phagocytosis. Ingested microorganisms are then sequestered within phagocytic vacuoles (phagosomes) (Fig. 1).

PMN phagocytosis initiates microbicidal processes that involve production of superoxide, and exposure of bacteria to antimicrobial peptides, proteins, and degradative enzymes (Fig. 1). Superoxide is generated by a multicomponent, membrane-bound complex called NADPH-dependent oxidase [16]. Superoxide is converted to hydrogen peroxide and other secondarily-derived reactive oxygen species (ROS), which have potent bactericidal activity. PMN phagocytosis also causes fusion of cytoplasmic granules with the plasma or phagosome membrane, a process known as degranulation. The cytoplasmic granules and secretory vesicles contain a reservoir of lumenal and membrane proteins that contribute to host defense. For example, azurophilic granules contain α-defensins, which are important for host defense against bacteria. The combination of ROS and antimicrobial peptides are effective at killing ingested microbes.

The neutrophil transcriptome

Cells of the immune system originate from common hematopoietic progenitor cells in bone marrow. Granulocyte differentiation and development are

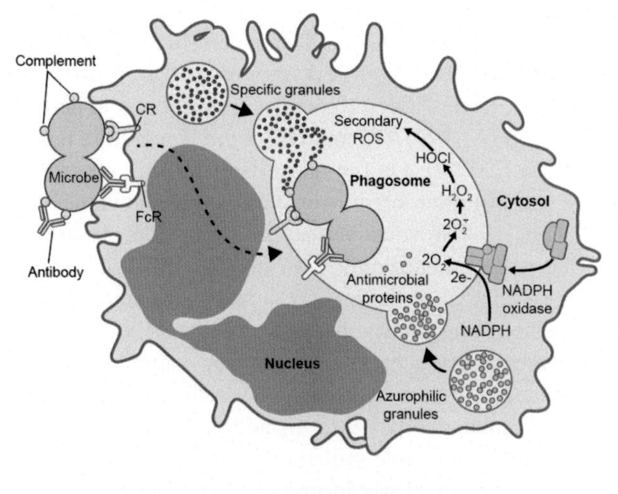

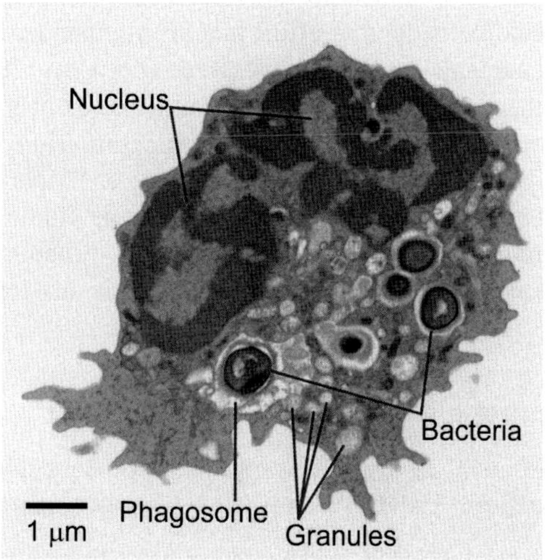

FIG. 1 Neutrophil phagocytosis and microbicidal processes. Top panel is a model of neutrophil phagocytosis and activation of microbicidal processes. Bottom panel is a transmission electron micrograph of a human neutrophil that has phagocytosed *Staphylococcus aureus* (bacteria).

influenced by the concerted activities of myeloid colony stimulating factors such as granulocyte colony-stimulating factor (G-CSF) and granulocyte-macrophage colony-stimulating factor (GM-CSF) [17]. During granulopoiesis, myeloid precursors sequentially acquire features necessary for microbicidal activity including receptors for phagocytosis and signaling, granule components, and NADPH oxidase proteins. Transcriptional processes occurring during granulopoiesis have been investigated intensely [18–20]. The transcriptional regulation of neutrophil development is mediated in part by the coordinated activities of several transcription factors and repressors. Transcription factors such as CCAAT/enhancer-binding protein alpha (C/EBPα), PU.1, RAR, CBF, and c-MYB are involved in early granulopoiesis; C/EBPε, PU.1, SP1, CDP, HOXA10, signal transducer and activator of transcription (STAT)1, STAT3, STAT5 and GFI-1 are involved in terminal neutrophil differentiation [21, 22]. On the other hand, mature neutrophils are able to phagocytose and kill microorganisms without initial synthesis of new proteins [23, 24]. Thus, it is provocative that the initial PMN transcriptome studies revealed that mature neutrophils also have significant capacity for gene transcription [25]. Several of these early studies used microarray analysis to gain insight into molecular processes that accompany PMN phagocytosis, and these findings were reviewed previously [3]. For example, neutrophil phagocytosis causes changes in transcripts that may reflect increased production of pro-inflammatory molecules [25–27]. In addition, PMN apoptosis (or programmed cell death) is a notable sequela of phagocytosis. The importance of neutrophil spontaneous apoptosis in maintenance of cellular homeostasis is well recognized, and considerable effort has been expended on determining the molecular mechanisms governing this essential process [28, 29]. However, the process of phagocytosis-induced cell death (PICD) differs from that of spontaneous neutrophil apoptosis [3]. Neutrophil transcriptome studies have been instrumental in re-shaping our view of the role of PMNs in the resolution of the inflammatory response. Notably, genome-wide studies have provided insight into key molecular determinants that regulate neutrophil PICD and have revealed mechanisms by which bacteria exploit this important process (Fig. 2).

Bacteria and neutrophil survival

PICD promotes removal of spent neutrophils, a process that facilitates the resolution of infection. The timely removal of spent neutrophils prevents

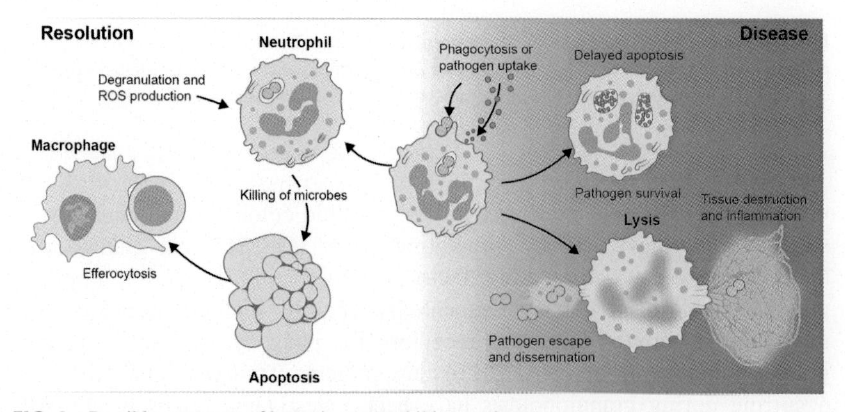

FIG. 2 Possible outcomes of bacteria-neutrophil interactions.

cytolytic release of cell contents, which includes molecules that can damage host tissues. Apoptotic PMNs are recognized and cleared by macrophages, thereby preventing excessive host tissue damage and limiting inflammation at sites of infection [30]. The process of phagocytosis significantly accelerates the rate of apoptosis in human PMNs. Similarly, neutrophil ingestion of *Escherichia coli, Streptococcus pneumoniae, Streptococcus pyogenes, Staphylococcus aureus, Mycobacterium tuberculosis, Burkholderia cepacia, Borrelia hermsii,* and *Listeria monocytogenes* significantly accelerates the rate of PMN apoptosis [25, 31–35]. PMN apoptosis plays a critical role in regulation of cell turnover and termination of the inflammatory process [30]. Inasmuch as the process of PMN apoptosis is essential for maintenance of host health, it is not surprising that several bacterial pathogens are capable of exploiting apoptosis/PICD as a potential mechanism of pathogenesis.

Recognition and ingestion of microorganisms results in a complex series of neutrophil signal transduction events including *Toll*-like receptor activation and the production of additional inflammatory mediators that are potentially pathogen specific. Bacteria are also capable of secreting molecules that alter function of host cells, including those of the innate immune system. Thus, the neutrophil response to bacterial pathogens involves complex signals induced by ligation of multiple receptors in addition to those participating in Fc- and complement-mediated phagocytosis. The PMN transcriptome has been queried following phagocytosis of many different microorganisms, including *Escherichia coli*, attenuated *Yersinia pestis*, *Staphylococcus aureus*, *Streptococcus pyogenes* (or group A *Streptococcus*, GAS), *Burkholderia cepacia, Listeria monocytogenes, Borrelia hermsii, Mycobacterium bovis, Candida albicans*, and *Anaplasma phagocytophilum* [26, 36–43]. Inasmuch as differences in pathogen-induced PMN transcript levels occur in those encoding signal transduction mediators and prominent transcription factors, it is not surprising that the PMN response is not entirely conserved.

Bacterial strategies that promote pathogenesis can involve mechanisms that delay neutrophil apoptosis or ultimately cause neutrophil lysis. Although the relatively short lifespan of neutrophils is generally not amenable to long-term survival strategies employed by most intracellular pathogens, some pathogens have adapted their niche to include PMNs as a host cell. *Anaplasma phagocytophilum*, the causative agent of human granulocytic anaplasmosis, was the first bacterial pathogen reported to delay PMN apoptosis, which ultimately promotes replication within an endosomal compartment [44]. Oligonucleotide microarrays have been used to investigate mechanisms that underlie survival of *Anaplasma phagocytophilum* within neutrophils [36, 44–46]. Infection of human PMNs with *Anaplasma phagocytophilum* delays up-regulation of transcripts that encode molecules involved in the inflammatory response [36]. Moreover, transcripts encoding molecules that block apoptosis are up-regulated in *Anaplasma phagocytophilum*-infected neutrophils, and there is a concomitant decrease in transcripts encoding molecules that promote apoptosis. Conversely, bacterial pathogens such as *Staphylococcus aureus* and GAS cause direct PMN lysis and/or accelerate bacteria-induced apoptosis to the point of secondary necrosis [26, 37]. It is likely that accelerated PMN lysis contributes to the overall levels of tissue necrosis associated with these infections. The ability of GAS to cause rapid PICD and ultimately PMN lysis is reflected by and/or results from the changes in neutrophil gene expression [26]. Interaction of PMNs with GAS elicits transcriptional changes in cell death pathways that include up-regulation of activator protein-1 (AP-1) complex-related transcripts such as *FOS*, *FOSL1*, *FOSB*, *JUNB*, and *TNFRSF5*, and down-regulation of transcripts encoding members of the NF-κB signal transduction pathway [26]. These findings indicate that the ability of GAS to exploit PMN fate pathways is an important component of streptococcal pathogenesis and a potential contributor to tissue destruction observed in invasive disease [47].

The ability of microbes to block PMN apoptosis or promote rapid cytolysis are potential components of virulence. Genome-wide analyses with human neutrophils and bacteria, coupled with the results of *in vitro* and *in vivo* studies, led to a hypothetical model of PMN-bacteria interactions (Fig. 2). On one hand, PMNs phagocytose and kill microorganisms, a process that leads to PICD and clearance of spent neutrophils by macrophages. On the other hand, pathogens are phagocytosed but not killed, and PMNs either lyse or survive to promote pathogen replication and subsequent disease.

Bacterial response to the host

The genomics era

As of 2017, complete genomes were available for at least 1708 bacterial species [48]. This wealth of genome sequence information has revolutionized our understanding of bacterial physiology and pathogenesis, and how microorganisms adapt to the host. It is now known that bacterial adaptation in general occurs by gain and/or loss of genes and changes in gene sequence or order [11]. Also, as new genes are identified, gene function or topology can be predicted, and novel vaccine approaches based upon this information can be implemented [49]. Early during the genomics era, microarrays based upon complete bacterial genomes were used successfully to compare strains and/or track the evolution of outbreaks [50–53]. Furthermore, genome-scale approaches such as DNA or oligonucleotide microarrays were routinely used to evaluate bacterial gene expression on a global scale and under a variety of conditions, including during interaction with the host or host cells (Table 1) [47, 107]. Inasmuch as DNA sequencing technologies have advanced significantly in recent years, microarray-based methods have been largely replaced by high-throughput genome sequencing approaches. For example, a DNA sequencing-based method known as RNA-seq is now used for comprehensive transcriptome analyses [108, 109]. RNA-seq utilizes cDNA derived from

TABLE 1 Selected studies of the bacterial transcriptome during host interaction

Pathogen	Host interaction	References
Actinobacillus pleuropneumoniae	St. Jude porcine lung cell line	[54]
Aeromonas hydrophila	RAW 264.7 murine macrophage-like cell line	[55]
Anaplasma phagocytophilum	HL60 human promyelocytic cell line and HMEC-1 human endothelial cell line, ISE6 tick cell line	[56]
Bacillus anthracis	RAW 264.7 murine macrophage-like cell line	[57]
Borrelia burgdorferi	H4 human neuroglial cell line	[58]
Brucella melitensis	HeLa human epithelial cell line	[59]
Chlamydia trachomatis	HeLa human epithelial cell line	[60, 61]
Clostridium difficile	Caco-2 human colon epithelial cell line	[62]
Escherichia coli	Human primarily neutrophils, human urinary tract	[63–66]
Francisella tularensis	Primary murine bone marrow-derived macrophages	[67]
Lactobacillus plantarum	Mouse caeca	[68]
Leptospira interrogans	J774A.1 murine macrophage-like cell line, THP-1 human macrophage-like cell line	[69]
Listeria monocytogenes	Mouse spleen, P388D1 murine macrophage-like cell line, Caco-2 human colon epithelial cell line	[70–73]
Mycobacterium avium subsp. *paratuberculosis*	Primary bovine monocyte-derived macrophages, bovine GI tract	[74, 75]
Mycobacterium tuberculosis complex	Primary human monocyte-derived macrophages, mouse lung, primary murine bone marrow-derived macrophages, THP-1 human macrophage-like cell line	[76–81]
Neisseria gonorrhoeae	A431 human epithelial cell line	[82]
Neisseria meningitidis	HeLa human epithelial cell line, human brain microvascular endothelial cell line, 16HBE14 human epithelial cell line	[83, 84]
Pseudomonas aeruginosa	Primary human lung epithelial cells	[85]

Continued

TABLE 1 Selected studies of the bacterial transcriptome during host interaction—cont'd

Pathogen	Host interaction	References
Rickettsia prowazekii	HMEC-1 human endothelial cell line	[86]
Rickettsia typhi	HeLa human epithelial cell line	[87]
Streptococcus pneumoniae	THP-1 human macrophage-like cell line, A549 human lung epithelial cell line	[88–90]
Streptococcus pyogenes	Human primary neutrophils, human blood, mouse skin infection, monkey pharyngitis, Detroit 562 human pharyngeal cells	[91–96]
Salmonella enterica serovar Typhimurium	HeLa human epithelial cell line	[97–99]
Salmonella enterica serovar Typhi	THP-1 human macrophage-like cell line	[100, 101]
Shigella flexneri	U937 human macrophage-like cell line, HeLa human epithelial cell line	[102]
Staphylococcus aureus	Human primary neutrophils, A549 human lung epithelial cells	[103, 104]
Vibrio cholera	Human vomitus and feces	[105]
Yersinia pestis	Flea, rat bubo	[106]

RNA and have several advantages over microarray-based methods. These advantages include the ability to detect and quantitate novel transcripts (no reference genome needed), low background signal, large dynamic range and high sensitivity, and less RNA is needed compared with microarray-based approaches [108]. RNA-seq has been used to better understand commensal interaction with the mammalian host, virulence and pathogenesis, and to elucidate potential targets for therapeutic maneuvers directed to prevent or treat infections.

Proteomics methods have also been used to study pathogen responses to the host on a genome-wide scale. For example, bacterial proteomics has been used successfully to better understand the interaction of *Staphylococcus aureus* with components of innate host defense in growth conditions that mimic specific host environments [110, 111], or to assess immune responses during infection [112]. Performing bacterial proteomics studies in the context of host-pathogen interaction is technically challenging; therefore, the vast majority of bacterial proteomics studies have been directed to identify proteins produced using specific growth conditions *in vitro*. Studies with *Staphylococcus aureus* have evaluated metabolism, post-translational modifications, stress and starvation responses, and virulence factors such as secreted molecules [113]. Proteomics approaches are essential for a comprehensive understanding of host-pathogen interactions and as such merit extensive discussion. However, there is insufficient space in this chapter for such a discussion, and we refer the reader to recent reviews on the topic [113–116]. Rather, this chapter will review our understanding of bacterial pathogenesis and/or host-pathogen interactions based upon knowledge gained from bacterial transcriptome studies, with an emphasis on selected pathogens.

Bacterial transcriptome studies *in vitro*

There have been many applications of bacterial transcriptome analyses *in vitro* (in the absence of the host or host cells). For example, such studies have been conducted to aid in understanding the role of gene regulatory molecules and/or the molecules controlled by global gene regulators, adaptation to the specific growth conditions or nutritional requirements (e.g., temperature, availability of iron or other metals), stress responses, biological processes such as biofilm formation or cell wall biosynthesis, impact of gene deletions, and responses to antimicrobial agents/antibiotics. Here we focus our discussion on a few selected transcriptome studies performed with *Staphylococcus aureus* and GAS, pathogens that are leading causes of human bacterial infections worldwide.

Some of the earliest transcriptome studies with GAS revealed that changes in culture temperature (those that are physiologically relevant) can have a profound impact on GAS gene expression *in vitro* [117]. These studies suggested that GAS alters gene expression depending on the site of infection (e.g., skin vs blood). Subsequent work identified virulence molecules or those important for infection that is controlled by GAS two-component gene regulatory systems [118–121]. For instance, an expression microarray study by Shelburne et al. identified a novel two-component gene regulatory system, *Spt*R/*Spt*S, that is important for GAS survival in human saliva [121]. Iterative microarray analyses using wild-type and *spt*R/*spt*S deletion mutants revealed the specific molecules controlled by *Spt*R/*Spt*S [121]. A similar approach identified a GAS survival response following exposure to neutrophil microbicides [120]. The GAS survival response includes differential regulation of molecules encoding cell wall metabolism, oxidative stress and virulence, and is controlled in part by a two-component gene regulatory system known as Ihk/Irr. These types of iterative microarray approaches are now widely used to identify molecules important for infection of the host.

Expression microarray work with *Staphylococcus aureus* has identified regulons for multiple global gene regulatory systems, such as accessory gene regulator (Agr) and staphylococcal accessory regulator (Sar) [122], vancomycin resistance-associated sensor/regulator (VraSR) [123], alternative sigma factor (*sigB* or σ^B) [124, 125], TcaR [126], ArlRS [127], MgrA [128], CidR [129], Aps [130], and SaeRS [131–134]. Collectively, these studies have provided information vital to our understanding of how *Staphylococcus aureus* regulates molecules important for interaction with the host.

The ability of *Staphylococcus aureus* to avoid or resist killing by phagocytic leukocytes—especially neutrophils—is key for its ability to cause disease in humans, and it is well known that individuals with defects in phagocyte function are highly susceptible to *Staphylococcus aureus* infection. Therefore, understanding the response of *Staphylococcus aureus* to microbicidal components of host phagocytes should provide insight into mechanisms used by the pathogen to survive in the human host. To that end, Richardson et al. used *Staphylococcus aureus* oligonucleotide microarrays to identify genes differently regulated following exposure to nitric oxide, an innate immune effector molecule produced by host cells [135]. Notably, the *Staphylococcus aureus* response to nitric oxide includes up-regulation of molecules involved in iron homeostasis, control of which is critical for bacterial survival in the host [136, 137]. Consistent with this finding, subsequent work by Palazzolo-Ballance et al. showed that exposure of strain MW2, the prototype CA-MRSA strain, to hydrogen peroxide caused up-regulation of transcripts involved in heme/iron uptake [138]. In addition, exposure of *Staphylococcus aureus* to neutrophil granule proteins, which occurs in the phagocytic vacuole, caused up-regulation of genes involved in stress response, including those that moderate ROS, and multiple toxins and hemolysins [138]. Subsequent DNA microarray studies indicated that *vraSR* is up-regulated by exposure of *Staphylococcus aureus* to cationic antimicrobial peptides [139], findings consistent with Li et al. [130] and Palazzolo-Ballance et al. and the general notion that VraSR is important for cell wall biosynthesis [123]. Such information has provided an enhanced view of how *Staphylococcus aureus* circumvents destruction by the innate immune system.

Antibiotic resistance in *Staphylococcus aureus* is a major problem in developed countries worldwide [140]. Although the mechanisms for some types of antibiotic resistance in *Staphylococcus aureus* are known and well characterized, such as resistance to β-lactam antibiotics, the mechanism for glycopeptide intermediate resistance is less well defined. Glycopeptide intermediate resistance is acquired gradually during prolonged treatment and involves multiple mutations. Several microarray or comparative whole genome sequence-based studies have investigated the molecular genetic basis for glycopeptide resistance in *Staphylococcus aureus* and identified genes linked to resistance, many with no previously characterized function [126, 141–144]. In aggregate, the findings indicate that glycopeptide intermediate resistance is due to altered cell wall metabolism and biosynthesis, which is at least partly controlled by the VraRS regulon and may include posttranslational protein processing.

Multiple gene expression profiling studies have provided exciting new insight into biofilm formation by staphylococci, such as the involvement of surfactant-like peptides known as phenol-soluble modulins [145] and extensive changes in metabolism [146]. This topic is important and merits more discussion, and we refer the reader to a comprehensive review on the subject [147].

Bacterial transcriptome analyses during host interaction

Elucidation of bacterial transcriptomes *in vitro* as described above is clearly a step forward in our efforts to generate a comprehensive view of bacterial physiology and adaptation in specific environments. Therefore, it is not unfounded to state that the ability to analyze global gene expression in bacterial pathogens during interaction with the host or during infection of the host *in vivo* is a major breakthrough for infectious disease research. Transcriptome studies of global gene expression during host or host cell interaction have been conducted for numerous bacterial pathogens (Table 1). Here we discuss a few representative studies and highlight the major findings and their implications.

Staudinger et al. were the first to evaluate the bacterial transcriptome during interaction with host cells [63]. The researchers investigated global gene expression in a urinary isolate of *Escherichia coli* following phagocytosis by human neutrophils from healthy individuals or those with chronic granulomatous disease (CGD). Phagocytes with CGD are incapable of generating reactive oxygen species, and thus acquire recurrent and severe bacterial and fungal infections. By comparing changes in transcripts following phagocytosis by healthy or CGD neutrophils, Staudinger et al. identified molecules used by *Escherichia coli* to moderate reactive oxygen species (e.g., the OxyR regulon), and the authors proposed that the bacterium responds to intraphagosomal hydrogen peroxide rather than superoxide [63].

Work by Voyich et al. immediately followed the Staudinger studies and evaluated the GAS transcriptome during phagocytic interaction with human neutrophils [91]. In addition to the identification of novel secreted molecules, one of the key findings was the discovery that *ihk/irr* is up-regulated following neutrophil phagocytosis and is essential for survival of the pathogen after uptake [91]. Iterative GAS microarray studies *in vitro* identified the Ihk/Irr regulon (described above), which engages a survival response in the pathogen [91]. Similar studies performed with hospital-associated MRSA and CA-MRSA strains identified SaeRS as potentially important for survival of *Staphylococcus aureus* after phagocytosis by human neutrophils [103]. Subsequent studies using animal infection models and an iterative microarray approach verified the importance of SaeRS for infection and identified molecules that contribute to *Staphylococcus aureus* survival after phagocytosis [131, 133].

Although technically challenging, some studies have evaluated bacterial transcriptomes during infection *in vivo*. For example, Talaat et al. reported global gene expression in *Mycobacterium tuberculosis* during intranasal infection of mice over the course of 28 days [76]. Notably, these researchers discovered genes whose expression is dictated by immune responses of the host and found that transcriptome data with macrophages *in vitro* aligned with their findings in the mouse *in vivo* after 21 days. These results provide an estimate of the immune response at this stage of infection.

Differential gene expression of *Listeria monocytogenes* recovered from the spleens of infected mice indicates that a significant component of the *Listeria* response to host infection is dedicated to virulence and subversion of the host immune system [70]. These *in vivo* transcriptome data demonstrated that molecules implicated in virulence were indeed up-regulated during infection. Further, the work identified PrfA and VirR as the two major regulators of *Listeria monocytogenes* virulence during infection *in vivo*.

Studies by Vadyvaloo and colleagues compared transcriptomes of *Yersinia pestis* during transit through the flea vector and those of the rat bubo [106]. Transit through the flea causes changes in gene expression that prepares the pathogen for enhanced survival following transmission to mammals. Some of this phenotype is related to induction of genes that confer resistance to phagocytosis by host leukocytes.

In a study of unprecedented magnitude, Shea et al. investigated the transcriptome of GAS during pharyngeal infection of monkeys and concomitantly identified host genes whose patterns of expression correlated with those of the pathogen [92]. The correlation of host and pathogen patterns of gene expression, dubbed the "interactome," revealed correlations between genes encoding GAS hyaluronic acid production and host cell endocytic vesicle formation, GAS mevalonic acid synthesis and host γδ T cells, and GAS responses to host phagocytes and host neutrophils. Notably, changes in the GAS transcriptome *in vivo* during the course of infection were associated with specific phases of pharyngitis and the host response. The interactome approach remains at the cutting edge of transcriptome analyses, although the methodology has changed.

More recently, a method based on RNA-seq—named dual RNA-seq—has been developed to evaluate the transcriptomes of host and microbe during interaction *in vitro* and/or infection *in vivo* [109, 148, 149]. Although there were technical caveats to overcome, the dual RNA-seq approach has been used to provide new insights into host-bacterial pathogen interaction (Table 2). For example, recent dual RNA-seq studies indicate that mouse (host) genetic background has a significant influence on the *Staphylococcus aureus* transcriptome, and most notably virulence factor expression [156]. Such studies underscore the potential complexity of therapeutic approaches that target bacterial virulence molecules.

Future perspective

Antibiotic resistance is arguably the single greatest problem for treatment of bacterial infections today. This is a long-term problem, and it will be imperative to develop new diagnostics, therapeutics and vaccines to treat and/or prevent

TABLE 2 Selected RNA-seq-based studies of bacterial pathogen and host interaction

Pathogen	Host interaction	References
Vibrio cholerae	Rabbit ceca and murine small intestine	[150]
Yersinia pestis	Murine lung	[151]
Haemophilus influenzae	Primary human bronchial epithelial cells	[152]
Salmonella enterica serovar Typhimurium	Human, murine, and porcine cell culture	[153]
Streptococcus pneumoniae	Human lung alveolar epithelial cells	[154]
Yersinia pseudotuberculosis	Murine Peyer's patches	[155]
Staphylococcus aureus	Murine kidney	[156]

severe bacterial infections. Genomics approaches such as duel RNA-seq are absolutely critical for these endeavors, which have been stymied in the past by a lack of information about host and bacterial pathogen responses during infection. Alternatively, information gained from host genomics and transcriptomics may ultimately be used to augment host defense and/or inform about appropriate prophylaxis to prevent infections. This is a reasonable goal, as such approaches have been successfully implemented for diagnosis of respiratory viral infections [157].

Acknowledgment

The authors are supported by the Intramural Research Program of the National Institute of Allergy and Infectious Diseases, National Institutes of Health.

References

[1] Lekstrom-Himes JA, Gallin JI. Immunodeficiency diseases caused by defects in phagocytes. N Engl J Med 2000;343(23):1703–14.

[2] Theilgaard-Monch K, Jacobsen LC, Borup R, Rasmussen T, Bjerregaard MD, Nielsen FC, et al. The transcriptional program of terminal granulocytic differentiation. Blood 2005;105(4):1785–96.

[3] Kobayashi SD, DeLeo FR. Role of neutrophils in innate immunity: a systems biology-level approach. Wiley Interdiscip Rev Syst Biol Med 2009;1(3):309–33.

[4] Theilgaard-Monch K, Porse BT, Borregaard N. Systems biology of neutrophil differentiation and immune response. Curr Opin Immunol 2006;18(1):54–60.

[5] Holland SM, DeLeo FR, Elloumi HZ, Hsu AP, Uzel G, Brodsky N, et al. STAT3 mutations in the hyper-IgE syndrome. N Engl J Med 2007;357(16):1608–19.
[6] Kobayashi SD, Voyich JM, Braughton KR, Whitney AR, Nauseef WM, Malech HL, et al. Gene expression profiling provides insight into the pathophysiology of chronic granulomatous disease. J Immunol 2004;172(1):636–43.
[7] Blattner FR, Plunkett 3rd G, Bloch CA, Perna NT, Burland V, Riley M, et al. The complete genome sequence of *Escherichia coli* K-12. Science 1997;277(5331):1453–62.
[8] Fleischmann RD, Adams MD, White O, Clayton RA, Kirkness EF, Kerlavage AR, et al. Whole-genome random sequencing and assembly of *Haemophilus influenzae* Rd. Science 1995;269(5223):496–512.
[9] Koonin EV, Mushegian AR, Rudd KE. Sequencing and analysis of bacterial genomes. Curr Biol 1996;6(4):404–16.
[10] Medini D, Serruto D, Parkhill J, Relman DA, Donati C, Moxon R, et al. Microbiology in the post-genomic era. Nat Rev Microbiol 2008;6(6):419–30.
[11] Pallen MJ, Wren BW. Bacterial pathogenomics. Nature 2007;449(7164):835–42.
[12] Perna NT, Plunkett 3rd G, Burland V, Mau B, Glasner JD, Rose DJ, et al. Genome sequence of enterohaemorrhagic *Escherichia coli* O157:H7. Nature 2001;409(6819):529–33.
[13] Tettelin H, Feldblyum T. Bacterial genome sequencing. Methods Mol Biol 2009;551:231–47.
[14] Wren BW. Microbial genome analysis: insights into virulence, host adaptation and evolution. Nat Rev Genet 2000;1(1):30–9.
[15] Hayashi F, Means TK, Luster AD. Toll-like receptors stimulate human neutrophil function. Blood 2003;102(7):2660–9.
[16] Quinn MT, Ammons MC, DeLeo FR. The expanding role of NADPH oxidases in health and disease: no longer just agents of death and destruction. Clin Sci (Lond) 2006;111(1):1–20.
[17] Dale DC, Liles WC, Llewellyn C, Price TH. Effects of granulocyte-macrophage colony-stimulating factor (GM-CSF) on neutrophil kinetics and function in normal human volunteers. Am J Hematol 1998;57(1):7–15.
[18] Ferrari F, Bortoluzzi S, Coppe A, Basso D, Bicciato S, Zini R, et al. Genomic expression during human myelopoiesis. BMC Genomics 2007;8:264.
[19] Martinelli S, Urosevic M, Daryadel A, Oberholzer PA, Baumann C, Fey MF, et al. Induction of genes mediating interferon-dependent extracellular trap formation during neutrophil differentiation. J Biol Chem 2004;279(42):44123–32.
[20] Theilgaard-Monch K, Knudsen S, Follin P, Borregaard N. The transcriptional activation program of human neutrophils in skin lesions supports their important role in wound healing. J Immunol 2004;172(12):7684–93.
[21] Friedman AD. Transcriptional regulation of granulocyte and monocyte development. Oncogene 2002;21(21):3377–90.
[22] McDonald PP. Transcriptional regulation in neutrophils: teaching old cells new tricks. Adv Immunol 2004;82:1–48.
[23] Kasprisin DO, Harris MB. The role of RNA metabolism in polymorphonuclear leukocyte phagocytosis. J Lab Clin Med 1977;90(1):118–24.
[24] Kasprisin DO, Harris MB. The role of protein synthesis in polymorphonuclear leukocyte phagocytosis II. Exp Hematol 1978;6(7):585–9.
[25] Kobayashi SD, Voyich JM, Buhl CL, Stahl RM, DeLeo FR. Global changes in gene expression by human polymorphonuclear leukocytes during receptor-mediated phagocytosis: cell fate is regulated at the level of gene expression. Proc Natl Acad Sci U S A 2002;99(10):6901–6.

[26] Kobayashi SD, Braughton KR, Whitney AR, Voyich JM, Schwan TG, Musser JM, et al. Bacterial pathogens modulate an apoptosis differentiation program in human neutrophils. Proc Natl Acad Sci U S A 2003;100(19):10948–53.

[27] Kobayashi SD, Voyich JM, Braughton KR, DeLeo FR. Down-regulation of proinflammatory capacity during apoptosis in human polymorphonuclear leukocytes. J Immunol 2003;170(6):3357–68.

[28] Serhan CN, Savill J. Resolution of inflammation: the beginning programs the end. Nat Immunol 2005;6(12):1191–7.

[29] Savill J. Apoptosis in resolution of inflammation. J Leukoc Biol 1997;61(4):375–80.

[30] Savill JS, Wyllie AH, Henson JE, Walport MJ, Henson PM, Haslett C. Macrophage phagocytosis of aging neutrophils in inflammation. Programmed cell death in the neutrophil leads to its recognition by macrophages. J Clin Invest 1989;83(3):865–75.

[31] DeLeo FR. Modulation of phagocyte apoptosis by bacterial pathogens. Apoptosis 2004;9(4):399–413.

[32] Coxon A, Rieu P, Barkalow FJ, Askari S, Sharpe AH, von Andrian UH, et al. A novel role for the beta 2 integrin CD11b/CD18 in neutrophil apoptosis: a homeostatic mechanism in inflammation. Immunity 1996;5(6):653–66.

[33] Gamberale R, Giordano M, Trevani AS, Andonegui G, Geffner JR. Modulation of human neutrophil apoptosis by immune complexes. J Immunol 1998;161(7):3666–74.

[34] Watson RW, Redmond HP, Wang JH, Condron C, Bouchier-Hayes D. Neutrophils undergo apoptosis following ingestion of *Escherichia coli*. J Immunol 1996;156(10):3986–92.

[35] Zhang B, Hirahashi J, Cullere X, Mayadas TN. Elucidation of molecular events leading to neutrophil apoptosis following phagocytosis: cross-talk between caspase 8, reactive oxygen species, and MAPK/ERK activation. J Biol Chem 2003;278(31):28443–54.

[36] Borjesson DL, Kobayashi SD, Whitney AR, Voyich JM, Argue CM, DeLeo FR. Insights into pathogen immune evasion mechanisms: *Anaplasma phagocytophilum* fails to induce an apoptosis differentiation program in human neutrophils. J Immunol 2005;174(10):6364–72.

[37] Kobayashi SD, Braughton KR, Palazzolo-Ballance AM, Kennedy AD, Sampaio E, Kristosturyan E, et al. Rapid neutrophil destruction following phagocytosis of *Staphylococcus aureus*. J Innate Immun 2010;2(6):560–75.

[38] Subrahmanyam YV, Yamaga S, Prashar Y, Lee HH, Hoe NP, Kluger Y, et al. RNA expression patterns change dramatically in human neutrophils exposed to bacteria. Blood 2001;97(8):2457–68.

[39] Mullick A, Elias M, Harakidas P, Marcil A, Whiteway M, Ge B, et al. Gene expression in HL60 granulocytoids and human polymorphonuclear leukocytes exposed to *Candida albicans*. Infect Immun 2004;72(1):414–29.

[40] Fradin C, Mavor AL, Weindl G, Schaller M, Hanke K, Kaufmann SH, et al. The early transcriptional response of human granulocytes to infection with *Candida albicans* is not essential for killing but reflects cellular communications. Infect Immun 2007;75(3):1493–501.

[41] Lee HC, Goodman JL. *Anaplasma phagocytophilum* causes global induction of antiapoptosis in human neutrophils. Genomics 2006;88(4):496–503.

[42] Suttmann H, Lehan N, Bohle A, Brandau S. Stimulation of neutrophil granulocytes with *Mycobacterium bovis* bacillus Calmette-Guerin induces changes in phenotype and gene expression and inhibits spontaneous apoptosis. Infect Immun 2003;71(8):4647–56.

[43] Zhang X, Kluger Y, Nakayama Y, Poddar R, Whitney C, DeTora A, et al. Gene expression in mature neutrophils: early responses to inflammatory stimuli. J Leukoc Biol 2004;75(2):358–72.

[44] Yoshiie K, Kim HY, Mott J, Rikihisa Y. Intracellular infection by the human granulocytic ehrlichiosis agent inhibits human neutrophil apoptosis. Infect Immun 2000;68(3):1125–33.

[45] Carlyon JA, Abdel-Latif D, Pypaert M, Lacy P, Fikrig E. *Anaplasma phagocytophilum* utilizes multiple host evasion mechanisms to thwart NADPH oxidase-mediated killing during neutrophil infection. Infect Immun 2004;72(8):4772–83.

[46] Mott J, Rikihisa Y. Human granulocytic ehrlichiosis agent inhibits superoxide anion generation by human neutrophils. Infect Immun 2000;68(12):6697–703.

[47] Musser JM, DeLeo FR. Toward a genome-wide systems biology analysis of host-pathogen interactions in group A *Streptococcus*. Am J Pathol 2005;167(6):1461–72.

[48] diCenzo GC, Finan TM. The divided bacterial genome: structure, function, and evolution. Microbiol Mol Biol Rev 2017;81(3). pii: e00019-17.

[49] Serruto D, Serino L, Masignani V, Pizza M. Genome-based approaches to develop vaccines against bacterial pathogens. Vaccine 2009;27(25–26):3245–50.

[50] Behr MA, Wilson MA, Gill WP, Salamon H, Schoolnik GK, Rane S, et al. Comparative genomics of BCG vaccines by whole-genome DNA microarray. Science 1999;284(5419):1520–3.

[51] Fitzgerald JR, Sturdevant DE, Mackie SM, Gill SR, Musser JM. Evolutionary genomics of *Staphylococcus aureus*: insights into the origin of methicillin-resistant strains and the toxic shock syndrome epidemic. Proc Natl Acad Sci U S A 2001;98(15):8821–6.

[52] Salama N, Guillemin K, McDaniel TK, Sherlock G, Tompkins L, Falkow S. A whole-genome microarray reveals genetic diversity among *Helicobacter pylori* strains. Proc Natl Acad Sci U S A 2000;97(26):14668–73.

[53] Tenover FC, McDougal LK, Goering RV, Killgore G, Projan SJ, Patel JB, et al. Characterization of a strain of community-associated methicillin-resistant *Staphylococcus aureus* widely disseminated in the United States. J Clin Microbiol 2006;44(1):108–18.

[54] Auger E, Deslandes V, Ramjeet M, Contreras I, Nash JH, Harel J, et al. Host-pathogen interactions of *Actinobacillus pleuropneumoniae* with porcine lung and tracheal epithelial cells. Infect Immun 2009;77(4):1426–41.

[55] Galindo CL, Sha J, Ribardo DA, Fadl AA, Pillai L, Chopra AK. Identification of *Aeromonas hydrophila* cytotoxic enterotoxin-induced genes in macrophages using microarrays. J Biol Chem 2003;278(41):40198–212.

[56] Nelson CM, Herron MJ, Felsheim RF, Schloeder BR, Grindle SM, Chavez AO, et al. Whole genome transcription profiling of *Anaplasma phagocytophilum* in human and tick host cells by tiling array analysis. BMC Genomics 2008;9:364.

[57] Bergman NH, Anderson EC, Swenson EE, Janes BK, Fisher N, Niemeyer MM, et al. Transcriptional profiling of *Bacillus anthracis* during infection of host macrophages. Infect Immun 2007;75(7):3434–44.

[58] Livengood JA, Schmit VL, Gilmore Jr. RD. Global transcriptome analysis of *Borrelia burgdorferi* during association with human neuroglial cells. Infect Immun 2008;76(1):298–307.

[59] Rossetti CA, Galindo CL, Lawhon SD, Garner HR, Adams LG. *Brucella melitensis* global gene expression study provides novel information on growth phase-specific gene regulation with potential insights for understanding *Brucella*:host initial interactions. BMC Microbiol 2009;9:81.

[60] Belland RJ, Nelson DE, Virok D, Crane DD, Hogan D, Sturdevant D, et al. Transcriptome analysis of chlamydial growth during IFN-gamma-mediated persistence and reactivation. Proc Natl Acad Sci U S A 2003;100(26):15971–6.

[61] Belland RJ, Zhong G, Crane DD, Hogan D, Sturdevant D, Sharma J, et al. Genomic transcriptional profiling of the developmental cycle of *Chlamydia trachomatis*. Proc Natl Acad Sci U S A 2003;100(14):8478–83.

[62] Janvilisri T, Scaria J, Chang YF. Transcriptional profiling of *Clostridium difficile* and Caco-2 cells during infection. J Infect Dis 2010;202(2):282–90.

[63] Staudinger BJ, Oberdoerster MA, Lewis PJ, Rosen H. mRNA expression profiles for *Escherichia coli* ingested by normal and phagocyte oxidase-deficient human neutrophils. J Clin Invest 2002;110(8):1151–63.

[64] Jandu N, Ho NK, Donato KA, Karmali MA, Mascarenhas M, Duffy SP, et al. Enterohemorrhagic *Escherichia coli* O157:H7 gene expression profiling in response to growth in the presence of host epithelia. PLoS One 2009;4(3):e4889.

[65] Kim Y, Oh S, Park S, Kim SH. Interactive transcriptome analysis of enterohemorrhagic *Escherichia coli* (EHEC) O157:H7 and intestinal epithelial HT-29 cells after bacterial attachment. Int J Food Microbiol 2009;131(2–3):224–32.

[66] Roos V, Klemm P. Global gene expression profiling of the asymptomatic bacteriuria *Escherichia coli* strain 83972 in the human urinary tract. Infect Immun 2006;74(6):3565–75.

[67] Wehrly TD, Chong A, Virtaneva K, Sturdevant DE, Child R, Edwards JA, et al. Intracellular biology and virulence determinants of *Francisella tularensis* revealed by transcriptional profiling inside macrophages. Cell Microbiol 2009;11(7):1128–50.

[68] Marco ML, Peters TH, Bongers RS, Molenaar D, van Hemert S, Sonnenburg JL, et al. Lifestyle of *Lactobacillus plantarum* in the mouse caecum. Environ Microbiol 2009;11(10):2747–57.

[69] Xue F, Dong H, Wu J, Wu Z, Hu W, Sun A, et al. Transcriptional responses of *Leptospira interrogans* to host innate immunity: significant changes in metabolism, oxygen tolerance, and outer membrane. PLoS Negl Trop Dis 2010;4(10):e857.

[70] Camejo A, Buchrieser C, Couve E, Carvalho F, Reis O, Ferreira P, et al. In vivo transcriptional profiling of *Listeria monocytogenes* and mutagenesis identify new virulence factors involved in infection. PLoS Pathog 2009;5(5):e1000449.

[71] Chatterjee SS, Hossain H, Otten S, Kuenne C, Kuchmina K, Machata S, et al. Intracellular gene expression profile of *Listeria monocytogenes*. Infect Immun 2006;74(2):1323–38.

[72] Hain T, Steinweg C, Chakraborty T. Comparative and functional genomics of *Listeria* spp. J Biotechnol 2006;126(1):37–51.

[73] Joseph B, Przybilla K, Stuhler C, Schauer K, Slaghuis J, Fuchs TM, et al. Identification of *Listeria monocytogenes* genes contributing to intracellular replication by expression profiling and mutant screening. J Bacteriol 2006;188(2):556–68.

[74] Janagama HK, Lamont EA, George S, Bannantine JP, Xu WW, Tu ZJ, et al. Primary transcriptomes of *Mycobacterium avium* subsp. paratuberculosis reveal proprietary pathways in tissue and macrophages. BMC Genomics 2010;11:561.

[75] Zhu X, Tu ZJ, Coussens PM, Kapur V, Janagama H, Naser S, et al. Transcriptional analysis of diverse strains *Mycobacterium avium* subspecies paratuberculosis in primary bovine monocyte derived macrophages. Microbes Infect 2008;10(12–13):1274–82.

[76] Talaat AM, Lyons R, Howard ST, Johnston SA. The temporal expression profile of *Mycobacterium tuberculosis* infection in mice. Proc Natl Acad Sci U S A 2004;101(13):4602–7.

[77] Cappelli G, Volpe E, Grassi M, Liseo B, Colizzi V, Mariani F. Profiling of *Mycobacterium tuberculosis* gene expression during human macrophage infection: upregulation of the alternative sigma factor G, a group of transcriptional regulators, and proteins with unknown function. Res Microbiol 2006;157(5):445–55.

[78] Dubnau E, Fontan P, Manganelli R, Soares-Appel S, Smith I. *Mycobacterium tuberculosis* genes induced during infection of human macrophages. Infect Immun 2002;70(6):2787–95.

[79] Fontan P, Aris V, Ghanny S, Soteropoulos P, Smith I. Global transcriptional profile of *Mycobacterium tuberculosis* during THP-1 human macrophage infection. Infect Immun 2008;76(2):717–25.

[80] Homolka S, Niemann S, Russell DG, Rohde KH. Functional genetic diversity among *Mycobacterium tuberculosis* complex clinical isolates: delineation of conserved core and lineage-specific transcriptomes during intracellular survival. PLoS Pathog 2010;6(7):e1000988.

[81] Rengarajan J, Bloom BR, Rubin EJ. Genome-wide requirements for *Mycobacterium tuberculosis* adaptation and survival in macrophages. Proc Natl Acad Sci U S A 2005;102(23):8327–32.

[82] Du Y, Lenz J, Arvidson CG. Global gene expression and the role of sigma factors in *Neisseria gonorrhoeae* in interactions with epithelial cells. Infect Immun 2005;73(8):4834–45.

[83] Dietrich G, Kurz S, Hubner C, Aepinus C, Theiss S, Guckenberger M, et al. Transcriptome analysis of *Neisseria meningitidis* during infection. J Bacteriol 2003;185(1):155–64.

[84] Grifantini R, Bartolini E, Muzzi A, Draghi M, Frigimelica E, Berger J, et al. Gene expression profile in *Neisseria meningitidis* and *Neisseria lactamica* upon host-cell contact: from basic research to vaccine development. Ann N Y Acad Sci 2002;975:202–16.

[85] Chugani S, Greenberg EP. The influence of human respiratory epithelia on *Pseudomonas aeruginosa* gene expression. Microb Pathog 2007;42(1):29–35.

[86] Bechah Y, El Karkouri K, Mediannikov O, Leroy Q, Pelletier N, Robert C, et al. Genomic, proteomic, and transcriptomic analysis of virulent and avirulent *Rickettsia prowazekii* reveals its adaptive mutation capabilities. Genome Res 2010;20(5):655–63.

[87] Ammerman NC, Rahman MS, Azad AF. Characterization of Sec-translocon-dependent extracytoplasmic proteins of *Rickettsia typhi*. J Bacteriol 2008;190(18):6234–42.

[88] Song XM, Connor W, Hokamp K, Babiuk LA, Potter AA. *Streptococcus pneumoniae* early response genes to human lung epithelial cells. BMC Res Notes 2008;1:64.

[89] Song XM, Connor W, Jalal S, Hokamp K, Potter AA. Microarray analysis of *Streptococcus pneumoniae* gene expression changes to human lung epithelial cells. Can J Microbiol 2008;54(3):189–200.

[90] Song XM, Connor W, Hokamp K, Babiuk LA, Potter AA. Transcriptome studies on *Streptococcus pneumoniae*, illustration of early response genes to THP-1 human macrophages. Genomics 2009;93(1):72–82.

[91] Voyich JM, Sturdevant DE, Braughton KR, Kobayashi SD, Lei B, Virtaneva K, et al. Genome-wide protective response used by group A *Streptococcus* to evade destruction by human polymorphonuclear leukocytes. Proc Natl Acad Sci U S A 2003;100(4):1996–2001.

[92] Shea PR, Virtaneva K, Kupko 3rd JJ, Porcella SF, Barry WT, Wright FA, et al. Interactome analysis of longitudinal pharyngeal infection of cynomolgus macaques by group A *Streptococcus*. Proc Natl Acad Sci U S A 2010;107(10):4693–8.

[93] Graham MR, Virtaneva K, Porcella SF, Barry WT, Gowen BB, Johnson CR, et al. Group A *Streptococcus* transcriptome dynamics during growth in human blood reveals bacterial adaptive and survival strategies. Am J Pathol 2005;166(2):455–65.

[94] Graham MR, Virtaneva K, Porcella SF, Gardner DJ, Long RD, Welty DM, et al. Analysis of the transcriptome of group A *Streptococcus* in mouse soft tissue infection. Am J Pathol 2006;169(3):927–42.

[95] Ryan PA, Kirk BW, Euler CW, Schuch R, Fischetti VA. Novel algorithms reveal streptococcal transcriptomes and clues about undefined genes. PLoS Comput Biol 2007;3(7):e132.

[96] Virtaneva K, Porcella SF, Graham MR, Ireland RM, Johnson CA, Ricklefs SM, et al. Longitudinal analysis of the group A *Streptococcus* transcriptome in experimental pharyngitis in cynomolgus macaques. Proc Natl Acad Sci U S A 2005;102(25):9014–9.

[97] Hautefort I, Thompson A, Eriksson-Ygberg S, Parker ML, Lucchini S, Danino V, et al. During infection of epithelial cells *Salmonella enterica* serovar Typhimurium undergoes a time-dependent transcriptional adaptation that results in simultaneous expression of three type 3 secretion systems. Cell Microbiol 2008;10(4):958–84.

[98] Sirsat SA, Burkholder KM, Muthaiyan A, Dowd SE, Bhunia AK, Ricke SC. Effect of sublethal heat stress on *Salmonella Typhimurium* virulence. J Appl Microbiol 2011;110(3):813–22.

[99] Wright JA, Totemeyer SS, Hautefort I, Appia-Ayme C, Alston M, Danino V, et al. Multiple redundant stress resistance mechanisms are induced in *Salmonella enterica* serovar Typhimurium in response to alteration of the intracellular environment via TLR4 signalling. Microbiology 2009;155 Pt 9: 2919–29.

[100] Faucher SP, Curtiss 3rd R, Daigle F. Selective capture of *Salmonella enterica* serovar typhi genes expressed in macrophages that are absent from the *Salmonella enterica* serovar Typhimurium genome. Infect Immun 2005;73(8):5217–21.

[101] Faucher SP, Porwollik S, Dozois CM, McClelland M, Daigle F. Transcriptome of *Salmonella enterica* serovar Typhi within macrophages revealed through the selective capture of transcribed sequences. Proc Natl Acad Sci U S A 2006;103(6):1906–11.

[102] Lucchini S, Liu H, Jin Q, Hinton JC, Yu J. Transcriptional adaptation of *Shigella flexneri* during infection of macrophages and epithelial cells: insights into the strategies of a cytosolic bacterial pathogen. Infect Immun 2005;73(1):88–102.

[103] Voyich JM, Braughton KR, Sturdevant DE, Whitney AR, Said-Salim B, Porcella SF, et al. Insights into mechanisms used by *Staphylococcus aureus* to avoid destruction by human neutrophils. J Immunol 2005;175(6):3907–19.

[104] Garzoni C, Francois P, Huyghe A, Couzinet S, Tapparel C, Charbonnier Y, et al. A global view of *Staphylococcus aureus* whole genome expression upon internalization in human epithelial cells. BMC Genomics 2007;8:171.

[105] Larocque RC, Harris JB, Dziejman M, Li X, Khan AI, Faruque AS, et al. Transcriptional profiling of *Vibrio cholerae* recovered directly from patient specimens during early and late stages of human infection. Infect Immun 2005;73(8):4488–93.

[106] Vadyvaloo V, Jarrett C, Sturdevant DE, Sebbane F, Hinnebusch BJ. Transit through the flea vector induces a pretransmission innate immunity resistance phenotype in *Yersinia pestis*. PLoS Pathog 2010;6(2):e1000783.

[107] Sturdevant DE, Virtaneva K, Martens C, Bozinov D, Ogundare O, Castro N, et al. Host-microbe interaction systems biology: lifecycle transcriptomics and comparative genomics. Future Microbiol 2010;5(2):205–19.

[108] Wang Z, Gerstein M, Snyder M. RNA-seq: a revolutionary tool for transcriptomics. Nat Rev Genet 2009;10:57.

[109] Westermann AJ, Gorski SA, Vogel J. Dual RNA-seq of pathogen and host. Nat Rev Microbiol 2012;10(9):618–30.

[110] Friedman DB, Stauff DL, Pishchany G, Whitwell CW, Torres VJ, Skaar EP. *Staphylococcus aureus* redirects central metabolism to increase iron availability. PLoS Pathog 2006;2(8):e87.

[111] Attia AS, Benson MA, Stauff DL, Torres VJ, Skaar EP. Membrane damage elicits an immunomodulatory program in *Staphylococcus aureus*. PLoS Pathog 2010;6(3):e1000802.

[112] Burlak C, Hammer CH, Robinson MA, Whitney AR, McGavin MJ, Kreiswirth BN, et al. Global analysis of community-associated methicillin-resistant *Staphylococcus aureus* exoproteins reveals molecules produced in vitro and during infection. Cell Microbiol 2007;9(5):1172–90.

[113] Hecker M, Becher D, Fuchs S, Engelmann S. A proteomic view of cell physiology and virulence of *Staphylococcus aureus*. Int J Med Microbiol 2010;300(2–3):76–87.

[114] Becher D, Hempel K, Sievers S, Zuhlke D, Pane-Farre J, Otto A, et al. A proteomic view of an important human pathogen—towards the quantification of the entire *Staphylococcus aureus* proteome. PLoS One 2009;4(12):e8176.

[115] Gotz F, Hacker J, Hecker M. Pathophysiology of staphylococci in the post-genomic era. Int J Med Microbiol 2010;300(2–3):75.

[116] Volker U, Hecker M. From genomics via proteomics to cellular physiology of the Gram-positive model organism *Bacillus subtilis*. Cell Microbiol 2005;7(8):1077–85.

[117] Smoot LM, Smoot JC, Graham MR, Somerville GA, Sturdevant DE, Migliaccio CA, et al. Global differential gene expression in response to growth temperature alteration in group A *Streptococcus*. Proc Natl Acad Sci U S A 2001;98(18):10416–21.

[118] Graham MR, Smoot LM, Migliaccio CA, Virtaneva K, Sturdevant DE, Porcella SF, et al. Virulence control in group A *Streptococcus* by a two-component gene regulatory system: global expression profiling and in vivo infection modeling. Proc Natl Acad Sci U S A 2002;99(21):13855–60.

[119] Sitkiewicz I, Musser JM. Expression microarray and mouse virulence analysis of four conserved two-component gene regulatory systems in group a streptococcus. Infect Immun 2006;74(2):1339–51.

[120] Voyich JM, Braughton KR, Sturdevant DE, Vuong C, Kobayashi SD, Porcella SF, et al. Engagement of the pathogen survival response used by group A *Streptococcus* to avert destruction by innate host defense. J Immunol 2004;173(2):1194–201.

[121] Shelburne 3rd SA, Sumby P, Sitkiewicz I, Granville C, DeLeo FR, Musser JM. Central role of a bacterial two-component gene regulatory system of previously unknown function in pathogen persistence in human saliva. Proc Natl Acad Sci U S A 2005;102(44):16037–42.

[122] Dunman PM, Murphy E, Haney S, Palacios D, Tucker-Kellogg G, Wu S, et al. Transcription profiling-based identification of *Staphylococcus aureus* genes regulated by the agr and/or sarA loci. J Bacteriol 2001;183(24):7341–53.

[123] Kuroda M, Kuroda H, Oshima T, Takeuchi F, Mori H, Hiramatsu K. Two-component system VraSR positively modulates the regulation of cell-wall biosynthesis pathway in *Staphylococcus aureus*. Mol Microbiol 2003;49(3):807–21.

[124] Bischoff M, Dunman P, Kormanec J, Macapagal D, Murphy E, Mounts W, et al. Microarray-based analysis of the *Staphylococcus aureus* sigmaB regulon. J Bacteriol 2004;186(13):4085–99.

[125] Pane-Farre J, Jonas B, Forstner K, Engelmann S, Hecker M. The sigmaB regulon in *Staphylococcus aureus* and its regulation. Int J Med Microbiol 2006;296(4–5):237–58.

[126] McCallum N, Spehar G, Bischoff M, Berger-Bachi B. Strain dependence of the cell wall-damage induced stimulon in *Staphylococcus aureus*. Biochim Biophys Acta 2006;1760(10):1475–81.

[127] Liang X, Zheng L, Landwehr C, Lunsford D, Holmes D, Ji Y. Global regulation of gene expression by ArlRS, a two-component signal transduction regulatory system of *Staphylococcus aureus*. J Bacteriol 2005;187(15):5486–92.

[128] Luong TT, Dunman PM, Murphy E, Projan SJ, Lee CY. Transcription profiling of the mgrA regulon in *Staphylococcus aureus*. J Bacteriol 2006;188(5):1899–910.

[129] Yang SJ, Dunman PM, Projan SJ, Bayles KW. Characterization of the *Staphylococcus aureus* CidR regulon: elucidation of a novel role for acetoin metabolism in cell death and lysis. Mol Microbiol 2006;60(2):458–68.

[130] Li M, Cha DJ, Lai Y, Villaruz AE, Sturdevant DE, Otto M. The antimicrobial peptide-sensing system aps of *Staphylococcus aureus*. Mol Microbiol 2007;66(5):1136–47.

[131] Voyich JM, Vuong C, DeWald M, Nygaard TK, Kocianova S, Griffith S, et al. The SaeR/S gene regulatory system is essential for innate immune evasion by *Staphylococcus aureus*. J Infect Dis 2009;199(11):1698–706.

[132] Rogasch K, Ruhmling V, Pane-Farre J, Hoper D, Weinberg C, Fuchs S, et al. Influence of the two-component system SaeRS on global gene expression in two different *Staphylococcus aureus* strains. J Bacteriol 2006;188(22):7742–58.

[133] Nygaard TK, Pallister KB, Ruzevich P, Griffith S, Vuong C, Voyich JM. SaeR binds a consensus sequence within virulence gene promoters to advance USA300 pathogenesis. J Infect Dis 2010;201(2):241–54.

[134] Liang X, Yu C, Sun J, Liu H, Landwehr C, Holmes D, et al. Inactivation of a two-component signal transduction system, SaeRS, eliminates adherence and attenuates virulence of *Staphylococcus aureus*. Infect Immun 2006;74(8):4655–65.

[135] Richardson AR, Dunman PM, Fang FC. The nitrosative stress response of *Staphylococcus aureus* is required for resistance to innate immunity. Mol Microbiol 2006;61(4):927–39.

[136] Pishchany G, McCoy AL, Torres VJ, Krause JC, Crowe Jr. JE, Fabry ME, et al. Specificity for human hemoglobin enhances *Staphylococcus aureus* infection. Cell Host Microbe 2010;8(6):544–50.

[137] Skaar EP. The battle for iron between bacterial pathogens and their vertebrate hosts. PLoS Pathog 2010;6(8):e1000949.

[138] Palazzolo-Ballance AM, Reniere ML, Braughton KR, Sturdevant DE, Otto M, Kreiswirth BN, et al. Neutrophil microbicides induce a pathogen survival response in community-associated methicillin-resistant *Staphylococcus aureus*. J Immunol 2008;180(1):500–9.

[139] Pietiainen M, Francois P, Hyyrylainen HL, Tangomo M, Sass V, Sahl HG, et al. Transcriptome analysis of the responses of *Staphylococcus aureus* to antimicrobial peptides and characterization of the roles of vraDE and vraSR in antimicrobial resistance. BMC Genomics 2009;10:429.

[140] Chambers HF, Deleo FR. Waves of resistance: *Staphylococcus aureus* in the antibiotic era. Nat Rev Microbiol 2009;7(9):629–41.

[141] Mwangi MM, Wu SW, Zhou Y, Sieradzki K, de Lencastre H, Richardson P, et al. Tracking the in vivo evolution of multidrug resistance in *Staphylococcus aureus* by whole-genome sequencing. Proc Natl Acad Sci U S A 2007;104(22):9451–6.

[142] Renzoni A, Barras C, Francois P, Charbonnier Y, Huggler E, Garzoni C, et al. Transcriptomic and functional analysis of an autolysis-deficient, teicoplanin-resistant derivative of methicillin-resistant *Staphylococcus aureus*. Antimicrob Agents Chemother 2006;50(9):3048–61.

[143] McCallum N, Karauzum H, Getzmann R, Bischoff M, Majcherczyk P, Berger-Bachi B, et al. In vivo survival of teicoplanin-resistant *Staphylococcus aureus* and fitness cost of teicoplanin resistance. Antimicrob Agents Chemother 2006;50(7):2352–60.

[144] Cui L, Lian JQ, Neoh HM, Reyes E, Hiramatsu K. DNA microarray-based identification of genes associated with glycopeptide resistance in *Staphylococcus aureus*. Antimicrob Agents Chemother 2005;49(8):3404–13.

[145] Yao Y, Sturdevant DE, Otto M. Genomewide analysis of gene expression in *Staphylococcus epidermidis* biofilms: insights into the pathophysiology of *S. epidermidis* biofilms and the role of phenol-soluble modulins in formation of biofilms. J Infect Dis 2005;191(2):289–98.

[146] Resch A, Rosenstein R, Nerz C, Gotz F. Differential gene expression profiling of *Staphylococcus aureus* cultivated under biofilm and planktonic conditions. Appl Environ Microbiol 2005;71(5):2663–76.

[147] Otto M. *Staphylococcus epidermidis*—the 'accidental' pathogen. Nat Rev Microbiol 2009;7(8):555–67.

[148] Westermann AJ, Barquist L, Vogel J. Resolving host-pathogen interactions by dual RNA-seq. PLoS Pathog 2017;13(2):e1006033.

[149] Wolf T, Kammer P, Brunke S, Linde J. Two's company: studying interspecies relationships with dual RNA-seq. Curr Opin Microbiol 2018;42:7–12.

[150] Mandlik A, Livny J, Robins WP, Ritchie JM, Mekalanos JJ, Waldor MK. RNA-Seq-based monitoring of infection-linked changes in *Vibrio cholerae* gene expression. Cell Host Microbe 2011;10(2):165–74.

[151] Stasulli NM, Eichelberger KR, Price PA, Pechous RD, Montgomery SA, Parker JS, et al. Spatially distinct neutrophil responses within the inflammatory lesions of pneumonic plague. MBio 2015;6(5): e01530-15.

[152] Baddal B, Muzzi A, Censini S, Calogero RA, Torricelli G, Guidotti S, et al. Dual RNA-seq of nontypeable *Haemophilus influenzae* and host cell transcriptomes reveals novel insights into host-pathogen cross talk. mBio 2015;6(6): e01765-15.
[153] Westermann AJ, Forstner KU, Amman F, Barquist L, Chao Y, Schulte LN, et al. Dual RNA-seq unveils noncoding RNA functions in host-pathogen interactions. Nature 2016;529(7587):496–501.
[154] Aprianto R, Slager J, Holsappel S, Veening JW. Time-resolved dual RNA-seq reveals extensive rewiring of lung epithelial and pneumococcal transcriptomes during early infection. Genome Biol 2016;17(1):198.
[155] Nuss AM, Beckstette M, Pimenova M, Schmuhl C, Opitz W, Pisano F, et al. Tissue dual RNA-seq allows fast discovery of infection-specific functions and riboregulators shaping host-pathogen transcriptomes. Proc Natl Acad Sci U S A 2017;114(5):E791–800.
[156] Thanert R, Goldmann O, Beineke A, Medina E. Host-inherent variability influences the transcriptional response of *Staphylococcus aureus* during in vivo infection. Nat Commun 2017;8:14268.
[157] Zaas AK, Chen M, Varkey J, Veldman T, Hero 3rd AO, Lucas J, et al. Gene expression signatures diagnose influenza and other symptomatic respiratory viral infections in humans. Cell Host Microbe 2009;6(3):207–17.

Chapter 6

Sepsis genomics and precision medicine

Hector R. Wong
Division of Critical Care Medicine, Cincinnati Children's Hospital Medical Center and Cincinnati Children's Research Foundation, Cincinnati, OH, United States, Department of Pediatrics, University of Cincinnati College of Medicine, Cincinnati, OH, United States

Introduction

The clinical challenges inherent to sepsis are in many ways ideally suited for the tenets of precision medicine. After decades of basic, translational, and clinical research seeking to modulate the host response to sepsis, therapy for sepsis is essentially limited to antibiotics and supportive care. There are no other sepsis-specific therapies approved currently, nor accepted universally. This remarkable set of circumstances reflects, in large part, the heterogeneous host response during sepsis, wherein any particular modulation of the host response is unlikely to benefit all patients. Moving forward, the hope is that precision medicine will enable the identification of patient subsets sharing biological commonalities and who are therefore more likely to benefit from new therapies targeting specific biological processes [1].

Enrichment

The concept of enrichment is fundamental to precision medicine [2]. At its most fundamental level, enrichment creates patient subsets based on shared clinical, physiological, and biological characteristics. Enrichment generally refers to the selection of a population of patients more likely to benefit from a given therapy, compared to an unselected population. More specifically, prognostic enrichment refers to the selection of a patient population more likely to have an outcome of interest, such as mortality. This type of enrichment is particularly important in the context of new therapies carrying more than minimal risk for adverse, therapy-related events. Such therapies are more likely to benefit patients at higher risk of mortality or other disease-related event.

Genomic and Precision Medicine. https://doi.org/10.1016/B978-0-12-801496-7.00006-X

Predictive enrichment is another form of enrichment, distinct from prognostic enrichment. Here, one selects a population more likely to benefit from a given therapy, but based on a biological or physiological mechanism. A recent example of this concept is embodied by the selection of oncology patients for anti-program cell death-1 (PD-1) therapy based on tumor positivity for PD-1 ligand [3].

Recently, several investigative teams have leveraged the discovery potential of genomics, particularly the sub-field of transcriptomics, to develop prognostic and predictive enrichment strategies for sepsis. These efforts, based on leukocyte-derived RNA, will be discussed in subsequent sections.

Historical examples of enrichment strategies for sepsis

In 2002, Annane and colleagues reported on the results of a trial testing the efficacy of adjunctive corticosteroids among adults with septic shock [4]. The primary endpoint for efficacy was survival at 28 days, and all study subjects were enrolled after undergoing corticotropin stimulation testing for relative adrenal insufficiency. There was no survival benefit attributable to corticosteroids for the entire cohort, but among the subset of patients who did not respond to the corticotropin stimulation test, corticosteroids were associated with a significant improvement in survival. Although this finding was not replicated in a subsequent trial [5], the approach based on a corticotropin stimulation test embodies the concept of predictive enrichment, wherein it is biologically plausible that patients who cannot mount an adequate cortisol response to corticotropin stimulation might have a greater likelihood of responding to exogenous corticosteroids.

Another example involves the Evaluating the Use of Polymyxin B Hemoperfusion in a Randomized controlled trial of Adults Treated for Endotoxemia and Septic shock (EUPHRATES) trial (clinicaltrials.gov #NCT01046669) [6]. Here, enrollment criteria were informed by measuring blood concentrations of endotoxin using a rapid assay. Only those patients with septic shock and having blood concentrations ≥0.6 endotoxin activity assay units were eligible for randomization to either extracorporeal hemoperfusion or sham perfusion. Although hemoperfusion did not improve 28-day survival [7], this study also represents a form of predictive enrichment, wherein it is biologically plausible that patients with a substantial burden of endotoxemia might have a greater likelihood of benefiting from an experimental therapy specifically designed for endotoxin removal.

Both examples represent biologically sound, but relatively simplistic, approaches to predictive enrichment. Simplicity is attractive in the setting of a sepsis clinical trial, which requires enrichment tools that can generate actionable data in relatively short time frame. While attractive and pragmatic, perhaps a primary lesson learned from these two well intended attempts at predictive enrichment is that future enrichment strategies are unlikely to capture the complexity of sepsis using a single variable or patient characteristic.

Prognostic enrichment strategies

The Pediatric Sepsis Biomarker Risk Model (PERSEVERE) incorporates five serum protein biomarkers (C-C chemokine ligand 3, interleukin-8, granzyme B, heat shock protein 70 kDa 1B, and matrix metallopeptidase 8) into a decision tree to estimate a range of baseline risk of mortality among children with septic shock [8]. As such, it represents a potential strategy for prognostic enrichment.

The PERSEVERE biomarkers were originally identified using discovery-oriented transcriptomic studies seeking to identify genes associated with poor outcome in this patient population [9]. From among 80 candidate genes, the PERSEVERE biomarkers were selected based on biological plausibility and the ability to measure the corresponding protein product in the blood compartment. PERSEVERE has been validated prospectively and recalibrated as PERSEVERE-II [10, 11], which incorporates platelet count along with the original PERSEVERE biomarkers. PERSEVERE-II appears to have broader applicability across diverse populations of children with septic shock. The PERSEVERE biomarkers have also been used to design an analogous model for adults with septic shock [12]. The most recent work in this area yielded PERSEVERE-XP, which uses a combination of mRNA-based biomarkers identified in the original transcriptomic studies, in combination with the PERSEVERE protein biomarkers [13].

In another comprehensive endeavor at prognostic enrichment, collaborating investigators recently catalogued all publically available transcriptomic data involving patients with sepsis [14]. The data included adults, children, and neonates. After normalization across the various datasets and analytical platforms, three different investigative groups set out to develop independent mortality prediction models for sepsis. Using different data-driven, machine learning approaches, each group developed robust models that performed well upon validation, and performance improved further when combined with existing illness severity scores.

In combination, the four models developed by this community approach yielded 65 candidate predictor genes. Among these 65 genes, 11 overlapped with the 80 genes identified during the initial discovery phase of the PERSEVERE biomarkers (Table 1). Based on Ingenuity Pathway Analysis, these 11 genes correspond to the gene network shown in Fig. 1. The gene network contains three highly connected central nodes: nuclear factor κB (NF-κB), tumor necrosis factor (TNF), and tumor protein 53 (TP53, a.k.a. p53), all of which have been linked to the pathobiology of sepsis.

The biological links among these 11 genes and the fact that they were identified independently across multiple investigative groups suggest that they are directly linked to the pathways that lead to poor outcomes in sepsis. Accordingly, the genes might represent more than simply biomarkers for prognostic enrichment. They might also serve as candidate therapeutic targets and pathways that can be readily interrogated experimentally.

TABLE 1 Genes common to a community approach for identification of prognostic enrichment genes and the original working gene list for the PERSEVERE biomarkers

Gene symbol	Description
CD24	CD24 molecule
CEACAM8	Carcinoembryonic antigen related cell adhesion molecule 8
CKS2	CDC28 protein kinase regulatory subunit 2
CX3CR1	C-X3-C motif chemokine receptor 1
DDIT4	DNA damage inducible transcript 4
EMR3	Adhesion G protein-coupled receptor E3
G0S2	G0/G1 switch 2
IL8	Interleukin 8, C-X-C motif chemokine ligand 8
MAFF	MAF bZIP transcription factor F
RGS1	Regulator of G protein signaling 1
TGFBI	Transforming growth factor beta induced

Predictive enrichment strategies

Several investigative groups have used whole genome expression data to identify gene expression-based subgroups of patients with sepsis. In general, these approaches involve unsupervised clustering algorithms based exclusively on commonalities of gene expression, and agnostic to patient characteristics, clinical trajectory, and outcome.

The Knight laboratory reported two gene expression-based subgroups of adults with sepsis, termed sepsis response signatures (SRS) 1 and 2 [15]. Patients allocated to SRS1 were characterized by a gene expression pattern consistent with endotoxin tolerance, T-cell exhaustion, and repression of human leukocyte antigen (HLA) class II. In addition, patients allocated to SRS1 had significantly greater 14-day mortality compared to patients allocated to SRS2. The existence of the SRS1 and 2 subgroups were prospectively validated in a follow up study and seem to be operative regardless of whether the source of sepsis is pulmonary or intra-abdominal [16].

The van der Poll laboratory identified four gene expression-based subgroups of adult patients with sepsis, termed molecular endotypes Mars1 to 4 [17]. Endotype assignment was associated with significant differences in 28-day mortality, with the Mars1 endotype being consistently associated with the highest risk of mortality. Similar to the SRS1 subgroup, the Mars1 endotype was associated with repression of genes corresponding to the innate and adaptive immune systems.

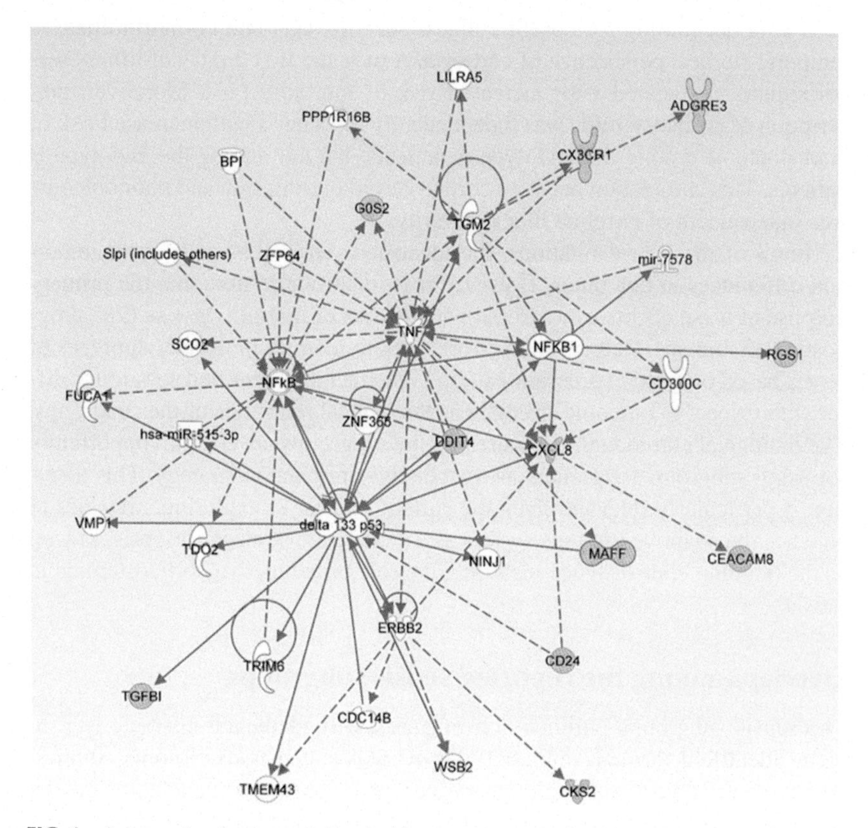

FIG. 1 A gene network represented by the 11 genes common to the 65 genes identified in the community approach to mortality risk prediction in sepsis, and the 80 genes identified during the initial discovery phase of the PERSEVERE biomarkers. The network was generated by uploading these 11 genes to the Ingenuity Pathway Analysis platform. The gene network contains three highly connected central nodes: nuclear factor κB (NF-κB), tumor necrosis factor (TNF), and tumor protein 53 (TP53, a.k.a. p53).

The Khatri laboratory recently used virtually all publically available transcriptomic data, including adults, children, and neonates, to uncover three clusters of patients with sepsis [18]. The three clusters were termed *inflammopathic*, *adaptive*, and *coagulopathic*, reflecting the biological functions of the cluster-defining genes. The coagulopathic cluster was associated with the highest mortality, whereas the adaptive cluster was associated with lower clinical severity and mortality.

Analogous studies were described in children with septic shock [19–23]. Here, the subgroups are termed pediatric septic shock endotypes A and B. Subjects allocated to endotype A were characterized by repression of genes corresponding to adaptive immunity and glucocorticoid receptor signaling. Allocation to endotype A was independently associated with increased mortality, even after accounting for baseline illness severity, age, and co-morbidities. In temporal studies, persistence of endotype A over the first 3 days of illness was particularly associated with increased risk of mortality [23]. Moreover, prescription of corticosteroids was independently associated with increased risk of poor outcome among the endotype A patients, but not among the endotype B patients. This association between corticosteroid prescription and poor outcome was independent of baseline illness severity.

In all of these investigations, the identified sepsis subgroups have important differences in outcomes. However, it is important to note that the primary purpose of these efforts is not to estimate the risk of mortality *per se* (i.e., prognostic enrichment). Rather, these efforts seek to identify molecular subtypes of sepsis based on gene expression patterns. The fact the identified subgroups differ with respect to outcome strengthens the clinical relevance of the subgroups. The biological commonalities shared by the subgroups provides an opportunity for using subgroup assignment as a predictive enrichment strategy. This assertion is particularly relevant given the current interest in exploring strategies to enhance the adaptive immune system as a therapeutic strategy in sepsis, as well as the ongoing controversies regarding the role of adjunctive corticosteroids in sepsis.

Overlaps among the reported sepsis subgroups

The sepsis subgroups outlined above share the fundamental characteristic of being identified through unbiased, discovery-oriented transcriptomic studies. As shown in Table 2, although they differ with respect to nomenclature and the subgroup-defining gene signatures, they generally share the common themes of differential gene expression corresponding to the innate and adaptive immune system, as well as the coagulation system in the work by the Khatri laboratory.

Nevertheless, it remains somewhat unclear whether the studies are essentially describing the same subgroups, albeit *via* different gene expression patterns, or unique subgroups that require further validation and eventual harmonization. To this end, the Knight laboratory investigated potential commonality between the SRS1 subgroup and pediatric septic shock endotype A using comparative differential gene expression analysis [16]. They found overlap with respect to pathway enrichment, but minimal overlap with respect to the specific subgroup-defining gene expression signatures. In contrast, the van der Poll laboratory reported overlap between the Mars3 endotype and the SRS2 subgroup [17], and the Khatri reported overlap between SRS1/2, pediatric septic shock endotypes A/B and the inflammopathic/adaptive clusters [18].

Recently, the cohorts reported by the Knight laboratory (median age 68 years) were directly assigned to pediatric endotypes A and B using the publically available gene expression data [24]. Seventy percent of the adult patients with sepsis were endotype B. This distribution is similar to that seen in children, but unlike the mortality differences observed in children, there was

TABLE 2 Commonalities across gene expression-based subgroups of sepsis

References	Nomenclature	Inferred biological features	Clinical association
Davenport et al. [15] Burnham et al. [16]	SRS1/2	Endotoxin tolerance, T-cell exhaustion, and repression of HLA class II in SRS1	↑ Mortality in SRS1
Scicluna et al. [17]	Mars1 to 4	Repression of genes corresponding to innate and adaptive immunity in Mars1	↑ Mortality in Mars1
Sweeney et al. [18]	Inflammopathic Adaptive Coagulopathic	Differential expression of genes corresponding to subgroup nomenclature	↑ Mortality in coagulopathic subgroup
Wong et al. [19–22]	Endotypes A/B	Repression of genes corresponding to glucocorticoid receptor signaling and adaptive immunity in endotype A	↑ Mortality in endotype A; corticosteroids associated with morality in endotype A

no difference in mortality when comparing adults with sepsis assigned to endotype A vs. endotype B. Further analyses revealed complex interactions between endotype, SRS assignment, and age. Most notably, the interaction between endotype A and younger age was associated with increased risk of mortality; the youngest patients co-assigned to SRS1 and endotype A had a >60% mortality risk. For the overall cohort, subjects co-assigned to SRS1 and endotype A had the highest mortality rate compared to the other three possible co-assignment groups.

These complex interactions between age, SRS assignment, and endotype highlight the importance of developmental age on the host response to sepsis. In particular, younger children and neonates appear to have unique gene expression patterns in response to sepsis [25–27]. This important biological variable has been corroborated experimentally [28, 29], and will be critical when developing predictive enrichment strategies and the corresponding experimental therapies.

Applying prognostic and predictive enrichment strategies

None of the aforementioned prognostic and predictive enrichment strategies have been used in a prospective manner to inform clinical trial enrollment nor allocation to a specific treatment regimen. This will require additional validation, and equally important, the development of rapid assay systems to meet the time sensitive demands surrounding decision making among critically ill patients with sepsis.

To date, the only demonstrations that these enrichment strategies can potentially improve patient care are limited to retrospective analyses of observational data. For example, stratification of children with septic shock using the PERSEVERE biomarkers demonstrated that the association between a positive fluid balance (fluid overload) and outcome is dependent on baseline mortality risk [30]. In this study, a positive fluid balance was associated with poor outcome among patients at low baseline risk of mortality, but not among those at intermediate to high risk. Similarly, a trial simulation of plasmapheresis among children with sepsis and thrombocytopenia-associated multiple organ failure demonstrated how PERSEVERE-based prognostic enrichment can potentially reduce the number of subjects needed to demonstrate efficacy [11].

The most recent observational study combined prognostic enrichment *via* PERSEVERE with endotype assignment to determine if there is a subgroup of children with septic shock more likely to benefit from adjunctive corticosteroids [31]. This study found that among endotype B patients at intermediate to high baseline risk of mortality, corticosteroid prescription was associated with a 15-fold decrease in the risk for poor outcome. Although intriguing, these data should be interpreted with caution because corticosteroid prescription was not standardized nor randomized in this observational cohort. However, the magnitude of the size effect warrants further investigation *via* a prospective study that specifically embeds these enrichment strategies into the trial design.

Considerations for the near future

The prognostic and predictive enrichment strategies described above, based on discovery-oriented transcriptomic studies, are bringing the field closer to embracing precision medicine, wherein new therapies will be best targeted to predefined subgroups based on mortality risk (prognostic enrichment) and common biological features (predictive enrichment). Bringing this to fruition will require alignment of multiple factors, all of which have relatively equivalent importance.

Opportunities exist at present to conduct *post hoc* analyses of failed clinical trials in sepsis using prognostic and predictive enrichment strategies. Leveraging these opportunities requires removing the often times idiosyncratic barriers to sharing the necessary biological specimens and the associated clinical data from previous trials. Recent successful examples include *post hoc*, enrichment-based analyses of previous trials testing the efficacy of anti-tumor necrosis factor-α [32] and recombinant interleukin-1 receptor antagonist [33] therapies for sepsis. While not definitive, these *post hoc* analyses provide a rationale for retesting previously failed therapies, but in the context of prognostic and predictive enrichment.

Collaborations will be required across multiple investigative groups in order to capture patient heterogeneity and the nuances of different analytical approaches. This is required in order to increase generalizability of the enrichment strategies. Collaborations will also be necessary to embed these enrichment strategies in future interventional trials, either for planned *post hoc* analyses of trial results, or to directly inform trial enrollment. The former seems more feasible at present given the current state of the candidate enrichment strategies. Funding of future clinical trials should include sufficient resources to bank biological specimens to enable the application of these enrichment strategies. Collaborations will also be necessary between academia and industry in order to bring these enrichment strategies to the bedside.

Unlike the oncology field where there is a relatively wide time frame for conducting laboratory testing for the purposes of enrichment, the time frame in the context of sepsis is substantially narrower given the acute nature and rapid progression of sepsis. Thus, all tests developed for prognostic and predictive enrichment in sepsis must eventually provide actionable data within a few hours in order to be useful. The technology exists at present to bring these tests to the bedside. Recognizing this need for rapid turnaround of results, all of the aforementioned investigative efforts have sought to reduce the respective gene expression signatures to the smallest number of genes or gene products possible [14–18, 34]. Rapid assay development will also require collaborations between academia and industry.

Precision critical care medicine for sepsis is within our reach. All future interventional trial designs must embrace this concept. Not doing so will lead to repetition of our past failures reflecting our inability to account for patient heterogeneity at both the clinical and biological levels. Prognostic and predictive enrichment can allow us to resolve this inherent heterogeneity of sepsis.

References

[1] Wong HR. Intensive care medicine in 2050: precision medicine. Intensive Care Med 2017;43(10):1507–9.

[2] Prescott HC, Calfee CS, Thompson BT, Angus DC, Liu VX. Toward smarter lumping and smarter splitting: rethinking strategies for sepsis and acute respiratory distress syndrome clinical trial design. Am J Respir Crit Care Med 2016;194(2):147–55.

[3] Sun C, Mezzadra R, Schumacher TN. Regulation and function of the PD-L1 checkpoint. Immunity 2018;48(3):434–52.

[4] Annane D, Bellissant E, Bollaert PE, Briegel J, Confalonieri M, De Gaudio R, et al. Corticosteroids in the treatment of severe sepsis and septic shock in adults: a systematic review. JAMA 2009;301(22):2362–75.

[5] Sprung CL, Annane D, Keh D, Moreno R, Singer M, Freivogel K, et al. Hydrocortisone therapy for patients with septic shock. N Engl J Med 2008;358(2):111–24.

[6] Klein DJ, Foster D, Schorr CA, Kazempour K, Walker PM, Dellinger RP. The EUPHRATES trial (evaluating the use of polymyxin B hemoperfusion in a randomized controlled trial of adults treated for endotoxemia and septic shock): study protocol for a randomized controlled trial. Trials 2014;15:218.

[7] Dellinger RP, Bagshaw SM, Antonelli M, Foster DM, Klein DJ, Marshall JC, et al. Effect of targeted polymyxin B hemoperfusion on 28-day mortality in patients with septic shock and elevated endotoxin level: the EUPHRATES randomized clinical trial. JAMA 2018;320(14):1455–63.
[8] Wong HR, Salisbury S, Xiao Q, Cvijanovich NZ, Hall M, Allen GL, et al. The pediatric sepsis biomarker risk model. Crit Care 2012;16(5):R174.
[9] Kaplan JM, Wong HR. Biomarker discovery and development in pediatric critical care medicine. Pediatr Crit Care Med 2011;12(2):165–73.
[10] Wong HR, Weiss SL, Giuliano Jr. JS, Wainwright MS, Cvijanovich NZ, Thomas NJ, et al. Testing the prognostic accuracy of the updated pediatric sepsis biomarker risk model. PLoS One 2014;9(1):e86242.
[11] Wong HR, Cvijanovich NZ, Anas N, Allen GL, Thomas NJ, Bigham MT, et al. Pediatric sepsis biomarker risk model-II: redefining the pediatric sepsis biomarker risk model with septic shock phenotype. Crit Care Med 2016;44(11):2010–7.
[12] Wong HR, Lindsell CJ, Pettila V, Meyer NJ, Thair SA, Karlsson S, et al. A multibiomarker-based outcome risk stratification model for adult septic shock*. Crit Care Med 2014;42(4):781–9.
[13] Wong HR, Cvijanovich NZ, Anas N, Allen GL, Thomas NJ, Bigham MT, et al. Improved risk stratification in pediatric septic shock using both protein and mRNA biomarkers. PERSEVERE-XP. Am J Respir Crit Care Med 2017;196(4):494–501.
[14] Sweeney TE, Perumal TM, Henao R, Nichols M, Howrylak JA, Choi AM, et al. A community approach to mortality prediction in sepsis via gene expression analysis. Nat Commun 2018;9(1):694.
[15] Davenport EE, Burnham KL, Radhakrishnan J, Humburg P, Hutton P, Mills TC, et al. Genomic landscape of the individual host response and outcomes in sepsis: a prospective cohort study. Lancet Respir Med 2016;4(4):259–71.
[16] Burnham KL, Davenport EE, Radhakrishnan J, Humburg P, Gordon AC, Hutton P, et al. Shared and distinct aspects of the sepsis transcriptomic response to fecal peritonitis and pneumonia. Am J Respir Crit Care Med 2017;196(3):328–39.
[17] Scicluna BP, van Vught LA, Zwinderman AH, Wiewel MA, Davenport EE, Burnham KL, et al. Classification of patients with sepsis according to blood genomic endotype: a prospective cohort study. Lancet Respir Med 2017;5(10):816–26.
[18] Sweeney TE, Azad TD, Donato M, Haynes WA, Perumal TM, Henao R, et al. Unsupervised analysis of transcriptomics in bacterial sepsis across multiple datasets reveals three robust clusters. Crit Care Med 2018;46(6):915–25.
[19] Wong HR, Cvijanovich N, Lin R, Allen GL, Thomas NJ, Willson DF, et al. Identification of pediatric septic shock subclasses based on genome-wide expression profiling. BMC Med 2009;7:34.
[20] Wong HR, Wheeler DS, Tegtmeyer K, Poynter SE, Kaplan JM, Chima RS, et al. Toward a clinically feasible gene expression-based subclassification strategy for septic shock: proof of concept. Crit Care Med 2010;38(10):1955–61.
[21] Wong HR, Cvijanovich NZ, Allen GL, Thomas NJ, Freishtat RJ, Anas N, et al. Validation of a gene expression-based subclassification strategy for pediatric septic shock. Crit Care Med 2011;39(11):2511–7.
[22] Wong HR, Cvijanovich NZ, Anas N, Allen GL, Thomas NJ, Bigham MT, et al. Developing a clinically feasible personalized medicine approach to pediatric septic shock. Am J Respir Crit Care Med 2015;191(3):309–15.
[23] Wong HR, Cvijanovich NZ, Anas N, Allen GL, Thomas NJ, Bigham MT, et al. Endotype transitions during the acute phase of pediatric septic shock reflect changing risk and treatment response. Crit Care Med 2018;46(3):e242–9.

[24] Wong HR, Sweeney TE, Hart KW, Khatri P, Lindsell CJ. Pediatric sepsis endotypes among adults with sepsis. Crit Care Med 2017;45(12):e1289–91.

[25] Wynn JL, Cvijanovich NZ, Allen GL, Thomas NJ, Freishtat RJ, Anas N, et al. The influence of developmental age on the early transcriptomic response of children with septic shock. Mol Med 2011;17(11–12):1146–56.

[26] Raymond SL, Lopez MC, Baker HV, Larson SD, Efron PA, Sweeney TE, et al. Unique transcriptomic response to sepsis is observed among patients of different age groups. PLoS One 2017;12(9):e0184159.

[27] Wynn JL, Guthrie SO, Wong HR, Lahni P, Ungaro R, Lopez MC, et al. Postnatal age is a critical determinant of the neonatal host response to Sepsis. Mol Med 2015;21:496–504.

[28] Atkinson SJ, Varisco BM, Sandquist M, Daly MN, Klingbeil L, Kuethe JW, et al. Matrix metalloproteinase-8 augments bacterial clearance in a juvenile sepsis model. Mol Med 2016;22:455–63.

[29] Wynn JL, Wilson CS, Hawiger J, Scumpia PO, Marshall AF, Liu JH, et al. Targeting IL-17A attenuates neonatal sepsis mortality induced by IL-18. Proc Natl Acad Sci U S A 2016;113(19):E2627–35.

[30] Abulebda K, Cvijanovich NZ, Thomas NJ, Allen GL, Anas N, Bigham MT, et al. Post-ICU admission fluid balance and pediatric septic shock outcomes: a risk-stratified analysis. Crit Care Med 2014;42(2):397–403.

[31] Wong HR, Atkinson SJ, Cvijanovich NZ, Anas N, Allen GL, Thomas NJ, et al. Combining prognostic and predictive enrichment strategies to identify children with septic shock responsive to corticosteroids. Crit Care Med 2016;44(10):e1000–3.

[32] Wong HR, Lindsell CJ. An enrichment strategy for sepsis clinical trials. Shock 2016;46(6):632–4.

[33] Meyer NJ, Reilly JP, Anderson BJ, Palakshappa JA, Jones TK, Dunn TG, et al. Mortality benefit of recombinant human interleukin-1 receptor antagonist for sepsis varies by initial interleukin-1 receptor antagonist plasma concentration. Crit Care Med 2018;46(1):21–8.

[34] Wong HR, Sweeney TE, Lindsell CJ. Simplification of a septic shock endotyping strategy for clinical application. Am J Respir Crit Care Med 2017;195(2):263–5.

Chapter 7

Sepsis: Metabolomics and proteomics

Yong B. Tan*, Raymond J. Langley†
**Department of Surgery, University of South Alabama College of Medicine, Mobile, AL, United States, †Department of Pharmacology, University of South Alabama College of Medicine, Mobile, AL, United States*

Introduction

Sepsis is an overarching term for an extremely heterogeneous clinical syndrome marked by life-threatening organ dysfunctions caused by a dysregulated host response to an initial infection. Sepsis remains one of the leading cause of death in the United States and the world [1]. Over 20 million patients in the world are treated for sepsis each year, and as many as 25% of those patients will die [2]. Given that we spend over $20 billion total hospital costs per year in the US on infection and sepsis treatment, there is an unstated urgency in developing accurate diagnostic and prognostic tools [1].

The common non-specific symptoms of sepsis include increased heart rate, fever, confusion, and increased respiratory rate. More severe symptoms include hypo-perfusion that will lead to eventual multiple organ dysfunction. Mortality has decreased in the last decade with the improvement in critical care protocols [1]. However, therapies that are optimized for individual patients and that target specific sepsis mechanisms remain difficult for implementation due to the delayed presentations, cryptic severity, and heterogeneity in clinical course. Two patients can arrive at an emergency room with similar mild symptoms and have diverging clinical courses. One patient can progress rapidly into sepsis and death while the other patient will retain a relatively benign course and be discharged home.

Advances in metabolomics and proteomics have great potential in identifying key microbial and host *perturbations* associated with sepsis outcome [3]. In this chapter, we will examine the literature to evaluate the usage of current biomarkers and the utilization of systems biology methods to investigate host response biomarkers for its utility in the determination of sepsis diagnosis and prognosis.

Genomic and Precision Medicine. https://doi.org/10.1016/B978-0-12-801496-7.00007-1

Current/past standards

For decades, the combination of an infection and the systemic inflammatory response syndrome (SIRS) criteria utilizing heart rate, white blood cell count, temperature, and respiratory rate were used to define sepsis. In 2016, the pressing need for an assessment score to identify early risk of sepsis has shifted from the usage of SIRS criteria to using a modified Sequential Organ Failure Assessment Score (qSOFA) [1]. The score is easily calculated based on three components: respiratory rate, altered mental status and systolic blood pressure. The modified SOFA score doesn't appear to perform any better than SIRS for predicting in-hospital mortality and additional biomarkers are still needed for better outcomes prediction [4, 5]. As mentioned previously, the clinical signs of sepsis are still relatively unspecific. Early goal-directed therapy can reduce mortality significantly if started within 6h, which highlights the need for highly sensitive and specific biomarkers for diagnosis and prognosis [6]. We will review the most popular array of biomarkers used in current medical practices that correlate with sepsis diagnosis and prognosis with mixed results.

Procalcitonin

In the 1990s, scientists noticed a rise in procalcitonin in response to an inflammatory stimulus [7]. Classified as an acute phase reactant, elevated serum procalcitonin levels were strongly associated with bacterial infection but it is less sensitive for viral and fungal infections. However, a meta-analysis of 18 studies found that procalcitonin was not able to distinguish between sepsis from nonseptic inflammation [8], which raises the concern for overaggressive antibiotic stewardship. However, Nobre et al. suggests that serial procalcitonin measurements allows for early termination of antibiotic use for sepsis without any adverse effects [9]. The results are affirmed in the PRORATA trial where a procalcitonin-guide antibiotic therapy to treat suspected bacterial infection in critically ill patients led to fewer days of antibiotic exposure without a significant change in mortality at 28 or 60 days [10].

C-reactive protein

C-reactive Protein (CRP) is an acute phase reactant produced in the liver in response to inflammation and participates in apoptotic cell removal through the complement system. Baseline CRP levels differs based on age, sex, and race [11]. Upon arrival to a medical facility, initial peak CRP levels are not associated with treatment failure and 28-day mortality [12] but when used in combination with other biomarkers or APACHE II score, CRP has higher predictability for mortality in sepsis scores [13]. Serial measurements to calculate CRP clearance may be useful since it is significantly associated with treatment failure and marginally significant for 28-day mortality [12].

Lactate

Lactate remains one of the most useful biomarker for the diagnosis of septic shock and outcome prediction [1, 14]. Previously assumed to be solely produced due to anaerobic glycolysis from inadequate oxygenation, recent studies shows increased production from muscle in the absence of hypoperfusion. Currently, septic shock is diagnosed by lactate >2 mmol/L or mean arterial pressure less than <65 mmHG after adequate fluid resuscitation. >2 mmol/L was associated with increased risk but more recent data shows that peak lactate >4 mmol/L was significantly associated with in-hospital mortality [15]. However, peak lactate alone is not diagnostic for sepsis. Furthermore, lactate clearance may play a larger role in sepsis prognosis. In patients with high levels of lactate, the inability to decrease lactate within 12 h was significantly associated with higher mortality [16].

High mobility group box 1

High mobility group box 1 (HMGB1) is a protein in the nucleus that interacts with the nucleosomes, transcription factors and histones [17]. Its main role is to bind and regulate DNA for transcription. Activated macrophages and monocytes will release HMGB1 into the extracellular space where it can act as an inflammatory mediator by decreasing the uptake of apoptotic cells by phagocytes [18]. Extracellular HMGB1 has the ability to bind to Toll-like receptor 4 and RAGE receptors to propagate the inflammatory pathway [18, 19]. The levels of HMGB1 is increased in sepsis, and its role as a cytokine mediator is frequently targeted in anti-inflammatory pharmacological therapies [20].

Utilization of systems biology for better disease biomarkers and understanding

There continues to be a need for better biomarkers for sepsis diagnosis and prognosis. The current biomarkers are agnostic to the biology of the disease and are often irrelevant to vital molecular pathways [21]. The heterogeneity of sepsis contributes to the complexity of detecting better biomarkers but the utilization of system biology to integrate and analyze large data sets may offer insight into the understanding of the disease susceptibility and development [3, 21]. Quantitative analysis of metabolites (metabolomics) and proteins (proteomics) may reveal the dynamic nature of the numerous pathways active or altered during the disease process. The results can provide an opportunity to identify clinical endpoints to improve clinical outcomes and to test for new interventions.

In the Community Acquired Pneumonia and Sepsis Outcome and Diagnostics (CAPSOD) study, plasma from patients that were enrolled with suspected community-acquired infections in the emergency department [22–24] was utilized in a broad spectrum analysis of metabolomic and proteomic changes. Six biochemical pathways could predict 28 day patient outcomes at presentation. The authors noted decreased acyl-glycerophosphocholines (acyl-GPCs); increased substrates for redox coenzyme nicotinamide adenine dinucleotide (NAD); increased tricarboxylic acid (TCA) intermediates; increased short-chain, medium-chain fatty acids and branched-chain amino acids (BCAA) bound to carnitine (also known as carnitine esters); sulfated steroids and increased bile acids. The changes in TCA, NAD+ and carnitine esters points toward an energy crisis in nonsurvivors.

Utilizing predictive modeling approaches, the authors determined that four carnitine esters, lactate, hematocrit and age predict sepsis patient outcomes in the ED better than lactate, SOFA and APACHEII [23–25]. The study was replicated in an ICU population of patients enrolled with sepsis (Registry of Critical Illness; RoCI) [24, 25, 23]. Once again, septic patients had significantly decreased arachidonoyl-GPC and increased carnitine esters. The results suggest that despite enrollment in the ED or the ICU, there was strong homogeneity in the two patient populations and that different metabolites could uncover novel biology. Similar findings were also noted in a non-human primate model of sepsis which strongly pointed to mitochondrial dysfunction as a driver of poor outcomes [25]. Many of these metabolites have been independently identified in a number of recent clinical ICU studies pointing to a common mechanism of metabolic dysfunction [26–29].

The results suggest that a bioenergetics crisis in sepsis is a cause of acute and long-term outcomes. The presence of metabolomics features consistent with mitochondrial dysfunction (e.g., dysregulation of the citric acid (TCA) cycle, β-oxidation, and nicotinamide adenine dinucleotide (NAD+) degradation) has been documented in critically ill human patients [3, 23–25].

Mitochondrial dysfunction

Mitochondrial dysfunction related to sepsis outcomes has been implicated for >20 years as changes in the mitochondria numbers and size are associated with poor outcomes (Fig. 1) [30]. Recently, a number of predictive biomarkers directly related to mitochondrial proteins and metabolites has been associated with patient outcomes. These include cardiolipin [31, 32], mitochondrial DNA (mtDNA) damage-associated molecular patterns (DAMPs) [33–35] and the previously described carnitine esters [23, 25]. Interestingly, mtDNA has been reported to induce immune paralysis in septic patients [36]. A single dose of mtDNA triggered the Toll-like Receptor 9 (TLR-9) immunosuppression response in adaptive T-cell cytotoxicity in wildtype mouse but not TLR-9 knockout mice.

The oxidative metabolism of glucose, lipids, and fat produces ATP in the mitochondria. A large number of reactive oxygen species (ROS) are produced as a by-product of this energy metabolism. In a pathological state like sepsis, the high ROS production provide the potential mechanism for cell damage by inhibiting cell respiration and inducing direct damage to mitochondrial protein

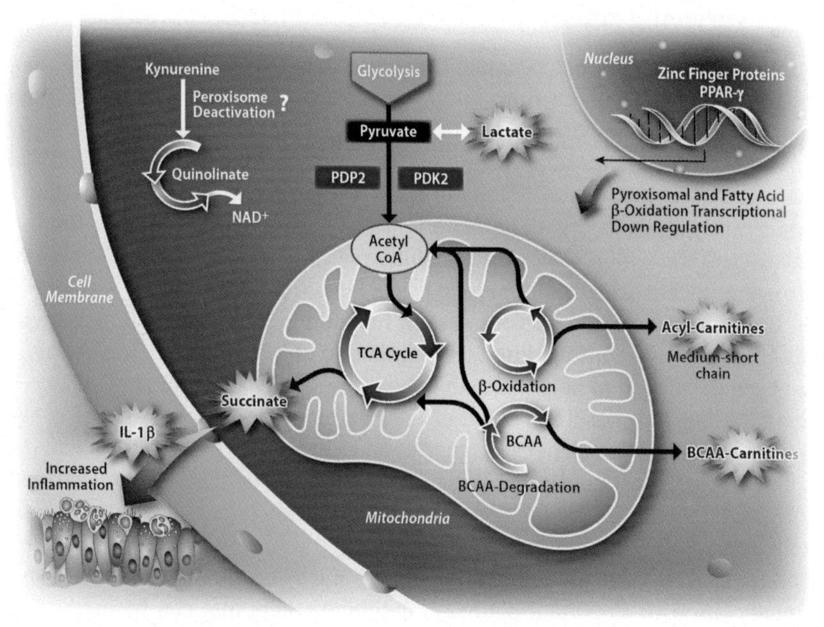

FIG. 1 Mitochondrial dysfunction is often predictive of patient outcomes in sepsis. The changes suggest a bioenergetics crisis in sepsis nonsurvivors. These changes appear to be regulated by mitochondrial biogenesis factors such as the PPARG coactivator 1 alpha (PPARGC1A) that lead to decreases in β-oxidation, checkpoint inhibition of the citric acid (TCA) cycle, and upregulation of inflammatory markers through succinate. Modulation of these pathways may lead to improved therapeutics.

and other structures like the cell membrane [37]. The ultimate effect is a decreased ATP production due to damage and decreased turnover of mitochondria [30]. The outcome of the patients could be dependent on whether these changes are permanent or temporary.

A Warburg-like phenotype (i.e., aerobic glycolysis) was found in monocytes with a decrease in β-oxidation and expression of the related genes, along with increased lactate and NAD+ production due to LPS treatment [38, 39]. In patients with sepsis, isolated monocytes demonstrated immunotolerance with decreased oxygen consumption. However, when these cell were treated with IFN-γ, cellular metabolism partially restored function leading to increased lactate production and upregulation of cytokines [40]. A similar finding was noted with the treatment of monocytes isolated from sepsis-patients with a sirtuin 1 (SIRT1) inhibitor EX-527 [41, 42]. Mice treated after 24h with EX-527 also induced a Warburg-like phenotype in monocytes, immune activation and improved survival.

The results demonstrate that the Warburg effect is critical for immune function in monocytes. These results point to promising new therapies that can reactivate the innate immune response in the later stages of sepsis potentially improving survival.

After surviving sepsis

Surviving sepsis may not be enough because of the negative sequela on quality of life once discharge from the hospital. One of the most daunting issues facing critical care medicine is impairment in the quality of function in sepsis survivors. Dramatic muscle wasting has been reported with as much as 25% of skeletal muscle lost within 1 week in the ICU [43] and duration of ICU stay is significantly associated with the severity of physical impairment [44]. In many cases, the loss of muscle tissue never fully recovers and may be related to permanent mitochondrial loss in muscle tissue [45, 46].

Baseline muscle and nervous system function in these critically ill patients has always been a confounding factor in that no objective testing methodology has been developed to help interpret the weakness seen at hospital discharge as being attributed to baseline deficit or attributable to the illness itself. It is also not clear whether the changes to muscles and the nervous system sustained due to the critical illness are permanent. The bioenergetics crisis associated with mitochondrial dysfunction is closely associated with poor 30-day patient outcomes [3, 23–27]. Could this crisis early in the disease syndrome also affect long-term quality of life? Once again, proteomics and metabolomics may potentially discover important pathways and biomarkers that can be predicative for patients that will have poor quality of life even a septic event.

Precision medicine in sepsis

Sepsis has had >30 years of failures when it comes to drug discovery. However, advances in omics has led to a better understanding of the pathophysiology of sepsis. Moreover, biomarkers are also identifying endotypes—defined as biological subtypes with distinct physiologic mechanisms within a phenotype that have clinically observed characteristics that identify a group of patients within a disease syndrome [47–50]. In the past, clinical trials in sepsis often failed to account for the vast heterogeneity of the patient population [50]. This has led to variation in treatment effect size nullifying potential benefit of new drug therapies that may be efficacious to subgroups of patient population. The utilizing of these new biomarkers that can predict outcomes as well accounting for endotypes should lead to vast improvement in drug discovery and treatment. For example, post-hoc analysis of the PROWESS study found that anti-thrombin therapy in patients that developed disseminated intravascular coagulation was beneficial, while providing no benefit in less severely ill patients [51]. Biomarkers that could quickly identify and subtype these patients could lead to more precise therapeutic strategies that maximize benefit for individual patients. Results from these biomarkers need to be rapid, accurate and relatively inexpensive for indicating appropriate drug interventions [52]. However, the increase in knowledge of biomarkers and unique endotypes is certain to improve patient care through precision medicine.

Conclusion

As highlighted numerous times, sepsis remains a highly heterogeneous disease. The need to accurately diagnosis sepsis in a timely fashion remains elusively with the current standard tests. Proteomics and metabolomics offer a unique methodology to look at the dynamic interactions that occur during sepsis. Furthermore, 'omics provide insight into vital molecular pathways that may become potential targets for pharmacological therapies.

References

[1] Singer M, Deutschman CS, Seymour CW, Shankar-Hari M, Annane D, Bauer M, et al. The third international consensus definitions for sepsis and septic shock (Sepsis-3). JAMA 2016;315(8):801–10. https://doi.org/10.1001/jama.2016.0287.

[2] Fleischmann C, Scherag A, Adhikari NK, Hartog CS, Tsaganos T, Schlattmann P, et al. Assessment of global incidence and mortality of hospital-treated Sepsis. Current estimates and limitations. Am J Respir Crit Care Med 2016;193(3):259–72. https://doi.org/10.1164/rccm.201504-0781OC.

[3] Langley RJ, Wong HR. Early diagnosis of Sepsis: is an integrated omics approach the way forward? Mol Diagn Ther 2017;21:525–37. https://doi.org/10.1007/s40291-017-0282-z.

[4] Maitra S, Som A, Bhattacharjee S. Accuracy of quick sequential organ failure assessment (qSOFA) score and systemic inflammatory response syndrome (SIRS) criteria for predicting mortality in hospitalized patients with suspected infection: a meta-analysis of observational studies. Clin Microbiol Infect 2018;24:1123–9. https://doi.org/10.1016/j.cmi.2018.03.032.

[5] Fernando SM, Tran A, Taljaard M, Cheng W, Rochwerg B, Seely AJE, et al. Prognostic accuracy of the quick sequential organ failure assessment for mortality in patients with suspected infection: a systematic review and meta-analysis. Ann Intern Med 2018;168(4):266–75. https://doi.org/10.7326/M17-2820.

[6] Rivers EP, Coba V, Visbal A, Whitmill M, Amponsah D. Management of sepsis: early resuscitation. Clin Chest Med 2008;29(4):689–704. ix–x, https://doi.org/10.1016/j.ccm.2008.06.005.

[7] Becker KL, Snider R, Nylen ES. Procalcitonin in sepsis and systemic inflammation: a harmful biomarker and a therapeutic target. Br J Pharmacol 2010;159(2):253–64. https://doi.org/10.1111/j.1476-5381.2009.00433.x.

[8] Tang BM, Eslick GD, Craig JC, McLean AS. Accuracy of procalcitonin for sepsis diagnosis in critically ill patients: systematic review and meta-analysis. Lancet Infect Dis 2007;7(3):210–7. https://doi.org/10.1016/S1473-3099(07)70052-X.

[9] Nobre V, Harbarth S, Graf JD, Rohner P, Pugin J. Use of procalcitonin to shorten antibiotic treatment duration in septic patients: a randomized trial. Am J Respir Crit Care Med 2008;177(5):498–505. https://doi.org/10.1164/rccm.200708-1238OC.

[10] Bouadma L, Luyt CE, Tubach F, Cracco C, Alvarez A, Schwebel C, et al. Use of procalcitonin to reduce patients' exposure to antibiotics in intensive care units (PRORATA trial): a multicentre randomised controlled trial. Lancet 2010;375(9713):463–74. https://doi.org/10.1016/S0140-6736(09)61879-1.

[11] Woloshin S, Schwartz LM. Distribution of C-reactive protein values in the United States. N Engl J Med 2005;352(15):1611–3. https://doi.org/10.1056/NEJM200504143521525.

[12] Ryu JA, Yang JH, Lee D, Park CM, Suh GY, Jeon K, et al. Clinical usefulness of procalcitonin and C-reactive protein as outcome predictors in critically ill patients with severe sepsis and septic shock. PLoS One 2015;10(9):e0138150. https://doi.org/10.1371/journal.pone.0138150.

[13] Khan F, Owens MB, Restrepo M, Povoa P, Martin-Loeches I. Tools for outcome prediction in patients with community acquired pneumonia. Expert Rev Clin Pharmacol 2017;10(2):201–11. https://doi.org/10.1080/17512433.2017.1268051.

[14] Shankar-Hari M, Phillips GS, Levy ML, Seymour CW, Liu VX, Deutschman CS, et al. Developing a new definition and assessing new clinical criteria for septic shock: for the third international consensus definitions for sepsis and septic shock (Sepsis-3). JAMA 2016;315(8):775–87. https://doi.org/10.1001/jama.2016.0289.

[15] Casserly B, Phillips GS, Schorr C, Dellinger RP, Townsend SR, Osborn TM, et al. Lactate measurements in sepsis-induced tissue hypoperfusion: results from the surviving sepsis campaign database. Crit Care Med 2015;43(3):567–73. https://doi.org/10.1097/CCM.0000000000000742.

[16] Haas SA, Lange T, Saugel B, Petzoldt M, Fuhrmann V, Metschke M, et al. Severe hyperlactatemia, lactate clearance and mortality in unselected critically ill patients. Intensive Care Med 2016;42(2):202–10. https://doi.org/10.1007/s00134-015-4127-0.

[17] Gentile LF, Moldawer LL. HMGB1 as a therapeutic target for sepsis: it's all in the timing!. Expert Opin Ther Targets 2014;18(3):243–5. https://doi.org/10.1517/14728222.2014.883380.

[18] Banerjee S, Friggeri A, Liu G, Abraham E. The C-terminal acidic tail is responsible for the inhibitory effects of HMGB1 on efferocytosis. J Leukoc Biol 2010;88(5):973–9. https://doi.org/10.1189/jlb.0510262.

[19] Sims GP, Rowe DC, Rietdijk ST, Herbst R, Coyle AJ. HMGB1 and RAGE in inflammation and cancer. Annu Rev Immunol 2010;28:367–88. https://doi.org/10.1146/annurev.immunol.021908.132603.

[20] Huang W, Tang Y, Li L. HMGB1, a potent proinflammatory cytokine in sepsis. Cytokine 2010;51(2):119–26. https://doi.org/10.1016/j.cyto.2010.02.021.

[21] Singer M. Biomarkers in sepsis. Curr Opin Pulm Med 2013;19(3):305–9. https://doi.org/10.1097/MCP.0b013e32835f1b49.

[22] Glickman SW, Cairns CB, Otero RM, Woods CW, Tsalik EL, Langley RJ, et al. Disease progression in hemodynamically stable patients presenting to the emergency department with sepsis. Acad Emerg Med 2010;17(4):383–90.

[23] Langley RJ, Tsalik EL, Velkinburgh JC, Glickman SW, Rice BJ, Wang C, et al. An integrated clinico-metabolomic model improves prediction of death in sepsis. Sci Transl Med 2013;5(195):195ra95. https://doi.org/10.1126/scitranslmed.3005893.

[24] Rogers AJ, McGeachie M, Baron RM, Gazourian L, Haspel JA, Nakahira K, et al. Metabolomic derangements are associated with mortality in critically ill adult patients. PLoS One 2014;9(1):e87538. https://doi.org/10.1371/journal.pone.0087538.

[25] Langley RJ, Tipper JL, Bruse S, Baron RM, Tsalik EL, Huntley J, et al. Integrative "omic" analysis of experimental bacteremia identifies a metabolic signature that distinguishes human sepsis from systemic inflammatory response syndromes. Am J Respir Crit Care Med 2014;190(4):445–55. https://doi.org/10.1164/rccm.201404-0624OC.

[26] Seymour CW, Yende S, Scott MJ, Pribis J, Mohney RP, Bell LN, et al. Metabolomics in pneumonia and sepsis: an analysis of the GenIMS cohort study. Intensive Care Med 2013;39(8):1423–34. https://doi.org/10.1007/s00134-013-2935-7.

[27] Kamisoglu K, Haimovich B, Calvano SE, Coyle SM, Corbett SA, Langley RJ, et al. Human metabolic response to systemic inflammation: assessment of the concordance between ex-

perimental endotoxemia and clinical cases of sepsis/SIRS. Crit Care 2015;19:71. https://doi.org/10.1186/s13054-015-0783-2.

[28] Neugebauer S, Giamarellos-Bourboulis EJ, Pelekanou A, Marioli A, Baziaka F, Tsangaris I, et al. Metabolite profiles in sepsis: developing prognostic tools based on the type of infection. Crit Care Med 2016;44(9):1649–62. https://doi.org/10.1097/CCM.0000000000001740.

[29] Ferrario M, Cambiaghi A, Brunelli L, Giordano S, Caironi P, Guatteri L, et al. Mortality prediction in patients with severe septic shock: a pilot study using a target metabolomics approach. Sci Rep 2016;6:20391. https://doi.org/10.1038/srep20391.

[30] Singer M. The role of mitochondrial dysfunction in sepsis-induced multi-organ failure. Virulence 2014;5(1):66–72. https://doi.org/10.4161/viru.26907.

[31] Chen BB, Coon TA, Glasser JR, Zou C, Ellis B, Das T, et al. E3 ligase subunit Fbxo15 and PINK1 kinase regulate cardiolipin synthase 1 stability and mitochondrial function in pneumonia. Cell Rep 2014;7(2):476–87. https://doi.org/10.1016/j.celrep.2014.02.048.

[32] Ray NB, Durairaj L, Chen BB, McVerry BJ, Ryan AJ, Donahoe M, et al. Dynamic regulation of cardiolipin by the lipid pump Atp8b1 determines the severity of lung injury in experimental pneumonia. Nat Med 2010;16(10):1120–7. https://doi.org/10.1038/nm.2213.

[33] Nakahira K, Kyung SY, Rogers AJ, Gazourian L, Youn S, Massaro AF, et al. Circulating mitochondrial DNA in patients in the ICU as a marker of mortality: derivation and validation. PLoS Med 2013;10(12):e1001577. discussion. https://doi.org/10.1371/journal.pmed.1001577.

[34] Simmons JD, Freno DR, Muscat CA, Obiako B, Lee YL, Pastukh VM, et al. Mitochondrial DNA damage associated molecular patterns in ventilator-associated pneumonia: prevention and reversal by intratracheal DNase I. J Trauma Acute Care Surg 2017;82(1):120–5. https://doi.org/10.1097/TA.0000000000001269.

[35] Kuck JL, Obiako BO, Gorodnya OM, Pastukh VM, Kua J, Simmons JD, et al. Mitochondrial DNA damage-associated molecular patterns mediate a feed-forward cycle of bacteria-induced vascular injury in perfused rat lungs. Am J Physiol Lung Cell Mol Physiol 2015;308(10):L1078–85. https://doi.org/10.1152/ajplung.00015.2015.

[36] Schafer ST, Franken L, Adamzik M, Schumak B, Scherag A, Engler A, et al. Mitochondrial DNA: an endogenous trigger for immune paralysis. Anesthesiology 2016;124(4):923–33. https://doi.org/10.1097/ALN.0000000000001008.

[37] Duran-Bedolla J, Montes de Oca-Sandoval MA, Saldana-Navor V, Villalobos-Silva JA, Rodriguez MC, Rivas-Arancibia S. Sepsis, mitochondrial failure and multiple organ dysfunction. Clin Invest Med 2014;37(2):E58–69.

[38] Cheng SC, Quintin J, Cramer RA, Shepardson KM, Saeed S, Kumar V, et al. mTOR- and HIF-1alpha-mediated aerobic glycolysis as metabolic basis for trained immunity. Science 2014;345(6204):1250684. https://doi.org/10.1126/science.1250684.

[39] Tannahill GM, Curtis AM, Adamik J, Palsson-McDermott EM, McGettrick AF, Goel G, et al. Succinate is an inflammatory signal that induces IL-1beta through HIF-1alpha. Nature 2013;496(7444):238–42. https://doi.org/10.1038/nature11986.

[40] Cheng SC, Scicluna BP, Arts RJ, Gresnigt MS, Lachmandas E, Giamarellos-Bourboulis EJ, et al. Broad defects in the energy metabolism of leukocytes underlie immunoparalysis in sepsis. Nat Immunol 2016;17:406–13. https://doi.org/10.1038/ni.3398.

[41] Liu TF, Vachharajani V, Millet P, Bharadwaj MS, Molina AJ, McCall CE. Sequential actions of SIRT1-RELB-SIRT3 coordinate nuclear-mitochondrial communication during immunometabolic adaptation to acute inflammation and sepsis. J Biol Chem 2015;290(1):396–408. https://doi.org/10.1074/jbc.M114.566349.

[42] Vachharajani VT, Liu T, Brown CM, Wang X, Buechler NL, Wells JD, et al. SIRT1 inhibition during the hypoinflammatory phenotype of sepsis enhances immunity and improves outcome. J Leukoc Biol 2014;96(5):785–96. https://doi.org/10.1189/jlb.3MA0114-034RR.

[43] Puthucheary ZA, Rawal J, McPhail M, Connolly B, Ratnayake G, Chan P, et al. Acute skeletal muscle wasting in critical illness. JAMA 2013;310(15):1591–600. https://doi.org/10.1001/jama.2013.278481.

[44] Goligher EC, Doufle G, Fan E. Update in mechanical ventilation, sedation, and outcomes 2014. Am J Respir Crit Care Med 2015;191(12):1367–73. https://doi.org/10.1164/rccm.201502-0346UP.

[45] Rocheteau P, Chatre L, Briand D, Mebarki M, Jouvion G, Bardon J, et al. Sepsis induces long-term metabolic and mitochondrial muscle stem cell dysfunction amenable by mesenchymal stem cell therapy. Nat Commun 2015;6:10145. https://doi.org/10.1038/ncomms10145.

[46] Wang X, Gu H, Qin D, Yang L, Huang W, Essandoh K, et al. Exosomal miR-223 contributes to mesenchymal stem cell-elicited cardioprotection in polymicrobial sepsis. Sci Rep 2015;5:13721. https://doi.org/10.1038/srep13721.

[47] Sweeney TE, Shidham A, Wong HR, Khatri P. A comprehensive time-course-based multicohort analysis of sepsis and sterile inflammation reveals a robust diagnostic gene set. Sci Transl Med 2015;7(287):287ra71. https://doi.org/10.1126/scitranslmed.aaa5993.

[48] Sweeney TE, Wong HR, Khatri P. Robust classification of bacterial and viral infections via integrated host gene expression diagnostics. Sci Transl Med 2016;8(346):346ra91. https://doi.org/10.1126/scitranslmed.aaf7165.

[49] Jacobs L, Wong HR. Emerging infection and sepsis biomarkers: will they change current therapies? Expert Rev Anti Infect Ther 2016;14(10):929–41. https://doi.org/10.1080/14787210.2016.1222272.

[50] Rello J, van Engelen TSR, Alp E, Calandra T, Cattoir V, Kern WV, et al. Towards precision medicine in sepsis: a position paper from the European society of clinical microbiology and infectious diseases. Clin Microbiol Infect 2018;24:1264–72. https://doi.org/10.1016/j.cmi.2018.03.011.

[51] Meziani F, Gando S, Vincent JL. Should all patients with sepsis receive anticoagulation? Yes. Intensive Care Med 2017;43(3):452–4. https://doi.org/10.1007/s00134-016-4621-z.

[52] Prescott HC, Calfee CS, Thompson BT, Angus DC, Liu VX. Toward smarter lumping and smarter splitting: rethinking strategies for sepsis and acute respiratory distress syndrome clinical trial design. Am J Respir Crit Care Med 2016;194(2):147–55. https://doi.org/10.1164/rccm.201512-2544CP.

Chapter 8

Aspiring to precision medicine for infectious diseases in resource limited settings

Kevin L. Schully, Danielle V. Clark
Austere environments Consortium for Enhanced Sepsis Outcomes (ACESO), The Henry M Jackson Foundation, Bethesda, MD, United States

Introduction

Precision medicine is defined as "the tailoring of medical treatment to the individual characteristics of each patient ... to classify individuals into subpopulations that differ in their susceptibility to a particular disease or their response to a specific treatment ... [allowing] preventative or therapeutic interventions [to] be concentrated on those who will benefit, sparing expense and side effects for those who will not" [1]. Currently most patients with infectious diseases in low-resource settings are treated empirically, with limited options for laboratory, imaging, or in-depth clinical evaluation. We, as a global community, know very little about the distribution of infectious etiologies, determinants of disease severity, or how host-response profiles vary by geographic region.

The first step toward development of a precision medicine approach suitable for low-resource settings is, perhaps not surprisingly, to generate and rigorously evaluate data from low-resource settings. Initially the focus should be on understanding the underlying heterogeneity, with the intent to identify pathogenic biologic pathways and subpopulations of patients. The technologies for multiomic approaches to characterize the temporal dynamics of the host response to infection are rapidly evolving and are thoroughly reviewed elsewhere [2]. Well-characterized host-response profiles would then provide a metric by which patient status could be assessed, ideally using a minimum set of markers (demographic, clinical, imaging, laboratory, or other) measured at the point of care. The necessary technology, computational resources, and expertise are available and are currently being harnessed to address these questions.

To fully realize the potential of precision medicine, however, requires systems to track, monitor, predict, and learn, both at the individual patient and the

Genomic and Precision Medicine. https://doi.org/10.1016/B978-0-12-801496-7.00008-3

population levels. While full electronic medical records and integrated laboratory information management systems backed by artificial intelligence are likely unrealistic for these settings in the near future, development of streamlined systems to collect a minimum set of markers as discussed above, analyze and interpret the data, and produce results relevant for clinical management decision-making is feasible. Such systems could range from relatively simple systems for continuous monitoring of vital signs to more complex systems with multiple inputs and predictive analytics.

Limitations of current diagnostic approaches

Diagnostics can be an undervalued and often neglected component of a hospital's infrastructure in the developing world. Commitment of hospital resources to diagnostics can be as low as 6% of a hospital's overall health care expenditures [3]. Resources from international agencies focus on high-profile pathogens such as HIV leaving major gaps in diagnostic capacity. Diagnostic resources that do exist are often concentrated in the major cities leaving provinces with even less. One answer to this problem is the new construction of traditional laboratory infrastructure or the renovation of existing space on a case-by-case basis. However, the limitations for laboratories in resource-limited settings go far beyond space. These labs suffer from daily power interruptions with implications for both samples and supplies/reagents alike. Furthermore, retention of skilled staff is difficult; often plagued by high turnover precipitated by poor salaries. Continuous training is required to prevent skill atrophy. In addition to being burdensome on laboratory resources, this often results in a more marketable staff who often seek to improve their position. In end, the impact of the laboratory is measured by its use and effectiveness. Physicians in hospitals where the laboratory is not a priority often either do not trust the results or do not know how to appropriately implement the results. Consequently, laboratory facilities in resource-limited settings often fall short of their potential in both use and effectiveness.

Culture remains the diagnostic gold standard for many of the world's most consequential bacterial pathogens. Diagnostic microbiology presents a myriad of problems for low-resource settings. In addition to standard blood culture materials, arrays of selective and differential growth media followed by biochemical testing are required for proper identification of bacterial pathogens. Complicating matters, self-medication through the purchase of antibiotics is common in the developing world [4–6] causing low return on ordered blood cultures. The resulting uncertainty and diagnostic delay often requires syndromic management of infectious diseases. While this form of treatment can be highly effective when appropriately implemented, empiric treatment for sepsis often does not include drugs for many of the neglected tropical diseases [7]. Unfortunately, this practice inevitably leads to the unnecessary treatment of patients resulting in wasted resources and poor antibiotic stewardship.

Augmenting diagnostic microbiology, laboratories often employ microscopy, immunodiagnostics and molecular assays. In areas prone to Malarial and tuberculosis infections, microscopy is an essential component of a diagnostic lab. This capacity requires a sufficiently equipped microscope and a skilled staff to interpret results and maintain the equipment. Point of care (POC) assays such as rapid diagnostic tests (RDTs) are a convenient alternative to microscopy. However, RDTs are often poorly regulated, improperly validated and not standardized. A recent study of malaria patients in Tanzania showed that 51% deemed negative by microscopy and 54% diagnosed as negative by RDT were still treated with antimalarial drugs demonstrating that diagnostic tests had no impact on patient care [8]. Certain laboratories have employed molecular assays such as PCR-based diagnostics. The effectiveness of these assays are often limited by the presence and concentration of pathogen nucleic acid present in clinical samples, unreliable power, expense (equipment and reagents), regional specificity, etc.

While standard diagnostic laboratories are an invaluable resource capable of having an enormous impact on the health of a population, implementing standard diagnostics from the ground up in resource-limited settings may not be as impactful in practice as it is in principle. As multi-omic technologies improve our understanding of the host response to infections, the potential to leapfrog standard diagnostic practices in favor of cutting-edge precision medicine approaches has never been closer to reality. Diagnostics based on the host response to infection rather than directly detecting the afflicting pathogen represent the next-generation of diagnostics and have the potential to have a major impact globally.

Aspiring to precision medicine in resource-limited settings

Effective implementation of a host-based diagnostic will require an unprecedented understanding of geographical, ethnic and pathogenic diversity that make up the host response to infectious and non-infectious illnesses alike. For example, the highest burden of infectious (and non-communicable) diseases on earth and the greatest genetic diversity in the population are both present in Africa [9]. How can host-based diagnostics account for that level of diversity? Studies to fully characterize the dynamics of the host response to disease in diverse settings will be required to encompass the enormous diversity of the host response to disease. Government agencies possess the resources to develop novel diagnostic tools but are typically outpaced by private/commercial entities (i.e., human genome project, certain vaccines, etc.). Commercial entities develop diagnostics for well-resourced settings that will produce acceptable returns on their investment. Non-profit research consortiums partnered with government funding agencies with shared goals represent a novel diagnostic paradigm; precision medicine for resource-limited settings.

The Austere environments Consortium for Enhanced Sepsis Outcomes (ACESO) is a consortium consisting of government, non-profit and university partners dedicated to improving survival for patients with sepsis in resource limited settings. This will be accomplished by development of host-based diagnostic and prognostic assays combined with evidence-based clinical management guidelines. Achieving such goals will require an exhaustive characterization of the temporal dynamics of the host response to a variety of infectious and non-infectious causes of illness across diverse range of host backgrounds. Our ongoing initiative aims to generate a vast amount of multi-omic data that can be translated into bedside host-based clinical assays. The ACESO approach to biomarkers and precision medicine is a departure from the single biomarker/specific disease approach of the past. Rather, a panel of biomarkers which provide a constellation of results to provide generalizable information to (1) classify the disease state (i.e., bacterial vs viral); (2) guide the appropriate treatment trajectories (fluids, antibiotics, etc.); (3) evaluate the patients' response to the treatment (antibiotic resistance/sensitivity); and (4) provide an accurate prognosis to fine-tune each patient's treatment.

To that end, ACESO is engaged in observational studies of sepsis in North America (Duke University Hospital, Durham North Carolina), Cambodia (Takeo Provincial Hospital, Takeo Province Cambodia) [7], West Africa (Komfo Anokye Teaching Hospital, Kumasi, Ghana; Phebe Hospital, Bong County, Liberia) and East Africa (Fort Portal Regional Referral Hospital, Fort Portal Uganda). Patients presenting to participating hospitals with at least two SIRS criteria (e.g., thermodysregulation, tachypnea, heart rate > 90 bpm, etc.) and a suspected infection are considered for enrollment. Detailed patient histories are obtained and blood is collected for standard clinical analyses such as culture, comprehensive metabolic panel, blood gases, electrolytes and hematology. Due to cultural sensitivities, the number and frequency of repeated collections vary depending on the site. In all cases however, blood is scheduled for collection at enrollment (T0), 6 h and 24 h.

In addition to blood collection for patient care, specimens are collected to examine the molecular characteristics of each patient's host response to the cause of their hospitalization. Blood is drawn directly into PAXgene RNA stabilization matrix (Pre Analytix) for whole transcriptome analysis by RNA sequencing. Blood collected into EDTA tubes (7.2 mg of K_2 EDTA) is separated into its three components (RBCs, plasma, buffy coat). The plasma is subjected to a variety of analyses including multiplex Luminex assays targeting a variety of pathway-specific proteins, and metabolomics (Biocrates AbsoluteIDQ p180 kit) [10]. Buffy coats are lysed and analyzed on a reverse-phase protein micro array [11].

Data analysis

Identifying subpopulations of patients with potentially different predisposition to disease, varying prognostic profiles, or divergent probabilities of responding to a given treatment is a core component of precision medicine. Given the complexity of disease pathogenesis and the heterogeneity in patient populations, this task requires measurement across a wide spectrum of biologic inputs paired with mathematical methods to cluster, classify, and predict. While individual biomarkers may be useful for some conditions, most complex diseases require panels of biomarkers to accurately place a patient in the appropriate subpopulation category.

Capturing large amounts of data to characterize the state of the biological system as a whole and then using advanced mathematical analysis to identify networks and associations of factors may be a powerful tool in understanding pathogenesis and targets for intervention. However, the wealth of clinical and multi-omic data collected in these studies presents both an opportunity and a host of challenges. An integrated, systems based approach is needed to delineate pathogenic biologic pathways as well as generate clinically relevant patient profiles. Computational methods such as latent factor models and graph theory based approaches in combination with machine learning algorithms are powerful tools to reduce the dimensionality inherent in multi-omic datasets, address issues of noise and differences of scale, as well as nonlinear relationships in the data.

Classify the disease state

Accurate diagnostic microbiology and antibiotic stewardship are inextricably linked. A recent survey demonstrated that although 88% of physicians in Cambodia are aware that antibiotic resistance is a problem, 86% of physicians would prescribe antibiotics for an uncomplicated common cold. In fact, 93% of respondents reported difficulties in selecting the appropriate antibiotics [6]. And this problem is not unique to resource-limited settings. A recent survey of American health professionals found that 95% of respondents have prescribed antibiotics in situations that were not necessary [12]. In 2015, U.S. President Barack Obama directed the development of diagnostics that could differentiate between bacterial and viral infections. There is mounting evidence that host-encoded biomarkers will be capable of robust differentiation between infectious and noninfectious processes. Host-based diagnostics capable of identifying or ruling out a bacterial infection could stem the inappropriate use of antibiotics throughout the world.

Guiding the appropriate use of antibiotics is important to reduce antimicrobial resistance, but also to prevent the waste of limited resources. For example, Cambodia is an area endemic for *Burkholderia pseudomallei*, the causative agent of melioidosis [13]. Melioidosis is a major cause of morbidity and mortality in resource limited settings. In Cambodia, melioidosis is gravely underdiagnosed [14]. Complicating matters, *B. pseudomallei* is not susceptible to many of the empiric sepsis treatment algorithms and Ceftazidime is only provided to provincial hospitals in limited quantities on a quarterly basis making the limited resources available even more precious [7]. Host-based diagnostics are not only capable of differentiating bacterial infections from viral infections, but could potentially diagnose specific infections. In fact, multivariate analysis of Luminex data generated from a subset of hospitalized sepsis patients from Cambodia identified a 13 Marker classifier panel capable of distinguishing *B. pseudomallei* infection from other sepsis-causing agents with an area under the curve (AUC) of 0.812 (Fig. 1). Similarly, Topological Data Analysis (TDA, Ayasdi Inc., Menlo Park, California) identified host transcriptomic patterns unique to melioidosis patients. Models generated using logistic regression on a subset of 35 patients identified classifiers with an AUC of 0.808 (Fig. 2). While promising preliminary data exists, considerable work remains to refine the models and ensure specificity.

Prognosis

The initial clinical presentation for a given patient is not an accurate predictor for the patient's clinical course. This unfortunate reality can result in inappropriate decisions regarding the urgency of care required, as well as decisions regarding treatment strategy. Early identification of patients at high risk of

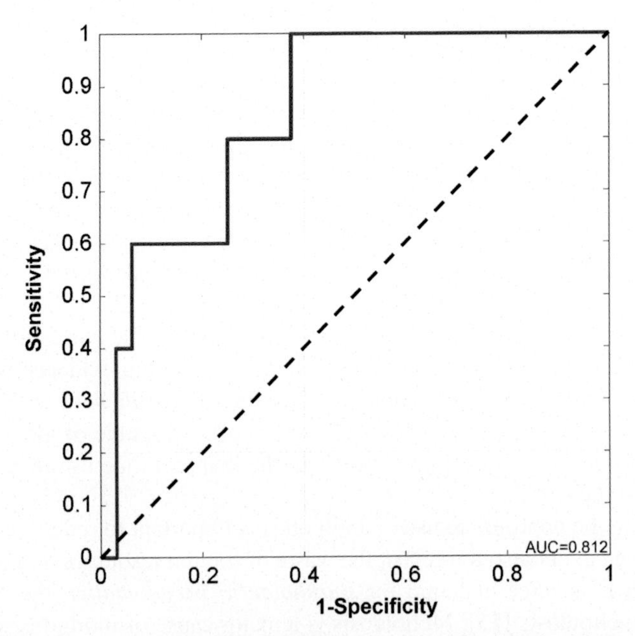

FIG. 1 Plasma biomarkers differentiate *Burkholderia pseudomallei* infections from other severe infections. Plasma protein biomarkers from 37 patients were measured at patient enrollment (Time 0) using multiplex Luminex assays. Sparse logistic regression models identified a panel of 13 biomarkers that was capable of differentiating melioidosis cases from other severe infections with a leave-one-out cross-validated AUC of 0.812.

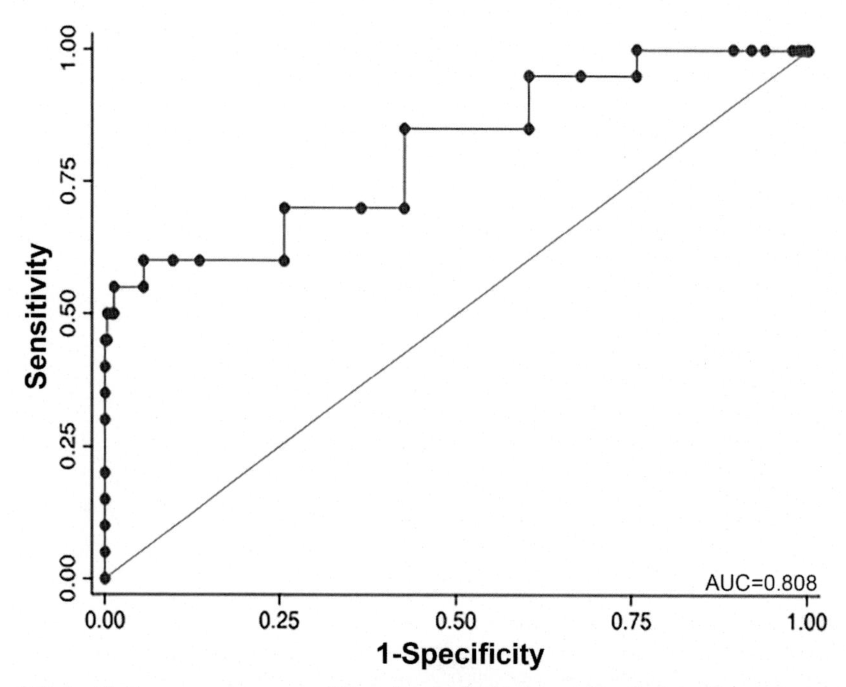

FIG. 2 The host transcriptome can differentiate *Burkholderia pseudomallei* infections from other infections. The host transcriptome of peripheral blood was characterized by RNAseq and analyzed by Topological Data Analysis (TDA, Ayasdi Inc., Menlo Park, California). Transcriptomic classifiers based on patient subgroups identified by TDA correctly classified 96.69% of patients with a specificity of 99.68%, sensitivity of 50% and leave-one-out cross-validated area under the curve of 0.808.

death would enable early initiation of care, ensure an appropriate level of care is obtained, and inform clinical management decisions. Accurate prognostic indicators would also allow for patient stratification in clinical trials evaluating interventions for these high-risk patients. We found that the host transcriptome (Fig. 3) can predict disease severity using 28-day mortality as the endpoint. Host-based biomarkers on a point of care platform would be an ideal approach to quickly screen patients as they present to the emergency department.

Biomarker-guided treatment

Predicting clinical outcome is not exclusively a feature of triage at presentation, rather a continual assessment throughout the course of treatment. Understanding if a patient's treatment regime is appropriate is an important feature of successful clinical management. Precision medicine based on the host response can be a powerful indicator of appropriate treatment. For example, recent studies to characterize the host-transcriptomic response to a variety of acute respiratory infections resulted in a host-based signature that could identify pre-symptomatic influenza infections [15–17]. Over the course of treatment with oseltamivir,

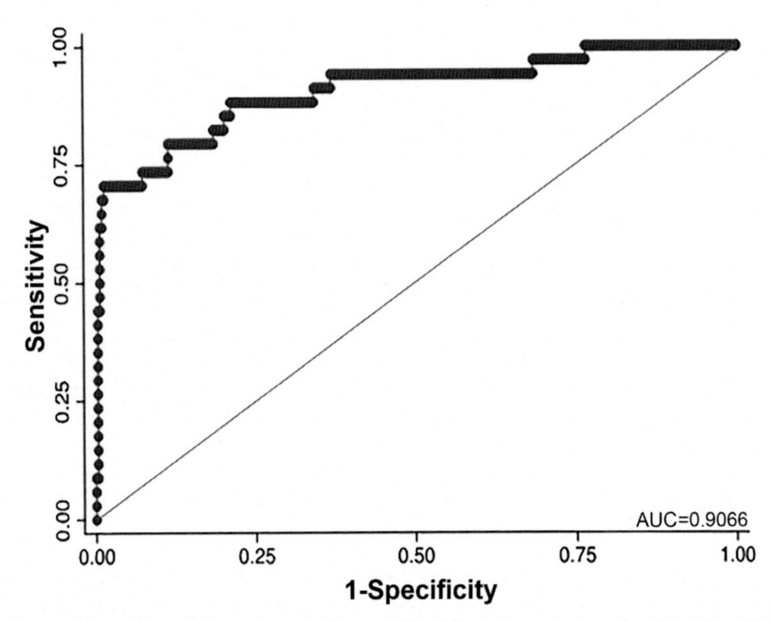

FIG. 3 The host transcriptome can predict mortality. The host transcriptome of peripheral blood was characterized by RNAseq and analyzed by Topological Data Analysis (TDA Ayasdi Inc., Menlo Park, California). The transcriptomic classifiers based on patient subgroups identified by TDA predicted mortality within 28 days with an accuracy of 95.78%, sensitivity of and specificity of 70.59% and 95.78%, respectively, and a leave-one-out cross-validated area under the curve of 0.9066.

the authors observed a return to baseline of the host transcriptomic signature demonstrating that the host response could serve as an indicator of treatment effectiveness [18]. Incorporating response to treatment POC assays into precision medicine algorithms could have a profound impact on a number of variables determining clinical outcomes. For example, identifying if an antibiotic is effective against the infection would not only improve the outcome for the patient but spare a potentially limited resource if an antibiotic is determined to be ineffective.

Precision medicine for infectious diseases in the developing world

The result of studies such as these have the potential to bring host-based assays for diagnosis and prognosis to the developing world. However, to be effective, host-based assays should still consider the WHO ASSURED guidelines for diagnostics in the developing world [19]. At first glance, these guidelines seem to exclude precision medicine as an option in resource limited settings; Affordable (to those at risk), Sensitive, Specific, User-friendly, Rapid, Equipment-free and Delivered to those who need it. However, as technology rapidly progresses, resource-limited settings may prove to be well suited for implementation of next-generation host-based diagnostics and prognostics. The cost of a host-based assay may be initially higher than diagnostic assays currently available in low-resource settings, but the cost of unnecessary treatments or interventions is most certainly higher. Developments in point of care assay technology are already moving toward low-cost, equipment-free systems capable of multiplexing multiple analytes with minimal trade-offs in sensitivity and specificity.

Finally, the obstacles encountered in resource-limited settings are not limited to the population of the developing world. There are many parallels between the limitations encountered in the medical facilities of forward deployed troops and those discussed above. Military personnel are at particular risk while deployed or in combat situations. Sepsis was the leading cause of death in the first 24 h after injury and the third leading cause overall during the Vietnam conflict [20]. In operations Iraqi and Enduring Freedom, 25% of deaths from otherwise survivable injuries could be attributed to sepsis [21]. Furthermore, distinguishing sepsis in severe combat injuries is complicated by nonspecific clinical presentations of systemic inflammatory responses which are present following trauma [22]. Implementation of host-based diagnostics by the Department of Defense could improve the force health protection and readiness of military forces around the world.

Conclusions

Early recognition and characterization of infectious disease threats currently rely on antiquated technologies throughout the world. Precision medicine has the potential to have a major impact on global health. Implementation of cutting-edge precision medicine approaches and technologies in resource-limited settings may at first seem contrary to conventional capacity building efforts. However, consider the explosion and exponential growth of other technologies such as cell phones in Africa. Traditional telecommunications infrastructure (i.e. landlines) is prohibitively expensive, with few countries willing or able to invest in such large infrastructure projects. When cell phones became available, low-resource countries "leap-frogged" over traditional landlines and recently became some of the fastest-growing mobile markets in the world (https://www.gsma.com/mobileeconomy/). We call on funding and research organizations alike to continue to engage in host-based studies that include appropriate levels of geographical, ethnic and pathogenic diversity. Investigators should foster collaboration to avoid duplication of efforts and encourage data sharing to maximize data availability. Commercial entities who develop diagnostics with global health in mind should be rewarded and new technologies and markers of disease should be integrated as they become available to produce the most robust diagnostics possible.

Disclaimers

The views expressed in this manuscript are those of the author and do not necessarily reflect the official policy or position of the Department of the Navy, Department of Defense, nor the U.S. Government.

Our study protocols are approved by the Naval Medical Research Center Institutional Review Board in compliance with all applicable Federal regulations governing the protection of human subjects.

This work was supported by work unit number A1316.

References

[1] President's Council of Advisors on Science and Technology (PCAST). Priorities for personalized medicine. President's Council of Advisors on Science and Technology, September 2008; Available at: www.whitehouse.gov/files/documents/ostp/PCAST/pcast_report_v2.pdf

[2] Yang WE, Woods CW, Tsalik EL. Host-based diagnostics for detection and prognosis of infectious diseases. In: Sails A, Tang Y-W, editors. Methods in microbiology. Academic Press; 2015. p. 465–500 [chapter 13].

[3] Peeling RW, Mabey D. Point-of-care tests for diagnosing infections in the developing world. Clin Microbiol Infect 2010;16(8):1062–9.

[4] Okeke IN, et al. Antimicrobial resistance in developing countries. Part I: recent trends and current status. Lancet Infect Dis 2005;5(8):481–93.

[5] Om C, et al. Pervasive antibiotic misuse in the Cambodian community: antibiotic-seeking behaviour with unrestricted access. Antimicrob Resist Infect Control 2017;6:30.

[6] Om C, et al. "If it's a broad spectrum, it can shoot better": inappropriate antibiotic prescribing in Cambodia. Antimicrob Resist Infect Control 2016;5:58.

[7] Schully KL, et al. Melioidosis in lower provincial Cambodia: a case series from a prospective study of sepsis in Takeo province. PLoS Negl Trop Dis 2017;11(9):e0005923.

[8] Reyburn H, et al. Rapid diagnostic tests compared with malaria microscopy for guiding outpatient treatment of febrile illness in Tanzania: randomised trial. BMJ 2007;334(7590):403.

[9] Mulder N. Development to enable precision medicine in Africa. Pers Med 2017;14(6):467–70.

[10] Schmerler D, et al. Targeted metabolomics for discrimination of systemic inflammatory disorders in critically ill patients. J Lipid Res 2012;53(7):1369–75.

[11] Paweletz CP, et al. Reverse phase protein microarrays which capture disease progression show activation of pro-survival pathways at the cancer invasion front. Oncogene 2001;20(16): 1981–9.

[12] Yox S. Too many antibiotics! Patients and prescribers speak up. 2014. https://www.medscape.com/features/slideshow/public/antibiotic-misuse#3.

[13] Limmathurotsakul D, et al. Predicted global distribution of *Burkholderia pseudomallei* and burden of melioidosis. Nat Microbiol 2016;1(1):15008.

[14] Suttisunhakul V, et al. Retrospective analysis of fever and sepsis patients from Cambodia reveals serological evidence of melioidosis. Am J Trop Med Hyg 2018;98:1039–45.

[15] Zaas AK, et al. A host-based rt-PCR gene expression signature to identify acute respiratory viral infection. Sci Transl Med 2013;5(203):203ra126.

[16] Zaas AK, et al. Gene expression signatures diagnose influenza and other symptomatic respiratory viral infections in humans. Cell Host Microbe 2009;6(3):207–17.

[17] Woods CW, et al. A host transcriptional signature for presymptomatic detection of infection in humans exposed to influenza H1N1 or H3N2. PLoS One 2013;8(1):e52198.

[18] McClain MT, et al. A genomic signature of influenza infection shows potential for presymptomatic detection, guiding early therapy, and monitoring clinical responses. Open Forum Infect Dis 2016;3(1):ofw007.
[19] Mabey D, et al. Diagnostics for the developing world. Nat Rev Microbiol 2004;2(3):231–40.
[20] Murray CK. Epidemiology of infections associated with combat-related injuries in Iraq and Afghanistan. J Trauma 2008;64(3 Suppl):S232–8.
[21] Holcomb J, et al. Causes of death in US special operations forces in the global war on terrorism: 2001–2004. US Army Med Dep J 2007;24–37.
[22] Hogan BK, et al. Correlation of American burn association sepsis criteria with the presence of bacteremia in burned patients admitted to the intensive care unit. J Burn Care Res 2012;33(3):371–8.

Chapter 9

Respiratory viral infections

Ann Regina Falsey
University of Rochester, Rochester, NY, United States, Rochester General Hospital, Rochester, NY, United States

Introduction

Respiratory viruses are a major cause of morbidity and mortality throughout the world and affect persons of all ages [1–4]. In addition to >100 million office visits for upper respiratory infections each winter, hospitals fill to capacity with admissions due to community acquired pneumonia, acute exacerbations of chronic obstructive pulmonary disease, asthma and bronchitis and many of these illnesses are due to viral infection. "Pneumonia and Influenza" consistently ranks as the fourth most common discharge diagnosis, and each year, 270,000 to 540,000 hospitalizations and 7600 to 72,000 deaths in the United States are attributable to influenza [3, 5–8]. Due to their epidemic nature influenza and RSV are widely recognized as pathogens in adults and children, respectively. However, the true burden of disease and the contributions of other viruses such parainfluenza viruses (PIV), human metapneumoviruses (hMPV), coronaviruses (CoV), and rhinoviruses (HRV) now being more fully recognized using modern molecular detection methods [9–13]. In addition to sensitive and rapid diagnostic testing, new molecular techniques allow an understanding of viral evolution, mechanisms and predictors of severe disease, interrogation of vaccine responses, improved bacterial and viral diagnostics and associations of viral infections with non-respiratory medical events. In this chapter the many ways molecular and precision medicine have impacted the field of respiratory viral disease will be reviewed.

Molecular virology of respiratory viruses

Viral diagnosis

In the past defining the epidemiology and impact of viral respiratory pathogens was significantly hampered by slow and/or insensitive diagnostic techniques such as cell culture and antigen detection [13, 14]. Polymerase chain reaction (PCR) has revolutionized the study of respiratory viruses and provides

Genomic and Precision Medicine. https://doi.org/10.1016/B978-0-12-801496-7.00009-5

extremely sensitive, specific and rapid means for the detection of fastidious and non-cultivatable respiratory viruses [13]. PCR based epidemiologic studies now provide a more complete understanding of the clinical spectrum and age ranges of populations affected [15–20]. In one study, conventional methods yielded a viral diagnosis in 14% of pneumonia cases, while use of PCR increased the yield to 56% [21]. Technology has rapidly evolved from single-plex PCR and gel electrophoresis to multiplex real time assays where products are detected by luminescent signals proportional to the target amplified [14]. There are currently a variety of commercially available assays that detect from 2 to 20 viral respiratory pathogens and maintain excellent sensitivity [14]. Many clinical microbiology laboratories are now moving to primarily molecular methods for viral detection and PCR formats are becoming increasingly simple so that nucleic acid extraction and PCR is fully automated with little operator input. Molecular point of care assays will soon be feasible [22].

Viral discovery

In addition to providing more sensitive means of detecting known viruses, molecular methods are extremely useful for viral discovery [23–25]. Over the past several decades a number of new respiratory viruses or variants have been identified including hMPV, novel strains of coronaviruses (HKU1, NL63, SARS-CoV, MERS-CoV), rhinovirus C, Human Boca virus, parechoviruses and new strains of avian influenza viruses. Molecular methods have been critical for the rapid identification of new viruses associated with dramatic lethal outbreaks but also for pathogen discovery for routine respiratory illnesses. Despite intensive investigation, in 12–62% of lower respiratory illnesses no pathogen can be identified suggesting additional agents may yet be discovered [26–28]. Several different genomic approaches for pathogen discovery have been used successfully and include random primer amplification, pan-viral DNA microarray and next generation sequencing (Fig. 1) [24].

If a viral class of an unknown pathogen or variant virus is suspected, consensus PCR using degenerate primers to detect sequences broadly conserved between members of a group can be used as was done to identify two new coronaviruses, HKU1 and MERS-CoV [29, 30]. Another technique for viral discovery is random primer amplification with conventional shotgun sequencing of PCR products [31–33]. Such was the case when Van den Hoogen discovered a new respiratory virus in 2001, in young children with bronchiolitis who tested negative for RSV [31]. After detecting paramyxovirus like particles in cell culture, RNA was subjected to random primer PCR and viral sequences were compared to all known pathogens. The new virus most closely aligned to avian pneumovirus but was determined to be a unique human pathogen and named human metapneumovirus (HMPV). Similarly, in 2003 Peiris identified a novel coronavirus as the cause of severe acute respiratory syndrome (SARS-CoV) using degenerate/random primers PCR

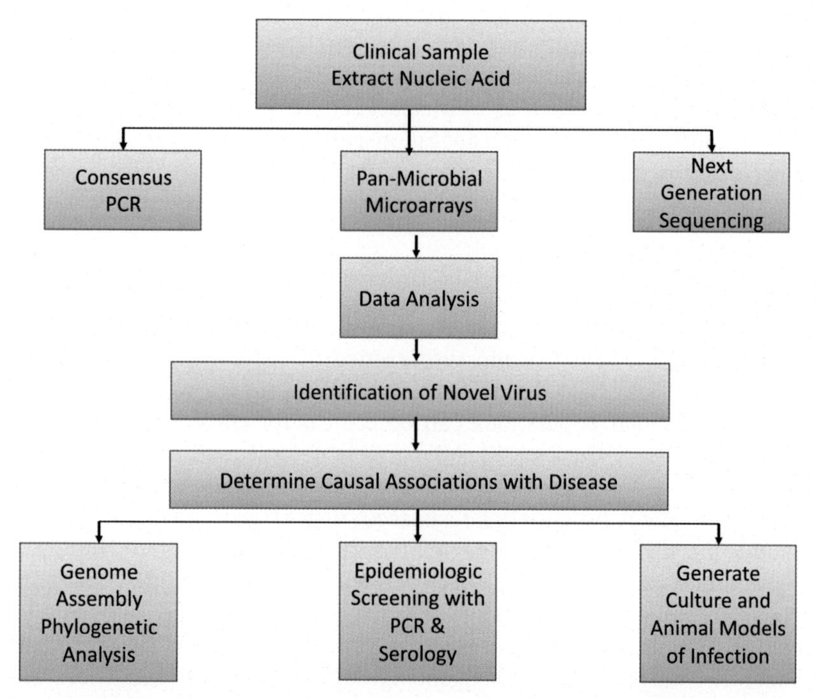

FIG. 1 Genomics approaches to viral discovery. Clinical material is subjected to extraction of nucleic acids, followed by genomic analysis and bioinformatics analysis. After novel pathogen is identified, additional studies are needed to determine causal relationship with disease. *(Adapted from reference Chiu CY. Viral pathogen discovery. Curr Opin Microbiol 2013;16(4):468–478.)*

amplification [32]. Using pan viral micro array investigators at the Center for Disease Control and Prevention independently identified the same SARS-CoV [34]. In this technique, after random primer amplification, PCR products are hybridized to microarrays consisting of 70mer oligonucleotides derived from every fully sequenced viral genome. Hybridized sequences are scraped from the microspot, amplified, cloned and finally sequenced [25]. Identification of completely novel infectious agents requires unbiased and sequence independent methods for universal amplification [23, 24, 35]. Conventional Sanger sequencing may have poor sensitivity for genomes at low quantity. Next Generation Sequencing (NGS) involves the analysis of millions of sequences and can detect small amounts of novel nucleic acid sequences in clinical samples. Continuous sequences are assembled, host sequences are subtracted and the residual sequences are analyzed for similarity to known microbial sequences. NGS has led to the discovery a numerous novel human and animal pathogens [24]. A recent study of nasopharyngeal aspirates of Thai children with respiratory illness using NGS identified a number of mammalian viral sequences belonging to newly described families of viruses such *Anelloviridae* as well as novel strains of HRV, enteroviruses and HBoV [35]. A critical step in viral discovery is the availability of bioinformatic tools to efficiently identify unique viral sequences in complex mixtures of host, bacterial and fungal sequences. New computational tools for analysis of the virome such as "VirusSeeker" are being developed [36]. Of note, detection does equate with causation and after discovery further studies are necessary to infer more than association.

Viral evolution

The genetic and antigenic evolution of error prone RNA respiratory viruses, particularly influenza, has been of interest for several decades [37, 38]. Understanding the selective pressure exerted by pre-existing immunity on viral evolution may help design more effective influenza vaccines and surveillance of animal populations can be critical for early identification of emerging influenza viruses [39]. Advances in deep sequencing make it possible to measure low frequency within host viral diversity and factors such as antigenic diversity, antiviral resistance, and tissue specificity can now be studied to understand the complexities of viral evolution [40]. Influenza evolution at a population level has been studied years, yet, new antigenic variants are initially generated and selected at the level of the individual infected host. Within a host, influenza viruses exist as a "swarm" of genetically distinct viruses [41]. Sanger sequencing defines consensus sequences and cannot resolve minority variants below 20% of the viral population. Deep sequencing has been used in natural infection and human challenge studies to characterize between and within host genetic diversity [41, 42]. The identification of low frequency mutations in the hemagglutinin (HA) antigenic sites or near the receptor-binding domain in vaccinated and unvaccinated influenza infected persons highlight viral evolution within a host due to selective immune pressure [41]. Similarly, NGS can reveal the rapid evolution of drug resistant variants during therapy [43]. Using samples collected over time, the mutational spectrum of H3N2 influenza A virus in an immunocompromised child was delineated [44]. Individual resistance mutations appeared weeks before they became dominant, evolved independently on cocirculating lineages. The within host evolution of antiviral resistance reflected a combination of frequent mutation, natural selection, and a complex pattern of segment linkage and reassortment. Within host sequencing diversity has also been examined in an infant with severe combined immune deficiency with persistent RSV infection [45]. NGS was performed on 26 samples obtained before and after bone marrow transplantation. The viral population appeared to diversify after engraftment with most variation occurring in the attachment protein (G). In addition, minority viral populations with palivizumab resistance mutations emerged after its administration. Deep sequencing of HRV during human challenge studies has shown that HRV generates new variants rapidly during the course of infection with accumulation of changes in "hot spots" in the capsid, 2C, and 3C genes [46].

Host genetic variation and genomic response to respiratory viral infection

Genome-wide association studies (GWAS)

A genome-wide association study (GWAS) involves rapidly scanning sets of DNA, or genomes, of many people to find genetic variations associated with a particular disease. Typically, the genomes of cases are compared to non-affected controls and search for single nucleotide polymorphisms (SNPS) or polygenic changes that are associated with risk or protection from susceptibility or severity of the condition. GWAS have been useful to find genetic variations and risk for asthma, cancer, diabetes, heart disease and autoimmune illnesses with relatively limited studies relating to infectious diseases [47, 48]. Recent studies examining host genetic factors conferring susceptibility to respiratory viruses such as pandemic H1N1 2009 influenza A, SARS-CoV and RSV now provide some insight into host genetic factors for respiratory viral infections [49–51]. Previously most influenza research focused on viral genetics of novel viruses, yet experience with H1N1 2009 and H5N1 clearly indicate host factors also influence disease severity [49, 50]. A number of candidate genes influencing respiratory virus susceptibility have been identified in animal and human studies and involve host virus interactions, innate immune signaling, interferon related pathways and cytokine responses (Table 1) [49–51, 69–75]. Over 20 studies have evaluated genetic polymorphisms associated with severe RSV disease and none demonstrates dramatic results [51]. Most focused on one or a few candidate genes resulting in only modestly increased odds ratios of severe illness. A relatively large study of almost 500 hospitalized children that examined 384 SNPS in 220 candidate genes demonstrated that susceptibility to RSV is complex with a several associations to a few innate immunity genes. These included a Vitamin D receptor gene associated with down regulating interleukin 12 (IL-12), gamma interferon (IFN-γ), nitrous oxide synthase (NOS2A), the JUN oncogene, an important transcriptional regulator for innate immune pathways, and IFN-α (IFNA5) an antiviral cytokine [68].

Host response to investigate mechanisms of disease

The host transcriptional response can be analyzed to investigate disease pathogenesis using a variety of methods including in-vitro studies of bronchial epithelial cells (BEC), animal models and infection both natural and experimental challenge [76–79]. In addition, two compartments, the respiratory epithelium and blood can be sampled in human studies and interrogated using different viruses or viral strains to develop gene signatures for prognosis, as indicators of severity and to identify potential therapeutic targets.

Mechanisms of disease

Most respiratory viral mechanistic studies have been performed using influenza viruses, RSV, HRV and coronaviruses [80–83]. Using BEC, the common and

TABLE 1 Genetic polymorphisms possibly associated with severity of respiratory viruses

Gene	Polymorphism	Significance	Virus	References
CCR5	CCR5Δ32	Increased allele frequency in Canadian ICU cases	H1N1,H5N1	[52, 53]
KIR	2DL2/L3	Increased allele frequency in Canadian ICU cases	H1N1	[54]
IFTIM 3	rs12252 altered splice receptor	Increased in hospitalized English and Scottish cases	H1N1	[55]
FcγRIIa	IGHG2 *n/*−n	IgG2 subclass deficiency	H1N1	[56, 57]
NLRP3	rs4612666(intron 7)	Dysregulation of inflammasome	H1N1	[58, 59]
	rs10754558 (3′UTR)	Alteration of NLRP3 m RNA stability and enhancer activity		
HLA	Various alleles	Influenza specific CTL	H1N1	[60]
MBL2	230G/A	Mannose-binding lectin	SARS	[61, 62]
MxA	-88G/T(rs2071430) -123C/A (rs17000900)	Encode IFN induced antiviral proteins	H5N1, SARS	[63, 64]
OAS1	rs2660(3′UTR A/G) rs3741981 (exon 3A/G) rs1077467	IFN induced antiviral proteins	H5N1,SARS, West Nile	[64–66]

TLR3	908T/C	Missense mutation in patient with encephalopathy	H5N1	[67]
VDR	Thr1met	Vitamin D receptor, downregulate IL-12 and IFNγ	RSV	[68]
JUN	G750A	Transcriptional regulator for innate immune pathways	RSV	[68]
IFNA5	C435T	Antiviral cytokines	RSV	[68]
NOS2A	G275A	Antimicrobial and anti-inflammatory	RSV	[68]
FCER1A	T-66C	Innate immunity	RSV	[68]

(Adapted from references Juno J, Fowke KR, Keynan Y. Immunogenetic factors associated with severe respiratory illness caused by zoonotic H1N1 and H5N1 influenza viruses. Clin Dev Immunol 2012;2012:797180; Keynan Y, Malik S, Fowke KR. The role of polymorphisms in host immune genes in determining the severity of respiratory illness caused by pandemic H1N1 influenza. Public Health Genomics 2013;16(1–2):9–16; Miyairi I, DeVincenzo JP. Human genetic factors and respiratory syncytial virus disease severity. Clin Microbiol Rev 2008;21(4):686–703.)

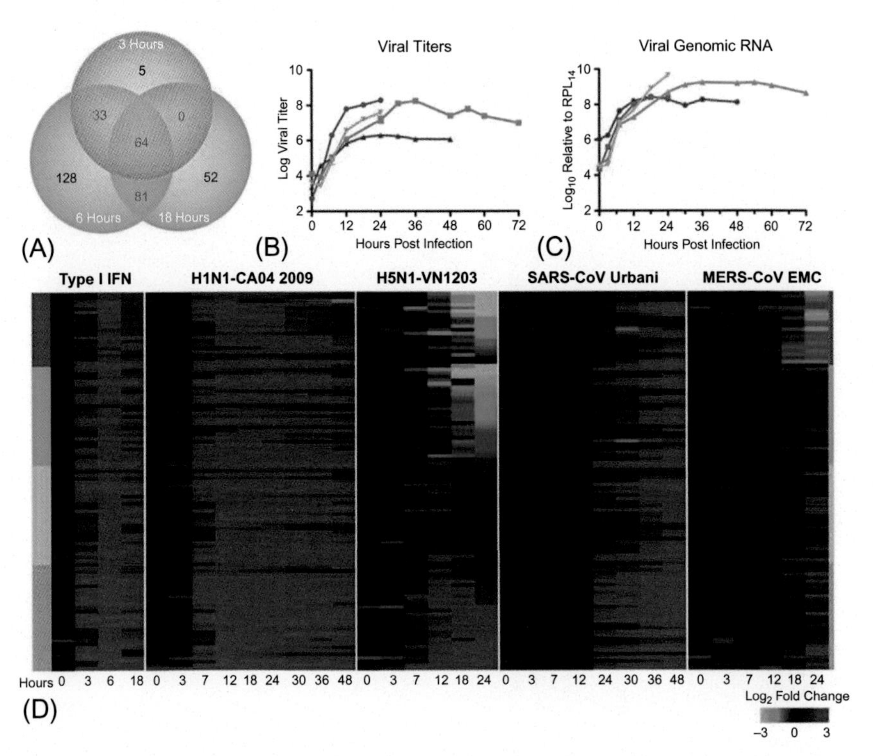

FIG. 2 Effect of type 1 interferon (IFN) treatment and viral infections on interferon stimulated gene (ISG) responses. (A) Total number of genes induced by type 1 IFN with a $\log_2$ fold change of >1.5 at indicated time points in Calu3 cells. (B) Viral titers. (C) Viral Genomic RNA with H5N1 in blue, H1N1 in red, SARS-CoV in green, or MERS-Co-V in orange. (D) Global ISG transcriptional response to IFN α treatment or infection. Genes ordered by MERS-CoV or H5N1 at 24 h and grouped into subsets based on fold change. Purple = downregulated in MERS-CoV and H5N1, blue = downregulated in H5N1, green = minimal stimulation. Increased expression compared to mock controls shown in red and decreased expression in green. *(Figure reproduced with permission from Menachery VD, Eisfeld AJ, Schafer A, Josset L, Sims AC, Proll S, et al. Pathogenic influenza viruses and coronaviruses utilize similar and contrasting approaches to control interferon-stimulated gene responses. MBio 2014;5(3):e01174-14. https://doi.org/10.1128/mBio.01174-14.)*

divergent pathways used by four virulent viruses (H1N1 2009, H5N1, SARS Co-V and MERS Co-V) to antagonize interferon stimulated genes (ISG) responses was demonstrated (Fig. 2) [84]. H5N1 exhibited early strong up and down regulation of ISG subsets, whereas, less virulent H1N1 did not. SARS Co-V and MERS Co-V also demonstrated delayed ISG allowing early viral replication. In a similar experiment Josset et al. infected BEC with different influenza viruses (H5N1, H7N7, H3N2 and H7N9) and analyzed cellular responses using microarray [83]. Common proinflammatory cytokines and antigen presentation were identified although each viral response was unique and notably, H7N9 responses were most similar to H3N2. The response of different clinical isolates of RSV in A549 cells, and monocyte derived human macrophages demonstrated that the pattern of innate immune activation was both host cell and viral strain specific [85]. Using RNA seq, differences in IL-6 and CCL5 were noted among the responses to different clinical isolates suggesting different RSV strains may vary in inherent virulence. Human studies have shown significant differences in the blood transcriptional profiles which change over time and differ depending on the infecting respiratory virus. Mejias and colleagues were able to differentiate RSV, HRV and influenza in young children based on the blood gene profile (Fig. 3). HRV infection exhibited the mildest innate and adaptive responses compared to RSV and influenza and neutrophil gene expression was greatest in RSV infection with marked suppression of B and T cell and lymphoid responses [79]. Notably, gene expression changes persisted up to 1 month after infection. Similarly, studies of H7N9 infected patients showed transcriptional profile changes persisting up to 1 month with a transition from innate to adaptive immunity [86].

Rhinovirus (HRV) association with asthma

Because of the association of HRV and exacerbations of asthma, the host response to HRV has been of particular interest [87–90]. Studies using BECs from asthmatic and healthy donors demonstrate different transcriptional profiles when infected with HRV [87]. HRV, similar to other picornaviruses induces gene expression down regulation by the 2A and 2C proteins. In both asthmatic and healthy control derived cells the majority of genes were down regulated after exposure to HRV. However, some significant expression differences in inflammatory, tumor suppressor, airway remodeling and metallopeptidase pathways have been noted in asthmatic derived cells. Asymptomatic HRV infection is quite common and its role in asthma pathogenesis has been questioned. Interestingly, Heinonen et al. did not find a difference in the blood transcriptome of asymptomatic HRV infected children compared to non-infected controls [91]. Whereas, Wesolowska-Anderson and colleagues demonstrated over 100 differentially expressed genes in the nasal epithelium of asymptomatic infected HRV patients [90]. Thus, the blood transcriptome may not be as informative as the nasal epithelial transcriptional response for asymptomatic HRV infection. Given the significant host response to asymptomatic infection, HRV may play a role in asthma exacerbations in the absence of clinically evident disease. Lastly, it may be possible to identify patients with asthma who are prone to frequent HRV related exacerbations by examining the gene expression response of their PBMCs stimulated with HRV [89].

Disease severity

Respiratory syncytial virus

Gene expression studies focusing on illness severity may enhance our understanding of disease pathogenesis, can identify potential therapies to modulate

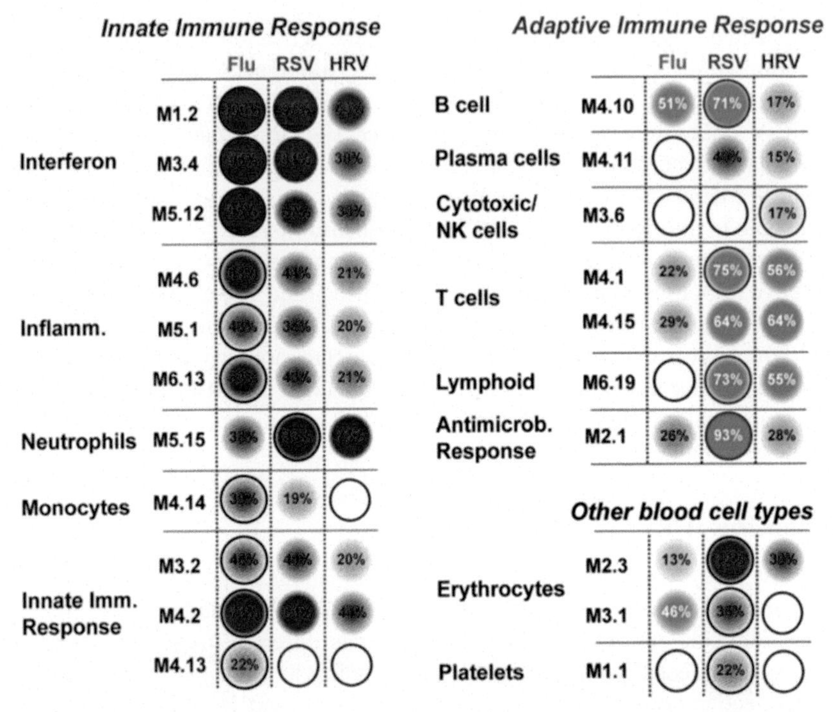

FIG. 3 Transcription profiles from blood samples of children with influenza, RSV and HRV Lower Respiratory Tract Infection (LRTI). 70 top ranked genes best discriminated influenza, RSV and HRV. Mean modular transcriptional fingerprint for influenza ($n=16$ and 10 matched controls), RSV ($n=44$ and 14 matched controls), and HRV LRTI ($n=30$ and 14 matched controls). The outer dark circles highlight the disease group (influenza, RSV, or HRV) with greater (red) or lower (blue) modular activation. Children with HRV infection demonstrated a milder activation of the innate and adaptive immune responses, compared with children with influenza or RSV infection. Children with influenza displayed a stronger activation of genes related to interferon (M1.2, M3.4, M5.12), inflammation (M4.6, M5.1, M6.13), monocytes (M4.14), and innate immune response (M3.2, M4.2, M4.13) compared with children with RSV or HRV. Several type I interferon and type II interferon genes were expressed only in influenza and RSV infection. In addition, the magnitude of the type I interferon and type II interferon response present was 2- to 22-fold higher in children with influenza compared with children with RSV or HRV. Similarly, genes related to inflammation, monocytes, and innate immune response were greatly overexpressed in children with influenza compared to children with RSV or HRV LRTI. Neutrophil-related genes (M5.15) were significantly overexpressed in RSV infection, followed by HRV infection and, at a lower level, influenza infection. The suppression of genes related to B cells (M4.10), T cells (M4.1, M4.15), lymphoid lineage (M6.19), and antimicrobial response (M2.1) observed in RSV infection was significantly milder or not present in children with influenza or HRV LRTI. *(Reproduced with permission from Mejias A, Dimo B, Suarez NM, Garcia C, Suarez-Arrabal MC, Jartti T, et al. Whole blood gene expression profiles to assess pathogenesis and disease severity in infants with respiratory syncytial virus infection. PLoS Med 2013;10(11):e1001549. https://doi.org/10.1371/journal.pmed.1001549.g005.)*

harmful host responses and can be used to develop biomarkers for predicting life threatening disease [79, 92–94]. A number of studies have been undertaken to understand the pathogenesis of severe RSV in young children and have identified a variety of gene expression patterns in blood including under expression of T cell cytotoxicity/NK cells and plasma cell genes, as well as upregulation of JAK/STAT, prolactin, IL-9 signaling, cell to cell signaling, and immune activation pathways [79, 92]. Using nasal epithelial gene expression analysis, van den Kieboom identified 5 differentially expressed genes in 30 children with mild, moderate and severe RSV infection [81]. Ubiquitin D, tetraspanin 8, mucin 13, β microseminoprotein, chemokine ligand 7 were up regulated and differentiated mild from severe illness. Lastly, nasal gene expression is complicated by interactions of the nasal microbiota and host cell gene responses [95]. In nasal samples from children with RSV infection, *H. influenzae* and *S. pneumoniae* dominated microbiota, Toll like receptors and neutrophil/macrophage signaling were over expressed and the presence of *H. influenzae* and *S. pneumoniae* along with age and sex were predictive of risk of hospitalization due to RSV.

Influenza

Transcriptional profiling related to severity has been analyzed in seasonal influenza as well as emerging avian pathogens with a recognition that disease is not only due to an infection with a novel virus in a non-immune host but may also be due to an exaggerated host immune response [78, 96]. In a study of primarily seasonal influenza (H1N1, H3N2), influenza infection was associated with a significantly stronger antiviral, cytokine, attenuation of T/NK cell response compared to patients with respiratory illnesses of unknown etiology regardless of severity [96]. Notably, IFN and ubiquitination was significantly down regulated in those with severe vs. mild to moderate disease. In a study of the lethality of 1918 H1N1 influenza and H5N1 Vietnam influenza virus in Macaques, upregulation of key components of the innate immune response and cell death pathways were noted were noted with 1918 H1N1 infection but were down regulated with H5N1 [78]. Early up regulation of the inflammasome likely resulted in some of the severe tissue damage noted with the 1918 H1N1 influenza infections.

Identification of potential therapeutic targets

In vitro, animal and human challenge studies have been used to identify new strategies control or prevent symptomatic or severe infection [82, 97]. In HRV challenge studies, virperin expression correlated with rhinorrhea and chilliness. Knockdown of expression resulted in increased viral replication in BECs suggesting virperin has antiviral actions and might have potential therapeutic use. Influenza challenge studies clearly show a definable transcriptomic profile in the blood prior to the onset of symptoms offering the possibility of earlier and more effective oseltamivir treatment [77, 98].

Associations with respiratory viral infection with non-respiratory medical events

Lastly, host gene expression studies may allow investigation into links between respiratory viral infections and specific non-respiratory events. There is ample epidemiologic evidence that influenza epidemics are linked with increased rates of strokes and myocardial infarction (MI) [99, 100]. Increased rates of falls and functional decline in nursing homes have also been associated with increased influenza activity [101, 102]. However, direct links of events to viral infection are scarce in part due the event of interest may follow the infection by several weeks when the virus is no longer detectable by traditional testing. Several gene profiling studies have identified viral infection signatures that may persist up to 1-month post infection [79, 86]. Thus, it might be possible to study patients with falls or cardiac events for evidence of recent viral infection using a host response viral signature. In addition, evaluating the host response can provide information on mechanisms of disease. A viral gene signature was used to evaluate patients undergoing cardiac catheterization [103]. Notably, 25% vs. 12%, $P=0.04$ of those with a viral gene signature present vs. those without viral signatures, suffered an MI. Furthermore, H1N1 infected patients showed an increased gene platelet expression signature providing insight into how infection may induce a prothrombotic state.

Host response for diagnosis

Diagnosis of viral infection based on host response

Given the availability of rapid and accurate Multiplex PCR for viral detection, host-based diagnostics might seem unnecessary. However, current PCR assays use conserved known viral sequences but can miss novel or significantly mutated viruses. This issue was seen in 2009 with pandemic H1N1 when influenza PCR assays had to be adapted to optimally detect the new influenza strain [104]. The emergence of novel respiratory viruses are a persistent threat and methods to detect a "viral signature" in the setting of clusters of severe pneumonia cases could be very useful. Zaas and colleagues developed an acute respiratory viral gene signature using microarray analysis of the blood from volunteers experimentally infected with influenza A, HRV or RSV [105]. The signature was subsequently 89% sensitive and 94% specific in classifying as viral 25 influenza and 3 HRV infected patients presenting to an emergency room. Additionally, a distinct blood transcriptome signature was noted in patients with severe H1N1 pneumonia [106]. Upregulated genes included those related to cell cycle, DNA damage, apoptosis, protein degradation, and T helper cells. Down regulated genes were primarily in immune response pathways suggesting immunosuppression as a mechanism of severe influenza pneumonia. Investigators developed a 29 gene classifier which predicted H1N1 influenza A regardless of concomitant bacterial infection and such a predicator could guide antiviral therapy in the face of negative pathogen detection methods.

Distinguishing viral and bacterial respiratory infections

In most cases of respiratory infection, the precise microbial etiology is unknown and antibiotics are frequently administered empirically [27, 107]. Although sensitive molecular diagnostics (PCR) now allow rapid diagnosis of a wide variety of respiratory viruses, their impact on patient management and antibiotic prescription has been modest primarily due to concern about bacterial co-infection [108–110]. Approximately 40% of adults hospitalized with a documented viral respiratory infection have evidence of concomitant bacterial infection and thus clinician concerns are reasonable [109]. Importantly, sensitive and specific diagnostic tests for bacterial lung infection are currently lacking [111, 112]. Although the site of infection is the respiratory tract, blood is a convenient sample comprised of components of the innate immune system (neutrophils, natural killer cells), as well as the adaptive immune system (B and T lymphocytes) [113]. Recent studies indicate that viral and bacterial infections trigger pathogen specific host transcriptional patterns in blood, yielding unique "bio-signatures" that may discriminate viral from bacterial causes of infection [114–117]. In the largest study to date, Tsalik et al. used gene expression in blood to discriminate bacterial from viral infection or non-infectious illness in 273 subjects with respiratory illness [118]. These investigators defined 130 predictor genes in a model with an accuracy of 87% to discriminate clinically adjudicated bacterial, viral, and non-infectious illness. Most studies to date have used micro array but recently RNAseq has been used to differentiate viral and bacterial respiratory illness and in one study 141 genes were noted to be differentially expressed [119]. Three pathways (lymphocyte, α-linoleic acid metabolism, IGF regulation pathways) which included 11 genes as predictors for bacterial infection from non-bacterial infection (naïve AUC = 0.94; nested CV-AUC = 0.86). To date, a number of gene expression studies of adults and children have developed predictors with similar accuracy (AUC ranging from 78% to 94%), yet there has been little overlap in classifying genes identified [105, 106, 114–116, 118–122]. Diverse populations, types of infection, plus alternate analytic tools used, likely explain the different genes identified. More work needs to be done to refine predictive gene sets including patients with mixed viral-bacterial respiratory tract infection. Most studies to date have focused on blood; however, analysis of the nasal respiratory epithelium which is the site of infection might offer advantages. Although data are limited, several recent papers demonstrate that nasopharyngeal host response can also be used as a diagnostic for respiratory viruses [93, 123, 124].

Influenza vaccine response

Immune response to influenza vaccine is variable and influenced by a variety of factors including prior vaccinations and infections, age, the presence of underlying conditions and the type of vaccine administered. Yet, even among a relatively homogeneous cohort of young healthy adults, antibody responses to vaccine can be variable [125]. Transcriptional profiling of whole blood provide insights into the mechanisms of variability, the effects of age, and vaccine types. The ability to predict vaccine response at baseline based on a transcriptomic signature would have significant clinical implications. To understand the biologic effects of live attenuated influenza vaccine (LAIV) compared to trivalent inactivated vaccine (TIV) blood transcription profiles from 85 young children were assessed by microarray at day 7 post vaccination [126]. Many more genes were differentially expressed in children receiving LAIV compared to TIV (245 vs. 49, respectively) and many modulated type 1 IFN. The efficacy of LAIV has been problematic in recent years and assessing stimulation of type 1 IFN genes could represent a potential biomarker for response to LAIV [126]. Bucasas and colleagues evaluated gene expression at multiple time points after vaccination of healthy young men with TIV [127]. They noted marked up regulation of gene expression of IFN signaling, IL-6 regulation, antigen processing and presentation genes within 24 h of vaccination and were able to define a 494 gene expression signature that correlated with the magnitude of antibody response. In another study, a gene profile predictive of antibody response 28 days after influenza vaccination of young and older adults was developed [128]. Notably, the predictive genes were the same in young and old as well as a subgroup of subjects with diabetes suggesting similar pathways were involved despite differences in age and underlying medical conditions. Additionally, transcriptional profiling has been used to signatures in blood associated with B cell memory responses to vaccination. In a study of 150 older and middle aged adults vaccinated with TIV including an H1N12009 antigen, metabolic, cell migration/adhesion, MAP kinase and NF-^{k}B cell genes correlated with peak memory B cell ELISPOTs [129]. Finally, in a study of over 500 subjects vaccinated over several seasons, a predictive signature of nine genes and three gene modules were significantly associated with the magnitude of the serum antibody response (Fig. 4) [130]. Interestingly and in contrast to a previous study, the signature was distinct to the younger cohort. For example, inflammatory genes were associated with better response in the young but a worse response in the elderly. In summary, gene expression studies could be used to evaluate new vaccines and develop predictors of vaccine response in different subgroups of patients based on age and disease state allowing for individualized vaccine regimens.

Conclusions

Molecular analysis of respiratory viruses and the host response to both infection and vaccination have transformed our understanding of these ubiquitous pathogens. The ability to accurately diagnosis viral infections has not only impacted patient care but also changed our perceptions of the burden of disease and populations effected. Transcriptional profiling of blood and nasal epithelium may provide therapeutic targets to prevent and ameliorate illness as well as offer predictors of severe disease.

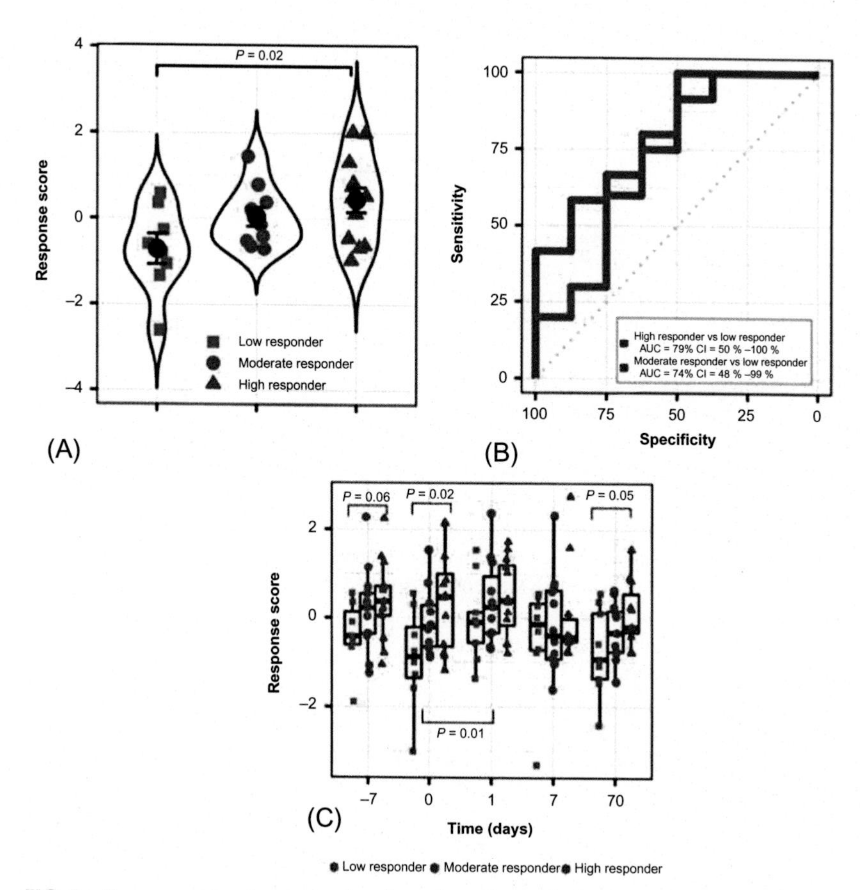

FIG. 4 Gene expression signatures at baseline as a predictor of influenza response in young adults. (A) The geographic mean of predictive genes (GRB2, ACTB, MVP, DPP7, ARPC4, PLEKHB2, ARRB1) z-scored expression values (response scores) was calculated for low, moderate and high responders. (B) Receiver Operator Characteristic (ROC) curves for classifiers designed to separate high from low to moderate responders. (C) Temporal behavior of response score in the validation cohort for low, moderate and high responders. Each point = individual subject and each group at a time point is summarized by a boxplot with significant *P* values for high vs. low responders above the date. *(Reproduced with permission from HIPC-CHI Signatures Project Team and HIPC-I Consortium. Multicohort analysis reveals baseline transcriptional predictors of influenza vaccination responses. Sci Immunol 2017;2(14).)*

References

[1] Bloom B, Cohen RA. Summary health statistics for U.S. children: National health interview survey. Vital Health Stat 10 2006; 2007. 234:1–79.

[2] Scheltema NM, Gentile A, Lucion F, Nokes DJ, Munywoki PK, Madhi SA, et al. Global respiratory syncytial virus-associated mortality in young children (RSV GOLD): a retrospective case series. Lancet Glob Health 2017;5(10):e984–91.

[3] Thompson WW, Shay DK, Weintraub E, Brammer L, Cox N, Anderson LJ, et al. Mortality associated with influenza and respiratory syncytial virus in the United States. JAMA 2003;289(2):179–86.
[4] Monto AS. Epidemiology of viral respiratory infections. Am J Med 2002;112(Suppl. 6A):4S–12S.
[5] Monto AS. Epidemiology of influenza. Vaccine 2008;26(Suppl. 4):D45–8.
[6] Fry AM, Shay DK, Holman RC, Curns AT, Anderson LJ. Trends in hospitalizations for pneumonia among persons aged 65 years or older in the United States, 1988-2002. JAMA 2005;294(21):2712–9.
[7] Zhou H, Thompson WW, Viboud CG, Ringholz CM, Cheng PY, Steiner C, et al. Hospitalizations associated with influenza and respiratory syncytial virus in the United States, 1993-2008. Clin Infect Dis 2012;54(10):1427–36.
[8] Thompson WW, Moore MR, Weintraub E, Cheng PY, Jin X, Bridges CB, et al. Estimating influenza-associated deaths in the United States. Am J Public Health 2009;99(Suppl. 2):S225–30.
[9] Oosterheert JJ, van Loon AM, Schuurman R, Hoepelman AI, Hak E, Thijsen S, et al. Impact of rapid detection of viral and atypical bacterial pathogens by real-time polymerase chain reaction for patients with lower respiratory tract infection. Clin Infect Dis 2005;41(10):1438–44.
[10] Walsh EE, Peterson DR, Falsey AR. Human metapneumovirus infections in adults: another piece of the puzzle. Arch Intern Med 2008;168(22):2489–96.
[11] Gorse GJ, O'Connor TZ, Hall SL, Vitale JN, Nichol KL. Human coronavirus and acute respiratory illness in older adults with chronic obstructive pulmonary disease. J Infect Dis 2009;199(6):847–57.
[12] Greenberg SB. Update on human rhinovirus and coronavirus infections. Semin Respir Crit Care Med 2016;37(4):555–71.
[13] Mahony JB. Detection of respiratory viruses by molecular methods. Clin Microbiol Rev 2008;21(4):716–47.
[14] Buller RS. Molecular detection of respiratory viruses. Clin Lab Med 2013;33(3):439–60.
[15] Falsey AR, Hennessey PA, Formica MA, Cox C, Walsh EE. Respiratory syncytial virus infection in elderly and high-risk adults. N Engl J Med 2005;352(17):1749–59.
[16] Johansson N, Kalin M, Tiveljung-Lindell A, Giske CG, Hedlund J. Etiology of community-acquired pneumonia: increased microbiological yield with new diagnostic methods. Clin Infect Dis 2010;50(2):202–9.
[17] van Elden LJ, van Kraaij MG, Nijhuis M, Hendriksen KA, Dekker AW, Rozenberg-Arska M, et al. Polymerase chain reaction is more sensitive than viral culture and antigen testing for the detection of respiratory viruses in adults with hematological cancer and pneumonia. Clin Infect Dis 2002;34(2):177–83.
[18] Jennings LC, Anderson TP, Beynon KA, Chua A, Laing RT, Werno AM, et al. Incidence and characteristics of viral community-acquired pneumonia in adults. Thorax 2008;63(1):42–8.
[19] Howard LM, Johnson M, Gil AI, Griffin MR, Edwards KM, Lanata CF, et al. Molecular epidemiology of rhinovirus detections in young children. Open Forum Infect Dis 2016;3(1):ofw001.
[20] Maitre NL, Williams JV. Human metapneumovirus in the preterm neonate: current perspectives. Res Rep Neonatol 2016;6:41–9.
[21] Johnstone J, Majumdar SR, Fox JD, Marrie TJ. Viral infection in adults hospitalized with community-acquired pneumonia: prevalence, pathogens, and presentation. Chest 2008;134(6):1141–8.
[22] Wang H, Deng J, Tang YW. Profile of the Alere i Influenza A & B assay: a pioneering molecular point-of-care test. Expert Rev Mol Diagn 2018;18(5):403–9.

[23] Lipkin WI, Firth C. Viral surveillance and discovery. Curr Opin Virol 2013;3(2):199–204.

[24] Chiu CY. Viral pathogen discovery. Curr Opin Microbiol 2013;16(4):468–78.

[25] Wang D, Urisman A, Liu YT, Springer M, Ksiazek TG, Erdman DD, et al. Viral discovery and sequence recovery using DNA microarrays. PLoS Biol 2003;1(2):E2.

[26] Iwane MK, Edwards KM, Szilagyi PG, Walker FJ, Griffin MR, Weinberg GA, et al. Population-based surveillance for hospitalizations associated with respiratory syncytial virus, influenza virus, and parainfluenza viruses among young children. Pediatrics 2004;113(6):1758–64.

[27] Jain S, Self WH, Wunderink RG, Team CES. Community-acquired pneumonia requiring hospitalization. N Engl J Med 2015;373(24):2382.

[28] Jartti T, Lehtinen P, Vuorinen T, Osterback R, van den Hoogen B, Osterhaus AD, et al. Respiratory picornaviruses and respiratory syncytial virus as causative agents of acute expiratory wheezing in children. Emerg Infect Dis 2004;10(6):1095–101.

[29] Woo PC, Lau SK, Chu CM, Chan KH, Tsoi HW, Huang Y, et al. Characterization and complete genome sequence of a novel coronavirus, coronavirus HKU1, from patients with pneumonia. J Virol 2005;79(2):884–95.

[30] Zaki AM, van Boheemen S, Bestebroer TM, Osterhaus AD, Fouchier RA. Isolation of a novel coronavirus from a man with pneumonia in Saudi Arabia. N Engl J Med 2012;367(19):1814–20.

[31] van den Hoogen BG, de Jong JC, Groen J, Kuiken T, de Groot R, Fouchier RA, et al. A newly discovered human pneumovirus isolated from young children with respiratory tract disease. Nat Med 2001;7(6):719–24.

[32] Peiris JS, Lai ST, Poon LL, Guan Y, Yam LY, Lim W, et al. Coronavirus as a possible cause of severe acute respiratory syndrome. Lancet 2003;361(9366):1319–25.

[33] Allander T, Tammi MT, Eriksson M, Bjerkner A, Tiveljung-Lindell A, Andersson B. Cloning of a human parvovirus by molecular screening of respiratory tract samples. Proc Natl Acad Sci USA 2005;102(36):12891–6.

[34] Ksiazek TG, Erdman D, Goldsmith CS, Zaki SR, Peret T, Emery S, et al. A novel coronavirus associated with severe acute respiratory syndrome. N Engl J Med 2003;348(20):1953–66.

[35] Prachayangprecha S, Schapendonk CM, Koopmans MP, Osterhaus AD, Schurch AC, Pas SD, et al. Exploring the potential of next-generation sequencing in detection of respiratory viruses. J Clin Microbiol 2014;52(10):3722–30.

[36] Zhao G, Wu G, Lim ES, Droit L, Krishnamurthy S, Barouch DH, et al. VirusSeeker, a computational pipeline for virus discovery and virome composition analysis. Virology 2017;503:21–30.

[37] Smith DJ, Lapedes AS, de Jong JC, Bestebroer TM, Rimmelzwaan GF, Osterhaus AD, et al. Mapping the antigenic and genetic evolution of influenza virus. Science 2004;305(5682):371–6.

[38] Bedford T, Suchard MA, Lemey P, Dudas G, Gregory V, Hay AJ, et al. Integrating influenza antigenic dynamics with molecular evolution. Elife 2014;3:e01914.

[39] Du Y, Chen M, Yang J, Jia Y, Han S, Holmes EC, et al. Molecular evolution and emergence of H5N6 avian influenza virus in Central China. J Virol 2017;91(12):e00143-17.

[40] Xue KS, Moncla LH, Bedford T, Bloom JD. Within-host evolution of human influenza virus. Trends Microbiol 2018;9:781–93.

[41] Dinis JM, Florek NW, Fatola OO, Moncla LH, Mutschler JP, Charlier OK, et al. Deep sequencing reveals potential antigenic variants at low frequencies in influenza A virus-infected humans. J Virol 2016;90(7):3355–65.

[42] Sobel Leonard A, McClain MT, Smith GJ, Wentworth DE, Halpin RA, Lin X, et al. Deep sequencing of influenza A virus from a human challenge study reveals a selective bottleneck and only limited intrahost genetic diversification. J Virol 2016;90(24):11247–58.

[43] Ghedin E, Holmes EC, DePasse JV, Pinilla LT, Fitch A, Hamelin ME, et al. Presence of oseltamivir-resistant pandemic A/H1N1 minor variants before drug therapy with subsequent selection and transmission. J Infect Dis 2012;206(10):1504–11.

[44] Rogers MB, Song T, Sebra R, Greenbaum BD, Hamelin ME, Fitch A, et al. Intrahost dynamics of antiviral resistance in influenza A virus reflect complex patterns of segment linkage, reassortment, and natural selection. MBio 2015;6(2):e02464-14.

[45] Grad YH, Newman R, Zody M, Yang X, Murphy R, Qu J, et al. Within-host whole-genome deep sequencing and diversity analysis of human respiratory syncytial virus infection reveals dynamics of genomic diversity in the absence and presence of immune pressure. J Virol 2014;88(13):7286–93.

[46] Cordey S, Junier T, Gerlach D, Gobbini F, Farinelli L, Zdobnov EM, et al. Rhinovirus genome evolution during experimental human infection. PLoS One 2010;5(5):e10588.

[47] Newport MJ, Finan C. Genome-wide association studies and susceptibility to infectious diseases. Brief Funct Genomics 2011;10(2):98–107.

[48] Khor CC, Hibberd ML. Host-pathogen interactions revealed by human genome-wide surveys. Trends Genet 2012;28(5):233–43.

[49] Juno J, Fowke KR, Keynan Y. Immunogenetic factors associated with severe respiratory illness caused by zoonotic H1N1 and H5N1 influenza viruses. Clin Dev Immunol 2012;2012:797180.

[50] Keynan Y, Malik S, Fowke KR. The role of polymorphisms in host immune genes in determining the severity of respiratory illness caused by pandemic H1N1 influenza. Public Health Genomics 2013;16(1–2):9–16.

[51] Miyairi I, DeVincenzo JP. Human genetic factors and respiratory syncytial virus disease severity. Clin Microbiol Rev 2008;21(4):686–703.

[52] Keynan Y, Juno J, Meyers A, Ball TB, Kumar A, Rubinstein E, et al. Chemokine receptor 5 big up tri, open32 allele in patients with severe pandemic (H1N1). Emerg Infect Dis 2009; 2010. 16(10):1621–2.

[53] Dawson TC, Beck MA, Kuziel WA, Henderson F, Maeda N. Contrasting effects of CCR5 and CCR2 deficiency in the pulmonary inflammatory response to influenza A virus. Am J Pathol 2000;156(6):1951–9.

[54] La D, Czarnecki C, El-Gabalawy H, Kumar A, Meyers AF, Bastien N, et al. Enrichment of variations in KIR3DL1/S1 and KIR2DL2/L3 among H1N1/09 ICU patients: an exploratory study. PLoS One 2011;6(12):e29200.

[55] Everitt AR, Clare S, Pertel T, John SP, Wash RS, Smith SE, et al. IFITM3 restricts the morbidity and mortality associated with influenza. Nature 2012;484(7395):519–23.

[56] Chan JF, To KK, Tse H, Lau CC, Li IW, Hung IF, et al. The lower serum immunoglobulin G2 level in severe cases than in mild cases of pandemic H1N1 2009 influenza is associated with cytokine dysregulation. Clin Vaccine Immunol 2011;18(2):305–10.

[57] Zuniga J, Buendia-Roldan I, Zhao Y, Jimenez L, Torres D, Romo J, et al. Genetic variants associated with severe pneumonia in A/H1N1 influenza infection. Eur Respir J 2012;39(3):604–10.

[58] Hitomi Y, Ebisawa M, Tomikawa M, Imai T, Komata T, Hirota T, et al. Associations of functional NLRP3 polymorphisms with susceptibility to food-induced anaphylaxis and aspirin-induced asthma. J Allergy Clin Immunol 2009;124(4). 779-85.e6.

[59] Verma D, Lerm M, Blomgran Julinder R, Eriksson P, Soderkvist P, Sarndahl E. Gene polymorphisms in the NALP3 inflammasome are associated with interleukin-1 production and severe inflammation: relation to common inflammatory diseases? Arthritis Rheum 2008;58(3):888–94.

[60] Boon AC, de Mutsert G, Graus YM, Fouchier RA, Sintnicolaas K, Osterhaus AD, et al. The magnitude and specificity of influenza A virus-specific cytotoxic T-lymphocyte responses in humans is related to HLA-A and -B phenotype. J Virol 2002;76(2):582–90.

[61] Ip WK, Chan KH, Law HK, Tso GH, Kong EK, Wong WH, et al. Mannose-binding lectin in severe acute respiratory syndrome coronavirus infection. J Infect Dis 2005;191(10): 1697–704.

[62] Zhang H, Zhou G, Zhi L, Yang H, Zhai Y, Dong X, et al. Association between mannose-binding lectin gene polymorphisms and susceptibility to severe acute respiratory syndrome coronavirus infection. J Infect Dis 2005;192(8):1355–61.

[63] Ching JC, Chan KY, Lee EH, Xu MS, Ting CK, So TM, et al. Significance of the myxovirus resistance A (MxA) gene -123C>a single-nucleotide polymorphism in suppressed interferon beta induction of severe acute respiratory syndrome coronavirus infection. J Infect Dis 2010;201(12):1899–908.

[64] He J, Feng D, de Vlas SJ, Wang H, Fontanet A, Zhang P, et al. Association of SARS susceptibility with single nucleic acid polymorphisms of OAS1 and MxA genes: a case-control study. BMC Infect Dis 2006;6:106.

[65] Hamano E, Hijikata M, Itoyama S, Quy T, Phi NC, Long HT, et al. Polymorphisms of interferon-inducible genes OAS-1 and MxA associated with SARS in the Vietnamese population. Biochem Biophys Res Commun 2005;329(4):1234–9.

[66] Lim JK, Lisco A, McDermott DH, Huynh L, Ward JM, Johnson B, et al. Genetic variation in OAS1 is a risk factor for initial infection with West Nile virus in man. PLoS Pathog 2009;5(2):e1000321.

[67] Hidaka F, Matsuo S, Muta T, Takeshige K, Mizukami T, Nunoi H. A missense mutation of the toll-like receptor 3 gene in a patient with influenza-associated encephalopathy. Clin Immunol 2006;119(2):188–94.

[68] Janssen R, Bont L, Siezen CL, Hodemaekers HM, Ermers MJ, Doornbos G, et al. Genetic susceptibility to respiratory syncytial virus bronchiolitis is predominantly associated with innate immune genes. J Infect Dis 2007;196(6):826–34.

[69] Stark JM, Barmada MM, Winterberg AV, Majumber N, Gibbons Jr. WJ, Stark MA, et al. Genomewide association analysis of respiratory syncytial virus infection in mice. J Virol 2010;84(5):2257–69.

[70] Larkin EK, Hartert TV. Genes associated with RSV lower respiratory tract infection and asthma: the application of genetic epidemiological methods to understand causality. Future Virol 2015;10(7):883–97.

[71] Hull J, Ackerman H, Isles K, Usen S, Pinder M, Thomson A, et al. Unusual haplotypic structure of IL8, a susceptibility locus for a common respiratory virus. Am J Hum Genet 2001;69(2):413–9.

[72] Hull J, Rowlands K, Lockhart E, Moore C, Sharland M, Kwiatkowski D. Variants of the chemokine receptor CCR5 are associated with severe bronchiolitis caused by respiratory syncytial virus. J Infect Dis 2003;188(6):904–7.

[73] Hull J, Thomson A, Kwiatkowski D. Association of respiratory syncytial virus bronchiolitis with the interleukin 8 gene region in UK families. Thorax 2000;55(12):1023–7.

[74] Wilson J, Rowlands K, Rockett K, Moore C, Lockhart E, Sharland M, et al. Genetic variation at the IL10 gene locus is associated with severity of respiratory syncytial virus bronchiolitis. J Infect Dis 2005;191(10):1705–9.

[75] Zhou J, To KK, Dong H, Cheng ZS, Lau CC, Poon VK, et al. A functional variation in CD55 increases the severity of 2009 pandemic H1N1 influenza A virus infection. J Infect Dis 2012;206(4):495–503.

[76] Statnikov A, Lytkin NI, McVoy L, Weitkamp JH, Aliferis CF. Using gene expression profiles from peripheral blood to identify asymptomatic responses to acute respiratory viral infections. BMC Res Notes 2010;3:264.

[77] Barton AJ, Hill J, Pollard AJ, Blohmke CJ. Transcriptomics in human challenge models. Front Immunol 2017;8:1839.

[78] Cilloniz C, Shinya K, Peng X, Korth MJ, Proll SC, Aicher LD, et al. Lethal influenza virus infection in macaques is associated with early dysregulation of inflammatory related genes. PLoS Pathog 2009;5(10):e1000604.

[79] Mejias A, Dimo B, Suarez NM, Garcia C, Suarez-Arrabal MC, Jartti T, et al. Whole blood gene expression profiles to assess pathogenesis and disease severity in infants with respiratory syncytial virus infection. PLoS Med 2013;10(11):e1001549. https://doi.org/10.1371/journal.pmed.1001549.g005.

[80] Reza Etemadi M, Ling KH, Zainal Abidin S, Chee HY, Sekawi Z. Gene expression patterns induced at different stages of rhinovirus infection in human alveolar epithelial cells. PLoS One 2017;12(5):e0176947.

[81] van den Kieboom CH, Ahout IM, Zomer A, Brand KH, de Groot R, Ferwerda G, et al. Nasopharyngeal gene expression, a novel approach to study the course of respiratory syncytial virus infection. Eur Respir J 2015;45(3):718–25.

[82] Gralinski LE, Bankhead 3rd A, Jeng S, Menachery VD, Proll S, Belisle SE, et al. Mechanisms of severe acute respiratory syndrome coronavirus-induced acute lung injury. MBio 2013;4(4):e00271-13.

[83] Josset L, Zeng H, Kelly SM, Tumpey TM, Katze MG. Transcriptomic characterization of the novel avian-origin influenza A (H7N9) virus: specific host response and responses intermediate between avian (H5N1 and H7N7) and human (H3N2) viruses and implications for treatment options. MBio 2014;5(1). e01102-13.

[84] Menachery VD, Eisfeld AJ, Schafer A, Josset L, Sims AC, Proll S, et al. Pathogenic influenza viruses and coronaviruses utilize similar and contrasting approaches to control interferon-stimulated gene responses. MBio 2014;5(3):e01174-14. https://doi.org/10.1128/mBio.01174-14.

[85] Levitz R, Gao Y, Dozmorov I, Song R, Wakeland EK, Kahn JS. Distinct patterns of innate immune activation by clinical isolates of respiratory syncytial virus. PLoS One 2017;12(9):e0184318.

[86] Guan W, Yang Z, Wu NC, Lee HHY, Li Y, Jiang W, Shen L, et al. Clinical correlations of transcriptional profile in patients infected with avian influenza H7N9 virus. J Infect Dis 2018;218:1238–48.

[87] Bochkov YA, Hanson KM, Keles S, Brockman-Schneider RA, Jarjour NN, Gern JE. Rhinovirus-induced modulation of gene expression in bronchial epithelial cells from subjects with asthma. Mucosal Immunol 2010;3(1):69–80.

[88] Gern JE. How rhinovirus infections cause exacerbations of asthma. Clin Exp Allergy 2015;45(1):32–42.

[89] Gardeux V, Berghout J, Achour I, Schissler AG, Li Q, Kenost C, et al. A genome-by-environment interaction classifier for precision medicine: personal transcriptome response to rhinovirus identifies children prone to asthma exacerbations. J Am Med Inform Assoc 2017;24(6):1116–26.

[90] Wesolowska-Andersen A, Everman JL, Davidson R, Rios C, Herrin R, Eng C, et al. Dual RNA-seq reveals viral infections in asthmatic children without respiratory illness which are associated with changes in the airway transcriptome. Genome Biol 2017;18(1):12.

[91] Heinonen S, Jartti T, Garcia C, Oliva S, Smitherman C, Anguiano E, et al. Rhinovirus detection in symptomatic and asymptomatic children: value of host transcriptome analysis. Am J Respir Crit Care Med 2016;193(7):772–82.
[92] Mariani TJ, Qiu X, Chu C, Wang L, Thakar J, Holden-Wiltse J, et al. Association of dynamic changes in the CD4 T-cell transcriptome with disease severity during primary respiratory syncytial virus infection in young infants. J Infect Dis 2017;216(8):1027–37.
[93] Do LAH, Pellet J, van Doorn HR, Tran AT, Nguyen BH, Tran TTL, et al. Host transcription profile in nasal epithelium and whole blood of hospitalized children under 2 years of age with respiratory syncytial virus infection. J Infect Dis 2017;217(1):134–46.
[94] Morrison J, Katze MG. Gene expression signatures as a therapeutic target for severe H7N9 influenza—what do we know so far? Expert Opin Ther Targets 2015;19(4):447–50.
[95] de Steenhuijsen Piters WA, Heinonen S, Hasrat R, Bunsow E, Smith B, Suarez-Arrabal MC, et al. Nasopharyngeal microbiota, host transcriptome, and disease severity in children with respiratory syncytial virus infection. Am J Respir Crit Care Med 2016;194(9):1104–15.
[96] Hoang LT, Tolfvenstam T, Ooi EE, Khor CC, Naim AN, Ho EX, et al. Patient-based transcriptome-wide analysis identify interferon and ubiquination pathways as potential predictors of influenza A disease severity. PLoS One 2014;9(11):e111640.
[97] Proud D, Turner RB, Winther B, Wiehler S, Tiesman JP, Reichling TD, et al. Gene expression profiles during in vivo human rhinovirus infection: insights into the host response. Am J Respir Crit Care Med 2008;178(9):962–8.
[98] McClain MT, Nicholson BP, Park LP, Liu TY, Hero 3rd AO, Tsalik EL, et al. A genomic signature of influenza infection shows potential for presymptomatic detection, guiding early therapy, and monitoring clinical responses. Open Forum Infect Dis 2016;3(1):ofw007.
[99] Warren-Gash C, Hayward AC, Hemingway H, Denaxas S, Thomas SL, Timmis AD, et al. Influenza infection and risk of acute myocardial infarction in England and Wales: a CALIBER self-controlled case series study. J Infect Dis 2012;206(11):1652–9.
[100] Kwong JC, Schwartz KL, Campitelli MA, Chung H, Crowcroft NS, Karnauchow T, et al. Acute myocardial infarction after laboratory-confirmed influenza infection. N Engl J Med 2018;378(4):345–53.
[101] Gozalo PL, Pop-Vicas A, Feng Z, Gravenstein S, Mor V. Effect of influenza on functional decline. J Am Geriatr Soc 2012;60(7):1260–7.
[102] Barker WH, Borisute H, Cox C. A study of the impact of influenza on the functional status of frail older people. Arch Intern Med 1998;158(6):645–50.
[103] Rose JJ, Voora D, Cyr DD, Lucas JE, Zaas AK, Woods CW, et al. Gene expression profiles link respiratory viral infection, platelet response to aspirin, and acute myocardial infarction. PLoS One 2015;10(7):e0132259.
[104] Zaas AK, Garner BH, Tsalik EL, Burke T, Woods CW, Ginsburg GS. The current epidemiology and clinical decisions surrounding acute respiratory infections. Trends Mol Med 2014;20(10):579–88.
[105] Zaas AK, Chen M, Varkey J, Veldman T, Hero 3rd AO, Lucas J, et al. Gene expression signatures diagnose influenza and other symptomatic respiratory viral infections in humans. Cell Host Microbe 2009;6(3):207–17.
[106] Parnell GP, McLean AS, Booth DR, Armstrong NJ, Nalos M, Huang SJ, et al. A distinct influenza infection signature in the blood transcriptome of patients with severe community-acquired pneumonia. Crit Care 2012;16(4):R157.
[107] Harris AM, Hicks LA, Qaseem A. Appropriate antibiotic use for acute respiratory tract infection in adults. Ann Intern Med 2016;165(9):674.

[108] Lautenbach E, Lee I, Shiley KT. Treating viral respiratory tract infections with antibiotics in hospitals: no longer a case of mistaken identity. LDI Issue Brief 2010;16(3):1–4.

[109] Falsey AR, Becker KL, Swinburne AJ, Nylen ES, Formica MA, Hennessey PA, et al. Bacterial complications of respiratory tract viral illness: a comprehensive evaluation. J Infect Dis 2013;208(3):432–41.

[110] Brendish NJ, Malachira AK, Armstrong L, Houghton R, Aitken S, Nyimbili E, et al. Routine molecular point-of-care testing for respiratory viruses in adults presenting to hospital with acute respiratory illness (ResPOC): a pragmatic, open-label, randomised controlled trial. Lancet Respir Med 2017;5:P401–11.

[111] Muller B, Harbarth S, Stolz D, Bingisser R, Mueller C, Leuppi J, et al. Diagnostic and prognostic accuracy of clinical and laboratory parameters in community-acquired pneumonia. BMC Infect Dis 2007;7:10.

[112] Tenover FC. Developing molecular amplification methods for rapid diagnosis of respiratory tract infections caused by bacterial pathogens. Clin Infect Dis 2011;52(Suppl 4):S338–45.

[113] Chaussabel D, Pascual V, Banchereau J. Assessing the human immune system through blood transcriptomics. BMC Biol 2010;8:84.

[114] Suarez NM, Bunsow E, Falsey AR, Walsh EE, Mejias A, Ramilo O. Superiority of transcriptional profiling over procalcitonin for distinguishing bacterial from viral lower respiratory tract infections in hospitalized adults. J Infect Dis 2015;212(2):213–22.

[115] Ramilo O, Allman W, Chung W, Mejias A, Ardura M, Glaser C, et al. Gene expression patterns in blood leukocytes discriminate patients with acute infections. Blood 2007;109(5):2066–77.

[116] Hu X, Yu J, Crosby SD, Storch GA. Gene expression profiles in febrile children with defined viral and bacterial infection. Proc Natl Acad Sci USA 2013;110(31):12792–7.

[117] Sweeney TE, Wong HR, Khatri P. Robust classification of bacterial and viral infections via integrated host gene expression diagnostics. Sci Transl Med 2016;8(346):346ra91.

[118] Tsalik EL, Henao R, Nichols M, Burke T, Ko ER, McClain MT, et al. Host gene expression classifiers diagnose acute respiratory illness etiology. Sci Transl Med 2016;8(322):322ra11.

[119] Bhattacharya S, Rosenberg AF, Peterson DR, Grzesik K, Baran AM, Ashton JM, et al. Transcriptomic biomarkers to discriminate bacterial from nonbacterial infection in adults hospitalized with respiratory illness. Sci Rep 2017;7(1):6548.

[120] Herberg JA, Kaforou M, Wright VJ, Shailes H, Eleftherohorinou H, Hoggart CJ, et al. Diagnostic test accuracy of a 2-transcript host RNA signature for discriminating bacterial vs viral infection in febrile children. JAMA 2016;316(8):835–45.

[121] Mahajan P, Kuppermann N, Mejias A, Suarez N, Chaussabel D, Casper TC, et al. Association of RNA biosignatures with bacterial infections in febrile infants aged 60 days or younger. JAMA 2016;316(8):846–57.

[122] Zaas AK, Burke T, Chen M, McClain M, Nicholson B, Veldman T, et al. A host-based RT-PCR gene expression signature to identify acute respiratory viral infection. Sci Transl Med 2013;5(203):203ra126.

[123] Yahya M, Rulli M, Toivonen L, Waris M, Peltola V. Detection of host response to viral respiratory infection by measurement of messenger RNA for MxA, TRIM21, and viperin in nasal swabs. J Infect Dis 2017;216(9):1099–103.

[124] Landry ML, Foxman EF. Antiviral response in the nasopharynx identifies patients with respiratory virus infection. J Infect Dis 2018;217(6):897–905.

[125] Franco LM, Bucasas KL, Wells JM, Nino D, Wang X, Zapata GE, et al. Integrative genomic analysis of the human immune response to influenza vaccination. Elife 2013;2:e00299.

[126] Zhu W, Higgs BW, Morehouse C, Streicher K, Ambrose CS, Woo J, et al. A whole genome transcriptional analysis of the early immune response induced by live attenuated and inactivated influenza vaccines in young children. Vaccine 2010;28(16):2865–76.

[127] Bucasas KL, Franco LM, Shaw CA, Bray MS, Wells JM, Nino D, et al. Early patterns of gene expression correlate with the humoral immune response to influenza vaccination in humans. J Infect Dis 2011;203(7):921–9.

[128] Nakaya HI, Wrammert J, Lee EK, Racioppi L, Marie-Kunze S, Haining WN, et al. Systems biology of vaccination for seasonal influenza in humans. Nat Immunol 2011;12(8):786–95.

[129] Haralambieva IH, Ovsyannikova IG, Kennedy RB, Zimmermann MT, Grill DE, Oberg AL, et al. Transcriptional signatures of influenza A/H1N1-specific IgG memory-like B cell response in older individuals. Vaccine 2016;34(34):3993–4002.

[130] HIPC-CHI Signatures Project Team, HIPC-I Consortium. Multicohort analysis reveals baseline transcriptional predictors of influenza vaccination responses. Sci Immunol 2017;2(14):1–28.

Chapter 10

Emerging viral infections

Daisy D. Colón-López, Christopher P. Stefan, Jeffrey W. Koehler
Diagnostic Systems Division, United States Army Medical Research Institute of Infectious Diseases, Fort Detrick, Frederick, MD, United States.

Introduction

Human populations living in close contact with their environment have continuously been exposed to emerging and reemerging infectious diseases. In many cases, localized outbreaks occur [1–4] that can lead to occasional, large scale epidemics and pandemics [5–8]. A well-known example of an emerging virus resulting in a pandemic is the human immunodeficiency virus (HIV-1). After jumping to humans from chimpanzees in Cameroon [9], this virus was able to maintain sustained, undetected person-to-person transmission in Kinshasa in the 1920s due to a variety of factors including high population densities [10]. The mobility of the population by well-connected railways allowed the virus to spread within Africa, and changing social behaviors such as injectable drug use and the sex trade allowed the virus to explode into a pandemic [10].

As mobile human populations become increasingly dense at this pathogen/host/environment interface, the risk of localized outbreaks becoming regional or world-wide epidemics is increasing. For instance, the 2014–2016 Ebola virus (Fig. 1) outbreak in West Africa originated in a rural region of Guinea [2]; travel, population density, and regional customs allowed the outbreak to become the largest known Ebola virus outbreak with multiple cases reaching other countries including Senegal [11], Nigeria [12], Mali [13], the United Kingdom [14], Italy [15], Spain [16], and the United States [17]. Of particular concern was a localized transmission chain in Lagos, Nigeria, that could have had an explosive impact in that country [12,18,19].

Similarly, Zika virus was first identified in 1947 [20] and caused limited infections [3,4] until its emergence in Micronesia in 2007 [5], French Polynesia in 2013 [6], and the Americas in 2015 [7,8]. In both of these cases, numerous factors had a role in enabling the rapid viral transmission of previously isolated infectious diseases. Besides increasing population density and ease of transportation, climate and ecological changes that expand a vector's range or habitat can lead to novel epidemics. For example, infections of Lassa virus, spread by

Genomic and Precision Medicine. https://doi.org/10.1016/B978-0-12-801496-7.00010-1

the *Mastomys natalensis* rat, are increased in West Africa during the dry seasons when the rats are more likely to be inside individuals' houses [21].

Fluctuations in climate resulting in increasing crop yields, and the associated rat population, followed by a severe drought can lead to increased cases of Lassa fever. Furthermore, climactic changes are altering vector-borne infectious disease risks due to expanded vector ranges. For example, there is an increased risk of Lyme disease in Canada due to tick range expansion [22], and climate change is expanding the mosquito range, and the associated infectious disease risk, of *Aedes albopictus* in the United States [23].

RNA viruses represent a significant source of viral outbreaks due partly to the higher mutation rate and error prone nature of RNA replication itself [24]. Newly introduced mutations into a viral genome can confer new pathogenic characteristics such as enabling transmission to new species or increased virulence. For instance, mutations within the severe acute respiratory syndrome coronavirus (SARS-CoV) genome allowed transmission from bats to humans [24,25]. The better understanding about the viral agents responsible of causing outbreaks in humans and the underlying factors driving emergence, the better the chances of preparedness and effective intervention for prevention of a new global epidemic (Figs. 1 and 2).

This chapter provides an overview of the genomics and precision medicine methodologies relevant to the topic of emerging and re-emerging viral threats. The chapter is divided up into two main sections: viral genomics and host genomics, and include examples of their applications in pathogen detection, population susceptibility, diagnostics, and outbreak response.

Viral genomics

The seminal investigations on the transmission of yellow fever virus by Walter Reed and colleagues in 1901 [27,28] ushered in an era of discovery and classification of a wide variety of viral pathogens including smallpox virus, poliovirus, and measles virus. Technological advancements including viral cell culture, plaque assays, enzyme-linked immunosorbent assays, and monoclonal antibodies pushed human virology forward. A second wave of viral discovery and characterization is currently ongoing, harvesting the power of next-generation sequencing to identify unknown viruses in acute febrile illness patients [29–31]. Increasingly, efforts are underway to identify the next emerging pathogen by sequencing and identifying the viruses circulating at the human/animal/pathogen interface [32,33]. While detection of viruses in wildlife does not predict transmission, identifying mutations or gene acquisitions through surveillance and continued discovery will provide valuable insight in the future.

Pathogen discovery

A major hindrance for precision medicine toward emerging zoonotic pathogens is the limited global characterization of potential threats. Taxonomic placement

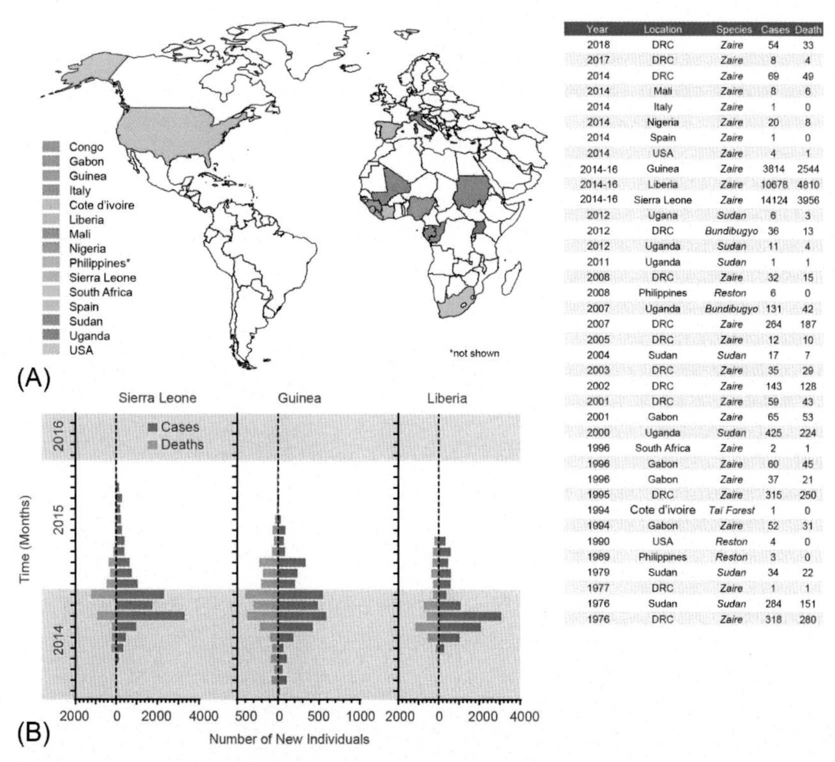

Year	Location	Species	Cases	Death
2018	DRC	*Zaire*	54	33
2017	DRC	*Zaire*	8	4
2014	DRC	*Zaire*	69	49
2014	Mali	*Zaire*	8	6
2014	Italy	*Zaire*	1	0
2014	Nigeria	*Zaire*	20	8
2014	Spain	*Zaire*	1	0
2014	USA	*Zaire*	4	1
2014-16	Guinea	*Zaire*	3814	2544
2014-16	Liberia	*Zaire*	10678	4810
2014-16	Sierra Leone	*Zaire*	14124	3956
2012	Ugana	*Sudan*	6	3
2012	DRC	*Bundibugyo*	36	13
2012	Uganda	*Sudan*	11	4
2011	Uganda	*Sudan*	1	1
2008	DRC	*Zaire*	32	15
2008	Philippines	*Reston*	6	0
2007	Uganda	*Bundibugyo*	131	42
2007	DRC	*Zaire*	264	187
2005	DRC	*Zaire*	12	10
2004	Sudan	*Sudan*	17	7
2003	DRC	*Zaire*	35	29
2002	DRC	*Zaire*	143	128
2001	DRC	*Zaire*	59	43
2001	Gabon	*Zaire*	65	53
2000	Uganda	*Sudan*	425	224
1996	South Africa	*Zaire*	2	1
1996	Gabon	*Zaire*	60	45
1996	Gabon	*Zaire*	37	21
1995	DRC	*Zaire*	315	250
1994	Cote d'ivoire	*Taï Forest*	1	0
1994	Gabon	*Zaire*	52	31
1990	USA	*Reston*	4	0
1989	Philippines	*Reston*	3	0
1979	Sudan	*Sudan*	34	22
1977	DRC	*Zaire*	1	1
1976	Sudan	*Sudan*	284	151
1976	DRC	*Zaire*	318	280

FIG. 1 Ebola virus outbreaks overtime. (A) Global map showing the countries where Ebola virus cases and deaths have been reported over the last 42 years. The information about each outbreak is shown on the right. As of November 2018, there is an ongoing Ebola virus outbreak in DRC and the data on the table does not reflect the final outcome. (B) The graphs show the number of new cases and deaths per month during the 2014 Ebola virus outbreaks in Sierra Leone, Guinea, and Liberia.*Panel (A): Data adapted from www.cdc.gov. Panel (B): Data adapted from www.who.int.*

of viruses is problematic due to the lack of universal genes, such as the bacterial 16S RNA gene, limiting classification to viruses with defined biological characteristics [34]. In fact a mere 5000 total species are currently classified by the International Committee on Taxonomy of Viruses (ICTV) which pales in comparison to the estimated 320,000 viral species infecting mammals alone [35]. Metagenomic sequencing has highlighted this significant gap in classified vs unknown viromes, and while biological characterization will not also always be defined, knowledge can be gained from genomic sequences including evolutionary relationships and genome structure [34]. Continuing to gather viral genomic information from environmental, plant, and animal samples will continue to fill gaps in our knowledge base and further prepare us for detecting and predicting emerging pathogens.

Metagenomic sequencing has been used for the discovery of previously unclassified human pathogens. In 2008 Lujo virus was discovered in South Africa

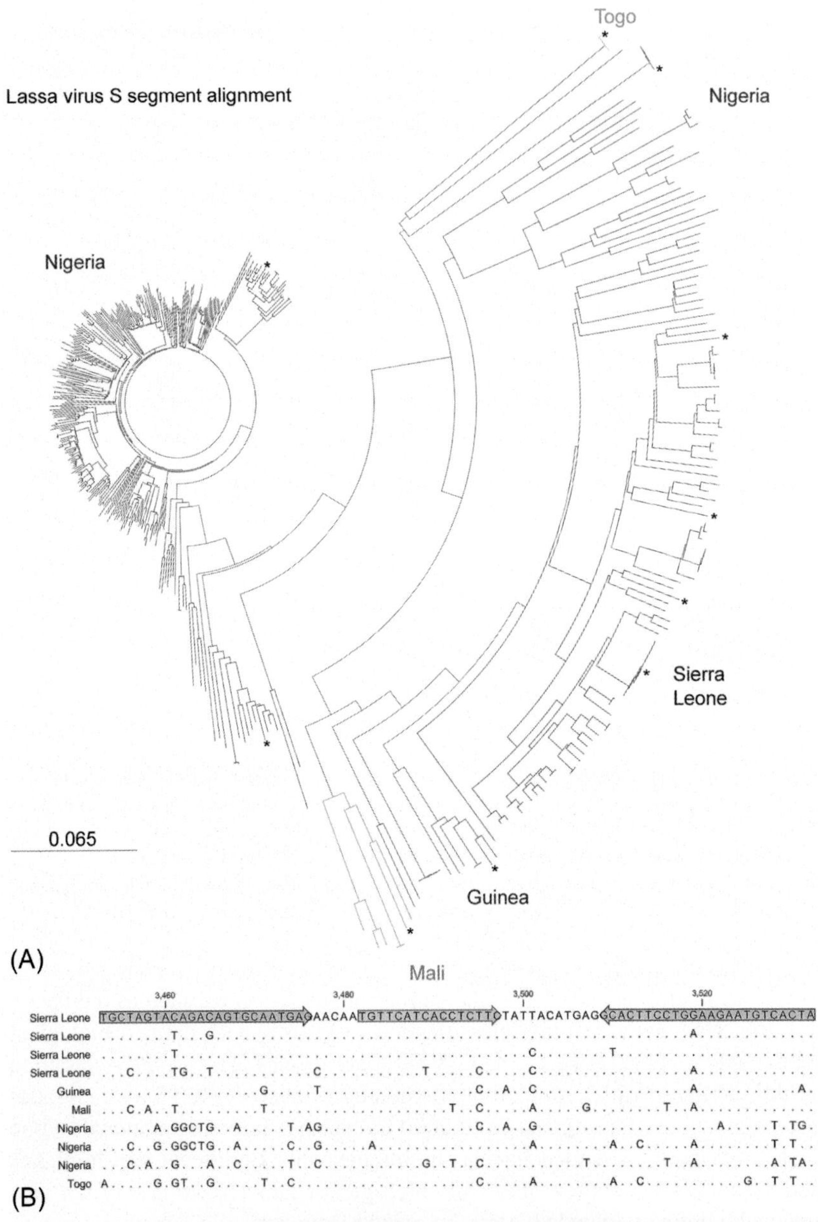

FIG. 2 Sequence diversity of Lassa virus complicates molecular diagnostic tool development. A phylogenetic tree (A) was generated using and alignment of full length and near full length Lassa virus S segment nucleic acid sequences available in GenBank. Diversity of Lassa virus is linked to geographic location as shown here for viruses sequenced from Nigeria, Sierra Leone, Guinea, Mali, and Togo. Selected sequences from each country [starred in (A)] were aligned in (B). The forward and reverse primers (red arrows) and the probe (green arrow) from a previously published Lassa virus Josiah real-time RT-PCR assay [26] are indicated. The primers and probe are exact matches to the Lassa virus Josiah sequence; however, contain mismatches within other Lassa virus isolates, including ones found in Sierra Leone.

[31]. Five patients with undiagnosed hemorrhagic fever were discovered after air transfer of a critically ill index case. The disease resulted in a case fatality rate of 80%. Unbiased pyrosequencing of RNA extracts from human samples identified approximately 50% of an arenavirus genome which was thereafter gap filled using primers designed off the sequenced genome. Phylogenetic analysis confirmed the identification of a novel Old World arenavirus [31]. Similarly, deep sequencing identified Bas-Congo virus, of a novel rhabdovirus, in three human cases in the Democratic Republic of Congo [30]. This virus was associated with high fever and rapid death and was found to have only 34% amino-acid identity with other rhabdoviruses. The discovery and characterization of these viruses has allowed novel diagnostics to be developed, including real-time PCR, leading to a better preparedness for future outbreaks.

Sequencing human samples from outbreaks has resulted in the discovery of several viruses; however, 70% of emerging human pathogens result from wildlife [36]. Analysis of samples from non-human hosts including arthropods, bats, rodents, and domestic animals offers potential insight into the transmission, evolution, and treatment of these pathogens [37]. Deep sequencing on more than 220 invertebrate species resulted in the discovery of 1445 phylogenetically distinct viromes filling in phylogenetic and evolutionary gaps [38]. Characterization of bacterial and viral relationships in mosquito arthropods demonstrated a symbiotic relationship between the bacterium and host, limiting dengue virus infection and potentially revealing new antiviral strategies [39,40]. The discovery of Middle-East Respiratory Syndrome-Coronavirus (MERS-CoV) in camels demonstrated these animals as a potential reservoir and host for transmission [41,42]. While detection of viruses in wildlife does not predict transmission, identifying mutations or gene acquisitions through surveillance and continued discovery will provide valuable insight in the future.

Epidemiology and outbreak response

In addition to pathogen discovery, metagenomic sequencing is increasingly used to monitor viral genomic changes as an outbreak is occurring [43–46]. The Ebola virus outbreak in West Africa resulted in 26,648 cases and 11,017 documented deaths, and genomic sequencing was applied in near real-time to provide information to aid in containing the outbreak [44,45]. Sequencing results early in the outbreak provided valuable insight into origination and transmission routes demonstrating the outbreak started from a single introduction in Guinea in December 2013 and was sustained by human to human transmissions [2]. Sequencing provided molecular evidence that Ebola virus was transmissible by sexual intercourse leading to changes in CDC recommendations for survivors [45,47] and establishing programs to support national testing of semen and other body fluids in male survivors [48]. Rapid outbreak sequencing allowed for the identification of transmission chains in sporadic clusters following the outbreak [46], further adding insight to end the epidemic.

Sequencing data from Sierra Leone early in the outbreak also found increased Ebola virus diversity that could have an impact on sequence-based therapeutics, vaccines, and diagnostic assays being fielded at the time [44]. Having such rapid sequencing information during an outbreak would inform responders about the potential efficacy of diagnostics and sequence-based countermeasures, either reassuring decision makers or informing them of the need to modify strategies based on the findings.

Diagnostics and therapeutic response

While sequencing has found a niche in the discovery and epidemiologic tracing of emerging infections, its roll for diagnostics is still in its infancy. Emerging pathogens often occur in austere environments or countries with limited infrastructure not conducive to sequencing technologies. Sequencing still has significant hurdles in decreasing the technical and mechanical requirements for routine clinical use. However, next-generation sequencing offers limitless potential due the necessity for little to no prior knowledge in sample composition. Newer technologies such as nanopore sequencing are looking to minimize these hurdles by creating smaller foot-prints and near real-time sequence analysis with minimal sample preparation [49]; however, difficult paths to their acceptance beyond laboratory derived tests to full regulatory approval remain.

Knowledge of viral genomic sequences in designing rapid point of care molecular diagnostics such as PCR is invaluable. The last decade has seen a significant increase in the use of PCR as a primary diagnostic due to its speed, sensitivity, and cost. However, due to the lack of available genomic sequences and high genetic variation within certain viral species, finding a conserved target can be difficult. The advent of multiplex PCR has allowed more targets to be captured in a single assay, lowering costs and increasing throughput. For example, this technique has been used to subtype influenza A and B viruses [50]. Other diagnostic devices such as the BioFire FilmArray can run multiple PCRs at once on a single device allowing the user to select between different panels [51–55]. Whether singleplex or multiplex, PCR remains the most rapid and sensitive method for the detection of viral genomes directly from clinical matrices.

Coordinating with detection, viral genome sequences are increasingly being used to predict and guide antiviral therapy. During the Ebola virus outbreak, sequence analysis of the viral genome over time demonstrated changes which could make the pathogen resistant to therapeutics such as siRNAs, phosphorodiamidate morpholino oligomers (PMOs), and antibodies [56]. The discovery of the CRISPR/Cas9 system and its ability to destroy dsDNA has brought to light its potential in mutating essential sites or removing proviral DNA from infected cells during and HIV infection CRISPR [57]. Similarly studies involving herpesviruses, large dsDNA viruses, have demonstrated the ability to destroy latent herpesvirus from infected cells [58,59]. These methods require prior knowledge of the viral genome promising a future where antiviral therapies are targeted specifically toward an individual's own specific infection.

Host genomics

Emerging and re-emerging viral diseases pose unique challenges to the use of omics and precision medicine tools. How does one predict when and where a pathogen will jump a species barrier or emerge into a new population that could rapidly spread into new geographic regions? To further complicate the scenario, a new human viral infection may not present with obvious signs of an infection. Such a virus may get introduced and transmitted across populations while remaining asymptomatic or unrecognized for long periods of times, like the hepatitis C virus (HCV) and HIV [10], for instance.

Viral infections and host responses

Viruses rely on their small-sized genomes to encode enough information to hijack the host's cellular machinery for target-cell recognition and entry, genome replication, protein synthesis, viral participle assembly, and propagation. Evolutionarily, hosts evolved conserved mechanisms to generate an immune response to detect and limit an infection and to prevent re-infection in the future. The early innate immune response functions to slow the infection once the immune system recognizes there is an infection (extensively reviewed in [60]) and guides the subsequent adaptive immune response to the pathogen.

Successful control of these immunological processes depend on tight control of the host's immunological and inflammatory pathways by regulating gene transcription. These processes result in measurable changes in the relative expression levels of coding and non-coding RNA and protein, even when an infection is asymptomatic [61–65] or difficult to diagnose [66]. The development of low-cost and less-time consuming genomics had facilitated the study of pathway-specific transcript alterations during viral infections. Consequently, various transcriptomic-based assays have been developed to study, characterize, and identify signs of infection using transcript signatures. For example, Zaas and colleagues identified a unique immunological gene expression profile capable of differentiating viral from bacterial respiratory infections in humans [64,65], laying to foundations for appropriate antibiotic use without direct pathogen identification.

Host-based diagnostics and genomics approaches

Traditionally, diagnosing viral infections is made based on clinical symptoms, PCR and serological testing, and virus isolation. While the world of omics is expanding with new and improved platforms, host transcriptome-based assays are becoming more feasible and widely available. Those of more relevance to this chapter are targeted and agnostic next-generation sequencing (NGS) platforms for genome-wide association studies (GWAS) and transcriptomics as well as microarray-based assays for amplification-independent profiling of transcripts. GWAS have direct applications to precision medicine whereas host-based transcriptomic assays have shown to have multiple applications for clinical diagnosis, and pathogen identification and characterization. These technologies were not specifically developed for emerging or re-emerging viruses but can be leveraged for these purposed based on their applications.

GWAS may provide insights into the genetic factors driving population susceptibility to viral infections [67,68]. This information can be utilized to identify populations within geographic regions at higher or lower risk of being infected with a new pathogen. Higher-risk populations may include those with increased direct exposure to animal reservoirs of zoonotic pathogens or their arthropod vectors. Other population subgroups without defined topographical relationships can be determined as well. A few examples of these are demographic, age and, gender groups that can be classified based on genetic polymorphisms and other susceptibility markers. In this setting, GWAS can provide guidance in outbreak preparedness and intervention.

Transcriptomic methods

The information provided by GWAS-based studies is not focused on the directed response to an infection but on the information already stored within the host's genome. On the other hand, transcriptomic approaches identifies the interactions between the pathogen and the host's genome by evaluating transcript levels, typically mRNAs. The most commonly used transcriptomic methods are RNA-seq [69,70] and microarrays [65,66]. RNA-seq is generally suitable for unbiased transcriptomic profiling and provides better understanding of global transcriptomic changes. This agnostic method is appropriate for identifying changes in the human transcriptome as a result of an emerging viral infection to show specific mechanisms of immune response evasion and other effects in the host's biology at the transcriptomic level. Conversely, targeted-approaches depend on previous knowledge of host-transcriptomics.

Studies with human cohorts and animal models have developed gene signatures that discriminate between viral and bacterial respiratory infections [61–65]. Other groups have developed sepsis classifiers to guide treatment options in clinical settings [71,72]. A goal of developing these disease classifiers is to use them to predict disease outcome prior to the onset of symptoms and to assist in decision making when a symptomatic disease with an unknown etiology is suspected. Effective and accurate discernment between a viral and a bacterial infection can make the difference in administering the appropriate therapeutics and potentially saving lives. When a specific viral species can't be identified, a gene classifier may be useful to determine potential disease outcomes such as lethality.

Closing remarks

As technology continues to advance, the linkages of genomics and precision medicine with emerging infectious diseases will continue to strengthen. Improvements with sequencing accuracy and speed are pushing pathogen-specific genomics to near real-time [73,74], allowing rapid access to critical information within a timeframe that can impact an outbreak response [44,45]. Confidence in the results from sequencing is leading to clinically actionable diagnostic patient testing [29], and it is reasonable to foresee rapid, pathogen agnostic diagnostic assays routinely utilized in the clinic.

Population growth, expansion into rural geographic regions, and rapid travel has greatly expanded the pathogen/host/environment interactome, putting a greater number of people at risk for formerly exotic infectious diseases such as Ebola virus or even the next unknown, soon-to-emerge pathogen. Applying genomics and the host transcriptome to generalize the response to a viral or bacterial pathogen would allow for more appropriate antibiotic use without requiring direct pathogen detection. Besides the benefits to antibiotic stewardship in an era where antimicrobial resistance is expanding rapidly, such an application can greatly improve clinical care in the event of an emerging, highly virulent unknown unknown organism.

Disclaimer

Opinions, interpretations, conclusions, and recommendations are those of the author and are not necessarily endorsed by the U.S. Army.

References

[1] Report of an International Commission. Ebola haemorrhagic fever in Zaire. Bull World Health Organ 1976;56(2):271–93. 1978, 307456. PubMed Central PMCID. PMC2395567.

[2] Baize S, Pannetier D, Oestereich L, Rieger T, Koivogui L, Magassouba N, et al. Emergence of Zaire Ebola virus disease in Guinea. N Engl J Med 2014;371(15):1418–25. https://doi.org/10.1056/NEJMoa1404505, 24738640.

[3] Macnamara FN. Zika virus: a report on three cases of human infection during an epidemic of jaundice in Nigeria. Trans R Soc Trop Med Hyg 1954;48(2):139–45, 13157159.

[4] Olson JG, Ksiazek TG, Suhandiman T. Zika virus, a cause of fever in Central Java, Indonesia. Trans R Soc Trop Med Hyg 1981;75(3):389–93, 6275577.

[5] Duffy MR, Chen TH, Hancock WT, Powers AM, Kool JL, Lanciotti RS, et al. Zika virus outbreak on Yap Island, federated states of Micronesia. N Engl J Med 2009;360(24):2536–43. https://doi.org/10.1056/NEJMoa0805715, 19516034.

[6] Cao-Lormeau VM, Roche C, Teissier A, Robin E, Berry AL, Mallet HP, et al. Zika virus, French polynesia, South pacific, 2013. Emerg Infect Dis 2014;20(6):1085–6. https://doi.org/10.3201/eid2006.140138, 24856001. PubMed Central PMCID. PMCPMC4036769.

[7] Campos GS, Bandeira AC, Sardi SI. Zika virus outbreak, Bahia, Brazil. Emerg Infect Dis 2015;21(10):1885–6. https://doi.org/10.3201/eid2110.150847, 26401719. PubMed Central PMCID. PMCPMC4593454.

[8] Hennessey M, Fischer M, Staples JE. Zika virus spreads to new areas—region of the Americas, may 2015–january 2016. MMWR Morb Mortal Wkly Rep 2016;65(3):55–8. https://doi.org/10.15585/mmwr.mm6503e1, 26820163.

[9] Keele BF, Van Heuverswyn F, Li Y, Bailes E, Takehisa J, Santiago ML, et al. Chimpanzee reservoirs of pandemic and nonpandemic HIV-1. Science 2006;313(5786):523–6. https://doi.org/10.1126/science.1126531, 16728595. PubMed Central PMCID. PMCPMC2442710.

[10] Faria NR, Rambaut A, Suchard MA, Baele G, Bedford T, Ward MJ, et al. HIV epidemiology. The early spread and epidemic ignition of HIV-1 in human populations. Science 2014;346(6205):56–61. https://doi.org/10.1126/science.1256739, 25278604. PubMed Central PMCID. PMCPMC4254776.

[11] Mirkovic K, Thwing J, Diack PA. Centers for disease C, prevention. Importation and containment of Ebola virus disease—Senegal, august–september 2014. MMWR Morb Mortal Wkly Rep 2014;63(39):873–4, 25275333. PubMed Central PMCID. PMCPMC4584878.

[12] Shuaib F, Gunnala R, Musa EO, Mahoney FJ, Oguntimehin O, Nguku PM, et al. Ebola virus disease outbreak—Nigeria, july–september 2014. MMWR Morb Mortal Wkly Rep 2014;63(39):867–72, 25275332. PubMed Central PMCID. PMCPMC4584877.

[13] Diarra B, Safronetz D, Sarro YD, Kone A, Sanogo M, Tounkara S, et al. Laboratory response to 2014 Ebola virus outbreak in Mali. J Infect Dis 2016;214(suppl 3):S164–8. https://doi.org/10.1093/infdis/jiw200, 27707892. PubMed Central PMCID. PMCPMC5050465.

[14] Gulland A. Second Ebola patient is treated in UK. BMJ 2014;349:g7861. https://doi.org/10.1136/bmj.g7861, 25552628.

[15] Castilletti C, Carletti F, Gruber CE, Bordi L, Lalle E, Quartu S, et al. Molecular characterization of the first Ebola virus isolated in Italy, from a health care worker repatriated from Sierra Leone. Genome Announc 2015;3(3):e00639-15. https://doi.org/10.1128/genomeA.00639-15, 26089420. PubMed Central PMCID. PMCPMC4472897.

[16] Lopaz MA, Amela C, Ordobas M, Dominguez-Berjon MF, Alvarez C, Martinez M, et al. First secondary case of Ebola outside Africa: epidemiological characteristics and contact monitoring, Spain, september to november 2014. Euro Surveill 2015;20(1):21003, 25613651.

[17] McCarthy M. Second US nurse with Ebola had traveled by plane. BMJ 2014;349:g6277. https://doi.org/10.1136/bmj.g6277. 25324208.

[18] Yusuf I, Adam RU, Ahmad SA, Yee PL. Ebola and compliance with infection prevention measures in Nigeria. Lancet Infect Dis 2014;14(11):1045–6. https://doi.org/10.1016/S1473-3099(14)70954-5. 25282666.

[19] Fasina FO, Shittu A, Lazarus D, Tomori O, Simonsen L, Viboud C, et al. Transmission dynamics and control of Ebola virus disease outbreak in Nigeria, july to september 2014. Euro Surveill 2014;19(40):20920. 25323076.

[20] Dick GW, Kitchen SF, Haddow AJ. Zika virus. I. Isolations and serological specificity. Trans R Soc Trop Med Hyg 1952;46(5):509–20. 12995440.

[21] Fichet-Calvet E, Lecompte E, Koivogui L, Soropogui B, Dore A, Kourouma F, et al. Fluctuation of abundance and Lassa virus prevalence in *Mastomys natalensis* in Guinea, West Africa. Vector Borne Zoonotic Dis 2007;7(2):119–28. https://doi.org/10.1089/vbz.2006.0520. 17627428.

[22] Kulkarni MA, Berrang-Ford L, Buck PA, Drebot MA, Lindsay LR, Ogden NH. Major emerging vector-borne zoonotic diseases of public health importance in Canada. Emerg Microbes Infect 2015;4:e33. https://doi.org/10.1038/emi.2015.33. 26954882. PubMed Central PMCID. PMCPMC4773043.

[23] Rochlin I, Ninivaggi DV, Hutchinson ML, Farajollahi A. Climate change and range expansion of the Asian tiger mosquito (*Aedes albopictus*) in Northeastern USA: implications for public health practitioners. PLoS One 2013;8(4):e60874. https://doi.org/10.1371/journal.pone.0060874. 23565282. PubMed Central PMCID. PMCPMC3614918.

[24] Howard CR, Fletcher NF. Emerging virus diseases: can we ever expect the unexpected? Emerg Microbes Infect 2012;1(12):e46. https://doi.org/10.1038/emi.2012.47. 26038413. PubMed Central PMCID. PMCPMC3630908.

[25] Bolles M, Donaldson E, Baric R. SARS-CoV and emergent coronaviruses: viral determinants of interspecies transmission. Curr Opin Virol 2011;1(6):624–34. https://doi.org/10.1016/j.coviro.2011.10.012. 22180768. PubMed Central PMCID. PMCPMC3237677.

[26] Trombley AR, Wachter L, Garrison J, Buckley-Beason VA, Jahrling J, Hensley LE, et al. Comprehensive panel of real-time TaqMan polymerase chain reaction assays for detection and absolute quantification of filoviruses, arenaviruses, and new world hantaviruses. Am J Trop Med Hyg 2010;82(5):954–60. https://doi.org/10.4269/ajtmh.2010.09-0636. 20439981. PubMed Central PMCID. PMC2861391s.

[27] Staples JE, Monath TP. Yellow fever: 100 years of discovery. JAMA 2008;300(8):960–2. https://doi.org/10.1001/jama.300.8.960. 18728272.

[28] Reed W, Carroll J, Agramonte A. The etiology of yellow fever: an additional note. JAMA 1901;36:431–40.

[29] Wilson MR, Naccache SN, Samayoa E, Biagtan M, Bashir H, Yu G, et al. Actionable diagnosis of neuroleptospirosis by next-generation sequencing. N Engl J Med 2014;370(25): 2408–17. https://doi.org/10.1056/NEJMoa1401268. 24896819. PubMed Central PMCID. PMC4134948.

[30] Grard G, Fair JN, Lee D, Slikas E, Steffen I, Muyembe JJ, et al. A novel rhabdovirus associated with acute hemorrhagic fever in central Africa. PLoS Pathog 2012;8(9):e1002924. https://doi.org/10.1371/journal.ppat.1002924. 23028323. PubMed Central PMCID. PMC3460624.

[31] Briese T, Paweska JT, McMullan LK, Hutchison SK, Street C, Palacios G, et al. Genetic detection and characterization of Lujo virus, a new hemorrhagic fever-associated arenavirus from southern Africa. PLoS Pathog 2009;5(5):e1000455. https://doi.org/10.1371/journal.ppat.1000455. 19478873. PubMed Central PMCID. PMCPMC2680969.

[32] Goldstein T, Anthony SJ, Gbakima A, Bird BH, Bangura J, Tremeau-Bravard A, et al. Author correction: the discovery of Bombali virus adds further support for bats as hosts of ebolaviruses. Nat Microbiol 2018;3(10):1084–9. https://doi.org/10.1038/s41564-018-0315-3. PubMed PMID. 30410089.

[33] Geldenhuys M, Mortlock M, Weyer J, Bezuidt O, Seamark ECJ, Kearney T, et al. A metagenomic viral discovery approach identifies potential zoonotic and novel mammalian viruses in Neoromicia bats within South Africa. PLoS One 2018;13(3):e0194527. https://doi.org/10.1371/journal.pone.0194527. 29579103. PubMed Central PMCID. PMCPMC5868816.

[34] Simmonds P, Adams MJ, Benko M, Breitbart M, Brister JR, Carstens EB, et al. Consensus statement: virus taxonomy in the age of metagenomics. Nat Rev Microbiol 2017;15(3):161–8. https://doi.org/10.1038/nrmicro.2016.177. 28134265.

[35] Anthony SJ, Epstein JH, Murray KA, Navarrete-Macias I, Zambrana-Torrelio CM, Solovyov A, et al. A strategy to estimate unknown viral diversity in mammals. MBio 2013;4(5):e00598-13. https://doi.org/10.1128/mBio.00598-13. 24003179. PubMed Central PMCID. PMCPMC3760253.

[36] Cutler SJ, Fooks AR, van der Poel WH. Public health threat of new, reemerging, and neglected zoonoses in the industrialized world. Emerg Infect Dis 2010;16(1):1–7. https://doi.org/10.3201/eid1601.081467. 20031035. PubMed Central PMCID. PMCPMC2874344.

[37] Temmam S, Davoust B, Berenger JM, Raoult D, Desnues C. Viral metagenomics on animals as a tool for the detection of zoonoses prior to human infection? Int J Mol Sci 2014;15(6):10377–97. https://doi.org/10.3390/ijms150610377. 24918293. PubMed Central PMCID. PMCPMC4100157.

[38] Shi M, Lin XD, Tian JH, Chen LJ, Chen X, Li CX, et al. Redefining the invertebrate RNA virosphere. Nature 2016;540(7634):539–43. https://doi.org/10.1038/nature20167. 27880757.

[39] Moreira LA, Iturbe-Ormaetxe I, Jeffery JA, Lu G, Pyke AT, Hedges LM, et al. A Wolbachia symbiont in *Aedes aegypti* limits infection with dengue, Chikungunya, and Plasmodium. Cell 2009;139(7):1268–78. https://doi.org/10.1016/j.cell.2009.11.042. 20064373.

[40] Bian G, Xu Y, Lu P, Xie Y, Xi Z. The endosymbiotic bacterium Wolbachia induces resistance to dengue virus in *Aedes aegypti*. PLoS Pathog 2010;6(4):e1000833. https://doi.org/10.1371/journal.ppat.1000833. 20368968. PubMed Central PMCID. PMCPMC2848556.

[41] Haagmans BL, Al Dhahiry SH, Reusken CB, Raj VS, Galiano M, Myers R, et al. Middle East respiratory syndrome coronavirus in dromedary camels: an outbreak investigation. Lancet Infect Dis 2014;14(2):140–5. https://doi.org/10.1016/S1473-3099(13)70690-X. 24355866.

[42] Briese T, Mishra N, Jain K, Zalmout IS, Jabado OJ, Karesh WB, et al. Middle East respiratory syndrome coronavirus quasispecies that include homologues of human isolates revealed through whole-genome analysis and virus cultured from dromedary camels in Saudi Arabia. MBio 2014;5(3):e01146-14. https://doi.org/10.1128/mBio.01146-14. 24781747. PubMed Central PMCID. PMCPMC4010836.

[43] Rohde H, Qin J, Cui Y, Li D, Loman NJ, Hentschke M, et al. Open-source genomic analysis of Shiga-toxin-producing *E. coli* O104:H4. N Engl J Med 2011;365(8):718–24. https://doi.org/10.1056/NEJMoa1107643. 21793736.

[44] Gire SK, Goba A, Andersen KG, Sealfon RS, Park DJ, Kanneh L, et al. Genomic surveillance elucidates Ebola virus origin and transmission during the 2014 outbreak. Science 2014;345(6202):1369–72. https://doi.org/10.1126/science.1259657. 25214632. PubMed Central PMCID. PMC4431643.

[45] Mate SE, Kugelman JR, Nyenswah TG, Ladner JT, Wiley MR, Cordier-Lassalle T, et al. Molecular evidence of sexual transmission of Ebola virus. N Engl J Med 2015;373(25):2448–54. https://doi.org/10.1056/NEJMoa1509773. 26465384. PubMed Central PMCID. PMCPMC4711355.

[46] Arias A, Watson SJ, Asogun D, Tobin EA, Lu J, Phan MVT, et al. Rapid outbreak sequencing of Ebola virus in Sierra Leone identifies transmission chains linked to sporadic cases. Virus Evol 2016;2(1):vew016. https://doi.org/10.1093/ve/vew016. 28694998. PubMed Central PMCID. PMCPMC5499387.

[47] Christie A, Davies-Wayne GJ, Cordier-Lassalle T, Blackley DJ, Laney AS, Williams DE, et al. Possible sexual transmission of Ebola virus—Liberia, 2015. MMWR Morb Mortal Wkly Rep 2015;64(17):479–81, 25950255.

[48] Purpura LJ, Soka M, Baller A, White S, Rogers E, Choi MJ, et al. Implementation of a National semen testing and counseling program for male Ebola survivors—Liberia, 2015–2016. MMWR Morb Mortal Wkly Rep 2016;65(36):963–6. https://doi.org/10.15585/mmwr.mm6536a5. 27632552.

[49] Heather JM, Chain B. The sequence of sequencers: the history of sequencing DNA. Genomics 2016;107(1):1–8. https://doi.org/10.1016/j.ygeno.2015.11.003. 26554401. PubMed Central PMCID. PMCPMC4727787.

[50] Stockton J, Ellis JS, Saville M, Clewley JP, Zambon MC. Multiplex PCR for typing and subtyping influenza and respiratory syncytial viruses. J Clin Microbiol 1998;36(10):2990–5. 9738055. PubMed Central PMCID. PMCPMC105099.

[51] Ruggiero P, McMillen T, Tang YW, Babady NE. Evaluation of the BioFire FilmArray respiratory panel and the GenMark eSensor respiratory viral panel on lower respiratory tract specimens. J Clin Microbiol 2014;52(1):288–90. https://doi.org/10.1128/JCM.02787-13. 24131685. PubMed Central PMCID. PMCPMC3911424.

[52] Gyawali P, Croucher D, Hewitt J. Preliminary evaluation of BioFire FilmArray((R)) gastrointestinal panel for the detection of noroviruses and other enteric viruses from wastewater and shellfish. Environ Sci Pollut Res Int 2018;25(27):27657–61. https://doi.org/10.1007/s11356-018-2869-2. 30083906.

[53] Pulido MR, Moreno-Martinez P, Gonzalez-Galan V, Fernandez Cuenca F, Pascual A, Garnacho-Montero J, et al. Application of BioFire FilmArray blood culture identification

panel for rapid identification of the causative agents of ventilator-associated pneumonia. Clin Microbiol Infect 2018;24(11):1213 e1–4. https://doi.org/10.1016/j.cmi.2018.06.001. 29906599.

[54] Soucek DK, Dumkow LE, VanLangen KM, Jameson AP. Cost justification of the BioFire FilmArray meningitis/encephalitis panel versus standard of care for diagnosing meningitis in a community hospital. J Pharm Pract 2019;32(1):36–40. https://doi.org/10.1177/0897190017737697. 897190017737697 29092659.

[55] Gay-Andrieu F, Magassouba N, Picot V, Phillips CL, Peyrefitte CN, Dacosta B, et al. Clinical evaluation of the BioFire FilmArray((R)) BioThreat-E test for the diagnosis of Ebola virus disease in Guinea. J Clin Virol 2017;92:20–4. https://doi.org/10.1016/j.jcv.2017.04.015. 28505570.

[56] Kugelman JR, Sanchez-Lockhart M, Andersen KG, Gire S, Park DJ, Sealfon R, et al. Evaluation of the potential impact of Ebola virus genomic drift on the efficacy of sequence-based candidate therapeutics. MBio 2015;6(1):e02227-14. https://doi.org/10.1128/mBio.02227-14. 25604787. PubMed Central PMCID. PMCPMC4313914.

[57] Soppe JA, Lebbink RJ. Antiviral goes viral: harnessing CRISPR/Cas9 to combat viruses in humans. Trends Microbiol 2017;25(10):833–50. https://doi.org/10.1016/j.tim.2017.04.005. 28522157.

[58] Wang J, Quake SR. RNA-guided endonuclease provides a therapeutic strategy to cure latent herpesviridae infection. Proc Natl Acad Sci U S A 2014;111(36):13157–62. https://doi.org/10.1073/pnas.1410785111. 25157128. PubMed Central PMCID. PMCPMC4246930.

[59] van Diemen FR, Kruse EM, Hooykaas MJ, Bruggeling CE, Schurch AC, van Ham PM, et al. CRISPR/Cas9-mediated genome editing of herpesviruses limits productive and latent infections. PLoS Pathog 2016;12(6):e1005701. https://doi.org/10.1371/journal.ppat.1005701. 27362483. PubMed Central PMCID. PMCPMC4928872.

[60] Samuel CE. Antiviral actions of interferons. Clin Microbiol Rev 2001;14(4):778–809. table of contents. Epub 2001/10/05, https://doi.org/10.1128/CMR.14.4.778-809.200111585785.

[61] McClain MT, Henao R, Williams J, Nicholson B, Veldman T, Hudson L, et al. Differential evolution of peripheral cytokine levels in symptomatic and asymptomatic responses to experimental influenza virus challenge. Clin Exp Immunol 2016;183(3):441–51. https://doi.org/10.1111/cei.12736. 26506932. PubMed Central PMCID. PMCPMC4750592.

[62] McClain MT, Nicholson BP, Park LP, Liu TY, Hero 3rd AO, Tsalik EL, et al. A genomic signature of influenza infection shows potential for presymptomatic detection, guiding early therapy, and monitoring clinical responses. Open Forum Infect Dis 2016;3(1):ofw007. https://doi.org/10.1093/ofid/ofw007. 26933666. PubMed Central PMCID. PMCPMC4771939.

[63] Woods CW, McClain MT, Chen M, Zaas AK, Nicholson BP, Varkey J, et al. A host transcriptional signature for presymptomatic detection of infection in humans exposed to influenza H1N1 or H3N2. PLoS One 2013;8(1):e52198. https://doi.org/10.1371/journal.pone.0052198. 23326326. PubMed Central PMCID. PMCPMC3541408.

[64] Zaas AK, Burke T, Chen M, McClain M, Nicholson B, Veldman T, et al. A host-based RT-PCR gene expression signature to identify acute respiratory viral infection. Sci Transl Med 2013;5(203):203ra126. https://doi.org/10.1126/scitranslmed.3006280. 24048524. PubMed Central PMCID. PMCPMC4286889.

[65] Zaas AK, Chen M, Varkey J, Veldman T, Hero 3rd AO, Lucas J, et al. Gene expression signatures diagnose influenza and other symptomatic respiratory viral infections in humans. Cell Host Microbe 2009;6(3):207–17. https://doi.org/10.1016/j.chom.2009.07.006. 19664979. PubMed Central PMCID. PMCPMC2852511.

[66] Kaforou M, Wright VJ, Oni T, French N, Anderson ST, Bangani N, et al. Detection of tuberculosis in HIV-infected and -uninfected African adults using whole blood RNA expression signatures: a case-control study. PLoS Med 2013;10(10):e1001538. https://doi.org/10.1371/journal.pmed.1001538. 24167453. PubMed Central PMCID. PMCPMC3805485.

[67] Zheng R, Li Z, He F, Liu H, Chen J, Chen J, et al. Genome-wide association study identifies two risk loci for tuberculosis in Han Chinese. Nat Commun 2018;9(1):4072. https://doi.org/10.1038/s41467-018-06539-w. 30287856. PubMed Central PMCID. PMCPMC6172286.

[68] Compaore TR, Soubeiga ST, Ouattara AK, Obiri-Yeboah D, Tchelougou D, Maiga M, et al. APOBEC3G variants and protection against HIV-1 Infection in Burkina Faso. PLoS One 2016;11(1):e0146386. https://doi.org/10.1371/journal.pone.0146386. 26741797. PubMed Central PMCID. PMCPMC4704832.

[69] Speranza E, Bixler SL, Altamura LA, Arnold CE, Pratt WD, Taylor-Howell C, et al. A conserved transcriptional response to intranasal Ebola virus exposure in nonhuman primates prior to onset of fever. Sci Transl Med 2018;10(434):eaaq1016. https://doi.org/10.1126/scitranslmed.aaq1016. 29593102.

[70] Speranza E, Altamura LA, Kulcsar K, Bixler SL, Rossi CA, Schoepp RJ, et al. Comparison of transcriptomic platforms for analysis of whole blood from Ebola-infected Cynomolgus Macaques. Sci Rep 2017;7(1):14756https://doi.org/10.1038/s41598-017-15145-7. 29116224. PubMed Central PMCID. PMCPMC5676990.

[71] Sweeney TE, Perumal TM, Henao R, Nichols M, Howrylak JA, Choi AM, et al. A community approach to mortality prediction in sepsis via gene expression analysis. Nat Commun 2018;9(1):694. https://doi.org/10.1038/s41467-018-03078-2. PubMed PMID 29449546. PubMed Central PMCID. PMCPMC5814463.

[72] Sweeney TE, Wong HR, Khatri P. Robust classification of bacterial and viral infections via integrated host gene expression diagnostics. Sci Transl Med 2016;8(346):346ra91. https://doi.org/10.1126/scitranslmed.aaf7165. 27384347. PubMed Central PMCID. PMCPMC5348917.

[73] Hoenen T, Groseth A, Rosenke K, Fischer RJ, Hoenen A, Judson SD, et al. Nanopore sequencing as a rapidly deployable Ebola outbreak tool. Emerg Infect Dis 2016;22(2):331–4. https://doi.org/10.3201/eid2202.151796. 26812583. PubMed Central PMCID. PMCPMC4734547.

[74] Walter MC, Zwirglmaier K, Vette P, Holowachuk SA, Stoecker K, Genzel GH, et al. MinION as part of a biomedical rapidly deployable laboratory. J Biotechnol 2017;250:16–22. https://doi.org/10.1016/j.jbiotec.2016.12.006. 27939320.

Chapter 11

Viral hepatitis

Nicholas A Shackel*,[†],[‡], Keyur Patel[§]
*University of New South Wales, Sydney, NSW, Australia, [†]Ingham Institute, Liverpool, NSW, Australia, [‡]Liverpool Hospital, Liverpool, NSW, Australia, [§]University Health Network, University of Toronto, Toronto, ON, Canada

Introduction

Viral hepatitis is a major cause of chronic liver disease and is associated with an increased risk of hepatocellular cancer (HCC). Hepatitis B (HBV) and hepatitis C viruses (HCV) result in chronic infection in >300 million people globally. Despite availability of HBV vaccine and effective antiviral therapy, screening practices and linkage to care is often inadequate, and most patients remain undiagnosed. Chronic hepatitis B (CHB) and C (CHC) will continue to pose a significant burden on health services due to the development of complications of end-stage liver disease, including hepatocellular cancer (HCC). Hepatitis Delta (HDV) is associated with HBV and results in an increased risk of progressive liver disease and HCC. Hepatitis A virus (HAV) infection usually follows a self-limiting course and is not complicated by significant clinical sequelae in most cases. Although Hepatitis E virus (HEV) infection is likely the most common global cause of acute viral hepatitis and jaundice, viral hepatitis research has mainly focused on HBV and HCV pathogenesis, mechanisms of viral persistence, and the development of HCC. Over the past decade, functional genomics approaches have significantly advanced our understanding of viral hepatitis pathogenesis and immune-mediated therapeutic responses.

Molecular virology of hepatitis A–E

Hepatitis A virus is a single-stranded RNA hepatovirus ~7.5 kb in length and member of the *Picornaviridae* family, with an estimated 126 million cases of acute HAV infection globally in 2005 [1]. HAV is mainly acquired through fecal-oral transmission resulting from exposure to contaminated water and food. HAV generally causes a self-limiting illness, but the clinical spectrum can range from asymptomatic infection to fulminant liver failure. The incidence of HAV has been significantly reduced in the western world due to improvement in hygiene, sanitation, public health measures, and availability of a vaccine. The spontaneous mutation rate of HAV is low, and a neutralizing IgG response to

Genomic and Precision Medicine. https://doi.org/10.1016/B978-0-12-801496-7.00011-3

closely-positioned epitopes in the highly conserved capsid proteins provides life-long protection against symptomatic reinfection [2]. Molecular detection methods and biomarkers are available for assessment of genetic relatedness that may be useful in epidemiological research or identifying food-contamination sources during outbreaks [3], but HAV RNA or nucleic acid sequencing tests are not routinely available.

Hepatitis B is a small, enveloped, primarily hepatotropic DNA virus belonging to the *Hepadnaviridae* family. There are an estimated 257 million persons living with HBV infection worldwide. The partially double-stranded HBV genome is 3.2 kb in length and is organized with multiple overlapping open reading frames (ORFs). HBV DNA incorporates into the host genome and forms a stable minichromosome, the closed circular covalent DNA (cccDNA), that provides a template for continued HBV RNA transcription in the hepatocyte nucleus. Long-term persistence of cccDNA continues despite viral suppression, and results in viral reactivation following cessation of currently available antiviral therapy [4]. The hepatitis B viral proteins include the surface (envelope) protein, core, polymerase, and X-proteins. Group-specific antigenic determinants on the HBV surface antigen protein give rise to four serotypes, and genetic divergence of the viral sequence provide 10 common HBV genotypes (A–J), and several subtypes.

Hepatitis C virus is an enveloped positive-sense virus classified in the *Hepacivirus* genus within the *Flaviviridae* family. The global prevalence of CHC infection is estimated at 1% or 75 million persons. There is considerable diversity in the HCV genome, with at least six distinct genotypes, and >50 subtypes. The high viral replication rate in hepatocytes, and absence of proof-reading by the viral RNA-dependent RNA polymerase, results in the presence of multiple closely related genetic variants. These "quasispecies" have probably developed through selection pressures but represent a critical life-cycle strategy that allows for host immune escape and viral persistence [5]. HCV consists of a single open reading frame (ORF) of 9.6 kb, flanked by 5′ and 3′ untranslated regions (UTR) that are necessary for viral translation and replication. The 5′ UTR is highly conserved among HCV isolates and contains the internal ribozyme entry site (IRES) that initiates translation of a 3000 AA polyprotein, which is subsequently processed by viral and host cellular proteases into 10 mature structural (core, E1, E2 and p7) and non-structural HCV proteins (NS2, NS3, NS4A, NS5A and NS5B). The non-structural proteins encode several key enzymes involved in viral replication [6].

Hepatitis Delta is a circular, single-stranded, RNA virus with an estimated 15–20 million people infected worldwide [7]. HDV requires hepatitis B surface antigen envelope proteins to form virus particles, and presents as a superinfection in inactive HBV carriers, or is acquired at the time of HBV exposure. Recognition of HDV is important as it can lead to fulminant liver failure, rapid progression to cirrhosis, and an increased risk of HCC.

Hepatitis E virus is a single-stranded, positive-sense RNA virus that is approximately 7.2 kb in length and was the first identified member of the

Hepeviridae family. HEV is a recognized zoonotic disease, transmitted through fecal-oral route, and is perhaps the most common cause of acute viral hepatitis and jaundice in developing countries. Acute HEV infection is associated with severe morbidity, and a reported mortality rate as high as 25% in pregnant women [8].

Genetic predisposition to hepatitis B and C

Common sources of HBV and HCV transmission include maternal-fetal vertical transmission or unsafe injection drug use. Genetic determinants of post-exposure outcomes include several human leukocyte antigen (HLA) alleles that are associated with both the persistence and clearance of both HBV and HCV (Table 1). Homozygosity for HLA class II locus increases the risk of HBV persistence. The HLA class II locus DRB1*1302 is associated with HBV clearance and the DQB1*0301 locus is associated with a self-limiting course of HCV infection. Non-HLA immune-associated genes are also implicated in viral hepatitis with tumor necrosis factor (TNF) promoter polymorphisms resulting in higher TNF secretion being associated with HBV clearance. The killer cell immunoglobulin like receptors (KIR) genes interact with HLA class I molecules and specific KIR heliotypes are associated with HCV clearance [9]. Many of the genetic predispositions linked to viral persistence or clearance can be directly implicated in the adaptive immune response. However, the currently documented genetic disease associations with viral hepatitis are of limited use in clinical practice. Future clinical practice is likely to see panels of genetic susceptibility markers being screened to determined prognosis, development of cancer, as well as the likelihood of CHB treatment response. The overall impact of genomic studies of hepatitis virus is considerable and impacts the stage of infection (acute and chronic), treatment responses, and consequences such as liver cancer.

Natural history of chronic hepatitis

Genomic research in viral hepatitis has mostly focused on chronic HBV and HCV infection, as HAV and HEV infections are almost always acute, and there is relatively limited natural history data, or available therapeutic options, in chronic HDV infection.

The natural history of CHB involves variable and complex interactions between virus and the host immune response that can result in risk of progressive liver injury and development of HCC. There are four main clinically relevant phases, and based on recently revised terminology, include HBeAg-positive chronic infection and hepatitis (immune tolerant and immune active phase) and HBeAg-negative chronic infection (inactive HbsAg carrier) and HBeAg-negative hepatitis. These phases are based on HBV DNA levels, serum alanine aminotransferase (ALT), HBeAg/HBeAb status, and severity of liver disease [10].

TABLE 1 Genetic susceptibility associations with hepatitis B and C

Allele/polymorphism	Hepatitis B	Hepatitis C	Comment
HLA-DRB1*1302 and HLA-DRB1*0301	Spontaneous elimination of infection		
HLA-DQA1*0501, DQB1*301, HLADRB1*1102 and HLADRB1*0301	Persistence of infection		
HLA-DRB1*0101, HLA-DRB1*0401, HLA-DRB1*15, HLA-DRB1*1101, HLA-DRB1*0301, HLA-A*2301, HLA-A*1101, HLA-A*03, HLA-B*57 and HLA-Cw*0102		Spontaneous elimination of infection	
HLA-DRB1*0701, HLA-A*01-B*08, Cw*07-DRB1*0301, DBQ1*0201, HLA-Cw*04, HLACw*04-B*53		Persistence of infection	
TNF promoter	Viral replication and clearance	Viral replication and clearance	Polymorphisms at −308 and −238 best characterized
Interleukin-10	Spontaneous elimination of infection		−1082 AA genotype
Vitamin D receptor	Control of viral replication		
GDNF family receptor alpha 1		Risk of HCC	
Chemokine (C-X-C motif) ligand 14		Risk of HCC	
IFN-lambda 4		– Spontaneous clearance – Virologic response to IFN-based therapy	ΔG variant in strong linkage disequilibrium with IFN-λ3 TT "poor response" predictor

TNF, Tumor necrosis factor; GDNF, glial cell line-derived neurotrophic factor.

Risk of progressive hepatic injury tends to occur during the HBeAg-positive or negative chronic hepatitis phase of infection during which HBV antigen-specific CD8+ cytotoxic T-cells target-infected hepatocytes. Low HBV titers and limited liver-injury progression are observed during HBeAg-negative chronic infection. HBV mutants are common during HbeAg-negative chronic hepatitis phase, driving chronic inflammation with flares of escape mutants followed by spikes of immune-mediated hepatocyte death. The presence of HBV escape mutants with PreS mutations (particularly A1762T and G1764A) predicts HCC risk with adjusted OR of 6.18 [11].

In contrast to the complex natural history of CHB, chronic HCV infection has a relatively segregated natural history, and only consists of acute and chronic viral persistence phases. Upon exposure, HCV replication continues at ~10^{12} virions per day followed by immune-mediated hepatocyte cell death. Chronic persistent hepatitis leads to progressive inflammatory injury leading to variable progression rate of liver fibrosis, and eventual development of cirrhosis in ~20% of patients.

Liver disease progression and hepatocellular cancer

The most important disease outcomes associated with chronic hepatitis virus infection are liver cirrhosis and HCC. Hepatocellular cancer is the sixth most common malignancy worldwide, and third common cause of cancer related mortality. HBV accounts for 75–80% of virus-associated HCCs, and HCV is responsible for 10–20%. The global burden prevalence of HCC is estimated to be 854,000 persons [12].

Single nucleotide polymorphisms (SNPs) in various genes have been shown to predict disease progression in both HBV and HCV infection (Table 2). HLA-associated SNPs play a significant role in viral hepatitis disease progression. SNPs in HLA loci have been shown to be associated with chronic HBV infection in Asian patients [13–16] However, in general HLA SNPs have been poorly predictive of progression of HCV-mediated liver injury [17]. HLA-associated SNPs are also associated with HBV-HCC in Japanese [13], Han Chinese [18–20], and Korean [20], but not in any other East-Asian cohorts [21]. A meta-analysis of eight case-control studies has shown that SNVs in HLA-DRB1*07, -DRB1*12, and -DRB1*15 are associated with increased risk of progression to viral-associated HCC (OR = ~2.88) for Asian populations and in DRB1*07 and DRB1*12 (OR = ~1.65) in multiple populations [22], regardless of etiology.

Further, SNPs in Interferon lambda 3 (*IFN-λ3*), patatin-like phospholipase domain-containing-3 (*PNPLA3*) and MHC class I polypeptide-related chain A (*MICA*) have been reported to predict disease progression in both chronic HBV and HCV infections. In Caucasian patients, *IFN-λ3* CC genotype (rs12979860) is associated with both HBV- and HCV-associated fibrosis progression (OR = 1.63 and 2.75, respectively) and inflammation (OR = 4.36 and 2.11, respectively) [23]. The rs738409 SNP in *PNPLA3* has been associated

TABLE 2 Gene polymorphisms associated with disease progression

Virus	Chromosome	Position	Gene	SNP	Risk allele
HBV	14	70,245,193	NTCP (SLC10A1)	rs2296651	C
	6		HLA Class I allele A*0301	Haplotype	–
	6	32,702,467	HLA-DQB1	rs2856718	AA
	6	32,762,235	HLA-DQB2	rs7453920	AA
	6	32,711,222	HLA-DQA2	rs9275572	A
	6	33,087,030	HLA-DPB1	rs9277534	G
	4	186,083,063	TLR3	rs3775290	C
	9	71,690,182	TMEM2	rs2297089	T
	9	21,384,972	IFNA2	rs146352658	T
	11	119,180,140	NLRX1	rs149770693	T
	11	110,413,469	Ferredoxin 1	rs2724432	C
	2	117,640,013	DDX18	rs2551677	A
HBV/HCV	6		HLA-DRB1*07 HLA-DRB1*12 HLA-DRB1*15	Haplotype	–
	19	39,248,147	IFN-λ4	rs12979860	CC
HCV	22	43,928,847	PNPLA3	rs738409	GC or GG
	17	34,320,812	MCP-2	rs1133763	C
	3	141,744,456	RNF7	rs16851720	AA
	2	112,013,193	MERTK	rs4374383	AG or GG
	6	35,534,425	TULP1	rs9380516	TT
	12	104,028,030	GLT8D2	rs2629751	GG
	9	90,419,249	LOC340515	rs883924	AA
	6		HLA Class I allele B*08	Haplotype	–
	6		HLA Class I allele B*44	Haplotype	–

with greater hepatic fibrosis (OR = 1.2) in Caucasian but not Asian populations [24]. This may be related to modulation of lipid metabolism pathways during the HCV replication cycle [25]. Further, genetic alterations in immune cell receptors TLR7 (c. 1–120T > G mutation, adjusted OR = 4.187) [26], and CCR5 (and its ligand MCP-2; rs333 and rs1133763, OR = 1.97 and 2.29) have been associated with greater fibrosis [27].

MICA is a ligand of the Natural Killer Group 2D Receptor (present on NK and T-cells) and mediates antiviral immune responses. The SNPs rs2596542 and rs2596538 in 5′ flanking region of MICA are linked to low serum levels of MICA and are associated with the progression of HCV-associated HCC (OR = ~1.4) in Japanese populations [28,29]. The rs2596542 has also been reported to associated with active HBV disease (OR = 1.31) [30] and increase the risk of HBV-associated HCC (OR = 1.2) [31].

Specific genomic mutations are different in HBV-HCC compared to non-HBV HCC. Inactivation of interferon regulatory factor 2 (*IRF2*) has been shown to be a rare, but enriched mutation in HBV-HCC, while mutations in β-catenin (*CTNNB1*) are generally absent [32]. Very few (if any) genomic features distinguish HCV-HCC from any non-HCV HCC. Precision therapy for HCC may be possible as exome sequencing has identified mutations that may be potentially targeted by compounds already in clinical development [33]. However, significant translational efforts are still required to develop molecular biomarkers of early disease stage, prognosis and therapeutic response in HCC [34]. Serum miRNAs continue to provide interest as biomarkers of liver disease progression and CHB associated HCC (miRNA-25, miRNA-375, and let-7f) [35], but none are available yet for routine clinical use.

Treatment response

Very few SNPs in either the host or HBV genome have been found to predict HBV response. Multiple SNPs in HLA loci (rs9276370 and rs7756516) have been associated with virologic response to oral nucleos(t)ide analogs and IFN treatment in a small cohort of Taiwanese–Han-Chinese patients, although this has yet to be verified in other populations [36]. While there was a positive relationship found with SNPs in the *IFN-λ3* gene (rs8099917) in a single early study, subsequent trials have found no correlation between *IFN-λ3* genotype and therapeutic outcome in chronic HBV infection [37,38]. A small study has shown an intrahepatic transcriptomic profile of 41 genes that predict combined response (HBeAg negativity, and HBV-DNA levels ≤2000 IU/mL) to Adefovir and IFN [39]. The majority of these were genes associated with MHC Class II presentation, chemokine activity, and IFN-stimulated genes.

Hepatitis C therapy is the quintessential success story for personalized pharmacogenomics. In 2009, three independent genome-wide association studies (GWAS) conducted North America, Japan, and Australia reported that single nucleotide polymorphism (SNPs) near the IFN-λ3 gene could be used to predict sustained virologic response (SVR) following treatment of CHC genotype 1 infection with Peg-IFN and Ribavirin [40–42], and subsequently validated in other cohorts [43,44]. The North American HCV study cohort identified rs12979860, a bi-allelic SNP (C/T) in chromosomal region 19q13.13 located ~3 kb upstream of IFN-λ3 (formerly *IL28B*) gene was associated with a two-to-threefold increase in the SVR rate. The frequency of the favorable response CC genotype was higher in Caucasians compared to that of African-Americans. This explained the significant differences in observed virologic response to IFN-based therapy between these two ethnic groups. A separate study noted association with *IFN-λ3* region polymorphisms and spontaneous clearance [45]. *IFN-λ3* is the gene coding for the protein IFN-λ3, which belongs to type III IFN, and acts via the up regulation of IFN stimulated genes (ISGs) to augment the natural killer cell and CD8+ T-cell cytotoxicity. The search for biological plausibility for a functional variant in the IFN-λ3 region led to the identification of a transiently induced region that harbors a dinucleotide variant rs368234815, originally designated as ss469415590, (TT or ΔG) upstream of *IFN-λ3*. This frame shift variant created a novel gene designated *IFN-λ4*, encoding the interferon-lambda 4 protein, also a type III IFN but in high linkage disequilibrium with rs12979860-TT. The IFN-λ4 ΔG functional variant controls the generation of IFN-λ4 and was strongly associated with spontaneous HCV clearance and virologic response following IFN-based therapy in individuals of African ancestry [46]. Ribavirin induced anemia was another common side effect of IFN-based therapy. The North American HCV cohort study also noted a strong association between SNP rs6051702 and Hgb reduction at week 4. This association was explained by two functional variants of the Inosine triphosphatase (ITPA) gene, a missense variant in exon 2 (rs1127354, P32T) and a splice-altering SNP in intron 2 (rs7270101) [47]. The rapid advent of non-IFN-based direct acting antiviral (DAA) therapy, with established safety and tolerability and SVR > 95% after 8–12 weeks of therapy, irrespective of race, viral genotype or disease severity, have now rendered the *IFN-λ3* association with CHC treatment into historical context. However, further delineating the functional role of IFN-λ4 in modulating innate and adaptive immune responses to other viruses and inflammatory states remains important [48].

Conclusions

Genomics studies have already made significant contributions to our understanding of viral hepatitis pathogenesis, but molecular pathways mediating acute HBV and HCV infection and the development of chronicity remain poorly understood. The identification of the role ISGs and the IFN- λ3 pathway in HCV infection led to one of the first examples of personalized therapy based on individual genomic variability, but in the current era of DAA therapy, the clinical relevance of this discovery for CHC patients is now limited. Future genomics studies assessing liver disease progression in CHB, or pathways to

HCC, will likely integrate transcriptomic and epigenomic profiles to stratify prognostic risks and individualize emerging therapeutic options. Genetic biomarkers and translated peptide identification appear promising as HCC prognostic and therapeutic biomarkers, but none are in clinical use, and there are still no reliable early diagnostic markers for HCC. Tagged proteomic approaches to assess subcellular protein interactions and quantify proteome evolution during HBV infection may help improve our understanding of the pathways to virus-mediated heptocarcinogenesis [49]. Defining the role of microbiota in HBV pathogenesis will continue to evolve [50]. Considering the complex interactions between host genetics, virus, and environmental factors for HCC, developing reliable molecular and genetic signatures for early diagnosis and prognosis in this population will likely continue to pose a considerable challenge over the next few years.

References

[1] World Health Organization. WHO position paper on hepatitis A vaccines—june 2012. Wkly Epidemiol Rec 2012;87(28/29):261–76.

[2] Lemon SM, et al. Type A viral hepatitis: a summary and update on the molecular virology, epidemiology, pathogenesis and prevention. J Hepatol 2018;68:167–84.

[3] Nainan OV, et al. Diagnosis of hepatitis A virus infection: a molecular approach. Clin Microbiol Rev 2006;19(1):63–79.

[4] Lucifora J, Protzer U. Attacking hepatitis B virus cccDNA—the holy grail to hepatitis B cure. J Hepatol 2016;64(1 Suppl):S41–8.

[5] Tsukiyama-Kohara K, Kohara M. Hepatitis C virus: viral quasispecies and genotypes. Int J Mol Sci 2018;19(1):23.

[6] Bartenschlager R, Lohmann V. Replication of hepatitis C virus. J Gen Virol 2000;81(Pt 7):1631–48.

[7] Wedemeyer H, Manns MP. Epidemiology, pathogenesis and management of hepatitis D: update and challenges ahead. Nat Rev Gastroenterol Hepatol 2010;7(1):31–40.

[8] Hoofnagle JH, Nelson KE, Purcell RH. Hepatitis E. N Engl J Med 2012;367(13):1237–44.

[9] Rehermann B, Nascimbeni M. Immunology of hepatitis B virus and hepatitis C virus infection. Nat Rev Immunol 2005;5(3):215–29.

[10] European Association for the Study of the Liver. *EASL* 2017 Clinical Practice Guidelines on the management of hepatitis B virus infection. J Hepatol 2017;67(2):370–98.

[11] Liu S, et al. Associations between hepatitis B virus mutations and the risk of hepatocellular carcinoma: a meta-analysis. J Natl Cancer Inst 2009;101(15):1066–82.

[12] Akinyemiju T, et al. The burden of primary liver cancer and underlying etiologies from 1990 to 2015 at the global, regional, and national level: results from the global burden of disease study 2015. JAMA Oncol 2017;3(12):1683–91.

[13] Kamatani Y, et al. A genome-wide association study identifies variants in the HLA-DP locus associated with chronic hepatitis B in Asians. Nat Genet 2009;41(5):591–5.

[14] Yan Z, et al. Relationship between HLA-DP gene polymorphisms and clearance of chronic hepatitis B virus infections: case-control study and meta-analysis. Infect Genet Evol 2012;12(6):1222–8.

[15] Hu Z, et al. New loci associated with chronic hepatitis B virus infection in Han Chinese. Nat Genet 2013;45(12):1499–503.

[16] Jiang DK, et al. Genetic variants in five novel loci including CFB and CD40 predispose to chronic hepatitis B. Hepatology 2015;62(1):118–28.
[17] Yee LJ. Host genetic determinants in hepatitis C virus infection. Genes Immun 2004;5(4): 237–45.
[18] Hu L, et al. Genetic variants in human leukocyte antigen/DP-DQ influence both hepatitis B virus clearance and hepatocellular carcinoma development. Hepatology 2012;55(5):1426–31.
[19] Li S, et al. GWAS identifies novel susceptibility loci on 6p21.32 and 21q21.3 for hepatocellular carcinoma in chronic hepatitis B virus carriers. PLoS Genet 2012;8(7):e1002791.
[20] Jiang DK, et al. Genetic variants in STAT4 and HLA-DQ genes confer risk of hepatitis B virus-related hepatocellular carcinoma. Nat Genet 2013;45(1):72–5.
[21] Sawai H, et al. No association for Chinese HBV-related hepatocellular carcinoma susceptibility SNP in other East Asian populations. BMC Med Genet 2012;13:47.
[22] Lin ZH, et al. Association between HLA-DRB1 alleles polymorphism and hepatocellular carcinoma: a meta-analysis. BMC Gastroenterol 2010;10:145.
[23] Eslam M, et al. Interferon-lambda rs12979860 genotype and liver fibrosis in viral and non-viral chronic liver disease. Nat Commun 2015;6:6422.
[24] Fan JH, et al. PNPLA3 rs738409 polymorphism associated with hepatic steatosis and advanced fibrosis in patients with chronic hepatitis C virus: a meta-analysis. Gut Liver 2016;10(3):456–63.
[25] Syed GH, Amako Y, Siddiqui A. Hepatitis C virus hijacks host lipid metabolism. Trends Endocrinol Metab 2010;21(1):33–40.
[26] Schott E, et al. A toll-like receptor 7 single nucleotide polymorphism protects from advanced inflammation and fibrosis in male patients with chronic HCV-infection. J Hepatol 2007;47(2):203–11.
[27] Hellier S, et al. Association of genetic variants of the chemokine receptor CCR5 and its ligands, RANTES and MCP-2, with outcome of HCV infection. Hepatology 2003;38(6):1468–76.
[28] Kumar V, et al. Genome-wide association study identifies a susceptibility locus for HCV-induced hepatocellular carcinoma. Nat Genet 2011;43(5):455–8.
[29] Lo PH, et al. Identification of a functional variant in the MICA promoter which regulates MICA expression and increases HCV-related hepatocellular carcinoma risk. PLoS One 2013;8(4):e61279.
[30] Al-Qahtani AA, et al. Genetic variation at −1878 (rs2596542) in MICA gene region is associated with chronic hepatitis B virus infection in Saudi Arabian patients. Exp Mol Pathol 2013;95(3):255–8.
[31] Tong HV, et al. Hepatitis B virus-induced hepatocellular carcinoma: functional roles of MICA variants. J Viral Hepat 2013;20(10):687–98.
[32] Guichard C, et al. Integrated analysis of somatic mutations and focal copy-number changes identifies key genes and pathways in hepatocellular carcinoma. Nat Genet 2012;44(6):694–8.
[33] Schulze K, et al. Exome sequencing of hepatocellular carcinomas identifies new mutational signatures and potential therapeutic targets. Nat Genet 2015;47(5):505–11.
[34] Zucman-Rossi J, et al. Genetic landscape and biomarkers of hepatocellular carcinoma. Gastroenterology 2015;149(5):1226–4.
[35] Li LM, et al. Serum microRNA profiles serve as novel biomarkers for HBV infection and diagnosis of HBV-positive hepatocarcinoma. Cancer Res 2010;70(23):9798–807.
[36] Chang SW, et al. A genome-wide association study on chronic HBV infection and its clinical progression in male Han-Taiwanese. PLoS One 2014;9(6):e99724.

[37] Galmozzi E, Vigano M, Lampertico P. Systematic review with meta-analysis: do interferon lambda 3 polymorphisms predict the outcome of interferon-therapy in hepatitis B infection? Aliment Pharmacol Ther 2014;39(6):569–78.

[38] Zhang Q, et al. IFNL3 (IL28B) polymorphism does not predict long-term response to interferon therapy in HBeAg-positive chronic hepatitis B patients. J Viral Hepat 2014;21(7):525–32.

[39] Jansen L, et al. An intrahepatic transcriptional signature of enhanced immune activity predicts response to peginterferon in chronic hepatitis B. Liver Int 2015;35(7):1824–32.

[40] Ge D, et al. Genetic variation in IL28B predicts hepatitis C treatment-induced viral clearance. Nature 2009;461(7262):399–401.

[41] Suppiah V, et al. IL28B is associated with response to chronic hepatitis C interferon-alpha and ribavirin therapy. Nat Genet 2009;41(10):1100–4.

[42] Tanaka Y, et al. Genome-wide association of IL28B with response to pegylated interferon-alpha and ribavirin therapy for chronic hepatitis C. Nat Genet 2009;41(10):1105–9.

[43] Rauch A, et al. Genetic variation in IL28B is associated with chronic hepatitis C and treatment failure: a genome-wide association study. Gastroenterology 2010;138(4):1338–45. [1345 e1-7].

[44] McCarthy JJ, et al. Replicated association between an IL28B gene variant and a sustained response to pegylated interferon and ribavirin. Gastroenterology 2010;138(7):2307–14.

[45] Thomas DL, et al. Genetic variation in IL28B and spontaneous clearance of hepatitis C virus. Nature 2009;461(7265):798–801.

[46] Prokunina-Olsson L, et al. A variant upstream of IFNL3 (IL28B) creating a new interferon gene IFNL4 is associated with impaired clearance of hepatitis C virus. Nat Genet 2013;45(2):164–71.

[47] Fellay J, et al. ITPA gene variants protect against anaemia in patients treated for chronic hepatitis C. Nature 2010;464(7287):405–8.

[48] Chinnaswamy S. Gene-disease association with human IFNL locus polymorphisms extends beyond hepatitis C virus infections. Genes Immun 2016;17(5):265–75.

[49] Douam F, Ploss A. Proteomic approaches to analyzing hepatitis C virus biology. Proteomics 2015;15(12):2051–65.

[50] Zhang Y, Lun CY, Tsui SK. Metagenomics: a new way to illustrate the crosstalk between infectious diseases and host microbiome. Int J Mol Sci 2015;16(11):26263–79.

Chapter 12

Vaccine development

Letitia D. Jones, Amelia B. Thompson, M. Anthony Moody
Duke Human Vaccine Institute and Department of Pediatrics, Duke University Medical Center, Durham, NC, United States

Introduction

Vaccination is one of the most cost-effective and efficient ways to prevent infectious diseases and to reduce their impact on the human population. Our understanding of immunology has paralleled vaccine development from early work using the principles of attenuation, to our current ability to quantify and manipulate host-pathogen interactions [1]. Despite our limited understanding of the complex mechanisms that comprise human immunity, vaccinology has allowed for significant public health achievements over the years. One well-known contribution was Edward Jenner's smallpox vaccine, which was developed in 1796 and widely distributed by 1800, leading to smallpox eradication by 1980 [2]. This vaccine was safer than variolation, a centuries old practice of inserting infectious material from a person with smallpox disease into a susceptible host to induce a milder protective infection [2]. By the end of the 19th century, Louis Pasteur's paradigm of attenuation became the foundation for vaccine development [3], before we understood or could measure immune correlates of protection [4,5].

While vaccination is crucial for public health, it also benefits the individual. In addition to helping to preserve one's health during productive years and reducing medical costs, vaccination improves the wellbeing of caregivers and relatives. Fig. 1 displays how vaccines fall along the continuum and shows how some currently used vaccines differ in their protection of individuals or the larger population. As we develop our understanding of genetic factors that contribute to the immune response, so too will the future of vaccine development evolve. From DNA sequencing technology in the 1970s and complete genome sequencing of *Haemophilus influenzae* in 1995, the development of genomic sequencing of pathogens, animal models and humans has resulted in a paradigm shift in vaccine development [6], and this overview will highlight examples of the current state of the field.

Genomic and Precision Medicine. https://doi.org/10.1016/B978-0-12-801496-7.00012-5

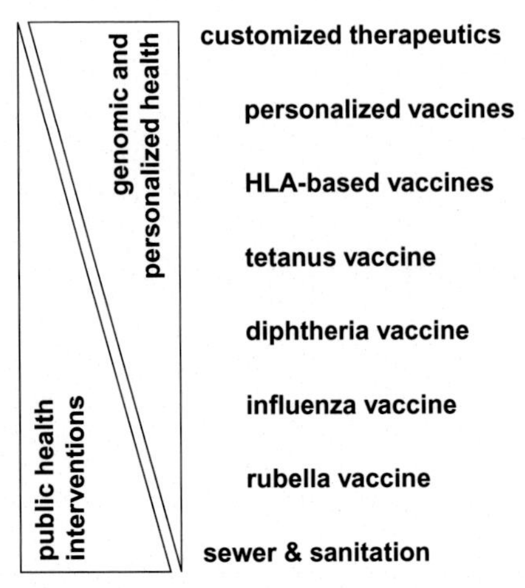

FIG. 1 Continuum of public and personalized health interventions. Public health interventions such as sewer and sanitation services reduce disease transmission by reducing hazards to all members of a community, while genomic and personalized medicine strategies like customized therapeutics are designed to benefit patients at the individual level. Vaccines reduce the transmission of infectious diseases and disease consequences for vaccinated persons. Current vaccine strategies are intended for deployment to large segments of the population while personalized and HLA-based vaccines would target individuals or small segments of the population. Examples of four vaccines currently in use show different models of protection. Tetanus vaccine provides personal protection by preventing disease caused by the soil organism *Clostridium tetani* but does not alter the likelihood of the organism growing in a dirty or deep wound. Diphtheria vaccine primarily prevents the toxin mediated effects of *Corynebacterium diphtheriae* and may affect person-to-person spread. Influenza vaccine reduces disease burden in the immunized and reduces transmission of influenza from vaccinated to susceptible persons. Rubella vaccine prevents the relatively mild disease German measles but is chiefly used to prevent congenital rubella syndrome by reducing transmission to susceptible pregnant women.

Measurement of vaccine response

Genomic variation and gene expression affect vaccine responses [7]. While genomic variation is largely fixed, gene expression is neither uniform across populations nor across the lifetime of an individual. Vaccine doses that are well tolerated in children may have more significant side effects in adults (e.g., diphtheria vaccine [8,9]) or be less immunogenic (e.g., varicella-zoster vaccine [10,11]). The resulting variability in efficacy, adverse effects and duration of protection in genetically diverse populations means that vaccines must be constantly re-evaluated as longitudinal data accumulate.

Vaccine success is typically based on two related criteria: protection from disease and immune correlates of protection [4,5]. Evaluating protection from disease is a simple concept, that is, comparing rates of disease in vaccinated to unvaccinated groups—e.g., a human challenge model to evaluate a cholera vaccine candidate [12]. However, conducting such studies in humans can be challenging and, in some cases, unethical. Difficulties include participant attrition, poor control of pathogen exposure prior to and after administration of a candidate vaccine, and self-reporting biases regarding risk and exposure. As a result, most of these studies are performed in animal models. While protection from disease is considered the *gold standard*, indirect methods of assessment are usually more practical and affordable (Table 1).

Vaccines often induce a measurable immunological response (e.g., antibody titer, T cell function) which can be compared to protection trials or used as endpoints themselves. If these immunological parameters are proportional to the observed protection and have a plausible mechanism of action against the targeted pathogen they are called correlates of protection. These correlates of protection then can be used to predict expected protection against a disease challenge [18,19]. For many pathogens, the vaccine-induced correlate of protection is identical or similar to the immune response found following recovery from disease (e.g., antibodies against influenza, Ebola hemorrhagic fever) [20].

The use of correlates of protection can be problematic. For example, in Dengue virus infection, anti-virus antibodies may be associated with protection [21] or, conversely, with the development of severe dengue hemorrhagic fever upon subsequent challenge [22,23]. Herpes virus infections may establish latency or persistent infection despite an appropriate immune response by the host [24]. Finally, in HIV-1, rapid mutations may permit escape from neutralizing antibody responses [25] while integration of viral genetic material into the host genome allows for a persistent reservoir of virus despite immune activation [26]. Efforts are ongoing to define immunologic and virologic correlates of HIV-1 prevention that may propel HIV-1 vaccine development forward.

TABLE 1 Examples of direct and indirect vaccine assessments

Vaccine against	Direct assessment	Indirect assessment	References
Influenza	Protection from disease	Mucosal and serum antibody (neutralizing and hemagglutination inhibition)	[5]
Polio	Protection from disease	Mucosal and serum antibody (neutralizing)	[13]
Smallpox	Protection from disease	Serum antibody titer (neutralizing)	[14,15]
Tetanus	Protection from disease	Animal protection after passive infusion	[16]
Tuberculosis	Protection from disease	IFN-γ release assays	[17]

The biology of immune response to vaccines

Various types of vaccines are currently in use, including live-attenuated, inactivated, subunit, toxoid, conjugate and recombinant vaccines. Experimental vaccines using genetic material and novel vectors to prevent infection with globally relevant pathogens are being studied [27,28]. These vaccines leverage several pathways to stimulate a humoral and cellular immune response [5,29–32], that is both durable and epitope-specific [33]. Novel technologies such as host genomic and transcriptomic analyses have allowed us to better characterize vaccine and adjuvant responses [34,35]

As with any challenge to the immune system, the body must first identify and classify a stimulus as a potential threat - in this case the vaccine antigen - and then mount a response. The initial reaction is typically mediated by the innate immune system, including the complement pathways and cells responsible for antigen-presentation [36]. The innate arm of the immune system performs critical functions for survival from infection through mechanisms such as phagocytosis by neutrophils and monocytes, and killing of infected targets by NK cells. Some of these functions are assisted by adaptive immunity (e.g., antibody-dependent cell-mediated cytotoxicity (ADCC) by NK cells [31,37], opsonization of bacterial [31] or viral [38] targets mediated by preformed antibodies), but these effector mechanisms also participate during primary infection. Therefore, the initial response to vaccines is dependent on innate immunity.

Vaccine responses may be further altered, as bacterial and viral components contained within vaccines can trigger pathogen-associated molecular pattern (PAMP) receptors [39], resulting in cytokines that modify B and T cell responses. Vaccine adjuvants (e.g., squalene-based emulsions containing bacterial products) can also trigger PAMP and damage-associated molecular pattern (DAMP) receptors, while other adjuvants can use alternative pathways, such as alum-adjuvanted hepatitis B vaccine that triggers the inflammasome via NLRP3 [40,41]. Thus, the genes and pathways that can modify vaccine response include those expressed in adaptive immune cells that directly determine specificity and genes expressed in other cell types that have indirect effects.

Vaccine-induced memory is mediated by the adaptive response, B lymphocytes (responsible for antibody production) and T lymphocytes (responsible for helper functions/regulation of B lymphocyte responses and for antigen-specific cell-mediated killing), which can be recruited upon subsequent vaccine or pathogen challenge. The immune response to vaccines involves variations in B and T lymphocyte gene expression, and the participation of non-lymphocyte cells, which affects vaccine performance [42]. Those other cells also provide cell-trafficking signals that facilitate the induction of immunological memory [43]. For example, the germinal center reaction requires interaction of multiple T and B lymphocytes with dendritic cells, stromal cells, monocytes, and intracellular matrix [33,44] and the cycling of B cells through that structure to select for the highest affinity responses [43].

Research has shown that people may be genetically predisposed to become symptomatic with infectious diseases [45], and given the homology between response to infections and vaccines, these genetic differences will likely modify vaccine response. Genetics can result in profound differences in immune response (e.g., Janus kinase 3 (Jak3) deficiency leading to severe combined immunodeficiency [46]) or subtle differences (e.g., inability to use immunoglobulin (Ig) gene *IGHV1–69* to make antibodies [47]), and these effects are probably not uniform for all vaccines. Differences in gene expression occur throughout the lifetime of the host, leading to declines in immune function as we age, termed immune senescence, that can increase our susceptibility to infectious diseases [48,49]. Despite advances in our understanding of immune senescence, vaccine development in the elderly is lacking [50], and it is hoped that transcriptomic and metabolomic techniques to identify metabolic correlates of vaccine immunity will allow for advancement in this area [51]. This area of research is highly active, and examples of pathways and gene candidates that contribute to differences in vaccine response are listed in Table 2.

Small animal models

At the start of the 20th century, studies of tumor rejection suggested a role for genetic differences in immune response. This work paralleled the development of well-characterized inbred strains of mice—the number of papers increased through the 1930s [70] as many factors were associated with rejection of transplanted tumors. In 1936, Gorer published two landmark papers [71,72] that provided the link between tumor rejection and blood group antigens, identifying what came to be known as the histocompatibility-2 (*H2*) gene complex [73]. Since then, the contributions of this complex to immune responses have been extensively studied (Table 3).

Because of low cost, ease of care, and small size, mice are common small animal models in immunology research. Most studies in mice use inbred strains, and thereby enforce genetic restriction based on the particular strain used. Those restrictions are sometimes inadvertent or based on convenience, while others are intentional comparisons. These studies have identified a number of genes associated with variations in vaccine response (examples in Table 3). The *H2* gene complex has been frequently implicated [75,86–91], and this is especially so for vaccines thought to be primarily T-cell-based or for studies that primarily measure T cell response. For vaccines with neutralizing or blocking antibody titer as the primary endpoint, the Ig gene complex has been identified [89,91]. However, in many cases additional genetic loci have been found to associate with vaccine response, either in genes linked to *H2* [86] or in distinct loci [89,91], and genes that contribute to vaccine responses in mice continue to be identified.

TABLE 2 Examples of human genes involved in vaccine response

Vaccine against	Associated genes	Observed phenomenon	References
Hepatitis B (subunit)	*IL10*	Antibody titer increased (ACC haplotype)	[52,53]
	HLA-DRA, FOXP1	Antibody titer (responder vs. non-responder)	[53,54]
HIV-1 (adenovirus vector)	*HLA-B* (various alleles)	gag-specific CD8+ T cell response increased	[53,55]
Influenza	*HLA-DQB1*	Hemagglutination inhibition titer increased	[56,57]
	IL6	Hemagglutination inhibition titer increased	[58]
Measles	*HLA-B*	Seroconversion after one immunization (high for *07, low for *08)	[59]
Mumps	*HLA-DQB1*0303*	Antibody titer decreased	[53]
	HLA-DRB1, HLA-DQA1, HLA-DQB2, IL10RA, IL12RB1, IL12RB2	Lymphoproliferation (various effects)	[60]
Pertussis (acellular)	*TLR4* (and downstream genes)	Antibody titer increased	[61]
Rubella	*IL12B, IFNGR1*	IFN-γ, IL-10 secretion (various effects)	[53,62]
	DDX58, RARB, TRIM5, TRIM22	Antibody titer (various effects)	[53,63]
	DDX58, MAVS, RARB, TLR3	IFN-γ, IL-6, TNF-α secretion (various effects)	[64]
Varicella	HCP5	IFN-γ secretion (various effects)	[65–67]
Yellow Fever Virus	*TLR2, TLR7, TLR8, TLR9*	IFN-γ, IFN-α/β secretion (various effects)	[68,69]

TABLE 3 Examples of genetic variation in vaccine response in small animal and non-human primate studies

Vaccine against	Animal model	Associated gene[a]	Phenomenon	References
Malaria (FALVAC-1A)	C57BL/10 and congenics	*H2*-associated loci	Highly immunogenic, recognize sporozoites and blood-stage parasites; adjuvant-based variability	[74]
Tuberculosis (Ad35-based)	BALB/c, C57BL/6	*H2*	BALB/c less responsive than C57BL/6, differential protection observed	[75]
Tuberculosis (BCG)	C57BL/10 and A congenics	*H2*	Delayed-type hypersensitivity response under *H2* control, protection correlated	[76]
Tuberculosis (BCG or CpG-C)	Humanized/NOG	*H2*	Protective immunity induced by cytokine secretions	[77]
Influenza (DNA)	BALB/c	*H2*	neuraminidase vaccine protected all strains, hemagglutinin vaccine only protected BALB/c	[78]
Influenza (DNA)	Ferret	*H1*	Protective immunity	[79, 80]
Rabies (G5 epitope)	C57BL/6, C3H	*H2*, Ig, additional loci	C3d-P28 adjuvant conferred enhanced protection	[81]
Pertussis	NHP	Ig	Increased antibody levels; protection from severe disease, not bacterial colonization	[82]
Group A streptococcal carbohydrate	A/J, BSVS	*H2*, Ig, additional locus	A/J strain produced higher titer antibodies vs. BSVS	[83]
Streptococcus pneumoniae 23F polysaccharide	BALB/c, CBA	*H2*	BALB/c strain produced higher titer antibodies	[84]
Human papillomavirus (HPV)	NHP	*IgG*	Elicited HPV specific IgG and neutralizing antibody titers	[85]

[a] *H2, histocompatibility-2 complex; Ig, immunoglobulin complex.*

Mice are not the only small animal models, and, in some cases, there are specific reasons to avoid using them. For example, HIV-1 cannot infect mice naturally and thus mutations must be made within the virus or the mice [92,93]. Also, some HIV-1 assays have poor performance in the presence of mouse serum; this discovery led to retraction of one vaccine candidate despite promising early results [94,95]. Other small animal models that have been used in vaccine trials include guinea pigs, hamsters, ferrets, rats, and rabbits. While these other small animal models have not been as extensively characterized, genetic contributions to vaccine response have been described in guinea pigs [96], rats [97], and in transgenic rabbits [98]. Rabbits, in particular, have been used as a model of systemic *Salmonella* infection [99,100] and to develop live-attenuated vaccines against *Salmonella typhi* [99,100].

Studies in non-human primates and humans

Genomics applied to vaccine development in humans and non-human primates (NHPs) has proceeded along three main tracks: (1) the use of known genetic variations to design vaccines that target a desired response (e.g., vaccines targeted to specific HLA types), (2) the use of genetics to dissect the variation in vaccine response among recipients, and (3) screening for genetic variants known to contribute to vaccine or pathogen response to improve study design. The first two have been applied to both NHP and human studies, while the last has not yet become routine in humans.

Vaccines are designed to elicit specific immune responses, and with each advance in our understanding of the adaptive immune response, there has been a corresponding wave of new vaccine approaches. For example, advances in our understanding of humoral immunity have paralleled antibody-based vaccine development. Once pathogen-specific antibodies were discovered to mediate protection from some infections, antisera and antitoxin products were developed, ultimately leading to the current tetanus and diphtheria vaccines. Most vaccines have anti-pathogen antibodies as their correlate of immunity [100] and given the diversity of the human antibody repertoire, genome level targeting, such as specific Ig genes, has not been required, although some families of Ig genes appear to be favored [101,102] or disfavored [103,104] in some immune responses. At this time, it is an open question as to whether vaccines that target particular gene families can selectively expand those cells after human vaccination.

Once the mechanisms of T cell function were determined, vaccine design began to target T cells in hopes of developing highly specific anti-virus vaccines. Unfortunately, such approaches applied to HIV-1 have so far been disappointing [105–108]. It is clear that T cell immunity is important for vaccine-mediated protection [109,110], and it is highly likely that some antibody correlates of protection are surrogate markers of a balanced immune response that includes T cell immunity. The integral role of T cells in the development of robust immune responses has been highlighted by work on follicular T helper (Tfh) cells [111] which are highly correlated with high affinity antibody responses. Whether vaccines that target specific Tfh cell populations are needed to develop vaccines against some pathogens is not yet known.

Another approach that has the potential to leverage advances in genomics is the pathogen focused approach of reverse vaccinology [112,113]. In this technique, pathogen whole-genome sequencing is used to determine the antigenic repertoire of an organism followed by bioinformatic techniques to select antigens with the greatest potential to elicit a protective immune response [114]. An example of this approach was a pan-genome analysis of *Leptospira*, a zoonotic bacterial infection recognized as an emerging disease with more than half a million affected persons reported annually [115]; in this case the researchers used negative selection to screen surface exposed proteins as potential vaccine candidates. Reverse vaccinology approaches using pathogen genome analyses are being applied to nosocomial pathogens with high rates of antimicrobial resistance [116–118]. An extension of this strategy, synthetic genomics, artificially creates organisms using genetic material from one pathogen expressed in a different organism [119,120]; for example, this technique has been used to generate synthetic influenza vaccine stocks using hemagglutinin (HA) and neuraminidase (NA) gene sequences [121], with the potential to accelerate manufacturing during future pandemics.

Complementing studies of pathogen genomics, host genetic approaches are being used to understand differential response to vaccines, with much of the work focused on vaccine non-responders. Twin-based studies have looked at whether specific alleles are associated with vaccine response [122] (Table 2). In other studies, specific HLA alleles were associated with lack of response to vaccines against influenza [56], tuberculosis [123], hepatitis B [124]. Importantly, this research has shown that lack of response is not an all-or-nothing phenomenon; for example, non-responders to standard hepatitis B vaccine had good response to an alternative vaccine [125]. This is encouraging from a public health perspective, since it appears that the use of customized vaccines could result in greater coverage of the total population, although such approaches will be costlier and more difficult to implement globally.

A third approach is the use of genetic information to select vaccine trial cohorts. This has been especially important in NHP trials of HIV-1 candidate vaccines due to the need to use protection from infection or disease progression as the marker of vaccine efficacy. In rhesus macaques, the HLA allele *Mamu-A*01* was found to protect against disease progression in simian immunodeficiency virus (SIV) infection [126]. As a result, researchers were able to characterize immune responses and vaccines based on antigens that are important to human disease processes. The rhesus macaque model is presently used to study the *Mamu-A*01* restricted T cell response for various infectious diseases [127]. As more markers associated with progression of infection emerge, such as the protective *TRIM5* polymorphism [128], additional screenings will be applied with the goal of reducing confounders and improving trial efficiency. To date, this approach has not been widely applied to human trials; ultimately, such screenings may be economically and politically difficult to implement.

Practical considerations

The number of genes known to be involved in response to vaccines is growing steadily (Tables 2 and 3). Currently, we are unable to use genetic testing in the clinic to predict vaccine responses, and genomic-based vaccine design and delivery methods are not ready for widescale implementation, but we are getting closer. Economic restrictions in development and manufacturing create an impediment to this progress. Unlike medicines for chronic illnesses such as hypertension and diabetes, vaccines are not highly profitable, and some manufacturers have abandoned the market [1]. Development of a vaccine effective in only a subset of the population reduces potential market share, and given the attacks leveled at some well-established vaccines (e.g., measles/MMR [129,130], hepatitis B [131–134], *Haemophilus influenzae* type B [135,136]) companies may be reluctant to pursue strategies with thin profit margins.

Perhaps more importantly, there are public health considerations that may not be initially apparent. Vaccines developed in industrialized countries are provided to less affluent nations through direct purchase, via non-governmental organizations (NGOs), and through the efforts of the World Health Organization (WHO). In many cases they have dramatically reduced disease burden [137]. In 1999, a rhesus-human reassortant rotavirus vaccine (RotaShield®) was withdrawn from the United States market after report of a small but statistically significant increased risk of intussusception following the first dose [138]. Following that withdrawal, there was active debate about the decision and whether the small risk was acceptable in the US or in any other country [139]. Analysis suggested that despite the apparent increased risk, overall rates of intussusception remained constant [140], and that providing the vaccine to the developing world could have saved hundreds of thousands of lives annually [141]. Ultimately, deployment of a vaccine deemed "not good enough" for the US market was unacceptable for the manufacturer, NGOs, WHO, and world governments [142]. It is likely that a vaccine used in conjunction with genomic testing in the US or other industrialized markets would have to be deployed using the same testing globally, even if a clear population-level benefit for use of the vaccine without testing could be shown.

As noted above, genetic testing is already being used to detect possible confounders of HIV-1 vaccine candidates in NHPs and at least one *post-hoc* analysis of a human trial has been performed [143]. Testing that is cost-effective and provided in "real time" could be used to improve the quality of vaccine trials, although ethical and political considerations may make implementation difficult. Regardless, none of these confounders are absolute contraindications to the use of genomics in vaccine development or deployment. Rather, with the application of appropriate scientific and technological advances, each of these hurdles can be overcome and the profound benefits of genomic approaches realized.

Conclusions

Genomic medicine approaches are improving vaccine development and manufacturing and may lead to personalized vaccines in the future. Currently, economic factors combined with biological complexity limit our ability to use genomics to individualize vaccines in a systematic way for individual patients, but the rapid evolution of the field and changes in our ability to identify relevant genes may soon make this a feasible option. Regardless, the use of genomics, in both animal models and humans, to improve vaccine trials by quantifying confounders has the potential for real, immediate benefits.

Acknowledgments

The authors report no conflicting financial interests related to this work.

References

[1] Baker JP, Katz SL. Childhood vaccine development: an overview. Pediatr Res 2004;55(2):347–56.

[2] Riedel S. Edward Jenner and the history of smallpox and vaccination. Proc (Baylor Univ Med Cent) 2005;18(1):21–5.

[3] Bambini S, Rappuoli R. The use of genomics in microbial vaccine development. Drug Discov Today 2009;14(5):252–60.

[4] Plotkin SA. Immunologic correlates of protection induced by vaccination. Pediatr Infect Dis J 2001;20(1):63–75.

[5] Plotkin SA. Vaccines: correlates of vaccine-induced immunity. Clin Infect Dis 2008;47(3):401–9.

[6] Fleischmann RD, Adams MD, et al. Whole-genome random sequencing and assembly of Haemophilus influenzae Rd. Science 1995;269(5223):496–512.

[7] Scepanovic P, Alanio C, et al. Human genetic variants and age are the strongest predictors of humoral immune responses to common pathogens and vaccines. Genome Med 2018;10(1):59.

[8] Galazka AM, Robertson SE. Immunization against diphtheria with special emphasis on immunization of adults. Vaccine 1996;14(9):845–57.

[9] Khetsuriani N, Zakikhany K, Fau - Jabirov S, et al. Seroepidemiology of diphtheria and tetanus among children and young adults in Tajikistan: nationwide population-based survey, 2010. Vaccine 2013;31:4917–22.

[10] Oxman MN, Levin Mj Fau - Johnson GR, et al. A vaccine to prevent herpes zoster and postherpetic neuralgia in older adults. NEJM 2005;352:2271–84. [s1533–4406 (Electronic)].

[11] Weinberger B. Vaccines for the elderly: current use and future challenges. Immun Ageing 2018;15:3.

[12] Shirley D-AT, McArthur MA. The utility of human challenge studies in vaccine development: lessons learned from cholera. Vaccine (Auckl) 2011;2011(1):3–13.

[13] Faden H, Modlin JF, et al. Comparative evaluation of immunization with live attenuated and enhanced-potency inactivated trivalent poliovirus vaccines in childhood: systemic and local immune responses. J Infect Dis 1990;162(6):1291–7.

[14] Sarkar JK, Mitra AC, et al. The minimum protective level of antibodies in smallpox. Bull World Health Organ 1975;52(3):307–11.
[15] Mack TM, Noble Jr. J, et al. A prospective study of serum antibody and protection against smallpox. Am J Trop Med Hyg 1972;21(2):214–8.
[16] Edsall G. Specific prophylaxis of tetanus. JAMA 1959;171:417–27.
[17] Fletcher HA. Correlates of immune protection from tuberculosis. Curr Mol Med 2007;7(3):319–25.
[18] Knight-Jones TJD, Edmond K, et al. Veterinary and human vaccine evaluation methods. Proc R Soc B Biol Sci 2014;281(1784).
[19] Swayne DE. Avian influenza vaccines and therapies for poultry. Comp Immunol Microbiol Infect Dis 2009;32:351–63. [1878–1667 (Electronic)].
[20] Sullivan N, Yang ZY, et al. Ebola virus pathogenesis: implications for vaccines and therapies. J Virol 2003;77(18):9733–7.
[21] Sabin AB. Research on dengue during world war II. Am J Trop Med Hyg 1952;1(1):30–50.
[22] Halstead SB, O'Rourke EJ. Dengue viruses and mononuclear phagocytes. I. Infection enhancement by non-neutralizing antibody. J Exp Med 1977;146(1):201–17.
[23] Halstead SB, Nimmannitya S, et al. Observations related to pathogenesis of dengue hemorrhagic fever. IV. Relation of disease severity to antibody response and virus recovered. Yale J Biol Med 1970;42(5):311–28.
[24] Daheshia M, Feldman LT, et al. Herpes simplex virus latency and the immune response. Curr Opin Microbiol 1998;1(4):430–5.
[25] Wei X, Decker JM, et al. Antibody neutralization and escape by HIV-1. Nature 2003;422(6929):307–12.
[26] Maldarelli F, Wu X, et al. HIV latency. Specific HIV integration sites are linked to clonal expansion and persistence of infected cells. Science 2014;345(6193):179–83.
[27] Liu W, Wong YC, et al. DNA prime/MVTT boost regimen with HIV-1 mosaic Gag enhances the potency of antigen-specific immune responses. Vaccine 2018;36(31):4621–32.
[28] Gao Q, Zhang NZ, et al. Immune response and protective effect against chronic Toxoplasma gondii infection induced by vaccination with a DNA vaccine encoding profilin. BMC Infect Dis 2018;18(1):117.
[29] Moylett EH, Hanson IC. Immunization. J Allergy Clin Immunol 2003;**111**(2):S754–65. 29.
[30] Moser M, Leo O. Key concepts in immunology. Vaccine 2010;**28**:C2–13. [1873-2518 (Electronic)].
[31] Pulendran B, Ahmed R. Immunological mechanisms of vaccination. Nat Immunol 2011;12:509.
[32] Plotkin SA. Complex correlates of protection after vaccination. Vaccine 2013;56:1458–65. [1537-6591 (Electronic)].
[33] Crotty S, Felgner P, Davies H, et al. Cutting edge: long-term B cell memory in humans after smallpox vaccination. J Immunol 2003;171:4969–73. [0022–1767 (Print)].
[34] O'Connor D, Pollard AJ. Characterizing vaccine responses using host genomic and transcriptomic analysis. Clin Infect Dis 2013;57(6):860–9.
[35] Santoro F, Pettini E, et al. Transcriptomics of the vaccine immune response: priming with adjuvant modulates recall innate responses After boosting. Front Immunol 2018;9:1248.
[36] Clem AS. Fundamentals of vaccine immunology. J Glob Infect 2011;3(1):73–8.
[37] Plotkin SA. Increasing complexity of vaccine development. Vaccine 2014;212:S12–6. [1537-6613 (Electronic)].
[38] Van Strijp JA, Van Kessel KP, et al. Complement-mediated phagocytosis of herpes simplex virus by granulocytes. Binding or ingestion. J Clin Invest 1989;84(1):107–12.

[39] Tukhvatulin A, Logunov D, et al. Toll-like receptors and their adapter molecules. Biochemistry (Mosc) 2010;75(9):1098–114.
[40] Kool M, Petrilli V, et al. Cutting edge: alum adjuvant stimulates inflammatory dendritic cells through activation of the NALP3 inflammasome. J Immunol 2008;181(6):3755–9.
[41] Li H, Willingham SB, et al. Cutting edge: inflammasome activation by alum and alum's adjuvant effect are mediated by NLRP3. J Immunol 2008;181(1):17–21.
[42] Dudley DJ. The immune system in health and disease. Baillieres Clin Obstet Gynaecol 1992;6(3):393–416.
[43] Mayer CT, Gazumyan A, et al. The microanatomic segregation of selection by apoptosis in the germinal center. Science 2017;358(6360).
[44] Allen CD, Okada T, et al. Germinal-center organization and cellular dynamics. Immunity 2007;27(2):190–202.
[45] Alcaïs A, Abel L, et al. Human genetics of infectious diseases: between proof of principle and paradigm. J Clin Invest 2009;119(9):2506–14.
[46] Notarangelo LD, Mella P, Fau - Jones A, et al. Mutations in severe combined immune deficiency (SCID) due to JAK3 deficiency. Hum Mutat 2001;18:255–63. [1098–1004 (Electronic)].
[47] Sasso EH, Johnson T, et al. Expression of the immunoglobulin VH gene 51p1 is proportional to its germline gene copy number. J Clin Invest 1996;97(9):2074–80.
[48] Frasca D, Blomberg BB. Aging affects human B cell responses. J Clin Immunol 2011;31(3):430–5.
[49] Montecino-Rodriguez E, Berent-Maoz B, et al. Causes, consequences, and reversal of immune system aging. J Clin Invest 2013;123(3):958–65.
[50] Dorrington MG, Bowdish DME. Immunosenescence and novel vaccination strategies for the elderly. Front Immunol 2013;4:171.
[51] Li S, Sullivan NL, et al. Metabolic phenotypes of response to vaccination in humans. Cell 2017;169(5):862–77. e17.
[52] Hohler T, Reuss E, et al. A functional polymorphism in the IL-10 promoter influences the response after vaccination with HBsAg and hepatitis A. Hepatology 2005;42(1):72–6.
[53] Castiblanco J, Anaya J-M. Genetics and vaccines in the era of personalized medicine. Curr Genomics 2015;16(1):47–59.
[54] Davila S, Froeling FE, et al. New genetic associations detected in a host response study to hepatitis B vaccine. Genes Immun 2010;11(3):232–8.
[55] Fellay J, Frahm N, et al. Host genetic determinants of T cell responses to the MRKAd5 HIV-1 gag/pol/nef vaccine in the step trial. J Infect Dis 2011;203(6):773–9.
[56] Gelder CM, Lambkin R, et al. Associations between human leukocyte antigens and nonresponsiveness to influenza vaccine. J Infect Dis 2002;185(1):114–7.
[57] Ovsyannikova IG, Poland GA. Vaccinomics: current findings, challenges and novel approaches for vaccine development. AAPS J 2011;13(3):438–44.
[58] Poland GA, Ovsyannikova IG, et al. Immunogenetics of seasonal influenza vaccine response. Vaccine 2008;26(Suppl. 4):D35–40.
[59] Jacobson RM, Poland GA, et al. The association of class I HLA alleles and antibody levels after a single dose of measles vaccine. Hum Immunol 2003;64(1):103–9.
[60] Ovsyannikova IG, Jacobson RM, et al. Human leukocyte antigen and cytokine receptor gene polymorphisms associated with heterogeneous immune responses to mumps viral vaccine. Pediatrics 2008;121(5):e1091–9.
[61] Kimman TG, Banus S, et al. Association of interacting genes in the toll-like receptor signaling pathway and the antibody response to pertussis vaccination. PLoS One 2008;3(11):e3665.

[62] Poland GA, Kennedy RB, et al. Vaccinomics, adversomics, and the immune response network theory: individualized vaccinology in the 21st century. Semin Immunol 2013;25(2):89–103.
[63] Ovsyannikova IG, Haralambieva IH, et al. Polymorphisms in the vitamin A receptor and innate immunity genes influence the antibody response to rubella vaccination. J Infect Dis 2010;201(2):207–13.
[64] Ovsyannikova IG, Dhiman N, et al. Rubella vaccine-induced cellular immunity: evidence of associations with polymorphisms in the Toll-like, vitamin A and D receptors, and innate immune response genes. Hum Genet 2010;127(2):207–21.
[65] Steain M, Sutherland JP, et al. Analysis of T cell responses during active varicella-zoster virus reactivation in human ganglia. J Virol 2014;88(5):2704–16.
[66] Grinde B. Herpesviruses: latency and reactivation—viral strategies and host response. J Oral Microbiol 2013;5. https://doi.org/10.3402/jom.v5i0.22766.
[67] Crosslin DR, Carrell DS, et al. Genetic variation in the HLA region is associated with susceptibility to herpes zoster. Genes Immun 2015;16(1):1–7.
[68] Douam F, Soto Albrecht YE, et al. Type III interferon-mediated signaling is critical for controlling live attenuated yellow fever virus infection in vivo. MBio 2017;8(4):e00819-17.
[69] Pulendran B. Learning immunology from the yellow fever vaccine: innate immunity to systems vaccinology. Nat Rev Immunol 2009;9:741.
[70] Bittner J. A review of genetic studies on the transplantation of tumours. J Genet 1935;31(3):471–87.
[71] Gorer P. The detection of antigenic differences in mouse erythrocytes by the employment of immune sera. Br J Exp Pathol 1936;17:42–50.
[72] Gorer P. The detection of a hereditary antigenic difference in the blood of mice by means of human group A serum. J Genet 1936;32:17–31.
[73] Klein J. Seeds of time: fifty years ago Peter A. Gorer discovered the H-2 complex. Immunogenetics 1986;24(6):331–8.
[74] Kaba SA, Price A, et al. Immune responses of mice with different genetic backgrounds to improved multiepitope, multitarget malaria vaccine candidate antigen FALVAC-1A. Clin Vaccine Immunol 2008;15(11):1674–83.
[75] Radošević K, Wieland CW, et al. Protective immune responses to a recombinant adenovirus type 35 tuberculosis vaccine in two mouse strains: CD4 and CD8 T-cell epitope mapping and role of gamma interferon. Infect Immun 2007;75(8):4105–15.
[76] Smith CM, Proulx MK, et al. Tuberculosis susceptibility and vaccine protection are independently controlled by host genotype. MBio 2016;7(5): e01516–16.
[77] Grover A, Troy A, et al. Humanized NOG mice as a model for tuberculosis vaccine-induced immunity: a comparative analysis with the mouse and guinea pig models of tuberculosis. Immunology 2017;152(1):150–62.
[78] de Vries RD, Rimmelzwaan GF. Viral vector-based influenza vaccines. Hum Vaccin Immunother 2016;12(11):2881–901.
[79] Robertson JS, Nicolson C, et al. The development of vaccine viruses against pandemic A(H1N1) influenza. Vaccine 2011;29(9):1836–43.
[80] Webster RG, Fynan EF, et al. Protection of ferrets against influenza challenge with a DNA vaccine to the haemagglutinin. Vaccine 1994;12(16):1495–8.
[81] Galvez-Romero G, Salas-Rojas M, et al. Addition of C3d-P28 adjuvant to a rabies DNA vaccine encoding the G5 linear epitope enhances the humoral immune response and confers protection. Vaccine 2018;36(2):292–8.

[82] Warfel JM, Zimmerman LI, et al. Acellular pertussis vaccines protect against disease but fail to prevent infection and transmission in a nonhuman primate model. Proc Natl Acad Sci U S A 2014;111(2):787–92.
[83] Rivera-Hernandez T, Pandey M, et al. Differing efficacies of lead group A streptococcal vaccine candidates and full-length M protein in cutaneous and invasive disease models. MBio 2016;7(3):e00618-16.
[84] Park C, Kwon EY, et al. Comparative evaluation of a newly developed 13-valent pneumococcal conjugate vaccine in a mouse model. Hum Vaccin Immunother 2017;13(5):1169–76.
[85] Gupta G, Giannino V, et al. Immunogenicity of next-generation HPV vaccines in non-human primates: measles-vectored HPV vaccine versus Pichia pastoris recombinant protein vaccine. Vaccine 2016;34(39):4724–31.
[86] Tian JH, Miller LH, et al. Genetic regulation of protective immune response in congenic strains of mice vaccinated with a subunit malaria vaccine. J Immunol 1996;157(3):1176–83.
[87] Apt AS, Avdienko VG, et al. Distinct H-2 complex control of mortality, and immune responses to tuberculosis infection in virgin and BCG-vaccinated mice. Clin Exp Immunol 1993;94(2):322–9.
[88] Brandler P, Saikh KU, et al. Weak anamnestic responses of inbred mice to *Yersinia* F1 genetic vaccine are overcome by boosting with F1 polypeptide while outbred mice remain nonresponsive. J Immunol 1998;161(8):4195–200.
[89] Briles DE, Krause RM, et al. Immune response deficiency of BSVS mice. Immunogenetics 1977;4(1):381–92.
[90] McCool TL, Schreiber JR, et al. Genetic variation influences the B-cell response to immunization with a pneumococcal polysaccharide conjugate vaccine. Infect Immun 2003;71(9):5402–6.
[91] Templeton JW, Holmberg C, et al. Genetic control of serum neutralizing-antibody response to rabies vaccination and survival after a rabies challenge infection in mice. J Virol 1986;59(1):98–102.
[92] Potash MJ, Chao W, et al. A mouse model for study of systemic HIV-1 infection, antiviral immune responses, and neuroinvasiveness. Proc Natl Acad Sci U S A 2005;102(10):3760–5.
[93] Jones LD, Jackson JW, et al. Modeling HIV-1 induced neuroinflammation in mice: role of platelets in mediating blood-brain barrier dysfunction. PLoS One 2016;11(3):e0151702.
[94] LaCasse RA, Follis KE, et al. Fusion-competent vaccines: broad neutralization of primary isolates of HIV. Science 1999;283(5400):357–62.
[95] Nunberg JH. Retraction. Science 2002;296(5570):1025.
[96] Cohen MK, Bartow RA, et al. Effects of diet and genetics on mycobacterium bovis BCG vaccine efficacy in inbred guinea pigs. Infect Immun 1987;55(2):314–9.
[97] Stankus RP, Leslie GA. Genetic influences on the immune response of rats to streptococcal A carbohydrate. Immunogenetics 1975;2(1):29–38.
[98] Hu J, Schell TD, et al. Using HLA-A2.1 transgenic rabbit Model to screen and characterize new HLA-A2.1 restricted epitope DNA vaccines. J Vaccines Vaccin 2010;1(1):1–10.
[99] Roland KL, Tinge SA, et al. Reactogenicity and immunogenicity of live attenuated Salmonella enterica serovar Paratyphi A enteric fever vaccine candidates. Vaccine 2010;28(21):3679–87.
[100] Panda A, Tatarov I, et al. A rabbit model of non-typhoidal Salmonella bacteremia. Comp Immunol Microbiol Infect Dis 2014;37(4):211–20.
[101] Corti D, Suguitan Jr. AL, et al. Heterosubtypic neutralizing antibodies are produced by individuals immunized with a seasonal influenza vaccine. J Clin Invest 2010;120(5):1663–73.

[102] Sui J, Hwang WC, et al. Structural and functional bases for broad-spectrum neutralization of avian and human influenza A viruses. Nat Struct Mol Biol 2009;16(3):265–73.
[103] Ditzel HJ, Itoh K, et al. Determinants of polyreactivity in a large panel of recombinant human antibodies from HIV-1 infection. J Immunol 1996;157(2):739–49.
[104] Berberian L, Goodglick L, et al. Immunoglobulin V_H3 gene products: natural ligands for HIV gp120. Science 1993;261(5128):1588–91.
[105] McElrath MJ, De Rosa SC, et al. HIV-1 vaccine-induced immunity in the test-of-concept step study: a case-cohort analysis. Lancet 2008;372(9653):1894–905.
[106] Gray GE, Allen M, et al. Safety and efficacy of the HVTN 503/Phambili study of a clade-B-based HIV-1 vaccine in South Africa: a double-blind, randomised, placebo-controlled test-of-concept phase 2b study. Lancet Infect Dis 2011;11(7):507–15.
[107] Buchbinder SP, Mehrotra DV, et al. Efficacy assessment of a cell-mediated immunity HIV-1 vaccine (the Step Study): a double-blind, randomised, placebo-controlled, test-of-concept trial. Lancet 2008;372(9653):1881–93.
[108] Hammer SM, Sobieszczyk ME, et al. Efficacy trial of a DNA/rAd5 HIV-1 preventive vaccine. N Engl J Med 2013;369(22):2083–92.
[109] Lam JH, Chua YL, et al. Dengue vaccine-induced CD8+ T cell immunity confers protection in the context of enhancing, interfering maternal antibodies. JCI Insight 2017;2(24).
[110] van Rooij EM, de Bruin MG, et al. Vaccine-induced T cell-mediated immunity plays a critical role in early protection against pseudorabies virus (suid herpes virus type 1) infection in pigs. Vet Immunol Immunopathol 2004;99(1–2):113–25.
[111] Crotty S. Follicular helper CD4 T cells (TFH). Annu Rev Immunol 2011;29:621–63.
[112] Bruno L, Cortese M, et al. Lessons from reverse vaccinology for viral vaccine design. Curr Opin Virol 2015;11:89–97.
[113] Seib KL, Zhao X, et al. Developing vaccines in the era of genomics: a decade of reverse vaccinology. Clin Microbiol Infect 2012;18(Suppl. 5):109–16.
[114] Pizza M, Scarlato V, et al. Identification of vaccine candidates against serogroup B meningococcus by whole-genome sequencing. Science 2000;287(5459):1816–20.
[115] Evangelista KV, Coburn J. Leptospira as an emerging pathogen: a review of its biology, pathogenesis and host immune responses. Future Microbiol 2010;5(9):1413–25.
[116] Ni Z, Chen Y, et al. Antibiotic resistance determinant-focused acinetobacter baumannii vaccine designed using reverse vaccinology. Int J Mol Sci 2017;18(2):E458.
[117] Rashid MI, Naz A, et al. Prediction of vaccine candidates against Pseudomonas aeruginosa: an integrated genomics and proteomics approach. Genomics 2017;109(3–4):274–83.
[118] Nandagopal N, Elowitz MB. Synthetic biology: integrated gene circuits. Science 2011;333(6047):1244–8.
[119] Endy D. Foundations for engineering biology. Nature 2005;438(7067):449–53.
[120] Pleiss J. The promise of synthetic biology. Appl Microbiol Biotechnol 2006;73(4):735–9.
[121] Dormitzer PR, Suphaphiphat P, et al. Synthetic generation of influenza vaccine viruses for rapid response to pandemics. Sci Transl Med 2013;5(185):185ra68.
[122] Kimman TG, Vandebriel RJ, et al. Genetic variation in the response to vaccination. Community Genet 2007;10(4):201–17.
[123] Newport MJ, Goetghebuer T, et al. Genetic regulation of immune responses to vaccines in early life. Genes Immun 2004;5(2):122–9.
[124] Kruskall MS, Alper CA, et al. The immune response to hepatitis B vaccine in humans: inheritance patterns in families. J Exp Med 1992;175(2):495–502.
[125] Zuckerman JN, Sabin C, et al. Immune response to a new hepatitis B vaccine in healthcare workers who had not responded to standard vaccine: randomised double blind dose-response study. BMJ 1997;314(7077):329–33.

[126] Zhang ZQ, Fu TM, et al. *Mamu-A*01* allele-mediated attenuation of disease progression in simian-human immunodeficiency virus infection. J Virol 2002;76(24):12845–54.

[127] Li J, Srivastava T, et al. Mamu-A*01/K(b) transgenic and MHC class I knockout mice as a tool for HIV vaccine development. Virology 2009;387(1):16–28.

[128] Lim SY, Rogers T, et al. TRIM5α modulates immunodeficiency virus control in Rhesus monkeys. PLoS Pathog 2010;6(1):e1000738.

[129] Wakefield AJ, Murch SH, et al. Ileal-lymphoid-nodular hyperplasia, non-specific colitis, and pervasive developmental disorder in children. Lancet 1998;351(9103):637–41.

[130] Murch SH, Anthony A, et al. Retraction of an interpretation. Lancet 2004;363(9411):750.

[131] Herroelen L, de Keyser J, et al. Central-nervous-system demyelination after immunisation with recombinant hepatitis B vaccine. Lancet 1991;338(8776):1174–5.

[132] Nadler JP. Multiple sclerosis and hepatitis B vaccination. Clin Infect Dis 1993;17(5):928–9.

[133] Zipp F, Weil JG, et al. No increase in demyelinating diseases after hepatitis B vaccination. Nat Med 1999;5(9):964–5.

[134] Ascherio A, Zhang SM, et al. Hepatitis B vaccination and the risk of multiple sclerosis. N Engl J Med 2001;344(5):327–32.

[135] Karvonen M, Cepaitis Z, et al. Association between type 1 diabetes and *Haemophilus influenzae* type b vaccination: birth cohort study. BMJ 1999;318(7192):1169–72.

[136] DeStefano F, Mullooly JP, et al. Childhood vaccinations, vaccination timing, and risk of type 1 diabetes mellitus. Pediatrics 2001;108(6):E112.

[137] Rogers KA, Scinicariello F, et al. IgG Fc Receptor III homologues in nonhuman primate species: genetic characterization and ligand interactions. J Immunol 2006;177(6):3848–56.

[138] Murphy TV, Gargiullo PM, et al. Intussusception among infants given an oral rotavirus vaccine. N Engl J Med 2001;344(8):564–72.

[139] Bines J. Intussusception and rotavirus vaccines. Vaccine 2006;24(18):3772–6.

[140] Simonsen L, Morens D, et al. Effect of rotavirus vaccination programme on trends in admission of infants to hospital for intussusception. Lancet 2001;358(9289):1224–9.

[141] Weijer C. The future of research into rotavirus vaccine. BMJ 2000;321(7260):525–6.

[142] Glass RI, Bresee JS, et al. The future of rotavirus vaccines: a major setback leads to new opportunities. Lancet 2004;363(9420):1547–50.

[143] Forthal DN, Landucci G, et al. FcγRIIa genotype predicts progression of HIV infection. J Immunol 2007;179(11):7916–23.

Chapter 13

HIV pharmacogenetics and pharmacogenomics: From bench to bedside

Sophie Limou[*,†,‡], Cheryl A. Winkler[‡], C. William Wester[§,¶,||]
[*]Institut de Transplantation Urologie-Néphrologie, CHU de Nantes, CRTI UMR1064, Inserm, Université de Nantes, Nantes, France, [†]Ecole Centrale de Nantes, Nantes, France, [‡]Frederick National Laboratory for Cancer Research, Frederick, MD, United States, [§]Vanderbilt University Medical Center (VUMC), Department of Medicine, Division of Infectious Diseases, Nashville, TN, United States, [¶]Vanderbilt Institute for Global Health (VIGH), Nashville, TN, United States, [||]Harvard School of Public Health, Department of Immunology and Infectious Diseases, Boston, MA, United States

Definitions

ADME genes: Genes affecting the absorption, distribution, metabolism, and excretion of drugs. These are critical in pharmacogenetics, as all of these factors affect the availability of the drug at the site of its activity.

Combined antiretroviral therapy (*ART*): Effective antiretroviral therapy generally combines three or four drugs from at least two different classes of antiretroviral medications to avoid the evolution of escape mutants and to fully suppress HIV replication. These are primarily reverse transcriptase inhibitors (NRTIs and NNRTIs) and protease inhibitors, but also entry/fusion inhibitors, co-receptor antagonists and integrase inhibitors; with the current standard being 2 NRTIs plus 1 integrase strand transfer inhibitor (INSTI) given as a single tablet fixed dose combination (FDC). NOTE: Integrase strand transfer inhibitors (INSTIs) are currently in the early phases of being /implemented in resource-constrained settings of the world, but due to some preliminary safety during early pregnancy concerns (*i.e.* higher than anticipated rates of neural tube defects being seen in a birth outcomes cohort in Botswana—for which additional data are needed and are forthcoming), the widespread implementation of the recommended first-line regimen in resource-constrained settings, namely tenofovir (TDF), lamivudine (3TC), plus the INSTI dolutegravir (DTG) (commonly referred to as TLD) is in many settings only being offered to males and females not having reproductive potential. ART is essential for the treatment of HIV/AIDS, since the high

Genomic and Precision Medicine. https://doi.org/10.1016/B978-0-12-801496-7.00013-7

mutation rate of HIV allows the virus to accumulate mutations that render a single drug ineffective; however, HIV is generally unable to overcome the multiple mutations needed to escape from multiple drugs simultaneously.

Drug disposition: The term refers to all aspects of absorption, distribution, metabolism, and excretion of drugs affecting drug concentrations in the appropriate target tissue(s).

Non-nucleoside reverse transcriptase inhibitor (*NNRTI*): Molecules that block the action of HIV reverse transcriptase by binding a pocket of the reverse transcriptase and distorting the shape of the molecule.

Nucleoside reverse transcriptase inhibitor (*NRTI*): Medications based on molecules that are converted in the cell to structures analogous to DNA bases, which are incorporated into the DNA strand transcribed by HIV polymerase (or reverse transcriptase) from HIV RNA, but block the DNA from further elongation.

Protease inhibitor (*PI*): A drug that inhibits the activity of the HIV protease, which is required by HIV to cleave the precursor HIV proteins produced by the infected cell to the smaller proteins required for the assembly of new HIV virions.

Pharmacogenetics is the study of inter-individual genetic variations in the response to and toxicity of medications–in short, it is how an individual's genetics influences the favorable or undesirable effects of a particular treatment.

Pharmacogenomics is the use of knowledge about a person's genetics, and how their unique genetic make-up influences the choice and/or prescribed dosage of a particular medication or class of medications for that individual. Whereas pharmacogenetics is based on the effects of ADME genes on drug disposition and toxicities, pharmacogenomics is a multi-genic approach. One significant benefit of pharmacogenomics includes the use of the most efficacious and well-tolerated medication for an individual or group of individuals sharing the same genetic characteristics—thus providing the basis of personalized medicine.

Introduction

The motivation for identifying genes associated with HIV/AIDS phenotypes has gained momentum as effective vaccines and sterilizing (curative) treatments have failed to materialize. Discovery of genes and pathways utilized by HIV in the course of its lifecycle provide new targets for drug development, and knowledge of the genetic correlates of immune regulation is crucial to the design of effective vaccines against HIV. A major challenge is to integrate the information from "-omic" technologies (genomics, proteomics, metabolomics, and epigenomics) to improve clinical care and long-term survival of the estimated 36.9 million persons living with HIV infection [1–3] and to translate that knowledge to improve therapeutic options and reduce toxicities associated with combined antiretroviral therapy (ART) [4]. Optimally, personalized medicine for HIV-positive individuals would incorporate genetic and biomarker profiling to: (1) guide the optimal timing of initiation of ART, (2) select the best combination of drugs to avoid toxicities and adverse reactions, thus maximizing adherence and efficacy, and (3) predict risk for non-AIDS-defining conditions (specifically cardiovascular, hepatic and renal diseases, as well as the development of non-AIDS-associated malignancies) related to HIV infection and treatment; this becomes increasingly important in aging populations on ART [5,6]. Despite increasing evidence that genetic variation, particularly in ADME (absorption, distribution, metabolism, and excretion) genes, affects efficacy, tolerability, and adherence, personalized HIV treatment utilizing pharmacogenetic screening is not a part of routine care [7,8]. Since common variants in ADME genes may differ in frequency among different populations or geographic regions, these variants have the potential to affect drug tolerance for large numbers of individuals within a given population influencing public health care decisions for national drug formularies [9]. Cost-benefit analyses and randomized clinical trials are required to bring personalized HIV care to the clinic. Here, we review the role of ADME and *HLA* genes on drug-related toxicities and their potential utility in personalized medicine [4].

Pharmacogenetics of combined antiretroviral therapy toxicity

The widespread availability of ART in the developed world has resulted in remarkable reductions in HIV-associated morbidity and mortality over the past two decades [10]. Significant progress has been made in this past decade in terms of making these potentially life-saving combination antiretroviral therapies available to persons residing in resource-limited settings of the world. Currently, >14 million HIV-positive persons are receiving ART in low- and middle-income countries, the vast majority of whom reside in sub-Saharan Africa [11]. Large numbers of national initiatives offering public non-nucleoside reverse transcriptase inhibitor (NNRTI)-based ART have commenced in the region, with preliminary data documenting impressive efficacy outcomes among the vast majority of ART-treated adults [12–16].

Antiretroviral medications (ARV) are also used to prevent HIV transmission or acquisition. Landmark clinical trials presented in Rome in July, 2011 [17–19] documented the impact of ARVs in preventing HIV sexual transmission. The HIV Prevention Trials Network HPTN052 study found that HIV-positive persons initiating ART with CD4+ cell counts between 350 and 550 cells/mm^3 had an unprecedented 96% reduced risk of transmitting HIV to uninfected, stable sexual partners [17]. Clinical and prevention synergy was confirmed when early ART also reduced disease progression in infected individuals. Knowledge of one's HIV status is critical for personal decision-making. If one's HIV-positive partner does not take ART, or an HIV-negative person is potentially exposed by a non-stable partner, ARV medications can be used orally or topically (microbicide) by HIV-negative, vulnerable persons themselves to prevent infection, *i.e.*, pre-exposure prophylaxis (PrEP). The PartnersPrEP trial demonstrated a greater than

60% reduction in heterosexual HIV acquisition among men and women in Kenya and Uganda, using either daily oral tenofovir plus emtricitabine (co-formulated as Truvada™) or daily oral tenofovir alone [18]. The TDF2 trial found similar protection with daily oral Truvada™ in Botswana [19]. PrEP efficacy in heterosexuals extends protective evidence from men who have sex with men [20] and in women who used topical intermittent 1% tenofovir vaginal gel (pre/post coitus) [21]. The use of antiretroviral therapy for mother and/or child to prevent mother-to-child transmission has greatly reduced maternal transmission to infants [22–25]. It is therefore likely that the number of ARV-treated individuals who are either HIV seronegative or who do not meet the "when to start" guidelines for ART will increase as pre-exposure prophylaxis and ARV treatment is increasingly used to reduce risk of HIV transmission and acquisition.

Effective therapies have promoted immunologic recovery, but adverse metabolic complications and negative health outcomes, including ARV medication-related toxicities, have emerged as major short- and longer-term health concerns among this rising number of ART-treated persons [26–30]. Recent studies showed that ARV treatment is characterized by varying rates of adverse events and responses [31]. Up to 45% of treatment-naïve patients change or discontinue treatment during their first year of ART, primarily due to poor tolerance or adverse drug reactions rather than suboptimal virologic response [32,33]. Given the recent widespread introduction of integrase strand inhibitors (i.e. dolutegravir, elvitegravir, etc.) for first-line ART (as well as other lines (i.e. 2nd/3rd line) ART regimens when applicable, it is anticipated that the prevalence of ARV-medication related toxicities will diminish substantially given the very favorable tolerability of ISIs. Genetic variation among human beings accounts for a substantial proportion of this variability [31]. A number of associations between human genetic variants and predisposition to adverse events have been described and for some ARV medications, a clear and causal genotype-phenotype correlation has already been established [31]. Here we highlight and review the most significant associations listed by major antiretroviral medication drug class that have been published to date. A list of antiretroviral medications, their side effects and associated genetic factors are shown in Table 1.

Established pharmacogenetic predictors in HIV treatment

Nucleoside (or nucleotide) reverse transcriptase inhibitors

Abacavir hypersensitivity reaction

Abacavir (ABC) is a guanosine nucleoside reverse transcriptase inhibitor utilized as a component in ART for the treatment of HIV-1 [57]. The main toxicity associated with abacavir treatment is a potentially life-threatening hypersensitivity reaction, commonly referred to as abacavir hypersensitivity reaction (ABC HSR), which has been documented to occur in approximately 5–8% of ABC-treated patients. Clinically, ABC HSR is characterized by a combination of the following symptoms: fever, rash, and the development of constitutional,

TABLE 1 Genetic associations with components of combination antiretroviral regimens

Generic name	Association	Gene	Function	
Nucleotide reverse transcriptase inhibitors (NRTI)				
3TC ZDV	Lamivudine Zidovudine	↑Intracellular exposure	*ABCC4* (3724G>A), (4131T>G)	Disposition [34]
ABC	Abacavir	Hypersensitivity reaction	*HLA-B*5701*[a]	Toxicity [35]
TDF	Tenofovir	Proximal renal tubulopathy	*ABCC2* CATC haplotype; 1249G>A	Toxicity [36]
		Peripheral neuropathy	Mitochondrial haplogroups T and L1c	Toxicity [37,38]
		Pancreatitis	*CFTR* I717-1G>A and *SPINK-1*-112C>T (also in general population)	Toxicity [39]
Non-nucleotide reverse transcriptase inhibitors (NNRTI)				
EFV NVP	Efavirenz Nevirapine	Increased exposure	*CYP2B6*6* (516G>T), *CYP2B6**18 (983T>G)	Disposition [6,40–42]
EFV	Efavirenz	CNS side effects, neuropathies	*CYP2B6*6* (516G>T), *CYP2B6**18 (983T>G)	Toxicity [40,41]
NVP	Nevirapine	Rash; Hypersensitivity reaction/rash; hepatotoxicity	*HLA-DRB1*0101*	Toxicity [43,44]

Continued

TABLE 1 Genetic associations with components of combination antiretroviral regimens

Generic name	Association	Gene	Function	
		Hypersensitivity reaction*	*HLA-Cw8-B14* (in some populations)	Toxicity [45]
EFV	Efavirenz	Reduced plasma exposure; increase in HDL cholesterol	*ABCB1* (*MDR1*) (3435C>T)	Disposition [6,46] Toxicity
Protease inhibitors (PI)				
SQV	Saquinavir	Faster oral clearance	*CYP3A5*1*, *CYP3A5**3	Disposition [47,48]
IDV	Indinavir	Faster oral clearance	*CYP3A5**3′, *CYP3A4**1B	Disposition [34,49]
ATV IDV	Atazanavir Indinavir	Unconjugated hyperbilirubinemia and jaundice	*UGT1A1**28 (*UGTA1*-TA_7), *ABCB1* (MDR1) 3435C>T	Disposition [34,50,51] Toxicity
LPV	Lopinavir boosted with ritonavir	Clearance rates	*SLO1B1**4*4, *SLO1B1**5, *CYP3A* rs6945984, *ABCC2* rs717620	Disposition [7,8,52]
	All	Hyperlipidemia	*Protease inhibitors* (*PI*)	Toxicity [6,53,54] Disposition

[a] *Information/warnings for abacavir and specifically the need for baseline (pre-initiation of abacavir) testing for the genetic variant HLA-B*5701 is contained in both clinical guidelines ("Guidelines for the use of antiretroviral agents in HIV-1-infected adults and adolescents." Department of Health and Human Services, January 10, 2011 [55] as well as the FDA drug label (effective July 18, 2008; http://www.accessdata.fda.gov/scripts/cder/drugsatfda/index.cfm) [56]).*

Table adapted from Tozzi V. Pharmacogenetics of antiretrovirals. Antiviral Res 2010;85:190-200.

gastrointestinal, and/or respiratory symptoms; appearing within the first 6 weeks of treatment, and becoming more severe with continued dosing [58–60]. Upon abacavir discontinuation, signs of ABC HSR rapidly reverse without sequelae. However, subsequent re-challenge in persons with suspected ABC HSR after initial exposure is contraindicated, since it may result in severe morbidity or even mortality [61,62].

Immunogenetic basis for abacavir hypersensitivity reactions

The immunological basis of ABC-HSRs is not fully understood. Early post-market reports suggested an immunogenetic basis, as the incidence was higher in individuals with European ancestry [63] and familial clustering was observed among family members [64]. In 2002, two independent groups reported an association between ABC HSR and the *HLA-B*5701* allele [35,65], which was then replicated by several groups of investigators [66–68]. *HLA-B*5701* is an allele of the multi-allelic *HLA* class I *HLA-B* locus coding for cell surface glycoproteins that present antigens to T-cell receptors to elicit immune responses.

False positive clinical diagnosis and patch testing

Early studies mistakenly suggested a lower sensitivity of *HLA-B*5701* for ABC HSR in non-Caucasian populations [66], but it became evident that this observation was driven by false positive clinical diagnoses resulting from nonspecific symptoms associated with ABC HSR. To reduce the overestimation of ABC HSR clinical diagnoses over the true immunological ABC HSR, epicutaneous patch testing was developed. Patch testing can discriminate between false positive clinically-diagnosed ABC HSRs and true immunologically mediated ABC HSRs among persons who have previously ingested the medication [29,30]. Patch testing cannot be used as a predictive screening tool for ABC HSR since prior exposure to abacavir is required for an immune reaction. To date, all clinically diagnosed patients exhibiting a positive result on patch testing carry *HLA-B*5701*, illustrating the robustness of the association between this *HLA* allele and ABC HSR [68–74].

*Utility of HLA-B*5701 screening and translation to clinical practice*

Before the development of patch testing, several observational studies had demonstrated that screening for *HLA-B*5701* prior to initiation of abacavir treatment significantly reduced the incidence of ABC HSR and all-cause abacavir discontinuations [75–80] (Fig. 1). A randomized clinical trial using patch testing to accurately diagnose ABC HSR clearly showed that *HLA-B*5701* screening prior to initiation of abacavir treatment fully eliminated immunological ABC HSR. It also significantly reduced clinically diagnosed ABC HSR, suggesting that the confidence in the *HLA-B*5701*-negative screening result impacts the management of patients' symptoms [81]. Some *HLA-B*5701*-positive patients

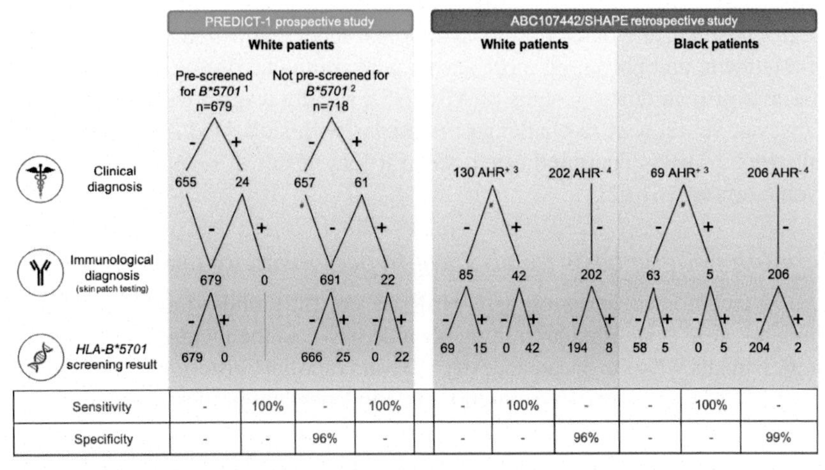

FIG. 1 Summary of the results obtained in the PREDICT-1 prospective study [70] (left, blue panel) and the SHAPE retrospective study [73] (right, red panel), focusing on the benefit of *HLA-B*5701* screening before abacavir treatment initiation for prevention of ABC-HSR.[1]*HLA-B*5701* screening before abacavir treatment initiation to exclude people suspected of developing AHR; [2]No *HLA-B*5701* screening before abacavir treatment initiation; [3]Retrospective identification of individuals who initiated abacavir treatment and who experienced a clinically diagnosed ABC-HSR; [4]Retrospective identification of individuals who initiated abacavir treatment and who did not experience any ABC-HSR; [#]Some individuals could not be tested for skin patch (5 in PREDICT-1, 3 in White-SHAPE and 1 in Black-SHAPE).

tolerate the ABC treatment and do not develop ABC HSR. Therefore, the clinical use of *HLA-B*5701* screening would exclude a subset of *HLA-B*5701*-positive individuals who would not display ABC HSR. However, it would also entirely prevent all ABC HSR (100% negative prediction), and it is therefore an ideal screening test to prevent these potentially fatal hypersensitivity reactions.

As shown in Fig. 1, *HLA-B*5701* is a sensitive (100%) marker for immunological ABC HSRs in both black and white populations in the United States [73], and potentially also in Hispanic and Thai patients [66,82]. Thus, even if the *HLA-B*5701* frequency varies across populations (Fig. 2) [83–86], the test should be generalizable. At a time of increasing migration and admixture of global populations, selective screening on the basis of race or ethnicity is not advisable. Screening for *HLA-B*5701* before the prescription of abacavir has been shown to have clinical utility—in 2008, the FDA approved changes to the drug's label, and *HLA-B*5701* screening is now recommended in clinical guidelines.

Tenofovir proximal tubulopathy

Tenofovir disoproxil fumarate (TDF) is a bioavailable ester pro-drug of tenofovir, an acyclic NRTI with activity against HIV-1, HIV-2, and hepatitis B virus [36]. The two main toxicities associated with TDF use are nephrotoxicity and bone mineral density (bone porosity) abnormalities. TDF is extensively

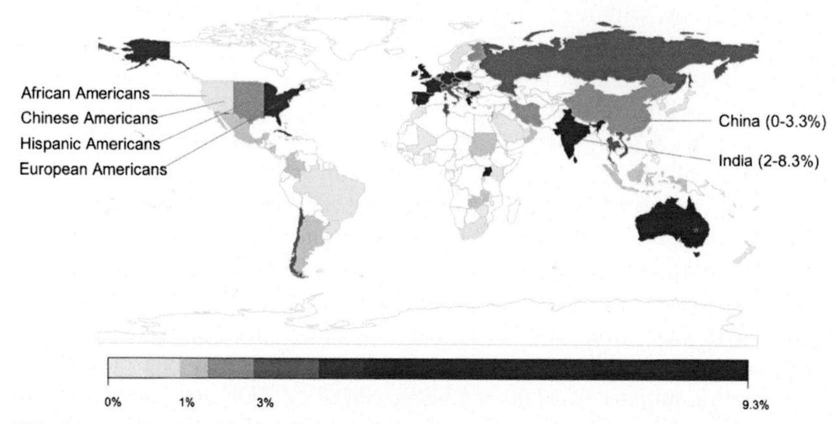

FIG. 2 *HLA-B*5701* frequency in global populations. *Australia and United Kingdom data are reported from Phillips and Mallal [60]. In allelefrequencies.net, United Kingdom is not represented, and the *B*5701* allele frequency in the Australia Aborigene (n = 103) and Caucasian (n = 134) populations is 1.5% and 0%, respectively. *All data except for Australia and United Kingdom were obtained from allelefrequencies.net (access on Feb.15, 2019).*

and rapidly excreted into the urine by the kidneys via glomerular filtration and tubular secretion [30,36]. Current routine practice is to calculate creatinine clearance (using one of the accepted formulas) and only initiate preferred TDF-containing ART if a person's creatinine clearance is greater than 50–60 mL/min, due to the potential of TDF-induced and/or exacerbated nephrotoxicity. Although large prospective trials, including accumulating data among TDF-treated adults residing in resource-limited settings of the world, have shown that TDF is very well-tolerated and relatively safe for the kidney, with a very low rate of renal insufficiency, cases of tubular dysfunction, including the development of Fanconi's syndrome, have been reported [87–90]. Renal abnormalities are relatively common in HIV-positive populations [87,91,92]. In addition, there is high inter-individual variability in the characteristics and severity of renal dysfunction associated with TDF use, which suggests an important underlying role of host genetics [87,93]. Carriers of high-risk *APOL1* genotype found only on African-ancestry chromosomes have an increased risk (odds ratio = 30) for HIV-associated nephropathy [94]; the potential interaction of *APOL1* variants with TDF or other ART components has not been investigated. Some have postulated that TDF-associated toxicity is mitochondrial in etiology [84], although TDF has not been shown to be as mitochondrially toxic as other earlier NRTIs that are no longer in routine clinical use (*e.g.*, stavudine (d4T), didanosine (ddI)). The pharmacogenetics of TDF has largely focused on the other possible mechanism of toxicity—interference with the normal function of tubular cells, specifically interference with key transporter proteins [87,93]. Multi-drug resistance (MDR) protein 2, coded by the *ABCC2* gene, is apically located in the cellular membrane of proximal renal tubular cells [50]. Several associations have been reported related to polymorphisms in transporter genes with renal clearance rates and TDF-associated tubulopathy/ renal toxicity; the *ABCC2* CATC haplotype (hallmarked by single nucleotide polymorphisms (SNPs) at positions −29, 1249, 3563, and 3972) is a determinant of proximal tubular damage, as is the *ABCC2*-24C allele [36,93,95]. In addition, the *ABCC4* (*MRP4*) 3463G allele has been associated with 35% higher intracellular concentrations of TDF compared to the wild-type allele [87,93,95], but variants in genes *SLCOA226* (*OAT1*) and *ABCCB1,* which encode other proteins implicated in TDF transport, have not shown any association with nephrotoxicity in TDF-treated patients [87,93]. Of note, older age and lower body weight have also been found to contribute to TDF tubulopathy [87]. As individuals age or lose body mass, the toxicities induced by genetic variants affecting drug metabolism, disposition, or plasma levels may be exacerbated. An increasing number of adults are being placed on TDF-containing ART in resource-replete as well as resource-limited settings of the world, including HIV-uninfected adults for the purposes of pre-exposure chemoprophylaxis (PrEP) [20]. Clearly, more research elucidating the underlying mechanism of TDF-induced renal dysfunction is needed, including investigation of gene-gene and gene-environment interactions that may contribute to TDF-induced renal dysfunction. It is critical to identify new variants and to determine the landscape of genetic variance and haplotype structure among ADME genes in diverse populations to assess the contribution of genetic and environmental factors associated with TDF-induced renal dysfunction.

Non-nucleoside reverse transcriptase inhibitors

Nevirapine-associated toxicities

The NNRTI nevirapine (NVP) is widely prescribed for the treatment of HIV-1, predominantly in resource-constrained settings of the world. NVP has been associated with two potentially serious treatment-related toxicities—cutaneous hypersensitivity reactions (seen in up to 6–10% of treated patients) and nevirapine-induced liver injury (DILI) [31,96–98]. The nevirapine-induced skin hypersensitivity reactions can manifest clinically in a myriad of ways from nevirapine-induced rash (without any systemic manifestations), to drug reaction with eosinophilia and systemic symptoms (DRESS), which is characterized by a combination of fever, rash and/or hepatitis and/or eosinophilia [99]; all the way to severe blistering skin reactions such as Stevens-Johnson syndrome and toxic epidermal necrolysis (1–2 per 1000 exposed individuals), which can be life-threatening [98]. [*Note*: Toxic epidermal necrolysis is a continuum/the most severe form of cutaneous sloughing/mucous membrane involvement and is characterized by >30% of total body surface area being involved]. Extracutaneous involvement typically manifests as hepatotoxicity and due to extensive liver involvement which typically manifests as hepatitis (*i.e.*, markedly elevated transaminases), we will refer to the NVP-induced hepatic insult as drug-induced liver involvement (DILI).

These toxicities typically occur during the first 6 weeks of therapy, especially in persons having higher baseline CD4+ cell counts. Nevirapine clearance is principally done via the cytochrome P450 pathway. Major metabolic pathways for NVP also involve hydroxylation by CYP2B6 and CYP3A4 [97]. Nevirapine induces *CYP2B6* and *CYP3A4* expression over a period of several weeks, increasing its own [97]. Genetic variants in the *ABCB1* gene, which encodes permeability glycoprotein 1 (Pgp), a multidrug resistance protein involved in removing xenobiotic substrates including antiretroviral medications from tissue and cellular compartments, have been associated with risk of NVP-induced hepatotoxicity [100–102].

Nevirapine-associated cutaneous hypersensitivity reactions

Nevirapine-treated persons may develop a cutaneous hypersensitivity reaction similar in scope and severity to ABC HSR, occurring in up to 5% of NVP-treated patients, typically within the first 6–8 weeks of treatment [31,96]. These Nevirapine-associated cutaneous hypersensitivity reactions (NVP HSR) are also typically more frequent and severe in uninfected persons who receive NVP for post-exposure prophylaxis [4,103]. Adult clinical trial data from a large cohort in Botswana (n=650) documented that 5.8% of NVP-treated adults developed treatment-modifying NVP HSR (*i.e.*, with extensive mucous membrane involvement—Stevens-Johnson syndrome), which suggests that rates of this potentially life-threatening toxicity appear similar or even slightly higher than rates reported in resource-replete settings [104].

Earlier studies in Thai populations reported an association between NVP-associated cutaneous hypersensitivity reactions and *HLA* class I alleles, specifically with *HLA-B*3505* and *HLA-Cw*0401* [105,106]. In studies evaluating 147 cases and 187 controls, Chantarangsu et al. identified an association between *HLA-B*3505* and NVP-HSR (odds ratio (OR) ~19); also predictive were *HLA-Cw*0401* (OR ~5) and the *HLA-B*3501:HLA-Cw*0401* haplotype (OR ~12) [105]. Similarly, Likanonsakul et al. reported an association between *HLA-Cw*04* and NVP-HSR, evaluating a smaller number of cases and controls [106]. Yuan et al. [97] supported these potential associations between *MHC* class I alleles *HLA-B*3505* and *HLA-Cw*0401* and NVP-HSR, evaluating a large series of patients having initial CD4+ cell counts of at least 150 cells/mm^3. In their multi-country study (n=175 cases with 2:1 matched controls), they found strong associations between *HLA-B*35*, *HLA-Cw*04* and NVP-associated cutaneous toxicities among Asians, particularly among those of Thai origin, and extended the association to African-ancestry and European-ancestry populations residing outside of Africa [97]. Results showed that among *HLA-Cw*04* carriers, NVP-HSR increased in persons concomitantly having the *CYP2B6 516G→T* polymorphism (rs3745274) [97]. In contrast, this study did not find any significant associations for *HLA-B*35* in African-ancestry and European-ancestry groups, which the authors suggest is possibly due to the relative infrequency of the *HLA-B*35* allele in these populations [97].

Recently published data [98] has shed new light onto host genetics and the risk for NVP HSR; the most severe form of hypersensitivity in terms of morbidity and mortality. These potentially life-threatening hypersensitivity cutaneous reactions manifest clinically as Steven's Johnson Syndrome (SJS) or the more extensive (in terms of body surface area) toxic epidermal necrolysis (TEN) when >30% body surface area is involved. The main finding by Carr et al. [98] was that *HLA-C*04:01* predisposes to SJS and all hypersensitivity phenotypes with the greatest risk being observed with SJS/TEN (OR=5.17 [95% CI, 2.39–11.18]). The risk associated with *HLA-B* 53:01:01/C*04:01* haplotype carriage was comparable [98]. Its sensitivity as a biomarker for SJS/TEN was 31.4%, compared to 63.9% for *HLA-C*04:01* alone, suggesting that this association may be driven by carriage of 1 risk allele at a single HLA locus [98]. This was supported by the haplotype analysis of the *HLA* locus in this particular Malawian population [98]. Of note, although *HLA-C*04* (along with *HLA-B*35*) has been associated with the risk for development of AIDS in Caucasians, no association between *HLA-C*04* and HIV has been reported in African populations or any other ethnic group [95]. The report by Carr et al. [98] of an association between nevirapine-induced SJS/TEN and *HLA-C*04:01* was the first in the literature, and this finding is consistent with prior published studies in black African (OR=5.17), Thai (OR=3.79), and Chinese (OR=3.23) populations that have reported an association with *HLA-C*04* nevirapine-cutaneous reactions. Nevirapine hypersensitivity reactions have been associated with multiple *HLA* class I and II alleles across different ethnicities [99]. The analyses revealed that, despite marked variation in the observed HLA allele repertoire across representative populations/ethnicities, the alleles associated with cutaneous NVP HSR share the structure of specific binding pockets within the antigen-binding groove [99]. While certain drug HSR syndromes show clear associations with only one specific allele, *i.e.*, abacavir with *HLA-B*57:01*, such single allele associations with 100% negative predictive values definitely appear to be the exception rather than the rule. As a result, Pavlos et al. [99] have proposed a novel analytic approach to better explore these multi-allelic associations, specifically via utilization of high-resolution typing coupled with a detailed analysis of peptide binding groove properties (including the peptide binding structure properties themselves). In a recent study [99], they performed multivariate modeling in order to examine the overall risk for cutaneous NVP HSR associated with the identified predisposing and protective clusters in both whole-cohort and ancestry-specific analyses. They found that the cluster effects remained strong when the models additionally account for the independent impact of NVP metabolism as conferred by *CYP2B6* genotype, noted previously (Table 2).

Nevirapine-associated drug-induced liver injury (DILI)

NVP-induced liver injury (DILI), which can be severe and potentially life-threatening, typically develops within the first 18 weeks following NVP initiation.

TABLE 2 Multivariate logistic regression model of predisposition to cutaneous NVP HSR, according to ancestry [99]

	ALL OR [95% CI]	Caucasian OR [95% CI]	Asian OR [95% CI]	African OR [95% CI]
*HLA-C*04:01*	4.06 [2.39–6.88]	3.14 [1.49–6.62]	7.30 [2.74–19.5]	5.60 [1.22–25.75]
Other risk F pocket (DNKLYLRNFYWTKW) *HLA-C *04:*(03/06/07), *-C*05:* (01/09), *-C*18*.01	2.91 [1.62–5.23]	3.92 [1.74–8.86]	1.64 [0.53–5.12]	3.18 [0.53–19.04]
Risk B07 alleles *HLA-B*35:05, -B*39:10. -B*51:*(*01*/02), *-B*54*:01, *HLA-B*55:*(*01*/02), *-B*56.* (01/04), *-B*67*.01, *HLA-B*78:01*	2.32 [1.42–3.79]	2.27 [1.06–4.85]	2.19 [1.10–4.35]	2.93 [0.23–37.10]
Risk non-B07 HLA-B alleles *HLA-B*13:02, B*38:*(01/*02*), *-B*39.* (*01*/05/06/09), *B*51:07*	1.76 [1.02–3.02]	1.87 [0.76–4.57]	1.50 [0.72–3.12]	5.54 [0.30–101.20]
Protective HLA-B alleles *HLA -B*15:* (01/12/24/25/27/32/35), *-B*52*:01	0.18 [0.07–0.46]	0.25 [0.06–1.15]	0.18 [0.05–0.61]	
Risk HLA-DRB1 alleles *HLA-DRB1*01:* (*01*/02/03), *-DRB1**04. (*04*/05/08/10)	2.00 [1.23–3.24]	1.34 [0.68–2.64]	3.23 [1.40–7.43]	4.77 [0.95–24.04]
SlowCYP2B6 metabolizer	2.07 [1.12–3.82]	3.19 [1.04–9.78]	1.66 [0.64–4.32]	1.81 [0.45–7.33]

Note: All models have been adjusted for ethnicity as appropriate.

It tends to occur with significantly higher frequency in the treatment of ART-naïve females having a baseline CD4+ cell count of greater than 250 cells/mm^3 and in the treatment of ART-naïve males having a baseline CD4+ cell count value of greater than 400 cells/mm^3. Patients with chronic B and/or C infection also appear to be at higher risk.

Genetic risk for NVP-associated hepatotoxicity, in contrast to NVP-associated cutaneous hypersensitivity reactions, appears to be mediated by *MHC* class II mechanisms, and is not under the influence of *CYP2B6* metabolism variants [94]. One study from Australia [4,68,97] documented *HLA-DRB*0101* as a risk for NVP-associated hepatotoxicity, whereas studies from Japan and Sardinia implicated *HLA-Cw*08* [45,107]. The Australian study [97,108] included 26 NVP-treated cases who experienced an increase in the liver enzyme alanine aminotransferase (ALT), fever and/or rash, and 209 controls. They found that *HLA-DRB1*0101* was significantly associated with hepatic/systemic reactions (OR=4.8), but not with isolated cutaneous toxicities. This elevated NVP-associated toxicity risk was only seen in persons having a CD4+ cell count percent of 25% or greater [97,108]. In their large, multi-country NVP genetic association study, Yuan et al. [97] confirmed an association between *HLA-DRB1*01* and NVP-associated hepatotoxicity among Whites [97], but they did not find any significant association with *HLA-DRB1*01* in Blacks and Asians [97]. They also found a possible association between *HLA-DQB*5* and NVP-associated hepatotoxicity in Whites, which they felt reflected a linkage between *HLA-DRB1*01* and *HLA-DQB*5* [97]. When evaluating only NVP-treated cases having isolated hepatotoxicity, without concomitant cutaneous and/or systemic involvement, they found that the association with *HLA-DQB*05* was no longer significant, but that the *HLA-DRB1*01* effect size increased from OR 3.0–3.6 [97].

The risk for NVP-associated hepatotoxicity is also influenced by its metabolism in hepatocytes *via* the cytochrome P450 system, specifically involving the transporter enzyme P-glycoprotein [4,31,100–102]. The 3435C>T variant in the *ABCB1* gene influences the risk of NVP-associated hepatotoxicity. Individuals with the *ABCB1* 3435C allele have a much higher risk of hepatotoxicity [101], especially among those co-infected with hepatitis B; in such cases, hepatotoxicity occurred 82% of the time [101]. Individuals with the T allele have a correspondingly lower risk of hepatotoxicity (OR=0.25; P=.021) [101]. Similar protective effects of the T allele (*ABCB1* 3435C>T) were also observed by Haas et al. [100] and Ciccacci et al. [102] in studies of HIV-positive adults receiving NVP-based ART in South Africa [100] and Mozambique [102], respectively. These resource-limited countries are in southern Africa, where HIV infection rates are among the highest in the world, and nevirapine is widely prescribed because of its low cost and efficacy (Table 3).

In summary, it appears that fundamentally different pathways are involved with the development of the most significant and potentially severe NVP-associated toxicities—cutaneous hypersensitivity reactions and hepatotoxicity [97].

TABLE 3 Nevirapine adverse effects studies

Genetic risk allele	Geographic region and continental ancestry	Risk group	References
Nevirapine hypersensitivity reaction			
*HLA-B*3505 & HLA-Cw*04*	Thailand	Asian	[64]
*HLA-B*3505*	Asians, African-descent (blacks) and European-descent (whites) enrolled from Argentina, Australia, Canada, France, Germany, Netherlands, Spain, Taiwan, Thailand, United Kingdom, and USA	Asian	[97]
*HLA-Cw*04*	Asians, African-descent (blacks) and European-descent (whites) enrolled from Argentina, Australia, Canada, France, Germany, Netherlands, Spain, Taiwan, Thailand, United Kingdom, and USA	All	[97]
*HLA-Cw*04*	Thailand [63]	Asian	[106]
*HLA-Cw*08*	Japan [64]	Asian	[107]
*HLA-Cw*08*	Sardinia [65]	European	[45]
CYP2B6 516T	Asians, African-descent (blacks) and European-descent (whites) enrolled from Argentina, Australia, Canada, France, Germany, Netherlands, Spain, Taiwan, Thailand, United Kingdom, and USA	All	[97]
*CYP2B6 516TT and HLA-Cw*04*	Asians, African-descent (blacks) and European-descent (whites) enrolled from Argentina, Australia, Canada, France, Germany, Netherlands, Spain, Taiwan, Thailand, United Kingdom, and USA	African descent	[97–99]
Nevirapine hepatotoxicity			
*HLA-DRB1*0101*	Australia	European	[44]
*HLA-DRB1*01*	France	European	[43]

Continued

TABLE 3 Nevirapine adverse effects studies—cont'd

Genetic risk allele	Geographic region and continental ancestry	Risk group	References
*HLA-DRB*01*	Asians, African-descent (blacks) and European-descent (whites) enrolled from Argentina, Australia, Canada, France, Germany, Netherlands, Spain, Taiwan, Thailand, United Kingdom, and USA	European descent	[97]
ABCB1 3435C>T	South Africa	Black Africans	[100]
ABCB1 3435C>T	Tennessee [66]	European and African Americans	[100]
ABCB1 3435C>T	Mozambique	Black Africans	[102]

Nevirapine-associated adverse reactions (cutaneous hypersensitivity and/or hepatic toxicities). Published with permission from author David W. Haas, MD (as presented at the "Pharmacogenomics: A Path Towards Personalized HIV Care" meeting; June 16th–17th, 2010 (convened in Rockville, MD, USA); sponsored by the National Institute of Allergy and Infectious Diseases (NIAID), National Institutes of Health, Bethesda, MD).

Both ADME genes and *HLA* class I and class II genes (Tables 1–3) affect risk of NVP-associated toxicity. Cutaneous hypersensitivity reactions seem to be primarily mediated *via* MHC class I mechanisms and influenced by NVP plasma levels regulated by *CYP2B6* 516G>T variants. On the other hand, hepatotoxicity is primarily mediated *via* MHC class II mechanisms [97] with NVP plasma levels or *CYP2B6* genotype showing little influence. However, in persons co-infected with the hepatitis B virus, the *ABCB1* 3435C>T variant is strongly associated with hepatotoxicity. Unfortunately, the presence or absence of these particular ADME and *HLA* variants lack sensitivity, and therefore have minimal clinical utility independently for persons initiating NVP-containing ART regimens [97], although it is feasible that by combining genetic and environmental risk factors, a more sensitive predictive risk score might be obtained [52]. Large proportions of NVP-treated persons developing cutaneous and/or hepatic toxicities carried no risk alleles for any of the genes, highlighting the need for in-depth genetics research to identify additional genetic factors associated with NVP toxicities.

Efavirenz clearance/drug levels

The NNRTI efavirenz (EFV) is widely prescribed for the treatment of HIV-1 infection, but central nervous system (CNS) toxicities occur commonly, and population-specific differences in pharmacokinetics and treatment response have been noted [46,72–75,109,110]. The enzyme CYP2B6 metabolizes EFV with minor involvement of CYP3A4 and CYP3A5, and since there is considerable variability in expression levels and function of *CYP2B6*, significant inter-individual variability exists in EFV exposure and the development of CNS abnormalities, its main treatment-related toxicity. The CNS toxicity manifests as abnormal and/or vivid dreams (or nightmares), dizziness, depression, feelings of intoxication/lethargy, confusion/abnormal thoughts, euphoria, aggression (and other forms of unusual behavior), as well as homicidal and suicidal ideation (rarely) [31,111]. As a result, *CYP2B6* polymorphisms and their association with EFV disposition have been extensively evaluated [31]; NCBI dbSNP and 1000 Genomes Project [112] list close to 500 missense or frame-shift variants in the gene encoding for *CYP2B6* to date, including >100 common genetic variants. The common CYP2B6*6G→T variant, which has the lowest functional enzyamatic activity, has a broad range of allele frequencies (~15–60%). In sub-Saharan Africa and in African Americans, ~20% of individuals are homozygous for the T allele [113].

Haas et al. [77], in their sub-study of HIV-positive patients participating in adult clinical trials group study 5097 (5097s), reported that the T allele of the *CYP2B6*6516*G→T polymorphism (Glu172His, rs3745274) was associated with greater EFV plasma exposure ($P<.0001$, Fig. 3A); the TT genotype most strongly associated with CNS toxicity is nearly seven-fold more frequent in African-Americans (20%) than in European-Americans (3%). Haas reported that EFV exposure was higher for carriers of the T allele with TT as slow EFV metabolizers; the area under the plasma concentration time curve (AUC) between 0 and 24h, according to G/G, G/T, and T/T genotype, was 44 (n=78),

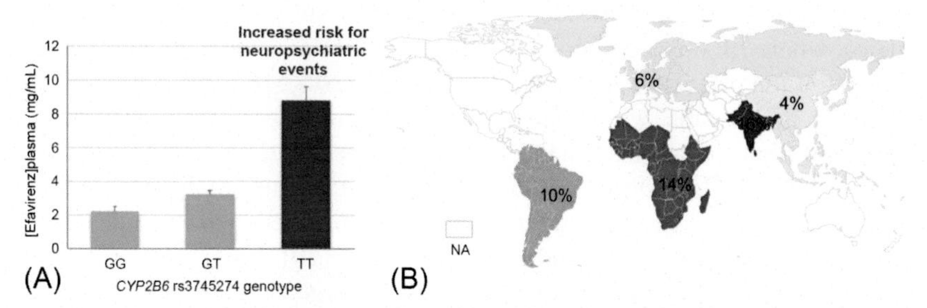

FIG. 3 Role for *CYP2B6* rs3745274 (*CYP2B6*6516*G→T) genotypes in Efavirenz plasma concentrations and adverse neuropsychiatric events. Higher Efavirenz plasma concentrations have been associated with an increased risk for adverse neuropsychiatric events. (A) Higher Efavirenz plasma concentrations are associated with slow metabolizing *CYP2B6* rs3745274 G to T allele (GG fast metabolizers, TT slow metabolizers and GT intermediate-to-fast phenotype). (B) Continental distribution of *CYP2B6* rs3745274 TT slow metabolizer genotype. *Panel (A): Reported data were adapted from Dalwadi DA, Ozuna L, Brian H. Harvey BH et al. Adverse neuropsychiatric events and recreational use of Efavirenz and other HIV-1 antiretroviral medications. Pharmacol Rev 2018;70:684–711; Panel (B): Data were obtained from gnomad.broadinstitute.org (access on Feb.15, 2019).*

60 (n=60), and 130 (n=14) μg/mL, respectively ($P < .0001$) [40]. Carriage of *CYP2B6* 516T, associated with elevated EFV exposure, was also associated with CNS symptoms at week 1 ($P = .036$) [40]. This common variant has a broad range of allele frequencies over the world (~4–60%, Fig. 3B). In regions of sub-Saharan Africa and in African Americans, up to ~15–20% of individuals are homozygous for the T allele.

Additional polymorphisms have been described that are associated with the loss or reduction in *CYP2B6* activity [4]. These include 983T→C (allele *CYP2B6*18*), 785A→G (allele *CYP2B6*16*), 593T→C (allele *CYP2B6*27*), and 1132C→T (allele *CYP2B6*28*), which, particularly in homozygous individuals, show evidence for heightened risk for development of high plasma EFV concentrations [4]. Although EFV is not a substrate of the permeability glycoprotein (P-glycoprotein) drug transporter encoded by *ABCB1*, Fellay et al. [114] have suggested that a polymorphism in the *ABCB1* gene may be associated with low plasma EFV concentrations.

The utility of EFV pharmacogenetics data in personalized treatment of HIV-1 was first shown by a small study where a patient's genotype was used to guide dosage of EFV. Lower doses (400 mg instead of the usual 600 mg) were administered to HIV-positive adults identified as being carriers of the *CYP2B6* 516T allele. Gatanaga et al. found that the majority (72%) of patients receiving lower EFV dosing had significantly reduced CNS symptoms without compromising virologic efficacy [115]. The randomized double-blind study ENCORE1 study reported similar efficacy of 400 mg versus 600 mg in a study of 630 treatment-naïve participants with only a modest reduction in CNS side effects [116]. As a consequence of concerns of CNS toxicities, USA guidelines no longer recommend EFV-based therapy for first-line treatment [117].The FDA also approved 400 mg Efavirenz (rather than 600 mg) together with lamivudine and tenofovir DR for use outside the United States in Association with the President's Emergency Plan [118]. In sub-Saharan Africa, where EFV is a cornerstone component of ART, there are conflicting reports on the effects of EFV clearance by *CYP2B6* 516T and CNS-related toxicities in Africans, with some studies finding that slow metabolizers experienced higher CNS disturbances while others found no [23–25]or fewer adverse events [119]. Differences in CNS adverse events among populations suggest that genetic background may affect *CYPB6*-mediated drug clearance and metabolite composition and abundance [119,120]. Large-scale randomized trials and cost-effectiveness analyses of targeted *CYP2B6* screening need to be performed in order to bring pharmacogenetics to clinical practice. Such studies will inform policymakers in regions of the world where higher proportions of those initiating ART may be at risk for these potentially debilitating EFV-related CNS toxicities that may compromise drug efficacy due to poor adherence. If unaddressed, poor adherence leading to a loss of virologic control may compromise the long-term success of such regimens and lead to unnecessarily high rates of circulating drug-resistant virus.

Protease inhibitors

Atazanavir-induced hyperbilirubinemia

Ritonavir-boosted atazanavir is one of the most commonly prescribed protease inhibitors for first-line ART. Atazanavir/ritonavir (ATV/r) is typically very well-tolerated, but the potential to develop hyperbilirubinemia is a cosmetically unappealing side effect. Early clinical trials suggested that ATV-associated hyperbilirubinemia occurred in 20–48% of patients. This abnormality results from the inhibition of the drug-transporter uridine diphosphate-glucuronosyltransferase 1A1, which is encoded by the gene *UGT1A1* [121–124]. This enzyme catalyzes the glucuronidation of bilirubin [46]. An increased number of thymine adenine (TA) repeats has been associated with diminished UGT1A1 activity, and persons homozygous for the 7 TA repeats (the *7/7* genotype) have chronic hyperbilirubinemia (Gilbert's syndrome) [121,122]. Promoters containing seven thymine adenine (TA) repeats (allele *UGT1A1*28*) have been found to be less active than the wild-type six repeats, resulting in lower gene expression and decreased UGT1A1 activity. Levels of UGT1A1 persons homozygous or even heterozygous for *UGT1A1**28 have been shown to be lower than those in persons with the wild-type six repeats [120,121]. UGT1A1 is important for the metabolism of numerous medications, and among patients treated for solid tumors, the homozygous *UGT1A1*28* genotype has been shown to predict toxicity (including diarrhea and neutropenia) to the anti-neoplastic medication irinotecan, a substrate of *UGT1A1* [122]. The homozygous *UGT1A1*28* genotype has also been associated with hyperbilirubinemia in 353 patients receiving the PI ATV in phase 1 clinical trials [122]. Among patients achieving therapeutic serum drug concentrations, the **28/*28* genotype was highly predictive of total serum bilirubin increases to greater than 2.5 mg/dL [122]. The *UGT1A1*28* genotype clearly appears to be associated with the development of ATV-induced hyperbilirubinemia. However, since ATV-induced hyperbilirubinemia is purely cosmetic and is not associated with any morbidity or mortality, genetic screening and/or testing for *UGT1A1*28* in HIV-positive individuals may have a low clinical priority. However, the development of stigmatizing, overt jaundice may lead to poor adherence or discontinuation of atazanavir.

Nephrolithiasis (kidney stones) is a rare but potentially serious toxicity of atazanavir. The stones from patients with ATV-related nephrolithiasis contain the drug, which is insoluble in alkaline fluids. It is well-known that adequate absorption of ATV requires an acidic gastric pH, therefore current treatment guidelines recommend avoiding the concomitant administration of atazanavir with proton-pump inhibitors (PPIs). This toxicity does not appear to have a pharmacogenetic association.

Lopinavir/ritonavir plasma concentrations

There is marked inter-individual variability in plasma concentrations of most, if not all, antiretroviral medications [125]. This is particularly true for PIs, for which therapeutic drug monitoring (TDM) should be considered for optimizing treatment in some circumstances [84]. Attaining and maintaining therapeutic plasma PI concentrations is critical for efficacy, as sub-therapeutic levels can lead to the emergence of drug resistance [125]. Ritonavir-boosted lopinavir (lopinavir/ritonavir; often referred to as LPV/r) is a commonly prescribed PI, especially in resource-limited settings of the world where it is typically the only PI available and is reserved for use in second-line ART regimens. In addition to ADME genes, age, gender and underlying synthetic liver function have all been shown to affect plasma drug concentrations [125].

PIs are primarily metabolized by the CYP3A enzyme system, but they are also substrates for the drug efflux transporters ABCB1, ABCC1, and ABCC2 [125]. These proteins affect integral components of PI drug metabolism, specifically absorption, distribution, and clearance [125]. In addition, many of the genes encoding these important efflux transporter proteins are highly variant, with functional polymorphisms [125]. Organic anion-transporting polypeptides (OATP), encoded by the *SLCO* genes, represent a family of membrane transport proteins involved in the influx of numerous chemical compounds [125]. These OATPs have varied substrate specificity and are often expressed simultaneously [125]. Numerous genetic polymorphisms have also been reported in the *SLCO1B1, SLCO1A2,* and *SLCO1B3* genes [125]. SNPs located within the trans-membrane domain of *OATP1B1* have been associated with decreased transport function, both *in vitro* and *in vivo* [125].

The *SLCO1B1* 521T→C (Val174Ala) (rs4149056) C allele is frequent in European populations (~15%) but is rare or near-absent in sub-Saharan Africa (3%) and Oceania [125,126]. Hartkoorn et al. [125] have recently shown that the C allele of *SLCO1B1* was significantly associated with higher plasma lopinavir levels [125]. The clinical utility of testing for *SLCO1B1* 521T→C is not apparent. The widespread use of ritonavir-boosted lopinavir is limited in resource-replete countries where the polymorphism is common. In sub-Saharan Africa, where increasing numbers of persons are being treated with this protease inhibitor as a second-line treatment, the polymorphism is rare-to-absent—genetic testing is therefore not required for dosage determinations. No associations with lopinavir plasma levels were observed for functional variants of *SLCO1A2* and *SLCO1B3* [125]. Additional studies are ongoing to assess the influence of other SNPs and functional variants within this important *SLCO* gene family [125].

Ritonavir-boosted protease inhibitors and lipodystrophy

Differential metabolic effects of various PIs have been reported [127]. A list of metabolic changes due to PIs is presented in Table 4, adapted from Flint et al. [128]. Metabolic complications of HIV infection and its treatment frequently present as lipodystrophy, an umbrella term encompassing loss of peripheral adipose tissue (lipoatrophy), redistribution of adipose tissue/central fat accumulation (*e.g.*, buffalo hump or visceral fat) (lipohypertrophy), and dyslipidemia or abnormalities in plasma lipids. These abnormalities may be associated with

TABLE 4 Multiple metabolic effects of protease inhibitors and possible consequences

Organ	Effect	Consequence
Adipose tissue	↓Glucose uptake	Lipoatrophy
	↓Triglyceride synthesis	Muscles, intracellular lipid levels; visceral fat disposition
Liver	↑Very low density lipid proteins	↑Triglycerides, Pancreatitis, arteriosclerosis
	↑ApoB[a]	Arteriosclerosis, Fatty liver (steatosis)
	↑Hepatic glucose production	Type 1 Diabetes
	↑SKEBP-1c[b]	Insulin resistance, atherosclerosis
Muscle	↓Glucose uptake	Glucose intolerance, Type 2 Diabetes
	↑Intramyocellular lipid levels	Increased visceral adiposity
Pancreas	↓Insulin secretion	Glucose intolerance, Type 2 Diabetes

[a] *Apolipoprotein B is the primary apolipoprotein of low density lipoproteins, responsible for cholesterol transport to tissues.*
[b] *SKEBP-1c is a transcription factor sterol regulatory element binding protein-1c inducing the expression of pathway genes for glucose utilization and fatty acid synthesis.*
Adapted from Flint OP, Noor MA, Hruz PW, et al. The role of protease inhibitors in the pathogenesis of HIV-associated lipodystrophy: cellular mechanisms and clinical implications. Toxicol Pathol 2009;37:65–77.

insulin resistance [129,130]. The biochemical disturbances observed in patients on PI-containing ART regimens resemble metabolic syndrome in the general population and may increase the risk of cardiovascular disease [131–133]. Table 3 provides a summary of the metabolic effects of HIV PIs and their potential end organ consequences [128]. Body composition changes due to HIV infection (*i.e.*, wasting syndrome) or due to PI components of ART (central fat accumulation in varying presentations) may be stigmatizing and affect treatment adherence [134].

Metabolic syndrome in the general population is also associated with cardiovascular diseases. Two large prospective studies comprising 23,437 and 2,386 patients respectively showed an increase in myocardial infarction correlated to PIs, but not to other non-PI ART components [135,136].

The effects of PIs on dyslipidemia (hypertriglyceridemia, elevated LDL cholesterol and decreased HDL cholesterol) are acute, manifesting within 4 weeks following initiation of ritonavir in HIV-positive persons [137].These effects are independent of underlying HIV-induced dyslipidemia, as shown by the observation that ritonavir-boosted lopinavir [138] administered short-term (4 weeks) to HIV-negative men also changes lipid profiles unfavorably. Lee et al. reported increases in fasting VLDL, free fatty acids and triglycerides, but no changes in fasting LDL, HDL, IDL glucose or insulin-mediated glucose disposal. Notably, there were no changes in weight, body fat, or abdominal adipose tissue by computed tomography. The effects of PIs on serum lipid levels are independent of their effects on HIV suppression, since not all PIs induce dyslipidemia in HIV-1-infected participants or uninfected patients [139,140].

Ritonavir is associated with acute insulin resistance at high therapeutic doses and causes body composition changes and insulin resistance. Ritonavir directly inhibits the GLUT4 insulin-regulated transporter, preventing glucose from entering fat and muscle cells, leading to insulin resistance [141]. Ritonavir-boosted tipranavir and lopinavir were shown to increase subcutaneous (limb) fat without evidence of increasing visceral fat or insulin in a 48-week study [142]. A study of healthy individuals showed differential responses to atazanavir (ATZ) and ritonavir-boosted lopinavir on insulin sensitivity [127]. After a 5-day course, ATZ did not affect insulin sensitivity, in contrast to ritonavir/lopinavir (LPV/r), which induced insulin resistance. The glycogen storage rate for LPV/r was also significantly lower compared to either ATZ (36%) or placebo [143]. However, other studies have reported no difference in insulin sensitivity after 4 weeks in healthy donors [143].

The MONARK trial evaluated the potential lipodystrophic effects of ritonavir-boosted lopinavir on ART-naïve HIV-positive patients. Those receiving LPV/r monotherapy had statistically significantly lower limb fat loss (median = −63 g) compared to those on LPV/r + ZDV/3TC combination therapy (−703 g). The proportion of patients having a greater than 20% loss of limb fat was also greater in the group on triple therapy (27.3%) compared to just 3% in the monotherapy arm. In both arms of the study, no changes in trunk fat were noted. These data suggest that LPV/r, and possibly other PIs, may not be the main contributors to lipoatrophy [6].

Because of the enormous health burden of metabolic disorders in the general population, numerous genome-wide association studies (GWAS) have been performed to identify genetic variants associated with various phenotypes associated with metabolic syndrome, type 2 diabetes, serum lipid levels, and body mass in non-HIV infected persons (an online catalog of GWAS genotype-phenotype associations is available at https://www.ebi.ac.uk/gwas/gwastudies [144]). The genetic factors identified to date as contributing to these disorders have had relatively small effect sizes (low penetrance), are multi-genic, and interact with environmental factors, including antiretroviral medications. Individually, small effect size markers are not good risk predictors, but by combining their additive effects, sensitivity and specificity may be improved, increasing clinical utility.

Using a pharmacogenomics approach, Rotger et al. tested 33 GWAS SNPs and nine candidate SNPs previously reported to contribute to serum lipid levels [6]. Patients with favorable genetic scores (32%), based on the number of dyslipidemia risk alleles, were found to have lower levels of low-density lipids (LDL (cholesterol)) compared to the 53% individuals with unfavorable scores who had higher levels of LDL. Low levels of high-density lipids (HDL (cholesterol)) were also associated with unfavorable scores in 42% of patients, compared to just 17% of participants with favorable scores for dyslipidemia risk alleles [6]. The study showed that a combination of genetic scores, based on additive scores or weighted for the effect size of the risk alleles, was equally predictive for identifying persons with sustained lipodystrophy. Genes significantly associated with non-HDL cholesterol, HDL cholesterol, and triglyceride levels are indicated in Table 1. Individuals receiving the most strongly dyslipidemic ART medications and carrying more risk alleles associated with dyslipidemia (Table 5) were much more likely to develop dyslipidemia, as indicated in Fig. 4. A combination of genetic risk score and antiretroviral treatment information provided the best risk prediction for identifying risk of lipodystrophy at the individual level. Notably, the proportion of the variance in lipid variation

TABLE 5 Antiretroviral regimens grouped according to their impact on serum lipid levels

Serum lipid analyzed	Group 1	Group 2	Group 3
HDL cholesterol	NNRTI	PI	No ART
			NRTI only
Non-HDL cholesterol	No ART	PI (except ATV/r)	NA
	NRTI only	NNRTI	
	Atazanavir boosted with ritonavir (ATV/r)		
Triglycerides	No ART	Single PI-containing ART (without ritonavir)	Ritonavir-containing ART (except ATV/r)
	NRTI only	ATV/r	
	Nevirapine-containing ART without PI	Efavirenz	
	Atazanavir unboosted		

Because <0.5% of lipid determinations were made during raltegravir, etravirine, or T20 exposure, these agents were not considered in the analysis. NA indicates not applicable; NNRTI, non-nucleoside reverse transcriptase inhibitor; NRTI, nucleoside reverse transcriptase inhibitor; PI, protease inhibitor [6].

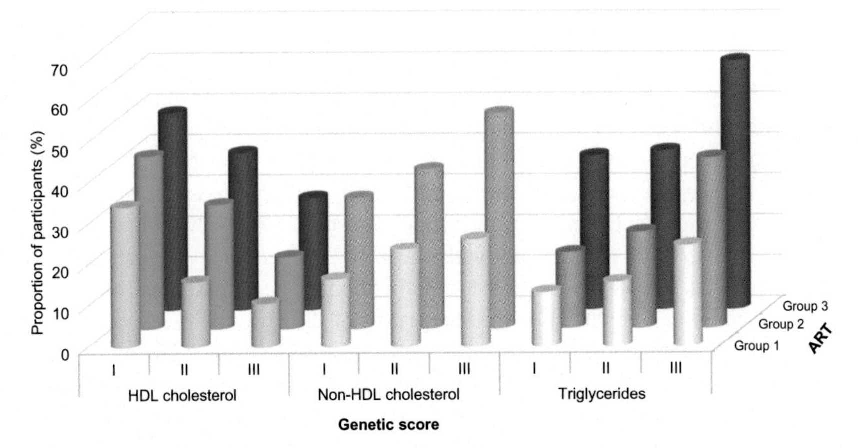

FIG. 4 Three-dimensional histogram representing the proportion of participants according to genetic score and ART group for each serum lipid levels analyzed: HDL cholesterol (blue gradient), non-HDL cholesterol (yellow gradient), and triglycerides (gray gradient). A description of ART groups is provided in Table 4. *Adapted from Rotger M, Bayard C, Taffe P, et al. Contribution of genome-wide significant single-nucleotide polymorphisms and antiretroviral therapy to dyslipidemia in HIV-positive individuals: a longitudinal study. Circ Cardiovasc Genet 2009;2:621–8.*

explained by genetic factors alone (~7.4%) or ART alone (~6.2%) was similar, indicating that genetic screening for lipid (or other drug-related toxicity) genetic risk factors may have clinical utility in the personalized selection of optimal ART regimens [5].

The future of HIV pharmacogenetics and pharmacogenomics the application of "pharmaco-omics" to personalized medicine

Although pharmacogenetics studies have increased our knowledge and understanding of antiretroviral therapy pharmacokinetics, efficacy and toxicity, very few associations will be readily translatable to clinical practice. To date, *HLA-B*5701* screening to prevent ABC HSR is the only example of pharmacogenetics translated to HIV clinical care. Because sustained HIV therapy with multiple drugs is required to avoid viral escape, and because of the complex interactions between genetic and other factors (*e.g.*, age, sex, body weight, drug-drug interactions, and co-morbidities), clinical management of HIV-positive patients is challenging. Extensive research efforts are required to discover new host factors and pathways that can be exploited to improve outcomes with antiretroviral treatment.

Hypothesis-free investigations of the genome have not been widely exploited in HIV pharmacogenetics. The major advance of the HapMap project has increased our knowledge of common genetic variations in the human genome, and combined with the development of SNP genotyping arrays, GWAS have provided significant advances in many pathologies, including HIV/AIDS [1,145] and AIDS-related complications (dyslipidemia, coronary artery disease, osteoporosis, obesity) [5]. A high-throughput pharmacogenetic association study (~1500 SNPs in ADME genes) targeting the pharmacokinetics of lopinavir co-formulated with ritonavir has revealed promising new pharmacokinetic risk alleles [8]. The next step is a shift from targeted genes (pharmacogenetics) to agnostic GWAS to identify all variants associated with drug-response phenotypes (pharmacogenomics). Pharmacogenomics has provided new insights in the management of commonly prescribed drugs for a number of disorders, including cardiovascular disease, mental illness, cancers and infectious diseases, as recently reviewed by Wang et al. [146].

Pharmacogenomic research should benefit from the additional advances offered by the large public genetic repositories, including the 1000 Genomes Project [112] and gnomad (https://gnomad.broadinstitute.org). These international projects are revealing a growing number of new sequence variants (rare to common) in human genomes, in populations representing distinct human ancestries. As this variation is revealed, it is publicly released and included in new versions of genotyping arrays, thus ensuring better coverage of human genetic diversity. Finally, the exploitation of next-generation sequencing platforms will allow rapid and full coverage of human genomic diversity and the discovery of rare or *de novo* mutations [147]. It is likely that rare variants will be associated with extreme but rare drug toxicities such as Stevens-Johnson syndrome, a major toxicity of nevirapine [148]. Rare and *de novo* variants have the potential to have larger effect sizes than common variants and may account for the missing heritability or unexplained variance seen with many common diseases and drug-related phenotypes [149,150].

Beyond the exploration of genetic variants impacting the response to HIV antiretroviral therapy, investigating transcriptomic profiles by *in vitro* and *in vivo* approaches has led to new perspectives in medications-induced toxicity [151]. Since the development of noninvasive biomarkers for clinical applications is one of the most important issues, blood is an excellent biological resource for this kind of investigation. As the study of gene expression patterns focuses only on mRNA, the results do not necessarily reflect the protein levels, and proteomics strategies could thus also participate in unraveling new candidates able to predict therapeutic responses. Interestingly, these technologies have been fruitfully exploited in rheumatoid arthritis response to treatment to identify new putative biomarkers [152].

Epigenetics is the study of heritable and somatic (non-heritable) changes in genes and gene expression that do not involved modifications of primary DNA nucleotide sequence. Epigenetic events include DNA methylation, histone modifications, and miRNA expression—consequences of epigenetic modification are changes in the regulation of gene expression and gene silencing. Recent studies have provided evidence of the important role of pharmacoepigenetics in survival and treatment response in heart disease and cancers [153–155].

Metabolomics is the study of small molecules in the blood or urine resulting from natural cellular metabolism. In the case of antiretroviral medications, biotransformation usually occurs *via* cytochrome P450 pathways. The study of metabolic profiles of antiretroviral medications is important for a number of reasons: (1) there is a potential for inhibitory or additive interactions among drug metabolites, (2) metabolites themselves may cause toxicities, (3) the metabolite profile provides an indication of clearance rates of the parent drug, and (4) alterations in metabolite levels may be biomarkers for toxicities. Also, antiretroviral medications may be metabolized by more than one metabolism enzymatic pathway, as is the case for efavirenz. The major pathway for efavirenz is *via* CYP2B6, but the minor pathways CYP3A4/3A5 are also utilized. Since the genes encoding these enzymes are extremely polymorphic, the gain or loss of function for any of these enzymes may affect metabolite profiles or clearance patterns, either or both of which may be associated with increased risk of toxicities. The role of metabolic profiles, clearance rates, and underlying genetic variation in key metabolizing genes on drug-induced toxicities has not been established for most antiretroviral medications, including the widely prescribed efavirenz [2,7,52,156,157].

The application of newer genotyping platforms that are more representative of global populations and contain 2.5–5 million rare-to-common SNPs should lead to the identification of new variant alleles associated with drug transporter genes. Although genetic variation in drug metabolism pathway genes has been well-investigated, much less is understood about the extent of variation in drug transporter genes and the potential for interactions with metabolizing variants and transporter genes on drug toxicities. Transporter genes and their proteins are critical because they modulate drug levels in cells and tissues. Thus drug transporters are key players in drug disposition and exposure in tissue compartments harboring HIV (*e.g.*, brain tissue). The *ABCB1* gene encoding the permeability glycoprotein (Pgp) is a member of the MDR subfamily. Pgp is an ATP-dependent drug efflux pump that removes drugs from the cell to the blood; Pgp also is a transporter in the blood-brain barrier. The *ABCB1* (*MDR1*) 3435C/T polymorphism predicts both PI nelfinavir plasma concentrations and CD4 T cell recovery [114]. In addition, poor-metabolizer alleles and extensive-metabolizing alleles of *CYP2D6,* encoding a cytochrome P450 isozyme, were also correlated with drug plasma concentrations in the same study, although they observed no changes in clinical CD4 cell count/immunologic recovery. Evidence of additive gene-gene interactions between transporter genes and metabolism genes in bioavailability and disposition of antiretroviral medications is not extensive. Gene-gene interactions between *CYP2B6* 516G>T and *ABCB1* 3435C>T have been associated with higher nevirapine plasma concentrations and hepatotoxicity in a population from Mozambique [102]. Other combinations of ultra-fast-metabolizing alleles (*CYP2D6*) and transporter genes (*ABCC2* rs717620) lead to sub-therapeutic levels of the PI ritonavir. Additive interactions between rapid or slow alleles of metabolizing genes and drug transporter genes have also been reported for premature discontinuation of therapy [52]. There is a reasonable likelihood that the rates of certain toxicities (*e.g.,* lactic acidosis) that apparently differ among geographic or ethnic population populations may be due to functional allele frequency differences in transporter and metabolizing genes. It is also likely that more population-specific alleles and rare variants remain to be discovered by GWAS using newer platforms. The challenge will be to identify single-gene effects as well as gene-gene and gene-environment interactions and translate these findings to the clinic to improve efficacy and reduce toxicities.

Barriers to personalized medicine in HIV-1 clinical care are not insurmountable. As potentially new genetically-predisposing hypersensitivity and other potentially serious adverse drug reactions emerge with the introduction of numerous new ARV medications (including new classes), these novel analytic approaches may be needed in order to guide clinical decision making for clinicians as they attempt to optimize ARV regimens from both a potency and tolerability perspective for their HIV-positive patients. Incomplete knowledge of drug metabolism and toxicity pathways limits the identification of specific and sensitive biomarkers that can be used as risk predictors. The integration of "pharmaco-omics" (genomics, proteomics, metabolomics, and epigenomics), using a systems-based approach, should lead to development of less toxic antiretroviral medications and to an increase in the number of informative biomarkers for risk prediction in personalized HIV-related health care. This comprehensive approach is critical as more people initiate treatment to prevent HIV transmission or acquisition, and as the population on ART ages [158].

A second barrier is that most common risk alleles have small effect sizes (OR < 1.5)—the corresponding lack of specificity and/or sensitivity makes the use of single-SNP markers for risk prediction problematic. Combining information from multiple small effect genetic risk factors into an individualized risk score has been shown to predict dyslipidemia [6,159] and premature discontinuation of ART due to toxicities [52]. These and other studies provide proof of principle that pre-screening with panels of ADME genes and other genes can be used to predict drug exposure and toxicities to optimize treatment. Determining the utility and cost-effectiveness of genetic screening to inform drug selection requires appropriately controlled, randomized clinical trials [160]. Before the expense of clinical trials can be justified, genetic factors associated with antiretroviral phenotypes must be securely validated by replication in diverse patient groups.

Application for resource-limited settings

Low-cost technology approaches for abacavir hypersensitivity genotyping

High-resolution *HLA* testing is needed to identify the *HLA-B*5701* allele, and to differentiate it from closely related alleles such as *HLA-B*5702*, *HLA-B*5703*, and *HLA-B*5801/5802*, which do not appear to be associated with ABC HSR. A high-resolution method (*i.e.*, DNA sequencing, which is expensive) may not be covered by health insurance and is not available in some settings. The development of feasible and cost-effective technologies, such as specific amplification or flow cytometry, has greatly helped to implement *HLA-B*5701* screening in clinical practice [108,161–164]. Previous genetic studies suggested absolute linkage disequilibrium (a measure of correlation between alleles of genes on the same chromosome) between the *HCP5* rs2395029 and *HLA-B*5701*. Studies with Europeans from Spain and Mexican Mestizos showed similar correlations between *HCP5* and *HLA-B*5701*, with all *HLA-B*5701* carriers also carrying the *HCP5* SNP. Since bi-allelic SNPs are more cheaply and easily genotyped than *HLA-B*5701*, *HCP5* genotyping could serve as a simple screening tool for ABC HSR, particularly in settings where sequence-based HLA typing is not available. However, further studies are required to determine if the strong correlation between *HLA-B*5701* and *HCP5* rs2395029 observed to date is generalizable to other populations.

Development of pre-treatment predictive scores

Another potential application of personalized genetic medicine in resource-limited settings may be the development of pre-treatment predictive scores. Such pre-treatment predictive scores have previously been proposed in the hepatitis C literature [165,166] by Abu Dayyeh et al. when risk stratifying patients for early and sustained virologic responses prior to antiviral therapy based on the presence (or absence) of the IL28B non-responder genotype (rs12979860), with maximal points (based on a 0, 3, and 6 point scale) being given to patients having the most favorable genotype [CC genotype as compared to having the CT or TT genotype] and/or the lowest pre-treatment hepatitis C RNA levels (<500,000 IU/mL) [165].

For example, if one were to identify specific risk alleles associated with the development of lactic acidosis, a potentially fatal toxicity, one could calculate a pre-treatment predictive score for an individual patient based on genetics (*i.e.*, the presence or absence of a deleterious risk allele) plus a key baseline or laboratory characteristic (*i.e.*, their baseline body mass index (BMI) and/or sex, as female sex and high BMI have been shown to be significant risk factors for the development of lactic acidosis [167]). Then, based on an individual's genetics, that is, if they had the risk allele, the intermediate (heterozygote) allele state, or did not have the risk allele, a predictive score would be developed. They could, for example, be given a score of 0 (no risk allele), 2 (intermediate), or 4 (presence of the risk allele). The same could be done for baseline BMI in a female patient, with overweight females being most at risk. For example, there could be scores of 0 (indicating normal BMI, 18.5–25), 2 (intermediate risk BMI, 25–30), and 4 (highest risk BMI, ≥30). Using statistics, one could then determine an appropriate pre-treatment risk score cutoff based on the negative predictive values (NPVs) of various possible values. If, for example, the NPV was >90% for a score of 4 or higher (out of a possible 8), then such recommendations could be incorporated into regional ART treatment guidelines, assuming that the cost of personalized genetics testing will decrease dramatically over the next few years and therefore be feasible for resource-limited settings. Such an example may not be as relevant now that the majority of adults initiating ART are receiving far less mitochondrially toxic NRTIs, namely tenofovir (TDF) plus emtricitabine (FTC). A more relevant example may be risk for tenofovir-associated tubulopathy and resultant renal insufficiency, which have been linked to a polymorphism in the *ABCC2* transporter [48], if a particular clinical and/or laboratory factor (*e.g.*, being above 40 years of age) was also associated with risk for this potentially untoward complication.

Conclusions

In summary, there are clinically relevant genetic predictors of ART efficacy and toxicity in addition to *HLA-B*5701*, which is the only genetic screening test that is currently standard practice in resource-replete settings. Many are of the opinion that the most important predictors of ART efficacy are yet to be discovered. There are some novel, low-cost technological approaches for existing and well-proven HIV pharmacogenetics testing, such as *HCP5* rs2395029 SNP genotyping as a proxy for *HLA-B*5701* screening, that may have substantial public health relevance and significance, and if implemented on a much larger global scale may result in substantial reductions in HIV treatment-associated morbidity and mortality. At the same time, it will be critically important to enhance pharmacovigilance in resource-limited settings by establishing regional ART outcomes databases, in order to systematically capture and longitudinally monitor these most-at-risk populations while monitoring them for the development of the most clinically relevant toxicities (*e.g.*, lactic acidosis, nevirapine cutaneous hypersensitivity and hepatotoxicity, tenofovir-associated renal toxicity, *etc.*). Pertinent pharmacogenetic discoveries, however, will only be attainable through the establishment of regional and openly collaborative DNA specimen repositories in resource-limited settings. Such an undertaking could be facilitated by the increasing collective insight and experience of clinicians, ethical review board committee members, and persons with bioinformatics expertise integrally involved in the establishment of such repositories elsewhere. While prioritizing what to study, researchers will need to carefully define their particular phenotype, and the proposed genetics screening recommendation that results from their findings will need to be easy to implement and cost-effective. The clinical benefit of genetic screening will only be determined through randomized clinical trials followed by cost-benefit analysis.

The future of HIV pharmacogenetics and specifically personalized medicine is promising, especially as the medical field is moving so rapidly toward individual genome testing. But because of the numerous variants and pathways simultaneously involved and the complexity of HIV disease and its treatment, the challenge is to integrate findings from pharmaco-omics and host genome-wide studies to inform and improve patient care. For both, the development of large, diverse cohorts is essential to confirm the associations previously suggested, to generalize the signals across populations, and to reveal new leads.

References

[1] An P, Winkler CA. Host genes associated with HIV/AIDS: advances in gene discovery. Trends Genet 2010;26:119–31.

[2] Arab-Alameddine M, Decosterd LA, Buclin T, Telenti A, Csajka C. Antiretroviral drug toxicity in relation to pharmacokinetics, metabolic profile and pharmacogenetics. Expert Opin Drug Metab Toxicol 2011;.

[3] Fellay J, Ge D, Shianna KV, et al. Common genetic variation and the control of HIV-1 in humans. PLoS Genet 2009;5:e1000791.

[4] Vidal F, Gutierrez F, Gutierrez M, et al. Pharmacogenetics of adverse effects due to antiretroviral medications. AIDS Rev 2010;12:15–30.

[5] Tarr PE, Telenti A. Genetic screening for metabolic and age-related complications in HIV-positive persons. F1000 Med Rep 2010;2:83.

[6] Rotger M, Bayard C, Taffe P, et al. Contribution of genome-wide significant single-nucleotide polymorphisms and antiretroviral therapy to dyslipidemia in HIV-positive individuals: a longitudinal study. Circ Cardiovasc Genet 2009;2:621–8.

[7] Lubomirov R, Csajka C, Telenti A. ADME pathway approach for pharmacogenetic studies of anti-HIV therapy. Pharmacogenomics 2007;8:623–33.

[8] Lubomirov R, di Iulio J, Fayet A, et al. ADME pharmacogenetics: investigation of the pharmacokinetics of the antiretroviral agent lopinavir coformulated with ritonavir. Pharmacogenet Genomics 2010;20:217–30.

[9] Nordling L. How the genomics revolution could finally help Africa. Nature 2017;544:20–2.

[10] Walensky RP, Paltiel AD, Losina E, et al. The survival benefits of AIDS treatment in the United States. J Infect Dis 2006;194:11–9.

[11] UNAIDS n.d.; http://www.unaids.org/sites/default/files/media_asset/UNAIDS_FactSheet_en.pdf

[12] Nglazi MD, Lawn SD, Kaplan R, et al. Changes in programmatic outcomes during 7 years of scale-up at a community-based antiretroviral treatment service in South Africa. J Acquir Immune Defic Syndr 2011;56:e1–8.

[13] Boulle A, Van Cutsem G, Hilderbrand K, et al. Seven-year experience of a primary care antiretroviral treatment programme in Khayelitsha, South Africa. AIDS 2010;24:563–72.

[14] Chi BH, Mwango A, Giganti M, et al. Early clinical and programmatic outcomes with tenofovir-based antiretroviral therapy in Zambia. J Acquir Immune Defic Syndr 2010;54:63–70.

[15] Bussmann H, Wester CW, Ndwapi N, et al. Five-year outcomes of initial patients treated in Botswana's National Antiretroviral Treatment Program. AIDS 2008;22:2303–11.

[16] Coetzee D, Hildebrand K, Boulle A, et al. Outcomes after two years of providing antiretroviral treatment in Khayelitsha, South Africa. AIDS 2004;18:887–95.

[17] Cohen MH. In: Antiretroviral treatment to prevent the sexual transmission of HIV-1: results from HPTN 052 multi-national randomized controlled trial. 6th international AIDS society (IAS) conference on HIV pathogenesis, treatment, and prevention. Rome, Italy; 2011.

[18] Baeten J, Celum C. In: Antiretroviral pre-exposure prophylaxis for HIV-1 prevention among heterosexual African men and women: the partners PrEP study. 6th IAS conference on HIV pathogenesis, treatment and prevention; 2011 July 17–20, 2011; Rome, Italy; 2011.

[19] Thigpen MC, Kebaabetswe PM, Smith DK, et al. In: Daily oral antiretroviral use for the prevention of HIV infection in heterosexually active young adults in Botswana: results from the TDF2 study. 6th IAS conference on HIV pathogenesis, treatment and prevention; 2011 July 17–22; Rome, Italy; 2011.

[20] Grant RM, Lama JR, Anderson PL, et al. Preexposure chemoprophylaxis for HIV prevention in men who have sex with men. N Engl J Med 2010;363:2587–99.
[21] Abdool Karim Q, Abdool Karim SS, Frohlich JA, et al. Effectiveness and safety of tenofovir gel, an antiretroviral microbicide, for the prevention of HIV infection in women. Science 2010;329:1168–74.
[22] Marazzi MC, Palombi L, Nielsen-Saines K, et al. Extended antenatal use of triple antiretroviral therapy for prevention of HIV-1 mother-to-child transmission correlates with favourable pregnancy outcomes. AIDS 2011;25:1611–8.
[23] de Vincenzi I. Triple antiretroviral compared with zidovudine and single-dose nevirapine prophylaxis during pregnancy and breastfeeding for prevention of mother-to-child transmission of HIV-1 (Kesho Bora study): a randomised controlled trial. Lancet Infect Dis 2011;11:171–80.
[24] Chasela CS, Hudgens MG, Jamieson DJ, et al. Maternal or infant antiretroviral medications to reduce HIV-1 transmission. N Engl J Med 2010;362:2271–81.
[25] WHO. In: Recommendations for use of antiretroviral medications for treating pregnant women and preventing HIV infection in infants. Guidelines on care, treatment and support for women living with HIV/AIDS and their children in resource-contrained settings; 2010.
[26] Brady MT, Oleske JM, Williams PL, et al. Declines in mortality rates and changes in causes of death in HIV-1-infected children during the HAART era. J Acquir Immune Defic Syndr 2010;53:86–94.
[27] Martinez E, Milinkovic A, Buira E, et al. Incidence and causes of death in HIV-positive persons receiving highly active antiretroviral therapy compared with estimates for the general population of similar age and from the same geographical area. HIV Med 2007;8:251–8.
[28] Palella Jr FJ, Baker RK, Moorman AC, et al. Mortality in the highly active antiretroviral therapy era: changing causes of death and disease in the HIV outpatient study. J Acquir Immune Defic Syndr 2006;43:27–34.
[29] Causes of death in HIV-1-infected patients treated with antiretroviral therapy. 1996–2006: collaborative analysis of 13 HIV cohort studies. Clin Infect Dis 2010;50:1387–96.
[30] Deeks SG, Phillips AN. HIV infection, antiretroviral treatment, ageing, and non-AIDS related morbidity. BMJ 2009;338:a3172.
[31] Tozzi V. Pharmacogenetics of antiretrovirals. Antivir Res 2010;85:190–200.
[32] Vo TT, Ledergerber B, Keiser O, et al. Durability and outcome of initial antiretroviral treatments received during 2000–2005 by patients in the Swiss HIV cohort study. J Infect Dis 2008;197:1685–94.
[33] Elzi L, Marzolini C, Furrer H, et al. Treatment modification in human immunodeficiency virus-infected individuals starting combination antiretroviral therapy between 2005 and 2008. Arch Intern Med 2010;170:57–65.
[34] Anderson PL, Lamba J, Aquilante CL, Schuetz E, Fletcher CV. Pharmacogenetic characteristics of indinavir, zidovudine, and lamivudine therapy in HIV-positive adults: a pilot study. J Acquir Immune Defic Syndr 2006;42:441–9.
[35] Mallal S, Nolan D, Witt C, et al. Association between presence of HLA-B*5701, HLA-DR7, and HLA-DQ3 and hypersensitivity to HIV-1 reverse-transcriptase inhibitor abacavir. Lancet 2002;359:727–32.
[36] Izzedine H, Hulot JS, Villard E, et al. Association between ABCC2 gene haplotypes and tenofovir-induced proximal tubulopathy. J Infect Dis 2006;194:1481–91.
[37] Canter JA, Robbins GK, Selph D, et al. African mitochondrial DNA subhaplogroups and peripheral neuropathy during antiretroviral therapy. J Infect Dis 2010;201:1703–7.

[38] Hulgan T, Haas DW, Haines JL, et al. Mitochondrial haplogroups and peripheral neuropathy during antiretroviral therapy: an adult AIDS clinical trials group study. AIDS 2005;19:1341–9.

[39] Felley C, Morris MA, Wonkam A, et al. The role of CFTR and SPINK-1 mutations in pancreatic disorders in HIV-positive patients: a case-control study. AIDS 2004;18:1521–7.

[40] Haas DW, Ribaudo HJ, Kim RB, et al. Pharmacogenetics of efavirenz and central nervous system side effects: an Adult AIDS Clinical Trials Group study. AIDS 2004;18:2391–400.

[41] Rotger M, Colombo S, Furrer H, et al. Influence of CYP2B6 polymorphism on plasma and intracellular concentrations and toxicity of efavirenz and nevirapine in HIV-positive patients. Pharmacogenet Genomics 2005;15:1–5.

[42] Tsuchiya K, Gatanaga H, Tachikawa N, et al. Homozygous CYP2B6 *6 (Q172H and K262R) correlates with high plasma efavirenz concentrations in HIV-1 patients treated with standard efavirenz-containing regimens. Biochem Biophys Res Commun 2004;319:1322–6.

[43] Vitezica ZG, Milpied B, Lonjou C, et al. HLA-DRB1*01 associated with cutaneous hypersensitivity induced by nevirapine and efavirenz. AIDS 2008;22:540–1.

[44] Martin AM, Nolan D, James I, et al. Predisposition to nevirapine hypersensitivity associated with HLA-DRB1*0101 and abrogated by low CD4 T-cell counts. AIDS 2005;19:97–9.

[45] Littera R, Carcassi C, Masala A, et al. HLA-dependent hypersensitivity to nevirapine in Sardinian HIV patients. AIDS 2006;20:1621–6.

[46] Ribaudo HJ, Liu H, Schwab M, et al. Effect of CYP2B6, ABCB1, and CYP3A5 polymorphisms on efavirenz pharmacokinetics and treatment response: an AIDS Clinical Trials Group study. J Infect Dis 2010;202:717–22.

[47] Mouly SJ, Matheny C, Paine MF, et al. Variation in oral clearance of saquinavir is predicted by CYP3A5*1 genotype but not by enterocyte content of cytochrome P450 3A5. Clin Pharmacol Ther 2005;78:605–18.

[48] Josephson F, Allqvist A, Janabi M, et al. CYP3A5 genotype has an impact on the metabolism of the HIV protease inhibitor saquinavir. Clin Pharmacol Ther 2007;81:708–12.

[49] Bertrand J, Treluyer JM, Panhard X, et al. Influence of pharmacogenetics on indinavir disposition and short-term response in HIV patients initiating HAART. Eur J Clin Pharmacol 2009;65:667–78.

[50] Rotger M, Taffe P, Bleiber G, et al. Gilbert syndrome and the development of antiretroviral therapy-associated hyperbilirubinemia. J Infect Dis 2005;192:1381–6.

[51] Rodriguez-Novoa S, Martin-Carbonero L, Barreiro P, et al. Genetic factors influencing atazanavir plasma concentrations and the risk of severe hyperbilirubinemia. AIDS 2007;21:41–6.

[52] Lubomirov R, Colombo S, di Iulio J, et al. Association of pharmacogenetic markers with premature discontinuation of first-line anti-HIV therapy: an observational cohort study. J Infect Dis 2011;203:246–57.

[53] Song Y, Stampfer MJ, Liu S. Meta-analysis: apolipoprotein E genotypes and risk for coronary heart disease. Ann Intern Med 2004;141:137–47.

[54] Talmud PJ, Hawe E, Martin S, et al. Relative contribution of variation within the APOBEC3/A4/A5 gene cluster in determining plasma triglycerides. Hum Mol Genet 2002;11:3039–46.

[55] Panel on Antiretroviral Guidelines for Adults and Adolescents. Guidelines for the use of antiretroviral agents in HIV-1-infected adults and adolescents. Department of Health and Human Services; 2011. p. 1–166. Available at http://www.aidsinfo.nih.gov/ContentFiles/AdultandAdolescentGL.pdf; (Accessed January 17, 2012) [pages 19–20].

[56] Food and Drug Administration (FDA)/Center for Drug Evaluation and Research, Office of Communications, Division of Information Services; Update Frequency: Daily; n.d. http://www.accessdata.fda.gov/scripts/cder/drugsatfda/index.cfm (the Drugs@fda website) (information for abacavir ("ZIAGEN (NDA # 020977)"); accessed January 2012.

[57] Thompson MA, Aberg JA, Cahn P, et al. Antiretroviral treatment of adult HIV infection: 2010 recommendations of the international AIDS society-USA panel. JAMA 2010;304:321–33.
[58] Chaponda M, Pirmohamed M. Hypersensitivity reactions to HIV therapy. Br J Clin Pharmacol 2011;71:659–71.
[59] Hetherington S, McGuirk S, Powell G, et al. Hypersensitivity reactions during therapy with the nucleoside reverse transcriptase inhibitor abacavir. Clin Ther 2001;23:1603–14.
[60] Phillips E, Mallal S. Successful translation of pharmacogenetics into the clinic: the abacavir example. Mol Diagn Ther 2009;13:1–9.
[61] Clay PG. The abacavir hypersensitivity reaction: a review. Clin Ther 2002;24:1502–14.
[62] Shapiro M, Ward KM, Stern JJ. A near-fatal hypersensitivity reaction to abacavir: case report and literature review. AIDS Read 2001;11:222–6.
[63] Symonds W, Cutrell A, Edwards M, et al. Risk factor analysis of hypersensitivity reactions to abacavir. Clin Ther 2002;24:565–73.
[64] Peyrieere H, Nicolas J, Siffert M, et al. Hypersensitivity related to abacavir in two members of a family. Ann Pharmacother 2001;35:1291–2.
[65] Hetherington S, Hughes AR, Mosteller M, et al. Genetic variations in HLA-B region and hypersensitivity reactions to abacavir. Lancet 2002;359:1121–2.
[66] Hughes AR, Mosteller M, Bansal AT, et al. Association of genetic variations in HLA-B region with hypersensitivity to abacavir in some, but not all, populations. Pharmacogenomics 2004;5:203–11.
[67] Hughes DA, Vilar FJ, Ward CC, et al. Cost-effectiveness analysis of HLA B*5701 genotyping in preventing abacavir hypersensitivity. Pharmacogenetics 2004;14:335–42.
[68] Martin AM, Nolan D, Gaudieri S, et al. Predisposition to abacavir hypersensitivity conferred by HLA-B*5701 and a haplotypic Hsp70-Hom variant. Proc Natl Acad Sci U S A 2004;101:4180–5.
[69] Phillips EJ, Sullivan JR, Knowles SR, Shear NH. Utility of patch testing in patients with hypersensitivity syndromes associated with abacavir. AIDS 2002;16:2223–5.
[70] Mallal S, Phillips E, Carosi G, et al. HLA-B*5701 screening for hypersensitivity to abacavir. N Engl J Med 2008;358:568–79.
[71] Phillips E, Rauch A, Nolan D, et al. Pharmacogenetics and clinical characterization of patch test confirmed patients with abacavir hypersensitivity. Rev Antivir Ther; 2006.
[72] Phillips E, Rauch A, Nolan D, et al. In: Genetic characterization of patients with MHC class I mediated abacavir hypersensitivity reaction [abstract 49]. The 4th IAS on HIV pathogenesis, Treatment and Prevention. Sydney; 2007.
[73] Saag M, Balu R, Phillips E, et al. High sensitivity of human leukocyte antigen-b*5701 as a marker for immunologically confirmed abacavir hypersensitivity in white and black patients. Clin Infect Dis 2008;46:1111–8.
[74] Shear NH, Milpied B, Bruynzeel DP, Phillips EJ. A review of drug patch testing and implications for HIV clinicians. AIDS 2008;22:999–1007.
[75] Rauch A, Nolan D, Martin A, et al. Prospective genetic screening decreases the incidence of abacavir hypersensitivity reactions in the Western Australian HIV cohort study. Clin Infect Dis 2006;43:99–102.
[76] Zucman D, Truchis P, Majerholc C, Stegman S, Caillat-Zucman S. Prospective screening for human leukocyte antigen-B*5701 avoids abacavir hypersensitivity reaction in the ethnically mixed French HIV population. J Acquir Immune Defic Syndr 2007;45:1–3.
[77] Waters LJ, Mandalia S, Gazzard B, Nelson M. Prospective HLA-B*5701 screening and abacavir hypersensitivity: a single centre experience. AIDS 2007;21:2533–4.

[78] Young B, Squires K, Patel P, et al. First large, multicenter, open-label study utilizing HLA-B*5701 screening for abacavir hypersensitivity in North America. AIDS 2008;22:1673–5.
[79] I. Reeves, D. Churchill and M. Fisher, Screening for HLA-B*5701 reduces the frequency of Abacavir hypersnsitivity reactions, In: Reeves I, Churchill D, Fisher M, editors. Screening for HLA-B*5701 reduces the frequency of Abacavir hypersnsitivity reactions [Abstract]. Antivir Ther 2006;11:L11, Abstract no. 14
[80] Trottier B, Thomas R, Nguyen VK, et al. In: How effectively HLA screening can reduce the early discontinuation of Abacavir in real life [abstract MOPEB002]. The 4th IAS Conference on HIV Pathogenesis, Treatment, and Prevention. Sydney; 2007.
[81] Phillips EJ, Wong GA, Kaul R, et al. Clinical and immunogenetic correlates of abacavir hypersensitivity. AIDS 2005;19:979–81.
[82] Mosteller M, Hughes A, Warren L, et al. In: Pharmacogenetic (PG) investigation of hypersensitivity to Abacavir [abstract 76]. The 16th international AIDS conference. Toronto; 2006.
[83] Sanchez-Giron F, Villegas-Torres B, Jaramillo-Villafuerte K, et al. Association of the genetic marker for abacavir hypersensitivity HLA-B*5701 with HCP5 rs2395029 in Mexican mestizos. Pharmacogenomics 2011.
[84] Poggi H, Vera A, Lagos M, et al. HLA-B*5701 frequency in Chilean HIV-positive patients and in general population. Braz J Infect Dis 2010;14:510–2.
[85] Orkin C, Wang J, Bergin C, et al. An epidemiologic study to determine the prevalence of the HLA-B*5701 allele among HIV-positive patients in Europe. Pharmacogenet Genomics 2010;20:307–14.
[86] Parczewski M, Leszczyszyn-Pynka M, Wnuk A, et al. Introduction of pharmacogenetic screening for the human leucocyte antigen (HLA) B*5701 variant in polish HIV-positive patients. HIV Med 2010;11:345–8.
[87] Rodriguez-Novoa S, Labarga P, Soriano V. Pharmacogenetics of tenofovir treatment. Pharmacogenomics 2009;10:1675–85.
[88] Nelson MR, Katlama C, Montaner JS, et al. The safety of tenofovir disoproxil fumarate for the treatment of HIV infection in adults: the first 4 years. AIDS 2007;21:1273–81.
[89] Barrios A, Garcia-Benayas T, Gonzalez-Lahoz J, Soriano V. Tenofovir-related nephrotoxicity in HIV-positive patients. AIDS 2004;18:960–3.
[90] Coca S, Perazella MA. Rapid communication: acute renal failure associated with tenofovir: evidence of drug-induced nephrotoxicity. Am J Med Sci 2002;324:342–4.
[91] Kopp J, Winkler CA. HIV-associated nephropathy in African Americans. Kidney Int 2003;65. S43-S9.
[92] Yanik EL, Lucas GM, Vlahov D, Kirk GD, Mehta SH. HIV and proteinuria in an injection drug user population. Clin J Am Soc Nephrol 2010;5:1836–43.
[93] Rodriguez-Novoa S, Labarga P, Soriano V, et al. Predictors of kidney tubular dysfunction in HIV-positive patients treated with tenofovir: a pharmacogenetic study. Clin Infect Dis 2009;48:e108–16.
[94] Kopp JB, Nelson GW, Sampath K, et al. APOL1 variants in focal segmental glomerulosclerosis and HIV-associated nephropathy. J Am Soc Nephrol 2011;22(11):2129–37.
[95] Kiser JJ, Aquilante CL, Anderson PL, et al. Clinical and genetic determinants of intracellular tenofovir diphosphate concentrations in HIV-positive patients. J Acquir Immune Defic Syndr 2008;47:298–303.
[96] Baylor MS, Johann-Liang R. Hepatotoxicity associated with nevirapine use. J Acquir Immune Defic Syndr 2004;35:538–9.
[97] Yuan J, Guo S, Hall D, et al. Toxicogenomics of nevirapine-associated cutaneous and hepatic adverse events among populations of African, Asian, and European descent. AIDS 2011;25:1271–80.

[98] Carr DF, Chaponda M, Jorgensen AL, Castro EC, van Oosterhout JJ. Association of human leukocyte antigen alleles and nevirapine hypersensitivity in a Malawian HIV-positive population. Clin Infect Dis 2013;56(9):1330–9.

[99] Pavlos R, McKinnon EJ, Ostrov DA, Peters B, Buus S, et al. Shared peptide binding of HLA Class I and II alleles associatewith cutaneous nevirapine hypersensitivity and identify novel risk alleles. Sci Rep 7: 8653 DOI:https://doi.org/10.1038/s41598-017-08876-0.

[100] Haas DW, Bartlett JA, Andersen JW, et al. Pharmacogenetics of nevirapine-associated hepatotoxicity: an Adult AIDS clinical trials group collaboration. Clin Infect Dis 2006;43:783–6.

[101] Ritchie MD, Haas DW, Motsinger AA, et al. Drug transporter and metabolizing enzyme gene variants and nonnucleoside reverse-transcriptase inhibitor hepatotoxicity. Clin Infect Dis 2006;43:779–82.

[102] Ciccacci C, Borgiani P, Ceffa S, et al. Nevirapine-induced hepatotoxicity and pharmacogenetics: a retrospective study in a population from Mozambique. Pharmacogenomics 2010;11:23–31.

[103] Patel SM, Johnson S, Belknap SM, et al. Serious adverse cutaneous and hepatic toxicities associated with nevirapine use by non-HIV-positive individuals. J Acquir Immune Defic Syndr 2004;35:120–5.

[104] Wester CW, Thomas AM, Bussmann H, et al. Non-nucleoside reverse transcriptase inhibitor outcomes among combination antiretroviral therapy-treated adults in Botswana. AIDS 2010;24(Suppl 1):S27–36.

[105] Chantarangsu S, Mushiroda T, Mahasirimongkol S, et al. HLA-B*3505 allele is a strong predictor for nevirapine-induced skin adverse drug reactions in HIV-positive Thai patients. Pharmacogenet Genomics 2009;19:139–46.

[106] Likanonsakul S, Rattanatham T, Feangvad S, et al. HLA-Cw*04 allele associated with nevirapine-induced rash in HIV-positive Thai patients. AIDS Res Ther 2009;6:22.

[107] Gatanaga H, Yazaki H, Tanuma J, et al. HLA-Cw8 primarily associated with hypersensitivity to nevirapine. AIDS 2007;21:264–5.

[108] Martin AM, Nolan D, Mallal S. HLA-B*5701 typing by sequence-specific amplification: validation and comparison with sequence-based typing. Tissue Antigens 2005;65:571–4.

[109] Gulick RM, Ribaudo HJ, Shikuma CM, et al. Three- vs four-drug antiretroviral regimens for the initial treatment of HIV-1 infection: a randomized controlled trial. JAMA 2006;296:769–81.

[110] Clifford DB, Evans S, Yang Y, et al. Impact of efavirenz on neuropsychological performance and symptoms in HIV-positive individuals. Ann Intern Med 2005;143:714–21.

[111] Mukonzo JK, Roshammar D, Waako P, et al. A novel polymorphism in ABCB1 gene, CYP2B6*6 and sex predict single-dose efavirenz population pharmacokinetics in Ugandans. Br J Clin Pharmacol 2009;68:690–9.

[112] Consortium TGP. A map of human genome variation from population-scale sequencing. Nature 2010;467:1061–73.

[113] Rajman I, Knapp L, Morgan T, Masimirembwa C. African genetic diversity: implications for cytochrome P450-mediated drug metabolism and drug development. EBioMedicine 2017;17:67–74.

[114] Fellay J, Marzolini C, Meaden ER, et al. Response to antiretroviral treatment in HIV-1-infected individuals with allelic variants of the multidrug resistance transporter 1: a pharmacogenetics study. Lancet 2002;359:30–6.

[115] Gatanaga H, Hayashida T, Tsuchiya K, et al. Successful efavirenz dose reduction in HIV type 1-infected individuals with cytochrome P450 2B6 *6 and *26. Clin Infect Dis 2007;45:1230–7.

[116] ENCORE1 Study Group, Carey D, Puls R, Amin J, Losso M, Phanupak P, et al. Efficacy and safety of efavirenz 400 mg daily versus 600 mg daily: 96-week data from the randomised, double-blind, placebo-controlled, non-inferiority ENCORE1 study. Lancet Infect Dis 2015;15(7):793–802.

[117] US Department of Health and Human Services. Guidelines for the use of antiretroviral agents in HIV-1-infected adults and adolescents. When to start: initial combination regimens for the antiretroviral-naïve patient. US Department of Health and Human Services; 2016.
[118] US Food and Drug Association. https://www.fda.gov/InternationalPrograms/PEPFAR/ucm119231.htm).
[119] Bellamy SL, Ratshaa B, Han X, Vujkovic M, Aplenc R, Steenhoff AP. CYP2B6 genotypes and early efavirenz-based HIV treatment outcomes in Botswana. AIDS 2017;31:2107–13.
[120] Tovar-y-Romo LB, Bumpus NN, Pomerantz D, et al. Dendritic spine injury induced by the 8-hydroxy metabolite of efavirenz. J Pharmacol Exp Ther 2012;343:696–703.
[121] Beutler E, Gelbart T, Demina A. Racial variability in the UDP-glucuronosyltransferase 1 (UGT1A1) promoter: a balanced polymorphism for regulation of bilirubin metabolism? Proc Natl Acad Sci U S A 1998;95:8170–4.
[122] Quirk E, McLeod H, Powderly W. The pharmacogenetics of antiretroviral therapy: a review of studies to date. Clin Infect Dis 2004;39:98–106.
[123] Haas DW, Zala C, Schrader S, et al. Therapy with atazanavir plus saquinavir in patients failing highly active antiretroviral therapy: a randomized comparative pilot trial. AIDS 2003;17:1339–49.
[124] Sanne I, Piliero P, Squires K, Thiry A, Schnittman S. Results of a phase 2 clinical trial at 48 weeks (AI424-007): a dose-ranging, safety, and efficacy comparative trial of atazanavir at three doses in combination with didanosine and stavudine in antiretroviral-naive subjects. J Acquir Immune Defic Syndr 2003;32:18–29.
[125] Hartkoorn RC, Kwan WS, Shallcross V, et al. HIV protease inhibitors are substrates for OATP1A2, OATP1B1 and OATP1B3 and lopinavir plasma concentrations are influenced by SLCO1B1 polymorphisms. Pharmacogenet Genomics 2010;20:112–20.
[126] Pasanen MK, Neuvonen PJ, Niemi M. Global analysis of genetic variation in SLCO1B1. Pharmacogenomics 2008;9:19–33.
[127] Noor MA, Parker RA, O'Mara E, et al. The effects of HIV protease inhibitors atazanavir and lopinavir/ritonavir on insulin sensitivity in HIV-seronegative healthy adults. AIDS 2004;18:2137–44.
[128] Flint OP, Noor MA, Hruz PW, et al. The role of protease inhibitors in the pathogenesis of HIV-associated lipodystrophy: cellular mechanisms and clinical implications. Toxicol Pathol 2009;37:65–77.
[129] Mulligan K, Tai VW, Schambelan M. Cross-sectional and longitudinal evaluation of body composition in men with HIV infection. J Acquir Immune Defic Syndr Hum Retrovirol 1997;15:43–8.
[130] Walli R, Herfort O, Michl GM, et al. Treatment with protease inhibitors associated with peripheral insulin resistance and impaired oral glucose tolerance in HIV-1-infected patients. AIDS 1998;12:F167–73.
[131] Carr A, Samaras K, Chisholm DJ, Cooper DA. Pathogenesis of HIV-1-protease inhibitor-associated peripheral lipodystrophy, hyperlipidaemia, and insulin resistance. Lancet 1998;351:1881–3.
[132] Carr A, Samaras K, Burton S, et al. A syndrome of peripheral lipodystrophy, hyperlipidaemia and insulin resistance in patients receiving HIV protease inhibitors. AIDS 1998;12:F51–8.
[133] Grundy SM, Brewer Jr HB, Cleeman JI, Smith Jr SC, Lenfant C. Definition of metabolic syndrome: Report of the National Heart, Lung, and Blood Institute/American Heart Association conference on scientific issues related to definition. Circulation 2004;109:433–8.
[134] Duran S, Saves M, Spire B, et al. Failure to maintain long-term adherence to highly active antiretroviral therapy: the role of lipodystrophy. AIDS 2001;15:2441–4.

[135] Friis-Moller N, Reiss P, Sabin CA, et al. Class of antiretroviral medications and the risk of myocardial infarction. N Engl J Med 2007;356:1723–35.
[136] Kaplan RC, Kingsley LA, Sharrett AR, et al. Ten-year predicted coronary heart disease risk in HIV-positive men and women. Clin Infect Dis 2007;45:1074–81.
[137] Purnell JQ, Zambon A, Knopp RH, et al. Effect of ritonavir on lipids and post-heparin lipase activities in normal subjects. AIDS 2000;14:51–7.
[138] Lee GA, Rao MN, Grunfeld C. The effects of HIV protease inhibitors on carbohydrate and lipid metabolism. Curr HIV/AIDS Rep 2005;2:39–50.
[139] Noor MA, Lo JC, Mulligan K, et al. Metabolic effects of indinavir in healthy HIV-seronegative men. AIDS 2001;15:F11–8.
[140] Jemsek JG, Arathoon E, Arlotti M, et al. Body fat and other metabolic effects of atazanavir and efavirenz, each administered in combination with zidovudine plus lamivudine, in antiretroviral-naive HIV-positive patients. Clin Infect Dis 2006;42:273–80.
[141] Rathbun RC, Rossi DR. Low-dose ritonavir for protease inhibitor pharmacokinetic enhancement. Ann Pharmacother 2002;36:702–6.
[142] Carr A, Ritzhaupt A, Zhang W, et al. Effects of boosted tipranavir and lopinavir on body composition, insulin sensitivity and adipocytokines in antiretroviral-naive adults. AIDS 2008;22:2313–21.
[143] Dube MP, Stein JH, Aberg JA, et al. Guidelines for the evaluation and management of dyslipidemia in human immunodeficiency virus (HIV)-infected adults receiving antiretroviral therapy: recommendations of the HIV Medical Association of the Infectious Disease Society of America and the Adult AIDS Clinical Trials Group. Clin Infect Dis 2003;37:613–27.
[144] Hindorff LA, Sethupathy P, Junkins HA, et al. Potential etiologic and functional implications of genome-wide association loci for human diseases and traits. Proc Natl Acad Sci U S A 2009;106:9362–7.
[145] Aouizerat BE, Pearce CL, Miaskowski C. The search for host genetic factors of HIV/AIDS pathogenesis in the post-genome era: progress to date and new avenues for discovery. Curr HIV/AIDS Rep 2011;8:38–44.
[146] Wang L, McLeod HL, Weinshilboum RM. Genomics and drug response. N Engl J Med 2011;364:1144–53.
[147] Metzker ML. Sequencing technologies—the next generation. Nat Rev Genet 2010;11:31–46.
[148] Metry DW, Lahart CJ, Farmer KL, Hebert AA. Stevens-Johnson syndrome caused by the antiretroviral drug nevirapine. J Am Acad Dermatol 2001;44:354–7.
[149] Hemminki K, Forsti A, Houlston R, Bermejo JL. Searching for the missing heritability of complex diseases. Hum Mutat 2011;32:259–62.
[150] Manolio TA, Collins FS, Cox NJ, et al. Finding the missing heritability of complex diseases. Nature 2009;461:747–53.
[151] Cui Y, Paules RS. Use of transcriptomics in understanding mechanisms of drug-induced toxicity. Pharmacogenomics 2010;11:573–85.
[152] Bansard C, Lequerre T, Daveau M, et al. Can rheumatoid arthritis responsiveness to methotrexate and biologics be predicted? Rheumatology (Oxford) 2009;48:1021–8.
[153] Mateo Leach I, van der Harst P, de Boer RA. Pharmacoepigenetics in heart failure. Curr Heart Fail Rep 2010;7:83–90.
[154] Dejeux E, Ronneberg JA, Solvang H, et al. DNA methylation profiling in doxorubicin treated primary locally advanced breast tumours identifies novel genes associated with survival and treatment response. Mol Cancer 2010;9:68.
[155] Ji J, Shi J, Budhu A, et al. MicroRNA expression, survival, and response to interferon in liver cancer. N Engl J Med 2009;361:1437–47.

[156] Markwalder JA, Christ DD, Mutlib A, Cordova BC, Klabe RM, Seitz SP. Synthesis and biological activities of potential metabolites of the non-nucleoside reverse transcriptase inhibitor efavirenz. Bioorg Med Chem Lett 2001;11:619–22.
[157] Mutlib AE, Gerson RJ, Meunier PC, et al. The species-dependent metabolism of efavirenz produces a nephrotoxic glutathione conjugate in rats. Toxicol Appl Pharmacol 2000;169:102–13.
[158] Arab-Alameddine M, Di Iulio J, Buclin T, et al. Pharmacogenetics-based population pharmacokinetic analysis of efavirenz in HIV-1-infected individuals. Clin Pharmacol Ther 2009;85:485–94.
[159] Arnedo M, Taffe P, Sahli R, et al. Contribution of 20 single nucleotide polymorphisms of 13 genes to dyslipidemia associated with antiretroviral therapy. Pharmacogenet Genomics 2007;17:755–64.
[160] Telenti A. Time (again) for a randomized trial of pharmacogenetics of antiretroviral therapy. Pharmacogenomics 2009;10:515–6.
[161] Kostenko L, Kjer-Nielsen L, Nicholson I, et al. Rapid screening for the detection of HLA-B57 and HLA-B58 in prevention of drug hypersensitivity. Tissue Antigens 2011;78:11–20.
[162] Martin AM, Krueger R, Almeida CA, et al. A sensitive and rapid alternative to HLA typing as a genetic screening test for abacavir hypersensitivity syndrome. Pharmacogenet Genomics 2006;16:353–7.
[163] Hammond E, Mamotte C, Nolan D, Mallal S. HLA-B*5701 typing: evaluation of an allele-specific polymerase chain reaction melting assay. Tissue Antigens 2007;70:58–61.
[164] Dello Russo C, Lisi L, Lofaro A, et al. Novel sensitive, specific and rapid pharmacogenomic test for the prediction of abacavir hypersensitivity reaction: HLA-B*5701 detection by real-time PCR. Pharmacogenomics 2011;12:567–76.
[165] Abu Dayyeh BK, Gupta N, Sherman KE, de PIW B, Chung RT, for the AIDS Clinical Trials Group A5178 Study Team. IL28B alleles exert an additive dose effect when applied to HCV-HIV Co-infected persons undergoing peginterferon and ribavirin therapy. PLoS One 2011;6(10):e25753.
[166] Haas DW, Kuritzkes DR, Ritchie MD, Amur S, Gage BF, et al. Pharmacogenomics of HIV therapy: summary of a Workshop Sponsored by the National Institute of Allergy and Infectious Diseases. HIV Clin Trials 2011;12(5):277–85.
[167] Wester CW, Okezie OA, Thomas AM, et al. Higher-than-expected rates of lactic acidosis among highly active antiretroviral therapy-treated women in Botswana: preliminary results from a large randomized clinical trial. J Acquir Immune Defic Syndr 2007;46:318–22.

Chapter 14

Genomics and precision medicine for malaria: A dream come true?

Desiree Williams, Karine G. Le Roch

Department of Molecular, Cell and Systems Biology, University of California Riverside, Riverside, CA, United States

Introduction

Malaria has been with us for the last 60,000 years, since the first human expansion and migration out of Africa [1]. Malaria is endemic to tropical and subtropical regions where the vector for transmission, the *Anopheles* mosquito, thrives. There are currently five species of *Plasmodium* parasites that can infect humans, *P. falciparum*, *P. vivax, P. ovale, P. malariae* and *P. knowlesi.* The most prevalent parasite is *Plasmodium falciparum* (99% of the cases in Africa) followed by *P. vivax* (the Americas, south-east Asia and some parts of Africa). *P. knowlesi* remains the only zoonotic parasite, where the natural hosts macaques in Southeast Asia, has most recently infected humans [2]. While efforts at reducing malaria such as physical barriers with insecticide treated nets and/or drug combinations with artemisinin (Artemisinin-based Combination Therapy (ACT)) have been effective, the infection rate is still high. In 2016, approximately 200 million infections and 445,000 malaria-related deaths were reported, an increased by 5 million from the previous year [3]. To identify innovative and long-lasting strategies to eradicate malaria, it is critical to understand both the pathogen and host responses to infection. The availability of the fully sequenced *Plasmodium* and human genomes, the rapid development of genomics, and the recent advent and easy access to next-generation sequencing technologies have started to complete and modernize our understanding of the malaria parasite and its interactions with its hosts. If well-funded and managed, these approaches could revolutionize the way we think about medicine and treatments to eradicate infectious diseases in developing countries. In this chapter, we present the recent genomic approaches that led to major advances in the understanding of the pathogen and its host as well

Genomic and Precision Medicine. https://doi.org/10.1016/B978-0-12-801496-7.00014-9

as their complex interactions. We discuss the impact of these approaches on the development of new therapeutic strategies in the near future including the development of method to reduce mosquito population, and possible long-term access to personalized medicine in developing countries.

Life cycle of the human malaria parasite

Malaria begins with the bite of an infected mosquito that transmit sporozoites (Fig. 1). The sporozoites are injected from the mosquito saliva into the bloodstream. The sporozoite rapidly reach the liver and invade hepatocytes. Within a week, the parasite develops into merozoites that bud out of hepatocytes in vesicles known as merosomes. These vesicles protect the merozoites from phagocytosis until they are released into the blood stream where they can invade red blood cells (RBCs) [4]. Once in a RBC, the parasite progresses through the intra-erythrocytic developmental cycle (IDC), consisting of three

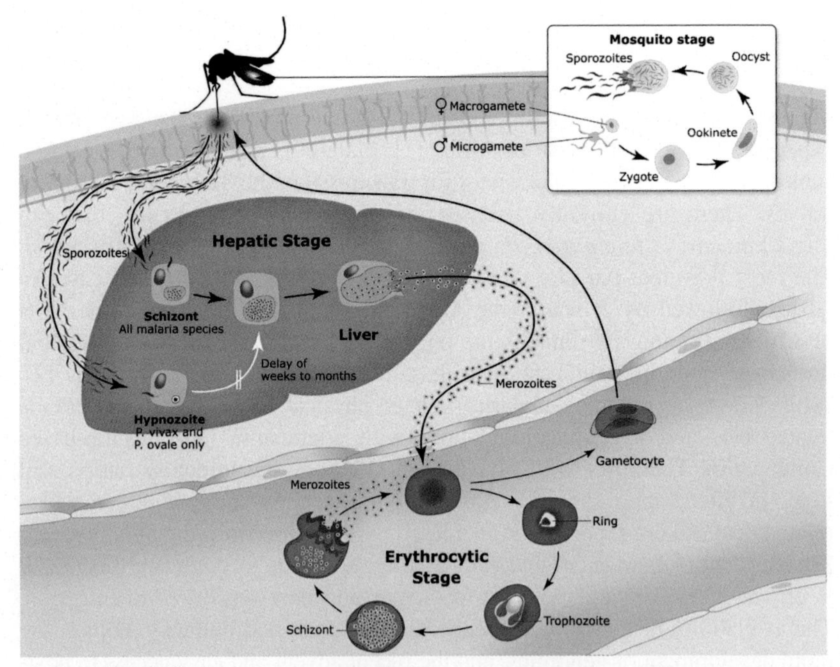

FIG. 1 The human malaria parasite life cycle. Infective sporozoites enter the bloodstream from the saliva of a female Anopheles mosquito and quickly invade liver cells. After ~10 days of differentiation and multiplication, merozoites are released into the bloodstream, where they invade circulating red blood cells. Over the next 48 h in the case of *P. falciparum*, parasites mature through ring, trophozoite and schizont stages; at this point, 16–32 new daughter cells are released to infect additional red blood cells. Exposure to stress, induces gametocyte development. A mosquito taking a blood meal from an infected person ingests these gametocytes, and sexual reproduction takes place in the mosquito midgut. Transmission into a new human host occurs when the mosquito feeds.

stages termed ring, trophozoite and schizont. During the IDC, 16–32 new merozoites are produced asexually every 48-h where they burst from the RBC and reinvade new healthy erythrocytes. Exposure to stress such as, fever or low-nutrient levels brought on by anemia can trigger the parasite to develop into a male or female gametocyte. If a mosquito bites a host and ingests mature gametocytes, the parasite reproduces sexually within the mosquito midgut. The fertilized female gametes develop into ookinetes that burrow through the mosquito's midgut wall and form oocysts. Inside the oocyst, thousands of active sporozoites develop. The oocyst eventually bursts, releasing sporozoites into the body cavity that travel to the mosquito's salivary glands to reinitiate disease transmission. It is worth denoting that two other malaria parasite species infecting humans, *P. vivax* and *P. ovale,* have the ability to develop dormant forms, hypnozoites, within hepatocytes for weeks or even years before an unknown mechanism reactivates parasite development. These dormant stages are associated with patients exhibiting multiple relapses and challenges significantly the malaria eradication program [5].

Parasite genomes and models

The availability of the complete sequences of the *P. falciparum*, *P. vivax*, *P. malariae, P. ovale* and *P. knowlesi* genomes [6–9], coupled with the exponential development of bioinformatics and computational biology, has revolutionized the malaria research field. The *P. falciparum* genome was the first to be sequenced with a genome size of 23 Mb, 14 chromosomes and close to 5300 coding genes [9]. The unusually high AT content of the *P. falciparum* genome (~80%) has hindered the application of reverse genetics approaches. As an alternative, the murine malaria species, *P. berghei and P. yoelii*, serve as a model for functional studies with the development of reverse genetics with its ease in transfection [10,11]. These murine malaria models are particularly relevant as comparative genome analysis of human and rodent parasites exhibit strong synteny in all chromosomes core with the exception of the subtelomeric regions [12]. Another critical model for *P. falciparum* has been developed with *P. knowlesi infection* in rhesus macaques. This model is particularly relevant to study antigenic variations. Antigenic variation of parasite-encoded proteins expressed at the surface of *Plasmodium* infected RBCs (iRBCs) is critical to parasite survival and the establishment and maintenance of chronic asymptomatic malaria infections. Malaria antigenic variation has been most prominently associated with the *P. falciparum* Erythrocyte Membrane Protein-1 (EMP-1) variant antigens, which are encoded by the large *var* multigene family with about 60 members dispersed throughout all 14 chromosomes, with the majority located near the telomeres [13–15]. In contrast to *P. falciparum* and *P. knowlesi*, the *var* gene family is not present in other human and murine malaria species. Necessity to understand the regulation of these gene family in vivo to control infection has led to an increase interest in using *P. knowlesi infection* as an ideal model in vivo [16].

Finally, the availability of these *Plasmodium* genome sequences and models provide the framework and possible comparison into their modes of infection, routes of drug resistance, and how the parasite can adapt to different host species.

Genomics for diagnosis

In the absence of a correct diagnosis and prompt treatment, severe malaria leading to death may occur. The first symptoms of malaria are attributed to the erythrocytic stage of the cycle. They are non-specific and resemble the ones provoked by viral infections such as flu viruses. Thus, diagnosis cannot be solely based on symptoms but is usually pronounced after microscopic examination of blood smears from infected patients (the *Gold Standard*). While the most common clinical detection of malaria relies mostly on visualizing infected erythrocytes within a blood smear under the microscope, rapid diagnostic tests (RDTs) are also convenient ways to detect the presence of infection. These tests use immune-antigen detection of parasite-specific proteins such as histidine-rich protein 2 (HRP2), lactate dehydrogenase or aldolase [17]. However, *P. falciparum* isolated in the amazon lacks the *pfhrp2* and *pfhrp3*, proteins structurally similar to hrp2, and thus evades RDTs [18]. Additionally, *hrp2* deletion has also been observed in malaria parasites isolated from patients in Africa and Asia. Thus, RDTs can often lead to false negative results if used to only detect one parasite antigen. False positive results can also occur as the parasite antigens persist weeks after clearance [17]. Furthermore, high parasitemia can also lead to false negative results due to the prozone effect where high antibody titer interferes with formation of antigen-antibody interactions [19]. Moreover, determination of the exact invading species in a patient, especially in the case of multi-infections, cannot be accomplished by such methods. Nonetheless, available genomic sequences and markers are now easily used to identify parasites in blood samples from patients. PCR-based methods have been extensively developed and optimized to complement standard malaria diagnostic tests and allow better diagnoses [20,21]. Because of their cost and requirement of heavy instrumentation and refrigeration of reagents, these methods are not readily available in the field. Development of more portable devices such as dielectrophoretic and magnetophoretic cartridges that can separate infected RBCs (iRBCs) from uninfected RBCs are now carefully investigated [22]. These devices increase sensitivity and specificity but have yet to be clinically tested.

Pathogenesis in the human host

Malaria involves a series of recurring episodes of chills, intense fever and sweating. Other symptoms include headache, malaise, fatigue, body aches, nausea, and/or vomiting. In some cases, mostly in young children and pregnant women, the disease can progress to "severe malaria" with complications such as seizures, severe anemia, respiratory distress, kidney and liver failure, cardiovascular collapse or cerebral malaria that can lead to coma and death. Individuals living in malaria endemic areas who are repeatedly exposed to parasites have reduced clinical complications [23]. However this phenomena can promote the transmission of the disease as individuals can have a high parasitemia without knowing they are infected [24]. A better understanding of the biological processes underlying the progression from infection to disease and protection is essential in order to reduce the morbidity and mortality associated with malaria as well as transmission of the disease.

In patients infected by *P. falciparum*, severity of the disease is linked to the ability of the parasite to evade the host immune system by changing the expression of parasite-specific surface antigens, PfEMP1 (*P. falciparum* erythrocyte membrane protein 1), on the surface of the infected red blood cell (iRBC) [25]. The parasite expresses one of these antigens at a time and is able to switch it to evade the host detection [26]. A major difficulty in studying virulence genes is the rapid adaptation of the model parasite strains used in the research labs, leading to important discrepancies when in vitro results are transposed to the field. The complete genome sequences for both *P. falciparum* and the human host coupled with the rapid development of affordable deep sequencing methods (Fig. 2) are tools to circumvent such limitations, as illustrated in a recent study where the transcription profiles of *var* genes were directly studied in *P. falciparum* patient isolates [27].

As described above, this antigenic variation phenomena is also present in *P. knowlesi* where it was first discovered and described as Schizont Infected Cell Agglutination (*SICA*) *var* gene family [28,29]. Publication of an improved *P. knowlesi* genome assembly [30] together with the annotation and characterization of all *SICA var* genes will most likely support a valuable understanding of the molecular mechanism and host factor regulating antigenic variation in vivo. While other *Plasmodium* species such as *P. vivax* have a different set of genes called variant interspersed repeats (*vir*), that are potentially involved in antigenic variation and pathogenicity, their exact function remain to be validated. Until now many vir proteins seem to lack the *Plasmodium* export signal element (PEXEL), a conserved motif required to direct mature proteins to the host cell [31]. It has also been demonstrated that as opposed to *var* genes in *P. falciparum,* several vir proteins can be expressed at the same time [31,32].

Host genetic factors and susceptibility/resistance to malaria

The idea of personalized medicine has been routinely taken into consideration for many years in the practice of medicine. For any type of disease, including infectious diseases, physicians seek information about the patient's environment, family history and ethnicity. The collected information is then used to determine more precisely the possible evolution of the pathology in the light of eventual relevant genetic factors to optimize the therapeutic care to be provided. Several independent genetic *loci* of susceptibility or resistance to diseases have emerged

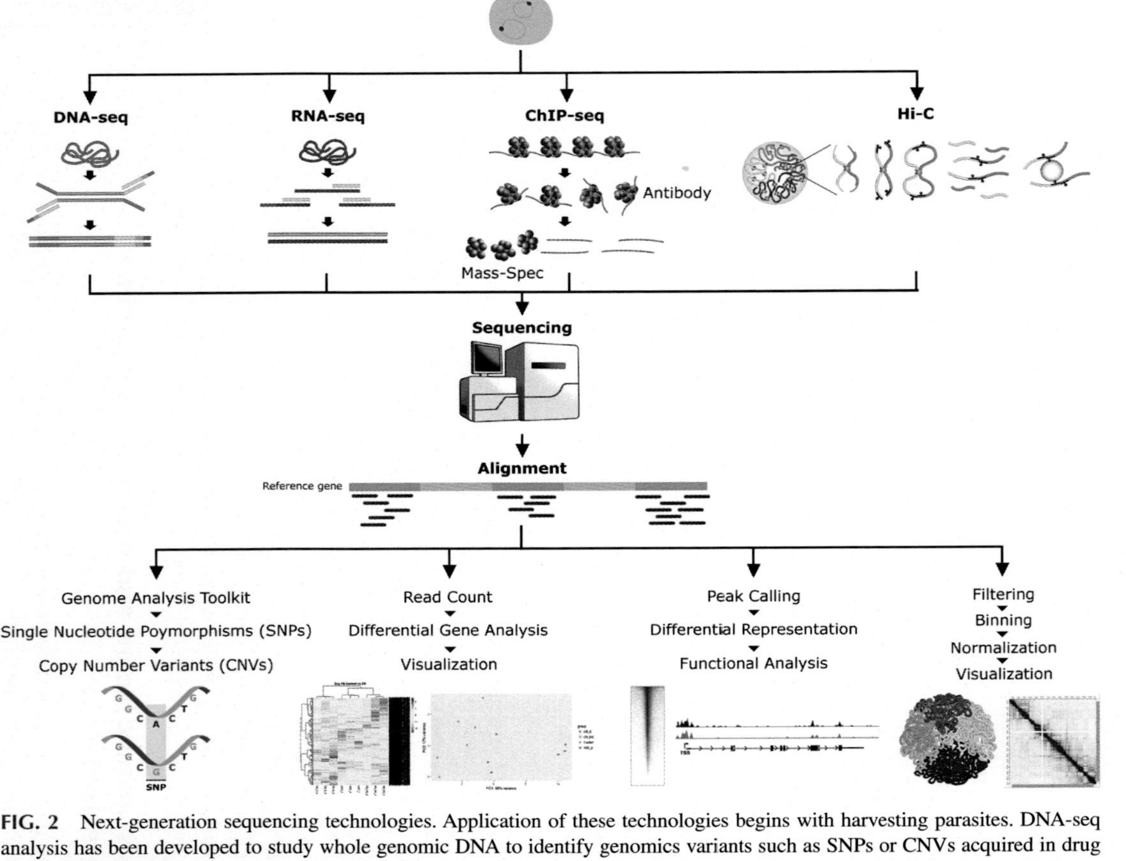

FIG. 2 Next-generation sequencing technologies. Application of these technologies begins with harvesting parasites. DNA-seq analysis has been developed to study whole genomic DNA to identify genomics variants such as SNPs or CNVs acquired in drug resistant strains. RNA-seq or Single-cell RNA-seq (scRNA-seq) is to measure the genome-wide expression of parasite population or single parasites, respectively. scRNA-seq is a valuable tool for field isolates or parasite species that are difficult to culture. ChIP-seq combines chromatin immunoprecipitation (*ChIP*) with massively parallel DNA sequencing to identify the binding sites of DNA-associated proteins. Chromosome conformation capture (Hi-C) provides 3D spatial relationship of chromosomes in the genome.

in the recent history of elucidation of the human genome. In the case of malaria, evidence has accumulated over the past few decades demonstrating strong genomic variations in susceptibility/resistance to malaria. Malaria is indeed the best-understood example of an infectious disease that has driven natural selection and human genome evolution. Sickle-cell anemia and various thalassemia disorders that result from a mutation in the *beta hemoglobin* (*HBB*) and/or in the *glucose-6-phosphate dehydrogenase* (*G6PD*) genes are genetic diseases that are highly prevalent in regions where malaria is endemic. In the heterozygous state, a mutation in Hbs provides protection against severe malaria as these individuals have a higher concentration of carbon monoxide (CO) in the blood, a by-product of heme breakdown. Higher CO levels inhibits heme release that occurs during parasite infection and prevented cerebral malaria in mouse models [33]. Furthermore, as sickled RBCs are also improperly formed due to a change in the beta hemoglobin amino acid, sickled cells are eliminated in high number by the spleen. Finally, because the cell membrane of the sickled RBC is stretched by its unusual shape, the cell "leaks" nutrients, such as potassium, that are vital for parasite survival. Recent genome-wide association (GWA) study identified 139 SNPs among the hemoglobin S (*HbS*) locus that were dispersed in 100 different susceptible *loci* in the genomes of 1060 Gambian children with severe malaria [34]. Other forms of natural immunity against malaria exist in the form of cell surface antigens. The Duffy antigen is expressed on the surface of immature RBCs (reticulocytes) where *P. vivax* and *P. knowlesi* bind to and invade [35]. In populations that are Duffy negative, such as the majority of sub-Saharan Africa, the incidence of *P. vivax* and *P. knowlesi* infections are rare [36]. Individuals in Sub-Saharan Africa harboring a newly identified variation of a RBC glycophorin receptor DUP4, displayed reduced risk of developing cerebral malaria and anemia [37].

Genomics for the development of new anti-malarial strategies

Target-based approaches for vaccine discovery

Development of vaccines has been difficult because of the plasticity of the parasite genome for genes involved in host-parasite interactions, and most importantly due to our inability to understand mechanisms regulating antigenic variation in *P. falciparum* and/or relapse infections in *P. vivax*. An ideal vaccine candidate would need to target various parasitic stages including the stages responsible for the transmission of the disease such as gametocytes and sporozoites. Liver stage parasites, or sporozoites, represent valuable candidate for vaccination, primarily due to the low parasite count at the time of infection and increased chances of parasite clearance. Vaccination with irradiated sporozoites or antigen based on pre-erythrocytic stages, targeting infected hepatocytes, is meant to block infection and has had some success in eliciting partial protection in human volunteers [38]. A whole parasite *P. falciparum* sporozoite (PfSPZ) vaccine has been recently developed, the Sanaria PfSPZ Vaccine, and is currently in phase I clinical trials [39]. Aside from the whole-cell vaccination approach, malaria vaccine development relies on the combination of parasite antigens with adjuvants designed to elicit the cell-mediated human immune response (CMI), which is often too weak to confer protection. The use of viral vectors to deliver and present the parasite's antigens can circumvent the problem. So far, the most advanced vaccine candidate is RTS, S/AS01 (Mosquirix Phase IV) developed by GlaxoSmithKline against a sporozoite surface protein, circumsporozoite protein (PfCSP), has shown to reduce clinical malaria by 51% after administering three doses [40]. RTS, S/AS01 has recently been approved for a pilot study to vaccinate infants and children in sub-Saharan Africa [41].

Whole sporozoite vaccines using genetically attenuated parasites (GAP), that have mutations within three genes that are expressed in pre-erythrocytic stages (Pf *p52*$^{-}$/*p36*$^{-}$/*sap1*$^{-}$), have shown to arrest liver stage of development and were effective in developing antibodies to sporozoites [42]. Development of GAP vaccines that arrest parasite development at later stages provides a wider range of immunity at lower doses compared to early liver stage parasites, as seen in rodent models [43]. However, these results have yet to be clinically tested.

Other possible vaccines are being investigated, targeting the blood stage parasites [44]. The most preponderant variant antigens produced by the malaria parasite are indubitably PfEMP1 encoded by the *Var* genes. However, the constant variant switching of the *Var* genes dramatically complicates the task of vaccination, against pregnancy-associated malaria (PAM) [45]. PAM is a severe form of malaria that causes maternal anemia and low birth weight. Numerous studies led to the identification of the receptor chondroitin sulfate A (CSA) on the placenta as the target of infected erythrocytes via the variant antigen PfEMP1, named *VAR2CSA* in the case of PAM [46]. Research for a vaccine against PAM has therefore been focusing on finding other genes that may be involved either in the regulation of *VAR2CSA* expression at the surface of infected erythrocytes or its binding to CSA rather than *VAR2CSA* itself. For RBC surface antigens, it has been difficult to generate antibodies as epitopes are only present for a short duration during the rapid asexual development. Because of this, high concentrations of antibodies are required for efficient antibody-antigen binding. So far, the best strategy for blood-stage vaccines has been to combine multiple antigens in a synergistic vaccine which is shown to be more effective than the use of a single antigen [47].

Merozoite surface proteins (MSP) and apical membrane antigen 1 (AMA1) also possess good antigenic properties. MSP1 and AMA1 have been particularly investigated as antigens for vaccine candidates (among many others) but do not seem to induce sufficient immunity to confer protection against clinical malaria, and research is ongoing to improve the adjuvant formulation that could elicit satisfactory levels of immune response. However, the high levels of genetic variations observed in MSP1 and AMA1 are challenging the viability of these vaccine candidates [48]. Screening for novel merozoite antigens from field isolates using patient sera from Mali has been beneficial to identify conserved antigens that are expressed within infected populations. The protein showing the highest parasite growth inhibition (~25%) was LSA3-C [49]. Another highly conserved merozoite antigen, *P. falciparum* reticulocyte-binding protein homolog 5 (PfRH5) is currently in Phase I/IIa clinical trials and has shown protection and inhibited growth (60%) of *P. falciparum* in vivo within *Aotus* monkeys [50]. Methods to improve the vaccine by studying the epitopes on PfRH5 and delivery of a soluble form of the protein for antibody production has been developed to aid in its clinical application [51,52]. Other approaches aim at transmission blocking vaccines, where ookinete and gametocyte surface antigens are used to prevent development within the mosquito vector and inhibit transmission. However, this would not protect an individual from infection but may be useful in asymptomatic carriers. The future of vaccines looks to be a multi-stage approach as studies show synergistic interactions between pre-erythrocytic and transmission blocking vaccines [53].

Target-based approaches to drug discovery

The most efficient antimalarials that have been successfully developed over the years are blood schizonticides (Table 1), that target the erythrocytic stages of the parasite. Quinine, one of the first compounds used to fight malaria, was isolated in South America in 1820 from the bark of a cinchona tree. Other compounds developed more recently, safer and less expensive to synthesize, such as chloroquine, amodiaquine, pyrimethamine, proguanil, sulfonamides, mefloquine, atovaquone, doxycycline, tetracycline are now far less efficient against infections as parasites have developed mechanisms of drug resistance to all of these compounds. More recently developed compounds such as artemisinin and its semi-synthetic artemisinin derivatives (e.g., artesunate, artemether), which are easier to use and have a very rapid mechanism of action on the vast majority of acute patients, are becoming less useful as parasites are also developing mechanism(s) of drug resistance.

Over the last 10 years, extensive efforts with the screening of several million chemicals have provided early leads for over 6000 potential anti-malarial compounds [54,55]. These studies have led to at least four new medications under development and clinical trials by drugmaker such as Novartis. One of the compounds now called cipargamin blocks a channel in the parasite's outer membrane, work faster than artemisinin and has also the advantage of preventing transmission [56].

To speed up the process of drug discovery against malaria parasites, over 6 million small molecules have been screened against the asexual and sexual stages of the parasite [57]. From these original screens, 400 compounds with confirmed antimalarial activity have been made available as open source to researchers who wish to study the mechanisms of action of these compounds.

TABLE 1 Antimalarial drugs and their associated markers of resistance.

Drug class	Drug	Resistance marker	Mode of action
Artemisinin derivative	Artesunate	*pfkelch13* [63]	Blood-stage schizonticide
Combination: hydroxynaphthoquinone—antifolate	Atovaquone-proguanil	*cytochrome b, pfdhfr* [74,75]	Blood-stage schizonticide and gametocyte
Aminoquinoline	Chloroquine	*pfcrt, pfmdr1* [65,66]	Blood-stage schizonticide
	Tafenoquine	unknown	Hypnozoiticide
	Primaquine	unknown	Hypnozoiticide
Quinoline derivative	Mefloquine	*pfmdr1* [72]	Blood-stage schizonticide
Antifolate	Sulfadoxine-pyrimethamine	*pfdhps, pfdhfr* [76]	Blood and liver schizonticide

crt, chloroquine-resistance transporter; *dhfr*, dihydrofolate reductase; *dhps*, dihydropteroate synthase; *mdr1*, multidrug resistance protein 1; *pfkelch13*, *P. falciparum* Kelch 13.

This set of compounds known as the "Malaria Box" is now available from Medicines for Malaria Venture, who distributes the selected compounds to researchers wanting to perform screens of their own to test for cardiotoxic effects, drug interactions and efficacy of the compounds throughout the parasite life cycle stages. A summary of combined screens from 55 different groups has identified nine compounds that are able to target the three main infectious stages of the parasite with low cytotoxicity [57]. An effective compound would target all stages of the parasite to reduce infectious rate, parasitemia, and increase patient survival. Future research should aim to identify the mechanism of action of these selected compounds to monitor and target the rise of potential drug resistance phenotypes. These open-source compounds are also available for screening other protozoa parasites as multiple compounds were able to target *Babesia, Toxoplasma* and *Wolbachia* [57]. Because these screenings are performed by multiple labs using different drug concentrations and assays, any overlap of positively identified compounds increases the chance of discovering effective drugs. The Malaria Box has boosted collaborative projects to support the identification of novel drugs and their targets.

The development of drug discovery projects within other *Plasmodium* species such as *P. vivax* to reduce latent or re-emergent infection, has been however more challenging. The hypnozoite stage has been arduous to study because of the lack of advanced liver culture systems and high-throughput assays to measure drug effects on *P. vivax* in human hepatocytes. Only recently has an in vitro method been developed to successfully culture *P. vivax* in primary human hepatocytes [58–60]. This new three dimensional (3D) culture system allowed the isolation of total RNA from infected cells and identification of the transcriptome expressed by the parasite throughout the different liver-infecting stages, including hypnozoites [58]. The transcripts expressed in hypnozoites were compared to the liver schizont stages of the parasite to uncover pathways specific for maintaining the quiescent state. This work will complement previous studies done in the nonhuman primate parasite *P. cynomolgi,* the closest relative to *P. vivax* [61] and will most likely lead to new therapeutic strategies.

Drug resistance

Efforts in genome research to better understand how resistance to drugs evolve in the parasite have been intense over the past years. Several genetic markers of resistance have been identified and seems to differ depending on the antimalarials (see Table 1). Kelch 13, involved in the turnover and proliferation of proteins, has been determined as the marker for artemisinin resistance in parasites in southeast Asia through genome-wide association studies (GWAS) [62,63]. A similar mechanism of resistance to chemotherapies has been observed in multiple cancers within the human homolog Keap1 [64]. Resistance markers to older antimalarials such a chloroquine have been identified within the multidrug resistance protein 1 (*Pfmdr1*) and/or the chloroquine resistance transporter (PfCRT) [65,66]. Both proteins are located on the membrane of the parasites' food vacuole and impact the ability to efflux chloroquine from the intracellular digestive vacuole. The use of combinatorial therapy such as Artemisinin-based combinatorial Therapy (ACTs) or Atovaquone-proguanil (Malarone), that exhibit different mechanisms of action, are now widely used as it reduces the likelihood of developing resistance. In areas where *P. vivax* and *P. ovale* are endemic, primaquine, the only drug known to eliminate hypnozoites is also co-administered to patients. However, side effects generated by the drug can be dangerous especially for glucose-6-phosphate-dehydrogenase (G6PD) deficiency patients as these patients are prone to hemolytic anemia when given primaquine [67]. In this particular condition testing for G6PD deficiency is essential. Additionally, patients with polymorphisms in cytochrome P-450 have reduced metabolism of primaquine which leads to reduced efficacy [68]. Alternatively, the longer-acting tafenoquine may be more effective, but can still result in hemolysis in patients with G6PD deficiency.

Improved functional genomics studies have allowed the identification of genetic variations contributing to the emergence of parasite resistance to old and new antimalarial drugs. This is particularly the case for a mutation in the membrane proton pump PfATP4 [69,70]. Because many drugs target ion channels, any genetic variations that may render the channel less active compared to the wild-type protein can contribute to the parasite drug resistance. The ability of the parasite to adapt its genome and circumvent antimalarial drugs has led to new approaches of identifying both drug targets and genes involved with resistance. The use of chemical screens to force the parasite to evolve resistance in vitro together with whole-genome sequencing techniques have revealed many novel drug targets, a methodology that has proved to be extremely powerful and time efficient compared to previous genetic crosses [71]. Genetic changes in the form of single nucleotide variants, copy number variants (CNVs) and insertions or deletions (indels) have been demonstrated to drive drug resistance mechanisms (Fig. 2). CNVs have been detected in *pfmdr1* where multiple copies of *pfmdr1* on chromosome 5 results in *Plasmodium* resistance to mefloquine [72]. To understand the entire "resistome" of the parasite, a recent screen using small molecules exhibiting different drug mechanisms of action, followed by whole genome sequencing, has been developed [73]. This assay led to the discovery of genetic changes and genes contributing to drug resistance phenotype. In this assay, mutations in 83 genes within the core genome were detected as genes contributing to the resistome of the parasite. Among the mutations detected were CNVs of *pfmdr1* and adenosine triphosphate (ATP)-binding cassette transporter I family member 1 (*pfabcI3*) and previously unknown resistance alleles within the chloroquine resistance transporter (*pfcrt*). These newly developed assays will help to direct discovery of small molecule inhibitors and advance our understanding of drug resistance mechanisms within the parasite.

Pharmacogenomics

Pharmacogenomics is the sine qua non condition of personalized medicine. It combines pharmacology and genomics to understand how genetics influences the human body's response to drugs (classically correlating gene expression or SNPs with drug efficacy). The ultimate goal of pharmacogenomics is to optimize drug and drug combination treatments to an individual's genetic makeup. Depending on the patient, the same drug can be toxic or not and effective or not. Antimalarial drugs are no exception [77–79]. Clinical trials revealed important variations of plasmatic concentrations of drugs, leading to lower effectiveness or increased toxicity, and pharmacogenomics would doubtlessly help to optimize anti-malaria chemotherapies and drug management. Polymorphisms in drug-metabolizing enzymes are responsible for such disparities, mainly various cytochrome P450 (CYP) and drug transporters. For example, plasma levels of proguanil and amodiaquine are linked to polymorphisms *CYP2C19* and *CYP2C8*, respectively. Polymorphisms in *CYP2C8* may also have important consequences for the use of chloroquine and dapsone, both substrates of the cytochrome [80]. Additionally, chlorcycloguanil metabolism is modulated by genetic variations at the *CYP2C19/CYP2C9* locus [81,82]. These examples are a sample of the published polymorphisms leading to different efficacies and toxicities of antimalarial drugs. The future should determine which populations are at greatest risk of potential treatment failures or side effects, and which drugs are most susceptible to genetic variation in metabolizing enzymes. These results should contribute to a pharmacogenomics paradigm for the treatment of malaria. When incorporated into clinical trials, the advent of pharmacogenomics for malaria should speed up the development of new drugs by facilitating sampling of patients with a certain genotype.

Genomics for the discovery of new drug targets

The role of functional genomics in drug discovery is to prioritize new drug targets from the genomics information with the ultimate goal to translate that knowledge into rational and reliable novel therapeutic agents (Fig. 3). Novel genome-wide screen assays to identify genes that are essential within the asexual blood stages have been recently conducted using saturation mutagenesis methodology [83]. By using the *piggyBac* transposon mutagenesis system, the entire *P. falciparum* genome has been successfully mutated to identify over 2680 essential genes. Furthermore this screen confirmed the essentiality of the genes involved in drug resistance (e.g., *Kelch13, DHFR* and *MDR)* and revealed a potential multistage antimalarial target, a cell-traversal protein for ookinetes and sporozoites (CelTOS), which was previously shown to inhibit sporozoite transmission [84]. Genes involved in lipid metabolisms were also identified as critical for parasite sexual development and were further validated as novel potential targets for transmission blocking strategies [85–87]. Overall, this study provided critical functional data for prioritizing high-value drug targets and

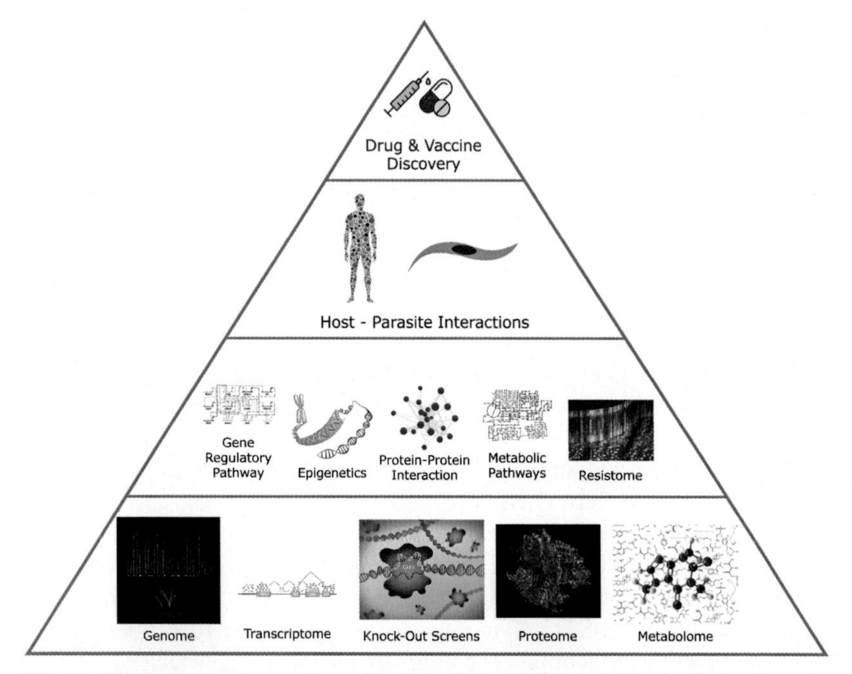

FIG. 3 Whole-genome approaches toward precision medicine against malaria. Genomics approaches are used to understand the intrinsic characteristics of the human host and *Plasmodium* involved in clinical malaria as well as the complex host-parasite interactions that occur. Ultimately, the integrated use of the genomics data generated by genomic analyses will help in developing precision medicine against malaria.

pathways such as genes involved in RNA metabolism and translation-related processes, supporting prior evidence for the importance of posttranscriptional and -translational control in the parasite life cycle [88–92]. Close to 2000 of the studied genes also have orthologs in other apicomplexan parasites such as *P. berghei* and *T. gondii* and *P. berghei*. The power of whole genome mutagenesis had previously been exploited in *T. gondii*. Using a genome-wide CRISPR Screen, researchers determined that approximately 60% of the genes were essential in the parasite. This latest screen identified and validated a novel claudin-like apicomplexan microneme protein (CLAMP), a protein involved in parasite invasion into its host cells, as a potential novel drug target in *T. gondii* and *P. falciparum* [93]. Altogether, the development of such genomic screening methodologies is rapid and cost-effective to discover novel targets that are opening new frontiers for antimalarial therapeutic research.

Genomics and parasite gene expression

Close to 50% of the genes encoded in the parasite genomes are still described as hypothetical proteins. Understanding when these genes are expressed, translated and how they are regulated may give us a clue about their function. Initial gene expression analysis using microarray technologies demonstrated that transitioning of the parasite through the different life cycle stages requires timely regulation of gene expression [94–98]. Understanding how this cascade of transcripts is regulated at the transcriptional level could lead to new ways to block the parasite development. Genome mining revealed that the parasites contain instructions for the proteins coding for the general transcription machinery such as the RNA polymerase II complex and the preinitiation complex [99]. However, the *Plasmodium* genome encodes very few specific transcription factors (TFs). In *P. falciparum* only 27 parasite-specific AP2 TFs have been identified and validated as regulating transcription. These TFs contain the AP2 domain found in plant TFs and are therefore an excellent lead for the development of therapeutic strategies [100]. Research into this family show that at least some of them regulate development of the parasite specific stages. This is particularly true for AP2-G that is essential for gametocyte commitment [101,102], AP2-Sp for sporozoite development [103] and AP2-L for development of liver-stages [104]. Considering the size of the *Plasmodium* genome and its adaptation to different host cells and environments, the small number of TFs identified in the genome is insufficient to regulate the timed expression of the 6000+ genes. A better understanding of the molecular components regulating the cascade of transcripts throughout the parasite life cycle is still needed.

While the first *P. falciparum* transcriptomes were analyzed at the population level using microarray [94,95], advanced methodologies using next-generation sequencing technology (RNA-seq) [96] or more recently, single cell RNA sequencing technology (scRNA-seq), has enabled transcriptome evaluation of the different parasite developmental stages more accurately (Fig. 2) [105,106]. Specific transcripts essential for stage's transition were successfully identified because of high-resolution at single cell level. ScRNA-seq can also be used to analyze transcriptomes of parasite extracted from the field that are harder or impossible to grow in culture (e.g., *P. vivax* patient samples). With the advancement of scRNA-seq technology, the Malaria Cell Atlas has been launched and will most likely assist researchers in improving our understanding of the parasite biology to identify more effectively novel therapeutic strategies. Additionally, research into epigenetic mechanisms and chromatin structure has started to reveal the complex mechanisms regulating transcription in the parasite and opened new ways of targeting parasite survival [107–113].

Epigenomics

Nuclear DNA is not free within the nucleus of eukaryotic cells. It is highly organized, condensed and wrapped in histones to form nucleosomes [114]. Occupancy of nucleosomes within the *Plasmodium* genome is similar to that seen in other eukaryotes. Active promoters are depleted of nucleosome to allow access for transcriptional machinery to enter and regulate gene expression.

To assist in maintaining DNA organization in the *P. falciparum* AT-rich genome, the parasite has developed specialized localization of histone variants, H2A.Z and H2A.B, that bind weakly to AT-rich intergenic regions [111,115,116]. This establishes a pattern in the parasite genome where GC content guides nucleosome formation and transcript levels, much like in other eukaryotes [117–119]. However, the exact mechanisms and proteins that regulate chromatin remodeling have yet to be identified.

Histone modifications, such as acetylation or methylation, direct the relaxation or compaction of chromatin. In *Plasmodium*, the vast majority of the genome is in an euchromatin (open) state maintained by active histone marks (H3K9ac), while subtelomeric regions are in a heterochromatin (closed) state, maintained by repressive histone marks (H3K9me3) and heterochromatin protein 1 (PfHP1) [107,120–122]. The heterochromatin cluster(s) of the parasite's genome harbor gene families encoding clonally variant antigens (e.g., *var*, rifin, stevor, and pfmc-2tm), invasion gene families (eba and clag), and a few other loci such as the gametocyte-specific transcription factor pfap2-g during the IDC [123–127]. The presence of these repressive marks suggests that mechanisms regulating transcription of these gene families may be more conserved with higher eukaryotes, than the rest of the *Plasmodium* genome. Strict control of these genes at the transcriptional level allows the parasite to escape the host immune system and adapt to environmental conditions for its survival. Mechanisms regulating *var* gene expression have emerged recently. Silent *var* genes are clustered to one or more repressive regions at the nuclear periphery, and are marked by H3K9me3 and PfHP1 [120,124–127]. The active *var* gene is positioned away from heterochromatin regions and contains H3K4me3 and H3K9Ac histone modifications at the 5′ end. When environmental cues or selective pressure are present, silent *var* genes are modified by histone acetyltransferases, which results in the relaxing of heterochromatin and the repositioning of the inactive gene to a transcriptionally permissive region [128]. Histone modifying proteins such as histone deacetylase (HDAC) silent information regulator 2 proteins, PfSIR2A and PfSIR2B, as well as the class II HDAC PfHDA2, are known to be critical in maintaining gene repression of these genes. Removal of these proteins results in loss of monoallelic *var* gene expression [129–131]. The methyltransferase, PfSET2, has also been demonstrated as essential for maintaining gene repression as knockout of PfSET2 results in reduced H3K36me3, a repressive histone mark, and leads to the expression of a majority of *var* genes [132]. These mechanisms coordinate mutually exclusive expression of antigenic variance genes [133]. Transcriptional regulation of genes involved with invasion of merozoites into new RBCs seems also to be mediated by chromatin modifying and remodeling complexes [34]. As an example, it has been demonstrated that positive regulation of *pfRH, a gene involved in invasion* was associated with the bromodomain protein (PfBDP1), where it binds to acetylated H3 (H3k9ac and H3K14ac) [134,135]. It is therefore likely that targeting histone-modifying proteins such as histone acetyltransferases (HATs) or HDACs with small molecule inhibitors could be promising antimalarial therapies. It is however important to highlight that the catalytic domains of these proteins are well conserved in eukaryotes and molecules targeting these proteins may also be toxic to the human host. Alternatives targeting the *Plasmodium* chromatin remodeling complex, SWI2/SNF2 (SWItch/Sucrose Non-Fermentable), should be more feasible as the parasite homolog of the protein contains an additional five PHD domains compared to the mammalian counterparts [136].

Another form of epigenetic regulation occurs through long non-coding RNAs (lncRNAs) [137]. These RNAs have been demonstrated to exhibit different mechanisms of regulation dependent on whether the transcript is in the sense or antisense orientation. The antisense lncRNA transcribed from the intron of a *var* gene has been shown to activate the silent gene [138]. While others such as the TARE-lncRNA have been shown to bind to subtelomeric regions to maintain *var* gene silencing [139]. Some additional lncRNAs have been implicated in the recruitment of proteins to telomeric regions to maintain heterochromatin structure [140]. Research into the regulatory roles of other lncRNAs have yet to be conducted as hundreds of lncRNAs have been recently identified in *P. falciparum's* genome [140] many of which could reveal novel antimalarial strategies.

Three-dimensional genome architecture

The importance of chromatin organization at the 3D level has recently emerged in all eukaryotic cells. Initial studies revealed that the multicellular eukaryotic genomes are compartmentalized into distinct open (actively transcribed) and compacted (repressed) regions [141]. Higher resolution of 3D architecture from in-depth sequencing studies collectively demonstrated that the genome is partitioned into topologically associated domains (TADs), that form loops between enhancer and promoter sequences, maintained by CCCTC-binding factor (CTCF) and cohesin [142,143]. Within mammalian cells these loops are cell type specific to regulate cell fate, and when disrupted result in pathologies [144–146]. In *Plasmodium*, the genome lacks the CTCF proteins known to be critical to tether TADs. Furthermore, the nuclear *lamins* that are major architectural fibrous proteins, essential to provide structure, function and transcriptional regulation in the cell nucleus are also missing in the parasite genome [147,148]. Proteins that assist chromosome assembly and segregation do however exist in the parasite genome. Structural maintenance of chromosome (SMC) protein, which are core subunits of cohesion and condensin [149]. However, their role in mediating chromosome structure has yet to be characterized. Additionally, the PfHP1 protein, a transcriptional repressor that directly binds to the methylated lysine 9 residue of histone H3 (H3K9me) in all eukaryotic cells, was demonstrated to be essential in *P. falciparum*, as knockdown of the protein resulted in loss of chromatin structure and parasite death [109,110].

Despite the lack of CTCF and lamin proteins, recent genome-wide chromosome conformation capture (Hi-C) experiments have revealed that the

Plasmodium genome is organized in the nucleus (Fig. 2). Chromosomes are anchored at the centromere and folded parallel to each other [107]. The subtelomeric *var* gene family add structural complexity as they form additional loops on chromosomes that possess internal *var* genes [107]. Furthermore, significant chromatin structural changes were observed during the transition stages [107,109] including regions of genes involved in pathogenesis, liver-cell invasion, erythrocyte remodeling, and regulation of sexual differentiation. It is now clear that genome organization adds complexity to gene regulation and transcriptional activity in the parasite. Targeting these features could disrupt parasite survival inside the host.

Post-transcriptional mechanisms of gene expression

In eukaryotes, gene expression is also regulated at the post-transcriptional level. Comparative analysis between steady-state mRNA, nascent RNA, polysome profiling, ribosome footprinting as well as proteomics data sets across the parasite life cycle [88–91,150–152] suggest that mechanisms regulating gene expression at the post-transcriptional level may also be essential in the parasite. Additional experiments demonstrate that translational downregulator PK4, a kinase targeting the eukaryotic translational initiation Factor 2 (eIF2), was essential to the parasite asexual cycle, validating further that control of gene expression at the translational level is critical in the parasite [153]. Within metazoans, mRNAs can be stabilized by proteins such as the DEAD-box RNA helicases, DDX6, until the mRNA is ready to be translated or processed in P-bodies where they are degraded by many enzymes involved in mRNA turnover [154]. In *Plasmodium* translational repression by the DOZI complex (homologous to DDX6) was identified in female *P. berghei* gametocytes. This complex was demonstrated as essential to store female gametocyte-specific mRNAs until required in fertilization for zygote development [155].

In *P. falciparum*, PfAlba1 protein was also identified as one of the post-transcriptional regulators in the parasite IDC. PfAlba1 binds to transcripts, repressing translation until the later stages in the IDC [156,157]. Knock-out or knock-down of PfAlba1 compromises the parasite survival indicating its necessity for parasite persistence and a clear mechanism regulating gene expression [156] at the post-transcriptional level. Since PfAlba1 is essential for parasite survival, designing a drug to inhibit transcript release at the later blood stages could serve as a potential antimalarial treatment [156,158]. Moreover, PfAlba1 is likely to be part of a team of RNA-binding proteins (RBPs) that collectively regulate translational repression. For example, CAF1 of the CCR4– Not complex, a key mediator of mRNA decay in eukaryotes, has been shown to regulate transcript abundance for approximately 20% of the parasite genome, including many proteins involved in egress and invasion [159]. Recent studies identifying and validating several *Plasmodium* RBPs have extended the number of proteins that could be targeted for novel antimalarial strategies [92,160].

Proteomics and metabolics parasite-host interaction

Soon after the release of the first *Plasmodium* genome, proteomics studies across the different life stages detected between 1800 and 2400 proteins further validating the accuracy of the genome annotation [151,152]. These studies were also able to identify novel vaccine targets that could prevent hepatocyte invasion and consequently the transmission of the parasite into the human host. More recent work has identified candidate vaccine proteins with post-translational modifications (PTMs) specifically exposed at the surface of *P. falciparum* and *P. vivax* sporozoites. The major vaccine candidates, the circumsporozoite protein (CSP) and the thrombospondin-related anonymous protein (TRAP) were identified as glycosylated [161,162], an information crucial for an improved and effective antibody design. Multiple studies targeting the identification of post-translational modifications (PTMs) such as phosphorylation demonstrated dynamic regulation of protein phosphorylation in the IDC [163–165]. Proteomic enrichment for specific PTMs such as phosphoproteome or lipid modifications can be challenging. Techniques to enrich for modified peptides or increase the specificity of protein pull down using biotin via click chemistry followed by immunoprecipitation of the biotinylated peptides were shown to be successful making mass spectrometry approaches more appealing to better understand the parasite biology, host-pathogen interactions and the identification of novel pathways that could be targeted for novel therapeutic intervention [166–170].

Metabolomic studies have also been performed recently to improve our understanding of the parasite biology and its interaction with its host. Experiments performed on infected blood samples have detected increased levels of bilirubin, a metabolite involved with the breakdown of heme [171]. Increased levels of bilirubin is known to suppress the immune response [172,173]. Therefore, dampening these metabolites could boost the immune system of infected patients. Other metabolic approaches using patient sera to measure metabolomes of patients with cerebral malaria demonstrate the presence of an increase level in arachidonic acid, a precursor for inflammation, and molecules implicated in seizures [174,175]. Other studies generated between patients with cerebral malaria and patients with sepsis, encephalitis or non-infected, identify malaria-specific perturbations such as decreased levels of glycoproteins [176]. Additional metabolome profiling is needed to confirm that these metabolites are parasite specific and are not a result of an immunologic response or normal physiological processes in the host. Still, metabolic profiling can be informative when used in combination with transcriptomic and proteomic data sets (Fig. 3).

Vector control

The Mosquito prevents itself from parasite infection through an innate immune response mechanism activated within its midgut. It has been proposed that invasion of parasite ookinetes in the midgut causes apoptosis of mosquito cells triggering a nitration response that leads to tagging of the parasites' surface with a thioester-containing protein 1 (TEP1) [177]. TEP1 is part of the mosquito complement immune system. This mechanism identifies and targets the ookinetes for cell lysis. However, some ookinetes survive this complement immune response and develop into sporozoites that infect human hepatocytes. Methods to suppress parasite infections in mosquitoes have been developed by generating transgenic mosquitoes that express the REL2 transcription factor, that activates expression of antimicrobial peptides within the mosquito gut cells [178]. This transgene expression leads to melanization, a process of arthropod innate immune response, whereby melanin triggers encapsulation of the parasite [179]. Coincidently, this transgene expression in the gut also influences their mating behavior, as wild-type mosquitoes prefer to mate with the transgenic REL2 mosquitoes [180]. This phenomenon helps to drive the transgene into the mosquito population. Unfortunately, these transgenic mosquitoes do not provide complete resistance to malaria. New methodologies aiming to suppress the mosquito population though the use of the bacterial adaptive immunity system CRISPR-Cas9 (clustered regularly interspaced palindromic repeats-CRISPR associated protein 9) are being developed. Specific genes reducing female fertility can be knocked out in the malaria mosquito, *Anopheles*, and have been shown to be able to propagate in the laboratory. However, after four generations of breeding selective pressure for resistance alleles are generated by the drive system through non-homologous end joining (NHEJ) during the repair process [181]. Efforts to reduce resistance, while still maintaining gene drive, can be achieved by reducing "leaky" expression of nuclease activity, which lowers the fitness of the mosquito. This has been demonstrated in *D. melanogaster* by using multiple guide RNAs that target the gene and its adjacent sites within the male germline [182–184]. This would reduce the chances of resistance, as multiple sites would need to evolve within the gene drive post-fertilization. Additionally, since the drive is within the male germline, paternally donated Cas9 proteins are not within the embryo reducing the chances of generating resistance alleles [185]. These combined strategies have yet to be implemented in mosquitoes as promoters for these drives need to be further optimized. Other CRISPR-Cas9 strategies to create mosquitoes resistant to malaria parasites have been tested, by knocking out fibrinogen-related protein 1 (FREP1) [186]. Silencing FREP1 can suppress *P. falciparum* development within the gut preventing transmission. However, knocking out FREP1 also led to lowered fitness costs as seen in reduced egg laying, delayed larval development and reduced propensity for blood-meals [186]. Strategies that combine robust gene drives that bias Mendelian inheritance and reduced allele resistance with less fitness costs are currently in the works by using tissue-specific promoters that are either germline-specific or gut tissue-specific. These strategies are more beneficial than the costly traditional vector control methods using insecticide bed nets, as these genetic population replacements are self-perpetuating. Combining this with the partial protection of barrier control could lead to reduced infection rates.

Precision medicine

Genomics has changed the study of malaria pathogenesis. Current and future research is not based on phenotypic changes anymore (such as the presence of sickle cells in a blood sample) but is based on the genotyping of thousands of individuals and the identification of malaria-resistance loci. These approaches, driven by the rapid evolution of high-throughput sequencing technologies, are the basis for genomic epidemiology of malaria. This term refers to the use of genomic tools in large-scale studies to detect and analyze genetic variants in the malaria parasite and its hosts to improve our understanding of the disease and develop efficient preventive and curative strategies. Assays to predict the susceptibility and severity of infection within patients can help tailor treatments for improved therapy. The current approach for identifying patient loci that are associated with severe malaria is mainly through genome-wide association studies (GWAS). For example, a screen in the Tanzanian population revealed single nucleotide polymorphisms (SNPs) in interleukin 23R (*IL-23R*), a pro-inflammatory cytokine, that may contribute to developing severe malarial anemia [187]. These large GWAS studies from diverse populations generate large-scale genomic data requiring the collaboration of many scientific groups. The establishment of large multi-center collaborations such as the Malaria Genomic Epidemiology Network or MalariaGEN (http://www.malariagen.net) aim to curate and make these large data sets publicly available. Ultimately, this will aid investigation on how these genetic variations effect the interactions between the host and parasite. The largest study to date has identified 70 genes with polymorphisms related to severe malaria with the most significant polymorphisms involved with the structure or function of red blood cells [188]. A benefit of large population screening is the access to field isolates of *Plasmodium* strains for monitoring drug resistance. In Cambodia such screens have been able to show *P. falciparum* gained two mutations within a year of introducing artemisinin combination therapy (dihydroartemisinin–piperaquine) [189]. The parasite gained two mutations that outcompeted all other resistant malaria parasites and led to the failure of the drug within western Cambodia. Increased genetic surveillance can monitor vulnerable populations and establish alternative drug combinations to decrease the fitness of such evolutionary changes within the parasite population.

A better understanding of the human genetic factors that influence susceptibility and response to both malaria and the drug/vaccine to treat it, as well as a better understanding of the parasite itself, is certainly necessary to eradicate malaria (Fig. 3). Genomic studies have already greatly improved our understanding of the parasite biology and its interaction with its host. It is very likely that it will change disease management in the near future. In this regard, governmental and non-governmental organizations, academic institutions, and individuals will have to be educated to ensure that the benefits of genomics are clearly understood and benefit everyone.

Acknowledgments

We would like to thank Zul Sabani for his talent in preparing all figures. This work was supported by the National Institutes of Allergy and Infectious Diseases and the National Institutes of Health (1 R01 AI06775-01 and 1 R01 AI136511 to K.L.R.) and the University of California, Riverside (NIFA-Hatch-225935).

References

[1] Tanabe K, Mita T, Jombart T, Eriksson A, Horibe S, Palacpac N, et al. Plasmodium falciparum accompanied the human expansion out of Africa. Curr Biol 2010;20(14):1283–9.

[2] Millar SB, Cox-Singh J. Human infections with Plasmodium knowlesi-zoonotic malaria. Clin Microbiol Infect 2015;21:640–8.

[3] WHO. World malaria report. World Health Organization; 20172017.

[4] Sturm A, Amino R, Van De Sand C, Regen T, Retzlaff S, Rennenberg A, et al. Manipulation of host hepatocytes by the malaria parasite for delivery into liver sinusoids. Science 2006;313(5791):1287–90.

[5] Battle KE, Karhunen MS, Bhatt S, Gething PW, Howes RE, Golding N, et al. Geographical variation in *Plasmodium vivax* relapse. Malar J 2014;13:144.

[6] Rutledge GG, Böhme U, Sanders M, Reid AJ, Cotton JA, Maiga-Ascofare O, et al. Plasmodium malariae and *P. ovale* genomes provide insights into malaria parasite evolution. Nature 2017;542:101–4.

[7] Pain A, Böhme U, Berry AE, Mungall K, Finn RD, Jackson AP, et al. The genome of the simian and human malaria parasite Plasmodium knowlesi. Nature 2008;455:799–803.

[8] Carlton JM, Adams JH, Silva JC, Bidwell SL, Lorenzi H, Caler E, et al. Comparative genomics of the neglected human malaria parasite *Plasmodium vivax*. Nature 2008;455:757–63.

[9] Gardner MJ, Hall N, Fung E, White O, Berriman M, Hyman RW, et al. Genome sequence of the human malaria parasite *Plasmodium falciparum*. Nature 2002;419:498–511.

[10] Mota MM, Thathy V, Nussenzweig RS, Nussenzweig V. Gene targeting in the rodent malaria parasite *Plasmodium yoelii*. Mol Biochem Parasitol 2001;113:271–8.

[11] Van Dijk MR, Waters AP, Janse CJ. Stable transfection of malaria parasite blood stages. Science 1995;268:1358–62.

[12] Otto TD, Böhme U, Jackson AP, Hunt M, Franke-Fayard B, WAM H, et al. A comprehensive evaluation of rodent malaria parasite genomes and gene expression. BMC Biol 2014;12:86.

[13] Leech JH, Barnwell JW, Aikawa M, Miller LH, Howard RJ. Plasmodium falciparum malaria: association of knobs on the surface of infected erythrocytes with a histidine-rich protein and the erythrocyte skeleton. J Cell Biol 1984;98:1256–64.

[14] Baruch DI, Pasloske BL, Singh HB, Bi X, Ma XC, Feldman M, et al. Cloning the *P. falciparum* gene encoding PfEMP1, a malarial variant antigen and adherence receptor on the surface of parasitized human erythrocytes. Cell 1995;82:77–87.

[15] Zhuan SX, Heatwole VM, Wertheimer SP, Guinet F, Herrfeldt JA, Peterson DS, et al. The large diverse gene family var encodes proteins involved in cytoadherence and antigenic variation of *Plasmodium falciparum*-infected erythrocytes. Cell 1995;82:89–100.

[16] Galinski MR, Lapp SA, Peterson MS, Ay F, Joyner CJ, Le Roch KG, et al. Plasmodium knowlesi: a superb in vivo nonhuman primate model of antigenic variation in malaria. Parasitology 2018;145(1):85–100.

[17] Cunningham J, Gatton ML, Kolaxzinski K. Malaria rapid diagnostic test performance: results of WHO product testing of malaria RDTs: round 7 (2015-2016) [Internet]. World Health Organization; 2017. Available from: https://eprints.qut.edu.au/111599/.

[18] Gamboa D, Ho MF, Bendezu J, Torres K, Chiodini PL, Barnwell JW, et al. A large proportion of *P. falciparum* isolates in the Amazon region of Peru lack pfhrp2 and pfhrp3: implications for malaria rapid diagnostic tests. PLoS One 2010;5(1):e8091.

[19] Luchavez J, Baker J, Alcantara S, Belizario V, Cheng Q, McCarthy JS, et al. Laboratory demonstration of a prozone-like effect in HRP2-detecting malaria rapid diagnostic tests: implications for clinical management. Malar J 2011;10:286.

[20] Cnops L, Jacobs J, Van Esbroeck M. Validation of a four-primer real-time PCR as a diagnostic tool for single and mixed Plasmodium infections. Clin Microbiol Infect 2011;17:1101–7.

[21] Zakeri S, Kakar Q, Ghasemi F, Raeisi A, Butt W, Safi N, et al. Detection of mixed *Plasmodium falciparum* & *P. vivax* infections by nested-PCR in Pakistan, Iran & Afghanistan. Indian J Med Res 2010;132:31–5.

[22] Kasetsirikul S, Buranapong J, Srituravanich W, Kaewthamasorn M, Pimpin A. The development of malaria diagnostic techniques: a review of the approaches with focus on dielectrophoretic and magnetophoretic methods. Malar J 2016;15.

[23] Bull PC, Lowe BS, Kortok M, Molyneux CS, Newbold CI, Marsh K. Parasite antigens on the infected red cell surface are targets for naturally acquired immunity to malaria. Nat Med 1998;4(3):358–60.

[24] Zhou Z, Mitchell RM, Kariuki S, Odero C, Otieno P, Otieno K, et al. Assessment of submicroscopic infections and gametocyte carriage of *Plasmodium falciparum* during peak malaria transmission season in a community-based cross-sectional survey in western Kenya, 2012. Malar J 2016;15(1):421.

[25] Howard RJ. Antigenic variation of bloodstage malaria parasites. Philos Trans R Soc Lond Ser B Biol Sci 1984. https://doi.org/10.1098/rstb.1984.0115.

[26] Scherf A, Lopez-Rubio JJ, Riviere L. Antigenic variation in Plasmodium falciparum. Annu Rev Microbiol 2008;62(2):445–70.

[27] Blomqvist K, Normark J, Nilsson D, Ribacke U, Orikiriza J, Trillkott P, et al. Var gene transcription dynamics in *Plasmodium falciparum* patient isolates. Mol Biochem Parasitol 2010;170:74–83.

[28] Brown KN, Brown IN. Immunity to malaria: antigenic variation in chronic infections of Plasmodium knowlesi. Nature 1965;1286–8.

[29] Howard RJ, Barnwell JW, Kao V. Antigenic variation of Plasmodium knowlesi malaria: identification of the variant antigen on infected erythrocytes. Proc Natl Acad Sci U S A 1983;80:4129–33.

[30] Lapp SA, Geraldo JA, Chien JT, Ay F, Pakala SB, Batugedara G, et al. PacBio assembly of a Plasmodium knowlesi genome sequence with Hi-C correction and manual annotation of the SICAvar gene family. Parasitology 2018;145:71–84. https://doi.org/10.1017/S0031182017001329.

[31] Singh V, Gupta P, Pande V. Revisiting the multigene families: plasmodium var and vir genes. J Vector Borne Dis 2014;51:75–81.

[32] Fernandez-Becerra C, Pein O, De Oliveira TR, Yamamoto MM, Cassola AC, Rocha C, et al. Variant proteins of *Plasmodium vivax* are not clonally expressed in natural infections. Mol Microbiol 2005;58:648–58.

[33] Ferreira A, Marguti I, Bechmann I, Jeney V, Chora Â, Palha NR, et al. Sickle hemoglobin confers tolerance to plasmodium infection. Cell 2011;145(3):398–409.

[34] Jallow M, Teo YY, Small KS, Rockett KA, Deloukas P, Clark TG, et al. Genome-wide and fine-resolution association analysis of malaria in West Africa. Nat Genet 2009;41:657–65.

[35] Cowman AF, Crabb BS. Invasion of red blood cells by malaria parasites. Cell 2006. https://doi.org/10.1016/j.cell.2006.02.006.

[36] Niangaly A, Gunalan K, Ouattara A, Coulibaly D, Sá JM, Adams M, et al. Plasmodium vivax infections over 3 years in Duffy Blood Group Negative Malians in Bandiagara. Mali Am J Trop Med Hyg 2017;97(3):744–52.

[37] Leffler EM, Band G, Busby GBJ, Kivinen K, Le QS, Clarke GM, et al. Resistance to malaria through structural variation of red blood cell invasion receptors. Science 2017;356 (6343):eaam6393. [Internet]. Available from http://www.sciencemag.org/lookup/doi/10.1126/science.aam6393.

[38] Draper Simon J, Sack BK, King CR, Nielsen CM, Rayner JC, Higgins MK, Long CA, et al. Malaria vaccines: recent advances and new horizons. Cell Host Microbe 2018;24(1):43–56. [Internet]. Available from https://doi.org/10.1016/j.chom.2018.06.008.

[39] Lyke KE, Ishizuka AS, Berry AA, Chakravarty S, De Zure A, Enama ME, et al. Attenuated PfSPZ Vaccine induces strain-transcending T cells and durable protection against heterologous controlled human malaria infection. Proc Natl Acad Sci U S A 2017;114:2711–6.

[40] RTS SCTP. Efficacy and safety of RTS,S/AS01 malaria vaccine with or without a booster dose in infants and children in Africa: final results of a phase 3, individually randomised, controlled trial. Lancet 2015;4:31–45. (London, England) [Internet]. 386(9988). Available from http://www.ncbi.nlm.nih.gov/pubmed/25913272.

[41] WHO. First malaria vaccine in Africa: a potential new tool for child health and improved malaria control. (No. WHO/CDS/GMP/2018.05) World Health Organization; 2018.

[42] Kublin JG, Mikolajczak SA, Sack BK, Fishbaugher ME, Seilie A, Shelton L, et al. Complete attenuation of genetically engineered *Plasmodium falciparum* sporozoites in human subjects. Sci Transl Med 2017;9. https://doi.org/10.1126/scitranslmed.aad9099.

[43] Sack BK, Keitany GJ, Vaughan AM, Miller JL, Wang R, SHI K. Mechanisms of stage-transcending protection following immunization of mice with late liver stage-arresting genetically attenuated malaria parasites. PLoS Pathog 2015;11:e1004855.

[44] Goodman AL, Draper SJ. Blood-stage malaria vaccines—recent progress and future challenges. Ann Trop Med Parasitol 2010;104:189–211.

[45] Staalsoe T, Shulman CE, Bulmer JN, Kawuondo K, Marsh K, Hviid L. Variant surface antigen-specific IgG and protection against clinical consequences of pregnancy-associated *Plasmodium falciparum* malaria. Lancet 2004;363:283–9.

[46] Fried M, Duffy PE. Adherence of *Plasmodium falciparum* to chondroitin sulfate A in the human placenta. Science 1996;1502–4.

[47] Bustamante LY, Powell GT, Lin Y-C, Macklin MD, Cross N, Kemp A, et al. Synergistic malaria vaccine combinations identified by systematic antigen screening. Proc Natl Acad Sci U S A. [Internet]. 2017;(10):201702944. Available from: http://www.pnas.org/lookup/doi/10.1073/pnas.1702944114.

[48] Kidgell C, Volkman SK, Daily J, Borevitz JO, Plouffe D, Zhou Y, et al. A systematic map of genetic variation in *Plasmodium falciparum*. PLoS Pathog 2006;2:e57.

[49] Morita M, Takashima E, Ito D, Miura K, Thongkukiatkul A, Diouf A, et al. Immunoscreening of *Plasmodium falciparum* proteins expressed in a wheat germ cell-free system reveals a novel malaria vaccine candidate. Sci Rep 2017;7:46086.

[50] Douglas AD, Baldeviano GC, Lucas CM, Lugo-Roman LA, Crosnier C, Bartholdson SJ, et al. A PfRH5-based vaccine is efficacious against heterologous strain blood-stage plasmodium falciparum infection in Aotus monkeys. Cell Host Microbe 2015;17:130–9.

[51] Wright KE, Hjerrild KA, Bartlett J, Douglas AD, Jin J, Brown RE, et al. Structure of malaria invasion protein RH5 with erythrocyte basigin and blocking antibodies. Nature 2014;515:427–30.

[52] Hjerrild KA, Jin J, Wright KE, Brown RE, Marshall JM, Labbé GM, et al. Production of full-length soluble Plasmodium falciparum RH5 protein vaccine using a Drosophila melanogaster Schneider 2 stable cell line system. Sci Rep 2016;6:30357.

[53] Sherrard-Smith E, Sala KA, Betancourt M, Upton LM, Angrisano F, Morin MJ, et al. Synergy in anti-malarial pre-erythrocytic and transmission-blocking antibodies is achieved by reducing parasite density. elife 2018.

[54] Plouffe D, Brinker A, McNamara C, Henson K, Kato N, Kuhen K, et al. In silico activity profiling reveals the mechanism of action of antimalarials discovered in a high-throughput screen. Proc Natl Acad Sci U S A 2008;105:9059–64.

[55] Gamo FJ, Sanz LM, Vidal J, De Cozar C, Alvarez E, Lavandera JL, et al. Thousands of chemical starting points for antimalarial lead identification. Nature 2010;465:305–10.

[56] White NJ, Pukrittayakamee S, Phyo AP, Rueangweerayut R, Nosten F, Jittamala P, et al. Spiroindolone KAE609 for falciparum and Vivax malaria. N Engl J Med 2014;371:403–10.

[57] Van Voorhis WC, Adams JH, Adelfio R, Ahyong V, Akabas MH, Alano P, et al. Open source drug discovery with the malaria box compound collection for neglected diseases and beyond. PLoS Pathog 2016;12(7):1–23.

[58] Gural N, Mancio-Silva L, Miller AB, Galstian A, Butty VL, Levine SS, et al. In vitro culture, drug sensitivity, and transcriptome of Plasmodium vivax hypnozoites. Cell Host Microbe 2018;23(3):395–406.e4. [Internet]. Available from https://doi.org/10.1016/j.chom.2018.01.002.

[59] Roth A, Maher SP, Conway AJ, Ubalee R, Chaumeau V, Andolina C, et al. A comprehensive model for assessment of liver stage therapies targeting Plasmodium vivax and Plasmodium falciparum. Nat Commun 2018;9(1):1–16. [Internet]. Available from https://doi.org/10.1038/s41467-018-04221-9.

[60] Rangel GW, Clark MA, Kanjee U, Lim C, Shaw-Saliba K, Menezes MJ, et al. Enhanced ex vivo *Plasmodium vivax* intraerythrocytic enrichment and maturation for rapid and sensitive parasite growth assays. Antimicrob Agents Chemother 2018;62. https://doi.org/10.1128/AAC.02519-17.

[61] van der Voorberg Wel A, Roma G, Gupta DK, Schuierer S, Nigsch F, Carbone W, et al. A comparative transcriptomic analysis of replicating and dormant liver stages of the relapsing malaria parasite *Plasmodium cynomolgi*. elife 2017;6:1–23.

[62] Haldar K, Bhattacharjee S, Safeukui I. Drug resistance in Plasmodium. Nat Rev Microbiol 2018;16(3):156–70. [Internet]. Available from https://doi.org/10.1038/nrmicro.2017.161.

[63] Ariey F, Witkowski B, Amaratunga C, Beghain J, Langlois AC, Khim N, et al. A molecular marker of artemisinin-resistant *Plasmodium falciparum* malaria. Nature 2014;505:50–5.

[64] No JH, Kim Y, Song YS. Targeting Nrf2 signaling to combat chemoresistance. J Cancer Prev 2014;19(2):111–7.

[65] Sidhu ABS, Verdier-Pinard D, Fidock DA. Chloroquine resistance in *Plasmodium falciparum* malaria parasites conferred by pfcrt mutations. Science 2002;298:210–3.

[66] Warhurst DC. A molecular marker for chloroquine-resistant falciparum malaria. N Engl J Med 2001.
[67] Chu CS, Bancone G, Moore KA, Win HH, Thitipanawan N, Po C, et al. Haemolysis in G6PD heterozygous females treated with primaquine for Plasmodium vivax malaria: a nested cohort in a trial of radical curative regimens. PLoS Med 2017;14:e1002224.
[68] Marcsisin SR, Reichard G, Pybus BS. Primaquine pharmacology in the context of CYP 2D6 pharmacogenomics: current state of the art. Pharmacol Ther 2016. https://doi.org/10.1016/j.pharmthera.2016.03.011.
[69] Rottmann M, McNamara C, Yeung BKS, Lee MCS, Zou B, Russell B, et al. Spiroindolones, a potent compound class for the treatment of malaria. Science 2010;329:1175–80.
[70] Spillman NJ, Allen RJW, McNamara CW, Yeung BKS, Winzeler EA, Diagana TT, et al. Na + regulation in the malaria parasite Plasmodium falciparum involves the cation ATPase PfATP4 and is a target of the spiroindolone antimalarials. Cell Host Microbe 2013;13:227–37.
[71] Peterson DS, Walliker D, Wellems TE. Evidence that a point mutation in dihydrofolate reductase-thymidylate synthase confers resistance to pyrimethamine in falciparum malaria. Proc Natl Acad Sci U S A 1988;85:9114–8.
[72] Cowman AF, Galatis D, Thompson JK. Selection for mefloquine resistance in *Plasmodium falciparum* is linked to amplification of the pfmdr1 gene and cross-resistance to halofantrine and quinine. Proc Natl Acad Sci U S A 1994;91(3):1143–7.
[73] Cowell AN, Istvan ES, Lukens AK, Gomez-Lorenzo MG, Vanaerschot M, Sakata-Kato T, et al. Mapping the malaria parasite druggable genome by using in vitro evolution and chemogenomics. Science 2018;359(6372):191–9. 12. [Internet]. Available from: http://www.sciencemag.org/lookup/doi/10.1126/science.aan4472.
[74] Parzy D, Doerig C, Pradines B, Rico A, Fusai T, Doury JC. Proguanil resistance in Plasmodium falciparum African isolates: assessment by mutation-specific polymerase chain reaction and in vitro susceptibility testing. Am J Trop Med Hyg [Internet] 1997;57(6):646–50. Available from: http://www.ajtmh.org/content/journals/10.4269/ajtmh.1997.57.646.
[75] Gil JP, Nogueira F, Strömberg-Nörklit J, Lindberg J, Carrolo M, Casimiro C, et al. Detection of atovaquone and Malarone™ resistance conferring mutations in Plasmodium falciparum cytochrome b gene (cytb). Mol Cell Probes 2003;17:85–39.
[76] Uhlemann AC, Krishna S. Antimalarial multi-drug resistance in Asia: mechanisms and assessment. Curr Top Microbiol Immunol 2005;295:39–53.
[77] Gil JP. Amodiaquine pharmacogenetics. Pharmacogenomics 2008;9:1385–90.
[78] Mehlotra RK, Henry-Halldin CN, Zimmerman PA. Application of pharmacogenomics to malaria: a holistic approach for successful chemotherapy. Pharmacogenomics 2009. https://doi.org/10.2217/14622416.10.3.435.
[79] Kerb R, Fux R, Mörike K, Kremsner PG, Gil JP, Gleiter CH, et al. Pharmacogenetics of antimalarial drugs: effect on metabolism and transport. Lancet Infect Dis 2009. https://doi.org/10.1016/S1473-3099(09)70320-2.
[80] Gil JP, Berglund EG. CYP2C8 and antimalaria drug efficacy. Pharmacogenomics 2007.
[81] Janha RE, Sisay-Joof F, Hamid-Adiamoh M, Worwui A, Chapman HL, Opara H, et al. Effects of genetic variation at the CYP2C19/CYP2C9 locus on pharmacokinetics of chlorcycloguanil in adult Gambians. Pharmacogenomics 2009;10:1423–31.
[82] Dagenais R, Wilby KJ, Elewa H, Ensom MHH. Impact of genetic polymorphisms on phenytoin pharmacokinetics and clinical outcomes in the Middle East and North Africa region. Drugs R D 2017;17(3):341–61. [Internet]. Available from: http://link.springer.com/10.1007/s40268-017-0195-7.

[83] Zhang M, Wang C, Otto TD, Oberstaller J, Liao X, Adapa SR, et al. Uncovering the essential genes of the human malaria parasite *Plasmodium falciparum* by saturation mutagenesis. Science 2018;360(6388). eaap7847.

[84] Espinosa DA, Vega-Rodriguez J, Flores-Garcia Y, Noe AR, Muñoz C, Coleman R, et al. The *Plasmodium falciparum* cell-traversal protein for ookinetes and sporozoites as a candidate for preerythrocytic and transmission-blocking vaccines. Infect Immun 2017;85. https://doi.org/10.1128/IAI.00498-16.

[85] Ikadai H, Shaw Saliba K, Kanzok SM, KJ ML, Tanaka TQ, Cao J, et al. Transposon mutagenesis identifies genes essential for *Plasmodium falciparum* gametocytogenesis. Proc Natl Acad Sci U S A 2013;110:E1676–84.

[86] Bobenchik AM, Witola WH, Augagneur Y, Nic Lochlainn L, Garg A, Pachikara N, et al. *Plasmodium falciparum* phosphoethanolamine methyltransferase is essential for malaria transmission. Proc Natl Acad Sci U S A 2013;110:18262–7.

[87] Brancucci NMB, Gerdt JP, Wang C, De Niz M, Philip N, Adapa SR, et al. Lysophosphatidylcholine regulates sexual stage differentiation in the human malaria parasite *Plasmodium falciparum*. Cell 2017;171:1532–44.e15.

[88] Caro F, Ahyong V, Betegon M, De Risi JL. Genome-wide regulatory dynamics of translation in the *Plasmodium falciparum* asexual blood stages. elife 2014;3. https://doi.org/10.7554/eLife.04106.

[89] Le Roch KG, Johnson JR, Florens L, Zhou Y, Santrosyan A, Grainger M, et al. Global analysis of transcript and protein levels across the Plasmodium falciparum life cycle. Genome Res 2004;14(11):2308–18.

[90] Foth BJ, Zhang N, Chaal BK, Sze SK, Preiser PR, Bozdech Z. Quantitative time-course profiling of parasite and host cell proteins in the human malaria parasite *Plasmodium falciparum*. Mol Cell Proteomics 2011;10:M110.006411.

[91] Bunnik EM, Chung D-WD, Hamilton M, Ponts N, Saraf A, Prudhomme J, et al. Polysome profiling reveals translational control of gene expression in the human malaria parasite Plasmodium falciparum. Genome Biol 2013;14(11):R128.

[92] Bunnik EM, Batugedara G, Saraf A, Prudhomme J, Florens L, Le Roch KG. The mRNA-bound proteome of the human malaria parasite Plasmodium falciparum. Genome Biol 2016;17(1):1–18. [Internet]. Available from: https://doi.org/10.1186/s13059-016-1014-0.

[93] Sidik SM, Huet D, Ganesan SM, Huynh MH, Wang T, Nasamu AS, et al. A genome-wide CRISPR screen in toxoplasma identifies essential apicomplexan genes. Cell 2016;166(6):1423–1435.e12. [Internet]. Available from https://doi.org/10.1016/j.cell.2016.08.019.

[94] Bozdech Z, Llinás M, Pulliam BL, Wong ED, Zhu J, De Risi JL. The transcriptome of the intraerythrocytic developmental cycle of *Plasmodium falciparum*. PLoS Biol 2003;1:E5.

[95] Le Roch KG, Zhou Y, Blair PL, Grainger M, Moch JK, Haynes JD, et al. Discovery of gene function by expression profiling of the malaria parasite life cycle. Science 2003;301:1503–8.

[96] Otto TD, Wilinski D, Assefa S, Keane TM, Sarry LR, Böhme U, et al. New insights into the blood-stage transcriptome of *Plasmodium falciparum* using RNA-Seq. Mol Microbiol 2010;76:12–24.

[97] Sorber K, Dimon MT, Derisi JL. RNA-Seq analysis of splicing in Plasmodium falciparum uncovers new splice junctions, alternative splicing and splicing of antisense transcripts. Nucleic Acids Res 2011;39:3820–35.

[98] Tuda J, Mongan AE, Tolba MEM, Imada M, Yamagishi J, Xuan X, et al. Full-parasites: database of full-length cDNAs of apicomplexa parasites, 2010 update. Nucleic Acids Res 2011;39:D625–31.

[99] Callebaut I, Prat K, Meurice E, Mornon JP, Tomavo S. Prediction of the general transcription factors associated with RNA polymerase II in *Plasmodium falciparum*: conserved features and differences relative to other eukaryotes. BMC Genomics 2005;6:100.

[100] Balaji S, Madan Babu M, Iyer LM, Aravind L. Discovery of the principal specific transcription factors of apicomplexa and their implication for the evolution of the AP2-integrase DNA binding domains. Nucleic Acids Res 2005;33:3994–4006.

[101] Kafsack BFC, Rovira-Graells N, Clark TG, Bancells C, Crowley VM, Campino SG, et al. A transcriptional switch underlies commitment to sexual development in malaria parasites. Nature 2014;507:248–52.

[102] Sinha A, Hughes KR, Modrzynska KK, Otto TD, Pfander C, Dickens NJ, et al. A cascade of DNA-binding proteins for sexual commitment and development in Plasmodium. Nature 2014;507:253–7.

[103] Yuda M, Iwanaga S, Shigenobu S, Kato T, Kaneko I. Transcription factor AP2-Sp and its target genes in malarial sporozoites. Mol Microbiol 2010;75:854–63.

[104] Iwanaga S, Kaneko I, Kato T, Yuda M. Identification of an AP2-family protein that is critical for malaria liver stage development. PLoS One 2012;7:e47557.

[105] Poran A, Nötzel C, Aly O, Mencia-Trinchant N, Harris CT, Guzman ML, et al. Single-cell RNA sequencing reveals a signature of sexual commitment in malaria parasites. Nature 2017;551(7678):95–9. [Internet]. Available from https://doi.org/10.1038/nature24280.

[106] Reid AJ, Talman AM, Bennett HM, Gomes AR, Sanders MJ, Illingworth CJR, et al. Single-cell RNA-seq reveals hidden transcriptional variation in malaria parasites. elife 2018;7:e33105. [Internet]. Available from: https://elifesciences.org/articles/33105.

[107] Ay F, Bunnik EM, Varoquaux N, Bol SM, Prudhomme J, Vert JP, et al. Three-dimensional modeling of the P. falciparum genome during the erythrocytic cycle reveals a strong connection between genome architecture and gene expression. Genome Res 2014;24(6):974–88.

[108] Lemieux JE, Kyes SA, Otto TD, Feller AI, Eastman RT, Pinches RA, et al. Genome-wide profiling of chromosome interactions in *Plasmodium falciparum* characterizes nuclear architecture and reconfigurations associated with antigenic variation. Mol Microbiol 2013;90:519–37.

[109] Bunnik EM, Cook KB, Varoquaux N, Batugedara G, Prudhomme J, Cort A, et al. Changes in genome organization of parasite-specific gene families during the Plasmodium transmission stages. Nat Commun 2018. [Internet]. Available from https://doi.org/10.1038/s41467-018-04295-5.

[110] Flueck C, Bartfai R, Volz J, Niederwieser I, Salcedo-Amaya AM, Alako BTF, et al. Plasmodium falciparum heterochromatin protein 1 marks genomic loci linked to phenotypic variation of exported virulence factors. PLoS Pathog 2009;5:e1000569.

[111] Bártfai R, Hoeijmakers WAM, Salcedo-Amaya AM, Smits AH, Janssen-Megens E, Kaan A, et al. H2A.Z demarcates intergenic regions of the *Plasmodium falciparum* epigenome that are dynamically marked by H3K9ac and H3K4me3. PLoS Pathog 2010;6:e1001223.

[112] Gupta AP, Chin WH, Zhu L, Mok S, Luah Y, Lim E-H, et al. Dynamic epigenetic regulation of gene expression during the life cycle of malaria parasite *Plasmodium falciparum*. PLoS Pathog 2013;9:e1003170.

[113] Karmodiya K, Pradhan SJ, Joshi B, Jangid R, Reddy PC, Galande S. A comprehensive epigenome map of Plasmodium falciparum reveals unique mechanisms of transcriptional regulation and identifies H3K36me2 as a global mark of gene suppression. Epigenetics Chromatin 2015;8:32.

[114] Khorasanizadeh S. The nucleosome: from genomic organization to genomic regulation. Cell 2004. https://doi.org/10.1016/S0092-8674(04)00044-3.

[115] Hoeijmakers WAM, Salcedo-Amaya AM, Smits AH, Françoijs K-J, Treeck M, Gilberger T-W, et al. H2A.Z/H2B.Z double-variant nucleosomes inhabit the AT-rich promoter regions of the Plasmodium falciparum genome. Mol Microbiol 2013;87:1061–73.

[116] Petter M, Selvarajah SA, Lee CC, Chin WH, Gupta AP, Bozdech Z, et al. H2A.Z and H2B.Z double-variant nucleosomes define intergenic regions and dynamically occupy var gene promoters in the malaria parasite *Plasmodium falciparum*. Mol Microbiol 2013;87:1167–82.

[117] Beh LY, Müller MM, Muir TW, Kaplan N, Landweber LF. DNA-guided establishment of nucleosome patterns within coding regions of a eukaryotic genome. Genome Res 2015;25:1727–38.

[118] Mavrich TN, Jiang C, Ioshikhes IP, Li X, Venters BJ, Zanton SJ, et al. Nucleosome organization in the Drosophila genome. Nature 2008;453:358–62.

[119] Lee CK, Shibata Y, Rao B, Strahl BD, Lieb JD. Evidence for nucleosome depletion at active regulatory regions genome-wide. Nat Genet 2004;36:900–5.

[120] Freitas-Junior LH, Bottius E, Pirrit LA, Deitsch KW, Scheidig C, Guinet F, et al. Frequent ectopic recombination of virulence factor genes in telomeric chromosome clusters of *P. falciparum*. Nature 2000;407:1018–22.

[121] Cui L, Miao J, Furuya T, Li X, Su XZ, Cui L. PfGCN5-mediated histone H3 acetylation plays a key role in gene expression in Plasmodium falciparum. Eukaryot Cell 2007;6(7):1219–27.

[122] Toenhake CG, Fraschka SAK, Vijayabaskar MS, Westhead DR, van Heeringen SJ, Bártfai R. Chromatin accessibility-based characterization of the gene regulatory network underlying *Plasmodium falciparum* blood-stage development. Cell Host Microbe 2018;23(4):557–569.e9.

[123] Salcedo-Amaya AM, van Driel MA, Alako BT, Trelle MB, van den Elzen AMG, Cohen AM, et al. Dynamic histone H3 epigenome marking during the intraerythrocytic cycle of Plasmodium falciparum. Proc Natl Acad Sci U S A 2009;106:9655–60.

[124] Crowley VM, Rovira-Graells N, de Pouplana LR, Cortés A. Heterochromatin formation in bistable chromatin domains controls the epigenetic repression of clonally variant *Plasmodium falciparum* genes linked to erythrocyte invasion. Mol Microbiol 2011;80:391–406.

[125] Freitas LH, Hernandez-Rivas R, Ralph SA, Montiel-Condado D, Ruvalcaba-Salazar OK, Rojas-Meza AP, et al. Telomeric heterochromatin propagation and histone acetylation control mutually exclusive expression of antigenic variation genes in malaria parasites. Cell 2005;121:25–36.

[126] Chookajorn T, Dzikowski R, Frank M, Li F, Jiwani AZ, Hartl DL, et al. Epigenetic memory at malaria virulence genes. Proc Natl Acad Sci U S A 2007;104:899–902.

[127] Lopez-Rubio JJ, Mancio-Silva L, Scherf A. Genome-wide analysis of heterochromatin associates clonally variant gene regulation with perinuclear repressive centers in malaria parasites. Cell Host Microbe 2009;5:179–90.

[128] Zhang Q, Huang Y, Zhang Y, Fang X, Claes A, Duchateau M, et al. A critical role of perinuclear filamentous actin in spatial repositioning and mutually exclusive expression of virulence genes in malaria parasites. Cell Host Microbe 2011;10:451–63.

[129] Duraisingh MT, Voss TS, Marty AJ, Duffy MF, Good RT, Thompson JK, et al. Heterochromatin silencing and locus repositioning linked to regulation of virulence genes in *Plasmodium falciparum*. Cell 2005;121:13–24.

[130] Tonkin CJ, Carret CK, Duraisingh MT, Voss TS, Ralph SA, Hommel M, et al. Sir2 paralogues cooperate to regulate virulence genes and antigenic variation in Plasmodium falciparum. PLoS Biol 2009;7:e84.

[131] Coleman BI, Skillman KM, Jiang RHY, Childs LM, Altenhofen LM, Ganter M, et al. A *Plasmodium falciparum* histone deacetylase regulates antigenic variation and gametocyte conversion. Cell Host Microbe 2014;16:177–86.

[132] Jiang L, Mu J, Zhang Q, Ni T, Srinivasan P, Rayavara K, et al. PfSETvs methylation of histone H3K36 represses virulence genes in Plasmodium falciparum. Nature 2013;499:223–7.
[133] Ukaegbu UE, Kishore SP, Kwiatkowski DL, Pandarinath C, Dahan-Pasternak N, Dzikowski R, et al. Recruitment of PfSET2 by RNA polymerase II to variant antigen encoding loci contributes to antigenic variation in *P. falciparum*. PLoS Pathog 2014;10:e1003854.
[134] Trelle MB, Salcedo-Amaya AM, Cohen AM, Stunnenberg HG, Jensen ON. Global histone analysis by mass spectrometry reveals a high content of acetylated lysine residues in the malaria parasite Plasmodium falciparum. J Proteome Res 2009;8(7):3439–50. [Internet]. Available from: http://pubs.acs.org/doi/pdfplus/10.1021/pr9000898.
[135] Josling GA, Petter M, Oehring SC, Gupta AP, Dietz O, Wilson DW, et al. A *Plasmodium falciparum* bromodomain protein regulates invasion gene expression. Cell Host Microbe 2015;17:741–51.
[136] Templeton TJ, Iyer LM, Anantharaman V, Templeton TJ, Iyer LM, Anantharaman V, et al. Comparative analysis of apicomplexa and genomic diversity in eukaryotes. Genome Res 2004;14:1686–95.
[137] Vembar SS, Scherf A, Siegel TN. Noncoding RNAs as emerging regulators of Plasmodium falciparum virulence gene expression. Curr Opin Microbiol 2014;20:153–61. https://doi.org/10.1016/j.mib.2014.06.013.
[138] Amit-Avraham I, Pozner G, Eshar S, Fastman Y, Kolevzon N, Yavin E, et al. Antisense long noncoding RNAs regulate var gene activation in the malaria parasite Plasmodium falciparum. Proc Natl Acad Sci U S A 2015;112(9):E982–91. [Internet]. Available from: http://www.ncbi.nlm.nih.gov/pubmed/25691743%5Cnhttp://www.pubmedcentral.nih.gov/articlerender.fcgi?artid=PMC4352787.
[139] Flueck C, Bartfai R, Niederwieser I, Witmer K, Alako BTF, Moes S, et al. A major role for the Plasmodium falciparum ApiAP2 protein PfSIP2 in chromosome end biology. PLoS Pathog 2010;6:e1000784.
[140] Broadbent KM, Park D, Wolf AR, Van Tyne D, Sims JS, Ribacke U, et al. A global transcriptional analysis of *Plasmodium falciparum* malaria reveals a novel family of telomere-associated lncRNAs. Genome Biol 2011;12:R56.
[141] Lieberman-Aiden E, van Berkum NL, Williams L, Imakaev M, Ragoczy T, Telling A, et al. Comprehensive mapping of long-range interactions reveals folding principles of the human genome. Science 2009;326(5950):289–93. 9. [Internet]. Available from: http://www.sciencemag.org/cgi/doi/10.1126/science.1181369.
[142] Dixon JR, Selvaraj S, Yue F, Kim A, Li Y, Shen Y, et al. Topological domains in mammalian genomes identified by analysis of chromatin interactions. Nature 2012;485:376–80.
[143] Rao SSP, Huntley MH, Durand NC, Stamenova EK, Bochkov ID, Robinson JT, et al. A 3D map of the human genome at kilobase resolution reveals principles of chromatin looping. Cell 2014;159:1665–80.
[144] Javierre BM, Sewitz S, Cairns J, Wingett SW, Várnai C, Thiecke MJ, et al. Lineage-specific genome architecture links enhancers and non-coding disease variants to target gene promoters. Cell 2016;167:1369–84.e19.
[145] Flavahan WA, Drier Y, Liau BB, Gillespie SM, Venteicher AS, Stemmer-Rachamimov AO, et al. Insulator dysfunction and oncogene activation in IDH mutant gliomas. Nature 2016;529:110–4.
[146] Gröschel S, Sanders MA, Hoogenboezem R, De Wit E, Bouwman BAM, Erpelinck C, et al. A single oncogenic enhancer rearrangement causes concomitant EVI1 and GATA2 deregulation in leukemia. Cell 2014;157:369–81.

[147] Batsios P, Peter T, Baumann O, Stick R, Meyer I, Gräf R. A lamin in lower eukaryotes? Nucleus 2012;3:237–43 (United States).

[148] Heger P, Marin B, Bartkuhn M, Schierenberg E, Wiehe T. The chromatin insulator CTCF and the emergence of metazoan diversity. Proc Natl Acad Sci U S A 2012;109:17507–12.

[149] Huang CE, Milutinovich M, Koshland D. Rings, bracelet or snaps: fashionable alternatives for Smc complexes. Philos Trans R Soc, B 2005. https://doi.org/10.1098/rstb.2004.1609.

[150] Lu XM, Batugedara G, Lee M, Prudhomme J, Bunnik EM, Le Roch KG. Nascent RNA sequencing reveals mechanisms of gene regulation in the human malaria parasite Plasmodium falciparum. Nucleic Acids Res 2017;45(13):7825–40.

[151] Florens L, Washburn MP, Raine JD, Anthony RM, Grainger M, Haynes JD, et al. A proteomic view of the *Plasmodium falciparum* life cycle. Nature 2002;419:520–6.

[152] Hall N, Karras M, Raine JD, Carlton JM, TWA K, Berriman M, et al. A comprehensive survey of the Plasmodium life cycle by genomic, transcriptomic, and proteomic analyses. Science 2005;307:82–6.

[153] Zhang M, Mishra S, Sakthivel R, Rojas M, Ranjan R, Sullivan WJ, et al. PK4, a eukaryotic initiation factor 2α(eIF2α) kinase, is essential for the development of the erythrocytic cycle of Plasmodium. Proc Natl Acad Sci U S A 2012;109:3956–61.

[154] Parker R, Sheth U. P bodies and the control of mRNA translation and degradation. Mol Cell 2007. https://doi.org/10.1016/j.molcel.2007.02.011.

[155] Mair GR, Braks JAM, Garver LS, Wiegant JCAG, Hall N, Dirks RW, et al. Regulation of sexual development of Plasmodium by translational repression. Science 2006;313(5787):667–9. [Internet]. Available from: http://www.sciencemag.org/cgi/doi/10.1126/science.1157880.

[156] Vembar SS, Macpherson CR, Sismeiro O, Coppée JY, Scherf A. The PfAlba1 RNA-binding protein is an important regulator of translational timing in *Plasmodium falciparum* blood stages. Genome Biol 2015;16(1):212.

[157] Chêne A, Vembar SS, Rivière L, Lopez-Rubio JJ, Claes A, Siegel TN, et al. PfAlbas constitute a new eukaryotic DNA/RNA-binding protein family in malaria parasites. Nucleic Acids Res 2012;40(7):3066–77. [Internet]. Available from: https://academic.oup.com/nar/article-lookup/doi/10.1093/nar/gkr1215.

[158] Bunnik EM, Le Roch KG. PfAlba1: master regulator of translation in the malaria parasite. Genome Biol 2015;16(1):4–7. [Internet]. Available from https://doi.org/10.1186/s13059-015-0795-x.

[159] Balu B, Maher SP, Pance A, Chauhan C, Naumov AV, Andrews RM, et al. CCR4-associated factor 1 coordinates the expression of Plasmodium falciparum egress and invasion proteins. Eukaryot Cell 2011;10:1257–63.

[160] Reddy BN, Shrestha S, Hart KJ, Liang X, Kemirembe K, Cui L, et al. A bioinformatic survey of RNA-binding proteins in Plasmodium. BMC Genomics 2015;16:890.

[161] Swearingen KE, Lindner SE, Flannery EL, Vaughan AM, Morrison RD, Patrapuvich R, et al. Proteogenomic analysis of the total and surface-exposed proteomes of *Plasmodium vivax* salivary gland sporozoites. PLoS Negl Trop Dis 2017;11:1–36.

[162] Swearingen KE, Lindner SE, Shi L, Shears MJ, Harupa A, Hopp CS, et al. Interrogating the Plasmodium sporozoite surface: identification of surface-exposed proteins and demonstration of glycosylation on CSP and TRAP by mass spectrometry-based proteomics. PLoS Pathog 2016;12(4):1–32.

[163] Lasonder E, Green JL, Grainger M, Langsley G, Holder AA. Extensive differential protein phosphorylation as intraerythrocytic *Plasmodium falciparum* schizonts develop into extracellular invasive merozoites. Proteomics 2015;15:2716–29.

[164] Treeck M, Sanders JL, Elias JE, Boothroyd JC. The phosphoproteomes of plasmodium falciparum and toxoplasma gondii reveal unusual adaptations within and beyond the parasites' boundaries. Cell Host Microbe 2011;10(4):410–9. [Internet]. Available from https://doi.org/10.1016/j.chom.2011.09.004.
[165] Pease BN, Huttlin EL, Jedrychowski MP, Talevich E, Harmon J, Dillman T, et al. Global analysis of protein expression and phosphorylation of three stages of *Plasmodium falciparum* intraerythrocytic development. J Proteome Res 2013;12:4028–45.
[166] Gisselberg JE, Zhang L, Elias JE, Yeh E. The prenylated proteome of *Plasmodium falciparum* reveals pathogen-specific prenylation activity and drug mechanism-of-action. Mol Cell Proteomics 2017;16:S54–64.
[167] Wright MH, Clough B, Rackham MD, Rangachari K, Brannigan JA, Grainger M, et al. Validation of N-myristoyltransferase as an antimalarial drug target using an integrated chemical biology approach. Nat Chem 2014;6:112–21.
[168] Jones ML, Collins MO, Goulding D, Choudhary JS, Rayner JC. Analysis of protein palmitoylation reveals a pervasive role in Plasmodium development and pathogenesis. Cell Host Microbe 2012;12:246–58.
[169] Anderson DC, Lapp SA, Akinyi S, EVS M, Barnwell JW, Korir-Morrison C, et al. *Plasmodium vivax* trophozoite-stage proteomes. J Proteome 2015;115:157–76.
[170] Anderson DC, Lapp SA, Barnwell JW, Galinski MR. A large scale *Plasmodium vivax- Saimiri boliviensis* trophozoite-schizont transition proteome. PLoS One 2017;12:e0182561.
[171] Gardinassi LG, Cordy RJ, Lacerda MVG, Salinas JL, Monteiro WM, Melo GC, et al. Metabolome-wide association study of peripheral parasitemia in *Plasmodium vivax* malaria. Int J Med Microbiol 2017;307:533–41.
[172] Hayashi S, Takamiya R, Yamaguchi T, Matsumoto K, Tojo SJ, Tamatani T, et al. Induction of heme oxygenase-1 suppresses venular leukocyte adhesion elicited by oxidative stress: role of bilirubin generated by the enzyme. Circ Res 1999;85:663–71.
[173] Tsai W-N, Wang Y-Y, Liang J-T, Lin S-Y, Sheu WH-H, Chang W-D. Serum total bilirubin concentrations are inversely associated with total white blood cell counts in an adult population. Ann Clin Biochem 2015;52:251–8.
[174] Gupta S, Seydel K, Miranda-Roman MA, Feintuch CM, Saidi A, Kim RS, et al. Extensive alterations of blood metabolites in pediatric cerebral malaria. PLoS One 2017;12:e0175686.
[175] Pappa V, Seydel K, Gupta S, Feintuch CM, Potchen MJ, Kampondeni S, et al. Lipid metabolites of the phospholipase A2 pathway and inflammatory cytokines are associated with brain volume in paediatric cerebral malaria. Malar J 2015.
[176] Sengupta A, Ghosh S, Das BK, Panda A, Tripathy R, Pied S, et al. Host metabolic responses to Plasmodium falciparum infections evaluated by 1H NMR metabolomics. Mol BioSyst 2016;12:3324–32.
[177] Castillo JC, Ferreira ABB, Trisnadi N, Barillas-Mury C. Activation of mosquito complement antiplasmodial response requires cellular immunity. Sci Immunol [Internet]. 2017;2(7):eaal1505. Available from http://immunology.sciencemag.org/lookup/doi/10.1126/sciimmunol.aal1505.
[178] Dong Y, Das S, Cirimotich C, Souza-Neto JA, McLean KJ, Dimopoulos G, Engineered Anopheles Immunity to Plasmodium Infection, Vernick KD. Engineered anopheles immunity to Plasmodium infection. PLoS Pathog 2011;7(12):e1002458. 22. [Internet]. Available from: https://doi.org/10.1371/journal.ppat.1002458.
[179] Christensen BM, Li J, Chen CC, Nappi AJ. Melanization immune responses in mosquito vectors. Trends Parasitol 2005;21(4):192–9.

[180] Pike A, Dong Y, Dizaji NB, Gacita A, Mongodin EF, Dimopoulos G. Changes in the microbiota cause genetically modified Anopheles to spread in a population. Science 2017;357(6358):1396–9. 29. [Internet]. Available from: http://science.sciencemag.org/content/357/6358/1396.full?utm_source=sciencemagazine&utm_medium=twitter&utm_campaign=6358issue-15492.

[181] Hammond A, Galizi R, Kyrou K, Simoni A, Siniscalchi C, Katsanos D, et al. A CRISPR-Cas9 gene drive system targeting female reproduction in the malaria mosquito vector *Anopheles gambiae*. Nat Biotechnol 2016;34(1):78–83.

[182] Esvelt KM, Smidler AL, Catteruccia F, Church GM. Concerning RNA-guided gene drives for the alteration of wild populations. *E*life 2014;3:e03401.

[183] Noble C, Olejarz J, Esvelt KM, Church GM, Nowak MA. Evolutionary dynamics of CRISPR gene drives. Sci Adv 2017;3:e1601964.

[184] Champer J, Buchman A, Akbari OS. Cheating evolution: engineering gene drives to manipulate the fate of wild populations. Nat Rev Genet 2016;17:146–59.

[185] Champer J, Liu J, Oh SY, Reeves R, Luthra A, Oakes N, et al. Reducing resistance allele formation in CRISPR gene drive. Proc Natl Acad Sci U S A 2018;115(21):5522–7.

[186] Dong Y, Simões ML, Marois E, Dimopoulos G. CRISPR/Cas9-mediated gene knockout of Anopheles gambiae FREP1 suppresses malaria parasite infection. PLoS Pathog 2018;14(3):e1006898.

[187] Ravenhall M, Campino S, Sepúlveda N, Manjurano A, Nadjm B, Mtove G, et al. Novel genetic polymorphisms associated with severe malaria and under selective pressure in North-eastern Tanzania. PLoS Genet 2018;14:e1007172.

[188] Ndila CM, Uyoga S, Macharia AW, Nyutu G, Peshu N, Ojal J, et al. Human candidate gene polymorphisms and risk of severe malaria in children in Kilifi, Kenya: a case-control association study. Lancet Haematol 2018;5:e333–45.

[189] Amato R, Pearson RD, Almagro-Garcia J, Amaratunga C, Lim P, Suon S, et al. Origins of the current outbreak of multidrug-resistant malaria in Southeast Asia: a retrospective genetic study. Lancet Infect Dis 2018;18:337–45.

Chapter 15

The role of disease surveillance in precision public health

David L. Blazes, Scott F. Dowell
Surveillance and Epidemiology, Bill and Melinda Gates Foundation, Seattle, WA, United States

Precision public health

Assuring the health of populations is a complex goal of most societies, dating back to when humans first started to live in communities. Public health practice has evolved over time from the anecdotal to a rigorous, evidence-based discipline. Applying precision to the practice of public health has been a continuously improving process, but recent advances in data access, molecular genomics and computing have vastly increased the potential impact of precision on public health.

Precision public health endeavors to use the best available data to efficiently target illnesses with interventions for those most in need [1,2]. Precision can be described in terms of several contexts, including spatial, temporal, and diagnostic. Spatial precision is perhaps the easiest to understand, and underlies the seminal public health contribution of John Snow in mapping the source of contaminated water that led to the control of cholera in London [3]. More recently, public health officials were able to map Zika-infected mosquito pools to the neighborhood level of resolution in Miami, thus allowing for the timely and accurate use of mosquito control methods that minimize costs while protecting the population (Fig. 1) [4]. Temporal precision includes identifying when a condition began as accurately as possible, the duration of the condition, and also how soon a definitive diagnosis can be made after the condition starts. For example, temporal precision becomes important when creating an epidemic curve with the appropriate time intervals so that the likely exposure period can be derived from known incubation periods [5]. Finally, diagnostic precision describes the need for accurate classification of a disease so that the correct therapeutic intervention may be applied. An example of diagnostic precision might be the paradigm of understanding a diarrheal outbreak as having broader epidemic potential or not, whether it is bacterial, viral or parasitic in nature, the exact species of pathogen involved and its antimicrobial susceptibility, and ultimately

Genomic and Precision Medicine. https://doi.org/10.1016/B978-0-12-801496-7.00015-0

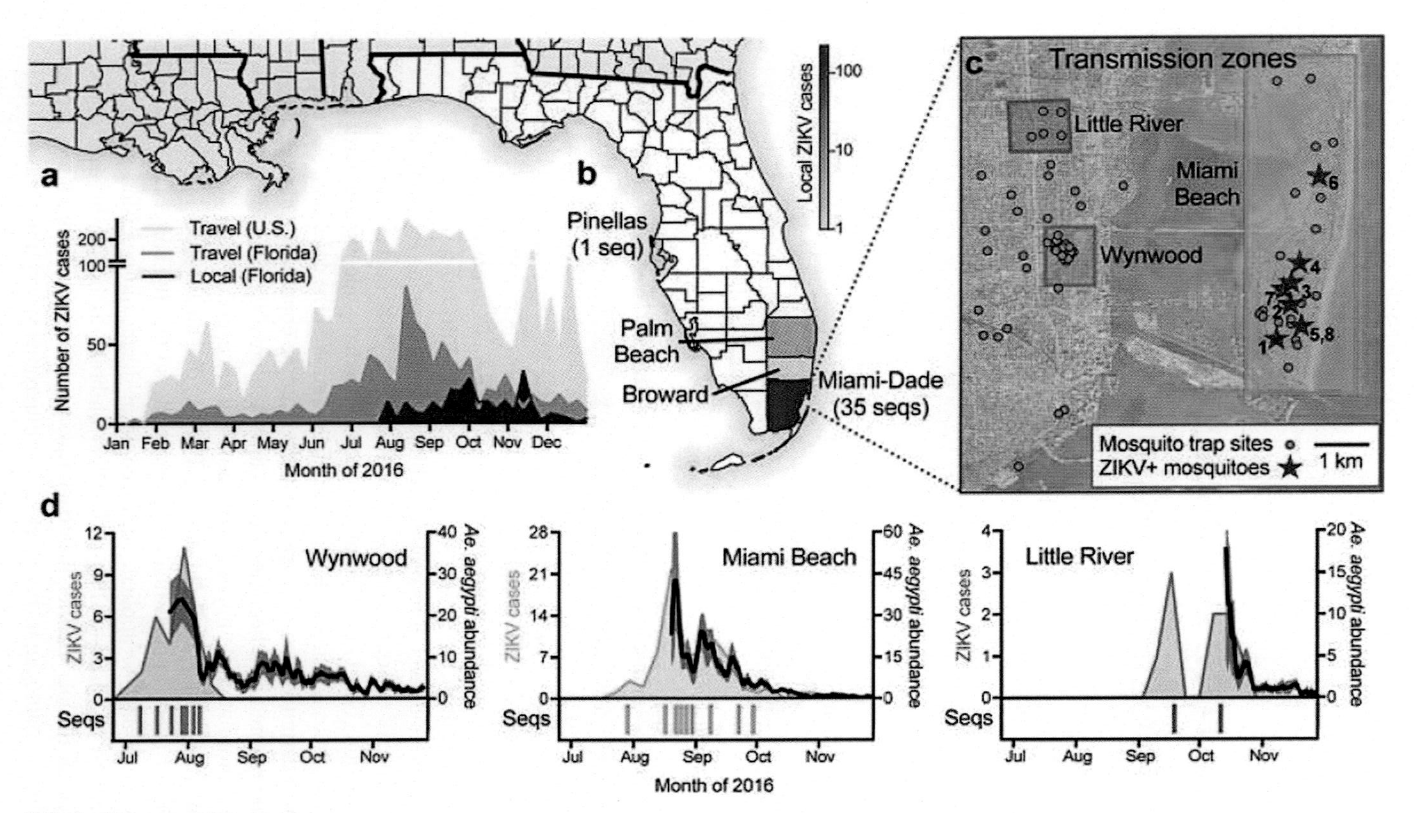

FIG. 1 Zika geospatial precision.

its genetic relatedness to other similar pathogens that have been collected previously. The most precise diagnostic test today can involve direct pathogen genetic sequencing, allowing for very rapid and accurate identification of the pathogen, but additionally, the chains of transmission and relatedness within a disease transmission network. With this increased precision of diagnosis, a more targeted intervention can then be applied.

Precision public health uses increasingly large and diverse datasets to more precisely guide interventions that benefit populations more efficiently. Included in the concept is leveraging advances in disease surveillance to provide ongoing information about the distribution of disease in vulnerable populations, increasing equity even as overall public health improves [1,6].

Progress in disease surveillance

Surveillance for infectious diseases is one of the most critical functions of a public health system, from the clinic or hospital level to the national and global levels. Surveillance can take many forms, but generally entails the continuous and systematic collection and analysis of data, as well as the subsequent reporting of any significant findings to effect positive change. There are many valid indications for implementing disease surveillance: establishing baseline levels of disease; defining hypotheses to investigate using further research; monitoring the effectiveness of interventions; and detecting outbreaks. Surveillance can be either active or passive. Active surveillance is often more sensitive but generally more expensive and labor intensive. Passive surveillance sometimes relies on the analysis of data collected for other purposes, and is often less sensitive, but usually more cost-effective.

Surveillance systems in the USA mostly follow a hierarchical reporting scheme, with local clinics, hospitals, or laboratories tabulating cases of reportable diseases and reporting them to the local public health authorities. The collated data is then passed to the state-level public health officials, who in turn report to the Centers for Disease Control and Prevention (CDC). The CDC reports to the World Health Organization (WHO) through the US national focal point, if necessary, about possible public health emergencies of international concern. Reports of outbreaks are updated by the CDC on Epi-X, https://emergency.cdc.gov/epix/index.asp, a secure web-based reporting system limited to public health officials, as well as in the Morbidity and Mortality Weekly Report (MMWR), which is available to the public electronically at http://www.cdc.gov/mmwr/ and in hard copy. An international treaty, the International Health Regulations governs what type of outbreak is reportable to WHO, as well as to whom and under what time requirements it must be reported [7]. Each signatory member state is responsible for controlling disease within its own borders. The IHR (2005) is designed to allow less developed countries to request assistance from the WHO and its response mechanism, the Global Outbreak and Alert Response Network (GOARN). The GOARN (https://www.who.int/ihr/alert_and_response/outbreak-network/en/) and WHO Outbreak News (https://www.who.int/csr/don/en/) websites provide public information about current outbreaks of potential global concern.

Technology is steadily improving disease surveillance, especially in the field of infectious diseases. Classic tools such as the outbreak line list and population serological surveys remain important for understanding and controlling disease, but increasingly, the power of "big data" is being leveraged to more rapidly and precisely address the challenges posed by infectious diseases.

Novel methods of disease surveillance are continuously being developed and refined. So-called event-based reporting systems, like ProMED mail, have been functioning since the 1990s. ProMED relies on a network of health care professionals and other interested parties (over 70,000 subscribers in 185 countries) to report outbreaks from around the world. This email and web-based system covers outbreaks affecting humans, animals and plants, and is moderated by a panel of public health experts who screen each report before release. Digital disease detection or surveillance is an approach that relies on collating and analyzing diverse streams of data that are increasingly available in machine-readable format. A prototypical digital disease detection platform is HealthMap (http://www.healthmap.org/en), which harnesses the broad reach of the internet by searching "on-line" news outlets and applying automated logic to sort "real" outbreaks from rumor [8].

Machine learning and related deep learning computer algorithms are synthesizing and analyzing previously unapproachably large data sets that surveillance systems can collect. This allows for aberration detection within surveillance data streams and may even lead to improved prediction of epidemiologic trends or forecasting of disease events. Better quality data also allow for more precise geospatial representation of disease, increasing the efficiency of identification, response and control of outbreaks.

It is important to remember the social, political and cultural context when thinking about precision. If there is no mechanism to implement an intervention, it may not be appropriate to spend scarce funds to improve surveillance. The level of precision in surveillance should be appropriate to the available response. In addition, the relative cost-effectiveness of potential diagnostics and interventions must also be considered in the calculus of how much to invest in precision surveillance activities.

Case studies

Surveillance systems are becoming increasingly precise, especially in higher income settings where data collection has become routine and standardized, tools for managing data are commonplace and automated, and accurate diagnosis can be assured. Key to well-performing surveillance systems are personnel trained in epidemiology, especially in bioinformatics, geospatial techniques and increasingly in molecular methods. Lesser developed settings have not been able to leverage these technological innovations for surveillance, and thus have not derived as much benefit from precision.

We will use the following two case studies on food-borne disease and influenza surveillance to demonstrate how precision is being increasingly applied in public health. While neither is comprehensive in terms of achieving all aspects of precision, each demonstrates key features and implications of a precision approach to disease surveillance.

Food-borne disease surveillance

Some surveillance systems attempt to identify specific diseases using defined populations (population-based surveillance). The Foodborne Diseases Active Surveillance Network (FoodNet) is part of CDC's Emerging Infections Program and conducts surveillance for enteric pathogens in 10 state health departments in collaboration with the US Department of Agriculture-Food Safety and Inspection Service (USDA-FSIS) and the Food and Drug Administration (FDA). The 10 sites cover approximately 49 million persons ($\approx$15% of the US population) and include Connecticut, Georgia, Maryland, Minnesota, New Mexico, Oregon, and Tennessee as well as selected counties in California, Colorado, and New York.

This system is used to monitor trends in incidence of lab-diagnosed infections caused by nine pathogens and has helped to identify numerous multi-state outbreaks of food-borne gastrointestinal infections [9]. FoodNet uses population estimates from the most recent census to estimate incidence for the following lab-diagnosed pathogens: *Campylobacter, Cryptosporidium, Cyclospora, Listeria, Salmonella,* Shiga-toxin producing *E. coli (*STEC)*, Shigella, Vibrio, and Yersinia species*. Overall incidence has declined for these pathogens, but one must be careful in interpreting the data because laboratory methods have become more precise over time, which must be weighed against the impact of a variety of control measures. For example, the addition of Culture Independent Diagnostic Tests (CIDT) has identified more cases of *Cyclospora, Yersinia* and *Vibrio* infections because of a much higher sensitivity of these tests when compared with older lab methods [9]. There are disadvantages to using DNA-based CIDTs, namely, that in many cases further characterization of pathogens (subtype or antimicrobial susceptibility) is not possible without actual isolates.

Another critical component of food-borne disease surveillance in the United States is PulseNet, which is a national laboratory network that is used to detect and investigate outbreaks by connecting cases through DNA fingerprinting of isolates [10]. There are 83 labs in the US network (local, state and federal labs) which have been used to investigate thousands of local and multi-state outbreaks since 1996. Over these two decades, lab techniques have evolved to increase precision, with whole genome sequencing (WGS) now allowing for rapid identification of the first cases of an outbreak and connecting cases spread out across multiple states. Critical to this system is the close collaboration and working with FDA to identify the food source and any gap in food safety systems. PulseNet International extends this paradigm to 88 member countries from 7 regions around the world, with similar techniques (WGS) and approach to connecting outbreaks [11].

Communication of critical information about food-borne outbreaks is also important. The CDC values transparency with respect to outbreaks, and keeps an updated list of outbreaks on the National Outbreak Reporting System (NORS) dashboard (https://wwwn.cdc.gov/norsdashboard/). Details on specific multi-state outbreaks are also available on-line at https://www.cdc.gov/foodsafety/outbreaks/multistate-outbreaks/outbreaks-list.html. Timely access to accurate information allows public health officials as well as the general public to make decisions regarding interventions, the ultimate goal of any surveillance system (Fig. 2)

Influenza surveillance

Traditional surveillance approaches for influenza varied according to the primary purpose for the system. For example, a nationwide network of sentinel providers in the United States helped to identify the onset and duration of the annual epidemic, a 122-city mortality tracking system helped to determining its severity, and a global network of laboratories acted to characterize the circulating viruses for vaccine strain selection.

In recent years innovations in surveillance and laboratory testing have made influenza surveillance increasingly efficient and precise. The characterization of influenza viruses, traditionally a meticulous process using antigenic approaches with panels of ferret antiserum, was gradually supplemented by genetic sequencing of a subset of viruses [12]. Most recently this process has been reversed, with sequencing of virus isolates as the initial step, and the more laborious antigenic characterization done on a minority of those sequenced [13].

The 122 cities system has been replaced by a more comprehensive system and supplemented by surveillance for pediatric deaths, giving a rapid and ongoing measure of severity throughout the season [14,15].

The use of big data to more rapidly identify the onset of the season has been attempted with varying degrees of success. A system known as Google flu trends serves as a cautionary tale. Initially successful in tracking the use of internet search engine terms such as "flu" or "sore throat and fever" to define the onset and magnitude of influenza season up to 2 weeks ahead of the CDC surveillance systems, the algorithms were not adequate for identifying the age-shifted pandemic of H1N1 influenza or the magnitude of subsequent epidemics, and the system has since been discarded [16]. The availability of big data sets and unprecedented analytic power does not consistently translate to better surveillance information or more reliable public health decisions.

Despite the occasionally stuttering progress, the improvements in influenza surveillance with the addition of big data, wide scale use of modeling, and the

FIG. 2 National Outbreak Reporting System: Campylobacter.

addition of virus genetic sequencing as a routine activity have resulted in a national and global surveillance system that can time the onset of influenza seasons in specific geographical areas such as states within days, accurately monitor the severity of the ongoing season within weeks, trace the genetic and inferred antigenic characteristics of circulating viruses in real time, and even predict with increasing accuracy the likely evolution of viruses for the subsequent season.

Conclusion

Improving precision in public health and specifically in disease surveillance is conceptually easy, but in practice, challenging to implement. Increased access to data allows for more precise public health actions, but also generates potential situations where individual privacy of sensitive information can be compromised. In addition, automated application of algorithms without appropriate oversight can perpetuate stereotypes and lead to incorrect decisions based on faulty assumptions. Increasing precision in disease surveillance has great potential to improve public health, as has been shown in the food-borne infection and influenza surveillance systems described above. However, these must be implemented carefully and with appropriate attention to unintended consequences for maximum utility to be realized.

References

[1] Dowell SF, Blazes D, Desmond-Hellmann S. Four steps to precision public health. Nature 2016;540:189–91. 08 December 2016, https://doi.org/10.1038/540189a.

[2] Horton R. Offline: in defense of precision public health. Lancet 2018;392(10157):1504. https://doi.org/10.1016/S0140-6736(18)32741-7.

[3] Johnson S. The ghost map: the story of London's most terrifying epidemic. Penguin; 2006.

[4] Grubaugh ND, Ladner JT, MUG K, Dudas G, et al. Genomic epidemiology reveals multiple introductions of Zika virus into the United States. Nature 2017;546(7658):401–5. https://doi.org/10.1038/nature22400. 2017 Jun 15.

[5] CDC Quick Lesson, n.d., Using an Epi curve to determine most likely time of exposure. In: Shiga toxin-producing *E. coli* cases by date of onset, Port Yourtown. https://www.cdc.gov/training/QuickLearns/exposure/. [Accessed 01 July 2019].

[6] Dolley S. Big data's role in precision public health. Front Public Health 2018;6:68. https://doi.org/10.3389/fpubh.2018.00068.

[7] WHO. International Health Regulations. 3rd ed. World Health Organization; 2005. https://www.who.int/ihr/publications/9789241580496/en/. Accessed 31 Dec 2018.

[8] Freifeld CC, Mandl KD, Reis BY, Brownstein JS. HealthMap: global infectious disease monitoring through automated classification and visualization of Internet media reports. J Am Med Inform Assoc 2008;15(2):150–7.

[9] Marder EP, Griffin PM, Cieslak PR, Dunn J, et al. Preliminary incidence and trends of infections with pathogens transmitted commonly through food—foodborne diseases active surveillance network, 10 U.S. Sites, 2006–2017. MMWR Morb Mortal Wkly Rep 2018;67(11):324–8. https://www.cdc.gov/mmwr/volumes/67/wr/mm6711a3.htm?s_cid=mm6711a3_w. Accessed 29 Dec 2018.

[10] About PulseNet, 2016. https://www.cdc.gov/pulsenet/about/index.html. (Accessed February 16, 2016).

[11] Nadon C, Van Walle I, Gerner-Smidt P, Campos J. FWD-NEXT expert panel PulseNet International: vision for the implementation of whole genome sequencing (WGS) for global food-borne disease surveillance. Eurosurveillance 2017;22(30):544. https://doi.org/10.2807/1560-7917.ES.2017.22.23.30544.

[12] Hampson A, Barr I, Cox N, Donis RO, et al. Improving the selection and development of influenza vaccine viruses—report of a WHO informal consultation on improving influenza vaccine

virus selection, Hong Kong SAR, China, 18–20 November 2015. Vaccine 2017;35(8):1104–9. https://doi.org/10.1016/j.vaccine.2017.01.018.

[13] Influenza virus genome sequencing and genetic characterization, 2017. https://www.cdc.gov/flu/professionals/laboratory/genetic-characterization.htm. (Accessed September 27, 2017).

[14] Notice to readers: update to reporting of pneumonia and influenza mortality. MMWR Morb Mortal Wkly Rep 2016;65(39):1088. https://doi.org/10.15585/mmwr.mm6539a8external.

[15] Biggerstaff M, Kniss K, Jernigan DB, Brammer L, et al. Systematic assessment of multiple routine and near real-time indicators to classify the severity of influenza seasons and pandemics in the United States, 2003–2004 through 2015–2016. Am J Epidemiol 2018;187(5):1040–50.

[16] Simonsen L, Gog JR, Olson D, Viboud C. Infectious disease surveillance in the big data era: towards faster and locally relevant systems. J Infect Dis 2016;214(Suppl. 4):S380–5.

Chapter 16

Multiple sclerosis

Athanasios Ploumakis, Howard L. Weiner, Nikolaos A. Patsopoulos
Department of Neurology, Brigham and Women's Hospital, Ann Romney Center for Neurological Diseases, Harvard Medical School, Boston, MA, United States

Introduction

Multiple sclerosis (MS) is a neurodegenerative disorder characterized by inflammation and loss of the insulating myelin sheath covering axons, resulting in lesions called plaques. MS affects >2 million people worldwide [1] and recent studies in the United States (US) alone have increased this estimate to at least 700,000 [2]. The prevalence of the disease differs in various countries, with estimates ranging from <1 in 100,000 in sub-Saharan African to over 200 in 100,000 in Canada and Scandinavian countries [3]. Diagnosis of MS is based on a combination of clinical, imaging, and paraclinical criteria, taking into account clinical history and other possible explanations besides MS [4]. MS exhibits a complex clinical profile, involving a wide range of symptom severity, regression, and relapse. The 2013 revision of MS classification proposed the following categories: clinically isolated syndrome (CIS), radiologically isolated syndrome (RIS) relapsing remitting MS (RRMS), and the progressive form of the disease (PMS) that can be primary or secondary to RRMS [5]. MS has severe negative health outcomes for patients, resulting in loss of physical functions, mental degradation [6] and even psychiatric problems [7]. To date there are dozens of treatments approved for the relapsing remitting form of the disease and one for the progressive form [8]. The cost of treatment is high, especially in the United States where the average annual cost for first-generation disease modifying drugs is more than $60,000 per year and newer generation ones are priced at 25–60% higher cost [9]. Given the heterogeneity of the presentation, clinical course, and potential treatment, MS is an excellent candidate for precision medicine approaches. Here, we present an overview of our current understanding of genetics and genomics in MS that could form the basis of future precision medicine applications.

Genetics

Multiple sclerosis has both a genetic and an environmental contribution, like any other common disease [10]. The heritable component of MS was established

Genomic and Precision Medicine. https://doi.org/10.1016/B978-0-12-801496-7.00016-2

early on. Family history of MS is present in about 15–20% of MS patients [11, 12]. Monozygotic or identical twins display significantly higher clinical concordance rate (25%–30%) compared to dizygotic or fraternal twins (3%–7%) [13]. The lifetime risk of MS in first-degree relatives of MS patients was estimated to be ~3%; more than three times higher than second- and third-degree relatives and 10 times higher than the general population [11]. Efforts to identify genetic drivers of the disease started early on [14–18] and besides the identification of strong and robust associations within the major histocompatibility complex (MHC) region the vast majority of these hypothesis-driven studies resulted in uncorroborated associations [19]. The MSgene database provides an exhaustive list of these studies [20].

The advent of genome-wide association studies (GWAS) and collaborative research allowed the agnostic interrogation of the whole-genome at large scale. The GWAS era was motivated by the common variant common disease (CVCD) hypothesis, which states that common diseases are mostly driven by several common genetic variants that have small effect size [21]. The application of the GWAS paradigm in MS resulted in the identification of several robustly associated genetics associations [22–29]. The most recent large-scale genetic study in MS analyzed 47,429 MS subjects and 68,374 controls and identified 233 genome-wide associated variants (P-value $<5\times10^{-8}$) [30]. Within the MHC region, 32 distinct genetic associations could be identified, replicating and extending findings from previous efforts, including many alleles of the human leukocyte antigen (HLA) system and non-HLA genes [31, 32]. The strongest association (odds ratio of 2.96) was reported to be driven by the well-established *HLA-DRB1*15:01* HLA Class II allele, followed by a protective association for the Class I *HLA-A*02:01*. Another 200 genetic associations where identified in the non-MHC autosomal genome and another in chromosome X. The effect sizes of the non-MHC associations were small, as predicted by the CVCD hypothesis, and ranged between odds ratios of 1.05 and 1.36. The large sample size allowed the identification of low frequency associated variants, implying that rare variants could be identified in MS. Another large-scale study that included 27,891 MS patients and 30,298 controls leveraged an array that targeted only exonic variants, the Exome Chip, replicated low frequency coding variant findings and reported another five coding rare variants, one of which has a frequency of 0.2% in the general population [33]. Overall, these large-scale genetic studies have allowed the mapping of about half of the genetic contribution to MS susceptibility [30].

The mapping of MS susceptibility's genetic architecture has been extremely successful; however, this success has not yet been repeated in other MS-related phenotypes. Current efforts to identify genetic associations of disease progression or severity have not resulted in any genome-wide variant [19, 28]. Magnetic resonance imaging (MRI) measurements have some potentially promising associations [34, 35] that will require further replication. A notable exception is the GWAS of intrathecal immunoglobulin G (IgG) in MS that resulted in a genome-wide associated hit in the immunoglobulin heavy chain (IGHC) locus [36] that was corroborated by another study [37]. Response to treatment is an extremely important phenotype that could have immediate translational impact. Several studies have performed genetic associations, either with targeted genetic variant panels or GWAS; however, no externally corroborated association has emerged to date [38]. Studies with large sample sizes of MS subphenotypes could potentially identify underlying genetic drivers, if any do exist.

A small percent of MS patients, about 5% in most recent estimates, exhibit symptoms at ages younger than 18 years old and are classified as pediatric MS [39, 40]. Given the low prevalence and resulting small sample size of pediatric MS, no large-scale and systematic genetic mapping has been performed to date. The most extensive genetic evaluation to date included 394 US and 175 Swedish pediatric MS patients [41], and tested 104 known non-MHC associations described in adult onset MS [26] and genetic variants spanning the MHC region. Despite the small sample size compared to equivalent studies of adult onset MS, 28 out of the 104 associated genetic variants and a weighted genetic risk score of the 104 variants were strongly associated in this pediatric onset MS study [41]. Further, *HLA–DRB1*15:01* and *HLA–A*02* were also replicated [41]. The findings of this study imply that the genetic architecture of pediatric and adult onset MS is similar, and if genetic drivers of early onset do exist these are yet elusive.

The genetic architecture of MS susceptibility poses a challenge on how this knowledge can be translated into precision medicine applications. The lack of robustly associated deterministic rare variants and the concrete evidence of hundreds common genetic associations with small effects, does not allow a single or few genetic variants to have diagnostic or prognostic utility. Instead, aggregation of several or all known genetic associations into genetic risk scores (GRSs) have been proposed to facilitate such translation. In MS, the relative success of GRS studies correlates with the current knowledge of MS genetics. One of the first investigations of weighted GRSs (wGRSs) in MS utilized the 19 genome-wide associated variants, that were known at that time, and reported an area under the curve (AUC) of 0.70 to predict MS status in an independent case-control data set [42]. In the same study, the wGRS could not predict transition from clinically isolated syndrome to MS. Another study of wGRS leveraged 16 known genetic variants and applied it in 1213 MS families [43]. They reported an increased risk of MS in siblings with high wGRS; however, they quantified an AUC of 0.57 [95% CI: 0.53–0.61] for the case-control status prediction of MS [43]. A Danish study with a wGRS based on 8 MHC and 109 non-MHC autosomal variants did not observe any association with overall disease activity or response to treatment with interferon-beta [44]. A wGRS study focusing only on HLA variants observed an association with younger age of onset and the atrophy of subcortical gray matter fraction in women with relapsing-onset MS [45]. They further reported that the *HLA-A*24:02-B*07:02-DRB1*15:01* haplotype was associated with subcortical gray atrophy, whereas *HLA-B*44:02* had a seemingly protective role [45]. A study from Belgium on 842 MS patients calculated wGRS based on four MHC and 106 non-MHC risk loci [46]. The MHC wGRS were associated with presence of oligoclonal bands, an increased immunoglobulin G index and female sex, whereas the non-MHC wGRS with increased relapse rate and shorter relapse-free intervals after disease onset [46].

Transcriptomics

The study of transcriptomics in multiple sclerosis has primarily concentrated in two main areas: identifying changes in gene expression in MS lesions and cells of the peripheral immune system. Early studies comparing lesion tissue and normal appearing white matter (NAWM) in secondary progressive MS used microarrays to compare the tissues, identifying 123 differentially expressed genes in lesions and 47 in NAWM compared to matched healthy controls [47]. Cellular immune response elements were notably deregulated in both lesions and NAWM alongside a reduction in expression of immunosuppressive elements, while lesions also specifically expressed higher levels of immunoglobulin and neuroglial differentiation associated genes, indicating that MS pathogenesis is likely a widespread process across the CNS. Additional studies have suggested that a number of other pathways are also affected by dysregulation of gene expression in MS, including ciliary neurotrophic factor signaling pathway [48], the mitochondrial respiratory chain [49], and heatshock proteins [50]. Comparison of actively demyelinating lesions with inflammation compared with inactive lesions in experimental autoimmune encephalomyelitis (EAE), the most prominent mouse model of MS, also demonstrated a similar pattern, with upregulation of gene transcripts encoding inflammatory cytokines and immunoglobulins [51, 52]. Evidence for promotion of inflammation by alteration of gene expression in acute MS was also demonstrated in both MS patient lesions and EAE models, notably including upregulation of osteopontin (*OPN*), which appeared to regulate Th1-mediated demyelination [53]. Indeed, anti-OPN autoantibodies are found in both EAE and MS and vaccination against OPN was shown to promote remission in EAE [54]. Newer studies comparing MS lesions and NAWM have suggested that T cell repertoires are variable among patients and show evidence of distinct clones being recruited to the same site by particular local antigens [55]. The boundary area around a demyelinating lesion will tend to have gene expression signatures similar to that of the expanding lesion in addition to the upregulation of genes involved in lipid binding, likely due to the presence of foamy macrophages in the area [56]. In terms of progression within the lesion, it would appear that the expression profiles of chronic and acute lesions show pathogenic modulation of similar gene sets but at different expression levels [57]. The location of the lesion also appears to play a role, as it has been found that gene expression of astrocytes in EAE follows distinctly region-specific signatures. Specifically, in the spinal cord, optic nerve and optic chiasm, decreased cholesterol synthesis was observed [58]. In the context of precision medicine this is important, as neuroprotective treatments would be more effective if specialized toward observed clinical manifestations compared to a general anti-inflammatory treatment.

Differences in expression patterns are not however limited to the neuronal and immune tissues. Differential gene expression comparing MS patients and non-MS affected individuals indicated that the expression profiles of the internal jugular vein were different [59]. MS patients demonstrated higher levels of immune related genes as well as *HOX* and other transcription factor genes. In contrast, expression of genes associated with cell adhesion and muscle contraction were downregulated, suggesting that MS associated dysregulation of gene expression is widespread in tissues [59]. Additionally, oligodendrocyte lineage cells have been shown to overexpress genes associated with antigen processing and presentation via the major histocompatibility complexes in both EAE mice and MS [60].

Transcriptomics have also been utilized to examine the progression of MS. In a 2008 study, gene expression in naïve CD4+ T cells from CIS patients was examined and used to construct a model of disease conversion [61]. It was found that a subset of 108 genes differentiated the group with the highest rate of progression to clinically definite MS (92% within 9 months), prominently including downregulation of the cell proliferation regulator transducer of ERBB2-1 (*TOB1*) [61]. On the other hand, conversion to clinically definite MS (CDMS) is more likely if overexpression of pSTAT3 in CD4+ T cells is detected [62]. Relapse episodes of MS have also been a target of investigation in transcriptomics. Initial examination had indicated that peripheral blood mononuclear cell (PBMC) transcriptional activity in RRMS can be used to deduce a transcriptional signature characteristic of relapse episodes [63]. In other studies, however, microarray-based comparison of cerebrospinal fluid (CSF) cells and PBMCs discovered that CSF cells, but not PBMCs, show differences in expression between MS patients and healthy controls; however, transcriptional changes in PBMCs were a better indicator of relapse episodes [64].

The response to therapy has been an area of significant investigation in the field of MS transcriptomics. In order to elucidate how interferon-beta (IFNb), a common treatment for MS, affects transcription, a 2004 study performed a time-course micro-array investigation of monocyte depleted PBMCs [65]. A wide range of immunological marker transcripts associated with immune response against viruses, IFNb targets, and lymphocyte activation markers were induced in a time-dependent manner. Later work, however, suggested the IFNb-regulated targets to be more variable on a per-patient basis [66]. Additional regulatory effects may also be provided by IFNb modulation of micro-RNAs (miRNAs), small nuclear RNAs (snRNAs), and miscellaneous short RNAs (misc-RNAs) [67]. The transcriptional signature to longitudinal treatment with three major IFNb formulations currently present in the market has recently been established [68]. These were found to involve quite similar expression patterns centered around IFNb dependent pathways but at different levels of expression

[68]. Dysregulation of the plasminogen activation cascade, an expression feature of MS resulting in increased expression of matrix metalloprotease 9, can also be corrected by IFNb administration [69]. Nevertheless, it should be noted that IFNb restores physiological levels of expression of only a subset of genes dysregulated during MS, at least in whole blood [70]. Studies on the effects of fingolimod, another approved MS treatment, indicated that lymphoid CD4 and CD8+ T cells show a repertoire of 900 deregulated genes mostly involved in the immune response, such as chemokine and Toll-like receptors [71] and population shifts in CD8+ T cell subsets [72]. The response to dimethyl fumarate has also been examined, which in PBMCs of RRMS patients consists of induction of nuclear factor (erythroid-derived 2)-like 2 (*Nrf2*) and inhibition of nuclear factor κB (*NFκB*) pathway transcripts [73].

Microglia, the resident innate immune cells of the CNS, have been the subject of some interest due to their importance in MS. In EAE models of MS, these have been shown to exhibit significant diversity and be either neuroprotective or neurotoxic. To characterize this diversity, Starossom et al. isolated subventricular zone microglia from acute and chronic EAE and analyzed their expression profiles using a combination of flow cytometry and expression microarrays [74]. Gene ontology and pathway analysis was further used to demonstrate that of the transcripts which could be sorted, EAE differential variants where enriched in inflammation, cell:cell communication and differentiation. Acute and chronic EAE also showed differences in the expression profile, with greater transcriptional activity found in acute EAE microglia. Further studies have utilized CD44-based flow cytometry to differentiate between microglia and monocyte-derived macrophages in terms of their transcriptomic profiles [75]. These have indicated the upregulation of common activation markers such as CD80 and CD86 in both cell populations but which are more highly expressed in monocyte derived macrophages. Additionally, monocyte derived macrophages show accessory markers such as *Sell*, *Cd69* and *Cd40*, not present in microglia, along with a more pro-proliferative expression profile. Microglia however show upregulation of gene expression of proteins involved in phagocytosis and phagoptosis, indicating their likely role in promoting debris clearance. The state of microglia activation appears to be dependent on the stage of lesion formation and remyelination. Specifically, it was found that M1 polarized proinflammatory microglia express the surface markers CD74, CD40, CD86, and CCR7, in contrast to M2 polarized anti-inflammatory microglia which express mannose receptor and CCL22. The polarization however can be switched, and NAWM resident microglia will exhibit different patterns of gene expression [76].

In T cells, dysregulation of early regulatory networks that could affect pathogenesis of MS is an important but difficult to answer question, as patients typically present when already symptomatic. To address this, Gustafsson et al. identified transcription factors enriched in MS associated polymorphisms, constructed a gene regulatory network by observing the time dependent in vitro differentiation of naïve CD4+ T cell into distinct subtypes and identified *GATA3*, *MAF*, and *MYB* as hub transcription factors. These were then validated as differentially expressed during asymptomatic stages of MS [77]. In B cells, analysis of the coding and non-coding transcriptome has been performed comparing single isolated B cells from MS patients not receiving therapy and sex/age matched control cohorts. The cells consistently show downregulation of Interferon Response Factor 1 (*IRF1*) and the chemokine receptor *CXCL10* [78], both of which have been implicated in MS [79, 80]. This reduction is likely mediated by upregulation of a miRNA targeting *IRF1*, shifting the B cells toward a pro-survival phenotype [78]. Studies on PBMCs have been used to demonstrate that MS patient PBMCs can be clustered into two distinct populations based on their RNA expression profile. One of these populations is characterized by higher levels of lymphocyte expression factors and was shown to be clinically characterized by increased probability of relapse even during treatment [81]. Similarly, monocytes isolated from PBMCs of people who fail to respond to IFNb therapy were found to overexpress type I interferon responsive genes compared to responders, possibly because the cells were already desensitized to chronically higher levels of Interferon (IFN) downstream factor expression [82].

Epigenetics

In recent years a growing body of evidence has been accumulated concerning the importance of epigenetics in the pathogenesis and progression of MS. Studies in this area have concentrated on modifications applied to nucleic acid sequences of genes pertinent to MS, histone modifications, and the non-coding RNA (ncRNA) based gene regulation.

Epidemiological evidence indicates the importance of factors other than genetics due to the low concordance rate of MS in twin studies [11, 19]. Although the epigenetic landscape of twins could start out virtually identical, environmental factors lead to epigenetic drift [83], which by the average age of MS onset may affect the probability of manifestation. However, no discernible DNA methylation differences were found in a 2010 twins' study [84]. Changes in methylation of promoter sites of genes pertinent to MS have been discovered in a number of studies. Specifically, increased promoter methylation of the anti-inflammatory Src homology region 2 domain-containing phosphatase-1 (*SHP-1*) in peripheral leucocytes was shown to result in increased expression of pro-inflammatory factors [85]. Conversely, decreased promoter methylation of the protein-arginine deiminase type-2 (*PAD-2*) results in upregulation of its expression and increased myelin citrullination, which may contribute to myelin autoreactivity. A similar pattern can be observed for the related protein-arginine deiminase type-4 (*PAD-4*), where citrullination of myelin nucleosome H3 histones likely contributes to reduced myelin expression in MS patients [86]. In EAE models, T cell factor-1 (TCF-1) (−/−) mice were shown to have exaggerated Th17 responses and increased disease susceptibility with increased IL-17 production, which could not be restored by *TCF-1* expression using mature T cells. Nucleosomal H3 histones at the *IL17* promoter site of these cells were found to have increased acetylation and Lys-4 trimethylation [87]. Additional studies have further implicated Th17 differentiation as being an important player in epigenetically determined EAE susceptibility. These involve reduced deacetylation of the interferon regulatory factor-1 (*IRF1*) promoter site by histone deacetylase sirtuin-1 in dendritic cells (DCs). This promoted Th17 differentiation, increasing EAE risk [88], and c-Jun N-terminal kinase (JNK) mediated CD27 signaling, resulting in epigenetic silencing of the *IL-17* gene through increased promoter site methylation [89].

Further evidence for the importance of epigenetic control via modulation of Th17 differentiation in EAE models of MS has also cropped up through examination of the Th17 miRNA repertoire. miR-326 overexpression has been found to result in promotion of Th17 differentiation and increased EAE disease severity, mediated by targeting of the negative regulator of Th-17 differentiation Ets-1 [90]. Additionally, it has been found that the transcription factor STAT3 promotes expression of miR-155 in Th17 cells by binding directly to the miR-155 locus, in turn promoting EAE onset [91]. In support of this, silencing of miR-155 results in amelioration of EAE symptoms [92]. Regulation of Th17 activity can also occur through miR-21, which targets the Mothers Against Decapentaplegic homolog 7 (*SMAD7*) transcript, alleviating negative regulation of TGF-beta signaling [93].

Other miRNAs have been shown to affect EAE by operating via different pathways. MiR-124, expressed in microglia but not PBMCs, has been shown to promote a quiescent microglial phenotype by inhibiting the transcription factor CCAAT/enhancer-binding protein-α (C/EBP-α) and its downstream target PU.1 [94]. Furthermore, it has been demonstrated that circulating miRNA populations are different when comparing both MS patients with healthy controls but also between RRMS and SPMS patients. Of the miRNAs identified, hsa-miR-454 and hsa-miR-145 could be used to differentiate between RRMS and SPMS, and RRMS from both SPMS and healthy controls, respectively. Hsa-miR-92a-1 appeared to be a general MS marker and its abundance was correlated with disability status, while other notable members included miRNAs targeting cell cycle and signaling, differentiation and T cell regulation [95]. The latter group included miRNAs of the *let* family, at least one member of which has been shown to be upregulated in MS in a separate study and suppresses regulatory T cell induction [96]. Finally, it is possible to associate MRI lesion phenotypes with specific miRNA profile repertoires, possibly based on the underlying changes in blood brain barrier pathology [97].

Proteomics

Proteomics studies have proven highly useful in the study of MS. Specifically, proteomic studies in MS have progressed our understanding of the development and biochemical manifestation of the disease; and in doing so have led to a list of candidate biomarkers. Numerous studies have generated a large number of exploratory biomarkers [98–105]. However, relatively few of these have been validated and fewer still have been translated into clinical practice [106], demonstrating a gap that needs to be addressed for precision medicine.

The traditional approach to MS proteomics is to perform 2D gel electrophoresis using patient derived biological fluid to obtain individual protein spots, which are then extracted and subject to tandem mass spectrometry. Newer approaches involve shotgun proteomics in which bulk protein content is isolated from the sample and digested with proteinases, and high pressure liquid chromatography (HPLC) is used to separate protein fragments, followed by tandem mass spectrometry. The latter approach has the advantage of being automatable and more sensitive [107] and is thus better suited for precision medicine practices. Protein arrays [108] and surface-enhanced laser desorption/ionization—time of flight (SELDI-TOF) mass spectrometry [109] have also been utilized, though not as commonly.

Proteomics have also proven very important in expanding our understanding of the pathways involved in MS. Brain tissue and CSF have been shown to exhibit increased level of numerous markers associated with inflammation. Notably, members of the complement system of proteins are consistently upregulated in CSF across multiple studies in humans suffering from MS [101] and in EAE mice [110, 111]. Of the complement proteins, four (C1inh, C1s, C5 and Factor H) can be used as biomarkers to distinguish MS from the related neuromyelitis optica [112]. Additionally, significant metabolomics changes take place, indicating a shift to catabolic state characterized by increased abundance of degradation enzymes, such as matrix metalloproteases [113] and oxidative burst activation in microglia [114, 115].

It is important to note, however, that the majority of these studies have been performed using blood samples. While blood may be a valuable source of information concerning alterations in the CNS when a blood brain barrier disruption occurs, as often in MS, [116] high abundance of "baseline" proteins such as serum albumin and immunoglobulins [117] can conceal the proteomic signal of the disease. To counter this, highly abundant proteins are often depleted in studies [118], which enhances detection but runs the risk of neglecting to observe effects in the high abundance proteins [118]. CSF of course represents the ideal substrate for MS proteomics, but lumbar punctures are invasive procedures and unlikely to be accepted for longitudinal studies. Therefore, in the scope of precision medicine, blood draws are likely to remain the biological fluid of choice for proteomics, though tears [119] [120] and saliva [121] have been proposed as possible alternatives. Fixed post-mortem tissue has also been examined as a source [103].

Antigen arrays

A powerful tool for the characterization of individualized MS phenotypes are antigen arrays, in which peptide fragments of autoantigens are assayed using an array-based approach. This allows for the in-depth profiling of auto-antibodies and has great potential for use in precision medicine. In particular, increased diversity of the autoantibody repertoire has been associated with more severe clinical manifestations in mice EAE models of MS [122]. The array results were then used to inform the choice of autoantigen cDNAs, which were cloned into expression plasmids and used to generate tolerizing vaccines. It was found that vaccines with greater diversity of tolerogenic myelin stimuli were more successful in treating established EAE [122]. Furthermore, autoantigen micro-array analysis examining the antibody reactivity of MS patient serum to central nervous system (CNS) and heatshock proteins, as well as lipid autoantigens, discovered that clinical forms of MS can be distinguished by unique autoantibody reactivity patterns. Heatshock and CNS antigens were primarily used in the characterization of the forms of MS, while lipid autoantigens could distinguish between pathologic subtypes of MS, defined by the type of demyelination observed [123, 124].

Conclusion

The complex interplay between genetics and environment creates a unique amalgam of disease manifestation in each MS patient. Our understanding of MS pathophysiology and the risk factors associated with it has greatly increased over the years; however, many components are yet to be uncovered. New technologies, including cell-specific and single-cell studies [125], will empower a deeper and refined understanding of disease-related biology. Integration of genetic, transcriptomic, epigenetic, proteomic, imaging, clinical, and other data will be a necessary step in order to facilitate a multilayered personalized disease model. This refined and granular information can lead to the identification of the most efficient treatment for each patient and perhaps, eventually, to the holy grail of MS cure.

References

[1] Zwibel HL, Smrtka J. Improving quality of life in multiple sclerosis: an unmet need. Am J Manag Care. 2011;17 Suppl. 5 Improving:S139–45. Epub 2011/07/27. PubMed PMID: 21761952.

[2] Wallin MT, Culpepper WJ, Campbell JD, Nelson LM, Langer-Gould A, Marrie RA, Cutter GR, Kaye WE, Wagner L, Tremlett H, Buka SL, Dilokthornsakul P, Topol B, Chen LH, LaRocca NG, Workgroup USMSP. The prevalence of MS in the United States: a population-based estimate using health claims data. Neurology. 2019;92(10):e1029-e40. https://doi.org/10.1212/WNL.0000000000007035. Epub 2019/02/17. PubMed PMID: 30770430.

[3] Browne P, Chandraratna D, Angood C, Tremlett H, Baker C, Taylor BV, Thompson AJ. Atlas of multiple sclerosis 2013: a growing global problem with widespread inequity. Neurology. 2014;83(11):1022–4. https://doi.org/10.1212/WNL.0000000000000768. Epub 2014/09/10. PubMed PMID: 25200713; PMCID: PMC4162299.

[4] Thompson AJ, Banwell BL, Barkhof F, Carroll WM, Coetzee T, Comi G, Correale J, Fazekas F, Filippi M, Freedman MS, Fujihara K, Galetta SL, Hartung HP, Kappos L, Lublin FD, Marrie RA, Miller AE, Miller DH, Montalban X, Mowry EM, Sorensen PS, Tintore M, Traboulsee AL, Trojano M, Uitdehaag BMJ, Vukusic S, Waubant E, Weinshenker BG, Reingold SC, Cohen JA. Diagnosis of multiple sclerosis: 2017 revisions of the McDonald criteria. Lancet Neurol. 2018;17(2):162–73. Epub 2017/12/26. https://doi.org/10.1016/S1474-4422(17)30470-2. PubMed PMID: 29275977.

[5] Lublin FD, Reingold SC, Cohen JA, Cutter GR, Sorensen PS, Thompson AJ, Wolinsky JS, Balcer LJ, Banwell B, Barkhof F, Bebo B, Jr., Calabresi PA, Clanet M, Comi G, Fox RJ, Freedman MS, Goodman AD, Inglese M, Kappos L, Kieseier BC, Lincoln JA, Lubetzki C, Miller AE, Montalban X, O'Connor PW, Petkau J, Pozzilli C, Rudick RA, Sormani MP, Stuve O, Waubant E, Polman CH. Defining the clinical course of multiple sclerosis: the 2013 revisions. Neurology. 2014;83(3):278–86. Epub 2014/05/30. https://doi.org/10.1212/WNL.0000000000000560. PubMed PMID: 24871874; PMCID: PMC4117366.

[6] Compston A, Coles A. Multiple sclerosis. Lancet. 2008;372(9648):1502–17. Epub 2008/10/31. https://doi.org/10.1016/S0140-6736(08)61620-7. PubMed PMID: 18970977.

[7] Chwastiak LA, Ehde DM. Psychiatric issues in multiple sclerosis. Psychiatr Clin North Am. 2007;30(4):803–17. Epub 2007/10/17. https://doi.org/10.1016/j.psc.2007.07.003. PubMed PMID: 17938046; PMCID: PMC2706287.

[8] Gholamzad M, Ebtekar M, Ardestani MS, Azimi M, Mahmodi Z, Mousavi MJ, Aslani S. A comprehensive review on the treatment approaches of multiple sclerosis: currently and in the future. Inflamm Res. 2019;68(1):25–38. Epub 2018/09/05. https://doi.org/10.1007/s00011-018-1185-0. PubMed PMID: 30178100.

[9] Hartung DM, Bourdette DN, Ahmed SM, Whitham RH. The cost of multiple sclerosis drugs in the US and the pharmaceutical industry: too big to fail? Neurology. 2015;84(21):2185–92. Epub 2015/04/26. https://doi.org/10.1212/WNL.0000000000001608. PubMed PMID: 25911108; PMCID: PMC4451044.

[10] Olsson T, Barcellos LF, Alfredsson L. Interactions between genetic, lifestyle and environmental risk factors for multiple sclerosis. Nat Rev Neurol. 2017;13(1):25–36. Epub 2016/12/10. https://doi.org/10.1038/nrneurol.2016.187. PubMed PMID: 27934854.

[11] Compston A, Coles A. Multiple sclerosis. Lancet. 2002;359(9313):1221–31. https://doi.org/10.1016/S0140-6736(02)08220-X. PubMed PMID: 11955556.

[12] Sawcer S, Franklin RJ, Ban M. Multiple sclerosis genetics. Lancet Neurol. 2014;13(7):700–9. https://doi.org/10.1016/S1474-4422(14)70041-9. PubMed PMID: 24852507.

[13] Dyment DA, Sadovnick AD, Ebers GC. Genetics of multiple sclerosis. Hum Mol Genet. 1997;6(10):1693–8. PubMed PMID: 9300661.

[14] Bertrams J, Kuwert E, Liedtke U. HL-A antigens and multiple sclerosis. Tissue Antigens. 1972;2(5):405–8. PubMed PMID: 4655776.

[15] Compston DA, Batchelor JR, McDonald WI. B-lymphocyte alloantigens associated with multiple sclerosis. Lancet. 1976;2(7998):1261–5. PubMed PMID: 63743.

[16] Jersild C, Fog T, Hansen GS, Thomsen M, Svejgaard A, Dupont B. Histocompatibility determinants in multiple sclerosis, with special reference to clinical course. Lancet. 1973;2(7840):1221–5. PubMed PMID: 4128558.

[17] Naito S, Namerow N, Mickey MR, Terasaki PI. Multiple sclerosis: association with HL-A3. Tissue Antigens. 1972;2(1):1–4. PubMed PMID: 5077731.

[18] Winchester R, Ebers G, Fu SM, Espinosa L, Zabriskie J, Kunkel HG. B-cell alloantigen Ag 7a in multiple sclerosis. Lancet. 1975;2(7939):814. PubMed PMID: 78174.

[19] Patsopoulos NA. Genetics of multiple sclerosis: an overview and new directions. Cold Spring Harb Perspect Med. 2018;8(7). Epub 2018/02/15. https://doi.org/10.1101/cshperspect.a028951. PubMed PMID: 29440325.

[20] Lill C, Roehr J, McQueen M, Bagade S, Schjeide B, Zipp F, Bertram L. The MSGene database. Alzheimer Research Forum. Available at http://wwwmsgeneorg/.

[21] Reich DE, Lander ES. On the allelic spectrum of human disease. Trends Genet. 2001;17(9):502–10. PubMed PMID: 11525833.

[22] Andlauer TF, Buck D, Antony G, Bayas A, Bechmann L, Berthele A, Chan A, Gasperi C, Gold R, Graetz C, Haas J, Hecker M, Infante-Duarte C, Knop M, Kumpfel T, Limmroth V, Linker RA, Loleit V, Luessi F, Meuth SG, Muhlau M, Nischwitz S, Paul F, Putz M, Ruck T, Salmen A, Stangel M, Stellmann JP, Sturner KH, Tackenberg B, Then Bergh F, Tumani H, Warnke C, Weber F, Wiendl H, Wildemann B, Zettl UK, Ziemann U, Zipp F, Arloth J, Weber P, Radivojkov-Blagojevic M, Scheinhardt MO, Dankowski T, Bettecken T, Lichtner P, Czamara D, Carrillo-Roa T, Binder EB, Berger K, Bertram L, Franke A, Gieger C, Herms S, Homuth G, Ising M, Jockel KH, Kacprowski T, Kloiber S, Laudes M, Lieb W, Lill CM, Lucae S, Meitinger T, Moebus S, Muller-Nurasyid M, Nothen MM, Petersmann A, Rawal R, Schminke U, Strauch K, Volzke H, Waldenberger M, Wellmann J, Porcu E, Mulas A, Pitzalis M, Sidore C, Zara I, Cucca F, Zoledziewska M, Ziegler A, Hemmer B, Muller-Myhsok B. Novel multiple sclerosis susceptibility loci implicated in epigenetic regulation. Sci Adv. 2016;2(6):e1501678. https://doi.org/10.1126/sciadv.1501678. PubMed PMID: 27386562; PMCID: PMC4928990.

[23] Australia and New Zealand Multiple Sclerosis Genetics Consortium. Genome-wide association study identifies new multiple sclerosis susceptibility loci on chromosomes 12 and 20. Nat Genet. 2009;41(7):824–8. https://doi.org/10.1038/ng.396. PubMed PMID: 19525955.

[24] Baranzini SE, Wang J, Gibson RA, Galwey N, Naegelin Y, Barkhof F, Radue EW, Lindberg RL, Uitdehaag BM, Johnson MR, Angelakopoulou A, Hall L, Richardson JC, Prinjha RK, Gass A, Geurts JJ, Kragt J, Sombekke M, Vrenken H, Qualley P, Lincoln RR, Gomez R, Caillier SJ, George MF, Mousavi H, Guerrero R, Okuda DT, Cree BA, Green AJ, Waubant E, Goodin DS, Pelletier D, Matthews PM, Hauser SL, Kappos L, Polman CH, Oksenberg JR. Genome-wide association analysis of susceptibility and clinical phenotype in multiple sclerosis. Hum Mol Genet. 2009;18(4):767–78. https://doi.org/10.1093/hmg/ddn388. PubMed PMID: 19010793; PMCID: PMC4334814.

[25] De Jager PL, Jia X, Wang J, de Bakker PI, Ottoboni L, Aggarwal NT, Piccio L, Raychaudhuri S, Tran D, Aubin C, Briskin R, Romano S, International MSGC, Baranzini SE, McCauley JL, Pericak-Vance MA, Haines JL, Gibson RA, Naeglin Y, Uitdehaag B, Matthews PM, Kappos L, Polman C, McArdle WL, Strachan DP, Evans D, Cross AH, Daly MJ, Compston A, Sawcer SJ, Weiner HL, Hauser SL, Hafler DA, Oksenberg JR. Meta-analysis of genome scans and replication identify CD6, IRF8 and TNFRSF1A as new multiple sclerosis susceptibility loci. Nat Genet. 2009;41(7):776–82. https://doi.org/10.1038/ng.401. PubMed PMID: 19525953; PMCID: PMC2757648.

[26] International Multiple Sclerosis Genetics Consortium, Beecham AH, Patsopoulos NA, Xifara DK, Davis MF, Kemppinen A, Cotsapas C, Shah TS, Spencer C, Booth D, Goris A, Oturai A, Saarela J, Fontaine B, Hemmer B, Martin C, Zipp F, D'Alfonso S, Martinelli-Boneschi F, Taylor B, Harbo HF, Kockum I, Hillert J, Olsson T, Ban M, Oksenberg JR, Hintzen R, Barcellos LF, Wellcome Trust Case Control C, International IBDGC, Agliardi C, Alfredsson L, Alizadeh M, Anderson C, Andrews R, Sondergaard HB, Baker A, Band G, Baranzini SE, Barizzone N, Barrett J, Bellenguez C, Bergamaschi L, Bernardinelli L, Berthele A, Biberacher V, Binder TM, Blackburn H, Bomfim IL, Brambilla P, Broadley S, Brochet B, Brundin L, Buck

D, Butzkueven H, Caillier SJ, Camu W, Carpentier W, Cavalla P, Celius EG, Coman I, Comi G, Corrado L, Cosemans L, Cournu-Rebeix I, Cree BA, Cusi D, Damotte V, Defer G, Delgado SR, Deloukas P, di Sapio A, Dilthey AT, Donnelly P, Dubois B, Duddy M, Edkins S, Elovaara I, Esposito F, Evangelou N, Fiddes B, Field J, Franke A, Freeman C, Frohlich IY, Galimberti D, Gieger C, Gourraud PA, Graetz C, Graham A, Grummel V, Guaschino C, Hadjixenofontos A, Hakonarson H, Halfpenny C, Hall G, Hall P, Hamsten A, Harley J, Harrower T, Hawkins C, Hellenthal G, Hillier C, Hobart J, Hoshi M, Hunt SE, Jagodic M, Jelcic I, Jochim A, Kendall B, Kermode A, Kilpatrick T, Koivisto K, Konidari I, Korn T, Kronsbein H, Langford C, Larsson M, Lathrop M, Lebrun-Frenay C, Lechner-Scott J, Lee MH, Leone MA, Leppa V, Liberatore G, Lie BA, Lill CM, Linden M, Link J, Luessi F, Lycke J, Macciardi F, Mannisto S, Manrique CP, Martin R, Martinelli V, Mason D, Mazibrada G, McCabe C, Mero IL, Mescheriakova J, Moutsianas L, Myhr KM, Nagels G, Nicholas R, Nilsson P, Piehl F, Pirinen M, Price SE, Quach H, Reunanen M, Robberecht W, Robertson NP, Rodegher M, Rog D, Salvetti M, Schnetz-Boutaud NC, Sellebjerg F, Selter RC, Schaefer C, Shaunak S, Shen L, Shields S, Siffrin V, Slee M, Sorensen PS, Sorosina M, Sospedra M, Spurkland A, Strange A, Sundqvist E, Thijs V, Thorpe J, Ticca A, Tienari P, van Duijn C, Visser EM, Vucic S, Westerlind H, Wiley JS, Wilkins A, Wilson JF, Winkelmann J, Zajicek J, Zindler E, Haines JL, Pericak-Vance MA, Ivinson AJ, Stewart G, Hafler D, Hauser SL, Compston A, McVean G, De Jager P, Sawcer SJ, McCauley JL. Analysis of immune-related loci identifies 48 new susceptibility variants for multiple sclerosis. Nat Genet. 2013;45(11):1353–60. https://doi.org/10.1038/ng.2770. PubMed PMID: 24076602; PMCID: PMC3832895.

[27] International Multiple Sclerosis Genetics Consortium, Hafler DA, Compston A, Sawcer S, Lander ES, Daly MJ, De Jager PL, de Bakker PI, Gabriel SB, Mirel DB, Ivinson AJ, Pericak-Vance MA, Gregory SG, Rioux JD, McCauley JL, Haines JL, Barcellos LF, Cree B, Oksenberg JR, Hauser SL. Risk alleles for multiple sclerosis identified by a genomewide study. N Engl J Med. 2007;357(9):851–62. https://doi.org/10.1056/NEJMoa073493. PubMed PMID: 17660530.

[28] International Multiple Sclerosis Genetics Consortium, Wellcome Trust Case Control Consortium, Sawcer S, Hellenthal G, Pirinen M, Spencer CC, Patsopoulos NA, Moutsianas L, Dilthey A, Su Z, Freeman C, Hunt SE, Edkins S, Gray E, Booth DR, Potter SC, Goris A, Band G, Oturai AB, Strange A, Saarela J, Bellenguez C, Fontaine B, Gillman M, Hemmer B, Gwilliam R, Zipp F, Jayakumar A, Martin R, Leslie S, Hawkins S, Giannoulatou E, D'Alfonso S, Blackburn H, Martinelli Boneschi F, Liddle J, Harbo HF, Perez ML, Spurkland A, Waller MJ, Mycko MP, Ricketts M, Comabella M, Hammond N, Kockum I, McCann OT, Ban M, Whittaker P, Kemppinen A, Weston P, Hawkins C, Widaa S, Zajicek J, Dronov S, Robertson N, Bumpstead SJ, Barcellos LF, Ravindrarajah R, Abraham R, Alfredsson L, Ardlie K, Aubin C, Baker A, Baker K, Baranzini SE, Bergamaschi L, Bergamaschi R, Bernstein A, Berthele A, Boggild M, Bradfield JP, Brassat D, Broadley SA, Buck D, Butzkueven H, Capra R, Carroll WM, Cavalla P, Celius EG, Cepok S, Chiavacci R, Clerget-Darpoux F, Clysters K, Comi G, Cossburn M, Cournu-Rebeix I, Cox MB, Cozen W, Cree BA, Cross AH, Cusi D, Daly MJ, Davis E, de Bakker PI, Debouverie M, D'Hooghe M B, Dixon K, Dobosi R, Dubois B, Ellinghaus D, Elovaara I, Esposito F, Fontenille C, Foote S, Franke A, Galimberti D, Ghezzi A, Glessner J, Gomez R, Gout O, Graham C, Grant SF, Guerini FR, Hakonarson H, Hall P, Hamsten A, Hartung HP, Heard RN, Heath S, Hobart J, Hoshi M, Infante-Duarte C, Ingram G, Ingram W, Islam T, Jagodic M, Kabesch M, Kermode AG, Kilpatrick TJ, Kim C, Klopp N, Koivisto K, Larsson M, Lathrop M, Lechner-Scott JS, Leone MA, Leppa V, Liljedahl U, Bomfim IL, Lincoln RR, Link J, Liu J, Lorentzen AR, Lupoli S, Macciardi F, Mack T, Marriott M, Martinelli V, Mason D, McCauley JL, Mentch F, Mero IL, Mihalova T, Montalban

X, Mottershead J, Myhr KM, Naldi P, Ollier W, Page A, Palotie A, Pelletier J, Piccio L, Pickersgill T, Piehl F, Pobywajlo S, Quach HL, Ramsay PP, Reunanen M, Reynolds R, Rioux JD, Rodegher M, Roesner S, Rubio JP, Ruckert IM, Salvetti M, Salvi E, Santaniello A, Schaefer CA, Schreiber S, Schulze C, Scott RJ, Sellebjerg F, Selmaj KW, Sexton D, Shen L, Simms-Acuna B, Skidmore S, Sleiman PM, Smestad C, Sorensen PS, Sondergaard HB, Stankovich J, Strange RC, Sulonen AM, Sundqvist E, Syvanen AC, Taddeo F, Taylor B, Blackwell JM, Tienari P, Bramon E, Tourbah A, Brown MA, Tronczynska E, Casas JP, Tubridy N, Corvin A, Vickery J, Jankowski J, Villoslada P, Markus HS, Wang K, Mathew CG, Wason J, Palmer CN, Wichmann HE, Plomin R, Willoughby E, Rautanen A, Winkelmann J, Wittig M, Trembath RC, Yaouanq J, Viswanathan AC, Zhang H, Wood NW, Zuvich R, Deloukas P, Langford C, Duncanson A, Oksenberg JR, Pericak-Vance MA, Haines JL, Olsson T, Hillert J, Ivinson AJ, De Jager PL, Peltonen L, Stewart GJ, Hafler DA, Hauser SL, McVean G, Donnelly P, Compston A. Genetic risk and a primary role for cell-mediated immune mechanisms in multiple sclerosis. Nature. 2011;476(7359):214–9. https://doi.org/10.1038/nature10251. PubMed PMID: 21833088; PMCID: PMC3182531.

[29] Patsopoulos NA, Bayer Pharma MS Genetics Working Group, Steering Committees of Studies Evaluating IFNβ-1b and a CCR1-Antagonist, ANZgene Consortium, GeneMSA, International Multiple Sclerosis Genetics Consortium, Esposito F, Reischl J, Lehr S, Bauer D, Heubach J, Sandbrink R, Pohl C, Edan G, Kappos L, Miller D, Montalban J, Polman CH, Freedman MS, Hartung HP, Arnason BG, Comi G, Cook S, Filippi M, Goodin DS, Jeffery D, O'Connor P, Ebers GC, Langdon D, Reder AT, Traboulsee A, Zipp F, Schimrigk S, Hillert J, Bahlo M, Booth DR, Broadley S, Brown MA, Browning BL, Browning SR, Butzkueven H, Carroll WM, Chapman C, Foote SJ, Griffiths L, Kermode AG, Kilpatrick TJ, Lechner-Scott J, Marriott M, Mason D, Moscato P, Heard RN, Pender MP, Perreau VM, Perera D, Rubio JP, Scott RJ, Slee M, Stankovich J, Stewart GJ, Taylor BV, Tubridy N, Willoughby E, Wiley J, Matthews P, Boneschi FM, Compston A, Haines J, Hauser SL, McCauley J, Ivinson A, Oksenberg JR, Pericak-Vance M, Sawcer SJ, De Jager PL, Hafler DA, de Bakker PI. Genome-wide meta-analysis identifies novel multiple sclerosis susceptibility loci. Ann Neurol. 2011;70(6):897–912. https://doi.org/10.1002/ana.22609. PubMed PMID: 22190364; PMCID: PMC3247076.

[30] Patsopoulos N, Baranzini SE, Santaniello A, Shoostari P, Cotsapas C, Wong G, Beecham AH, James T, Replogle J, Vlachos I, McCabe C, Pers T, Brandes A, White C, Keenan B, Cimpean M, Winn P, Panteliadis I-P, Robbins A, TFM A, Zarzycki O, Dubois B, Goris A, Bach Sondergaard H, Sellebjerg F, Soelberg Sorensen P, Ullum H, Wegner Thoerner L, Saarela J, Cournu-Rebeix I, Damotte V, Fontaine B, Guillot-Noel L, Lathrop M, Vukusik S, Berthele A, Biberacher V, Buck D, Gasperi C, Graetz C, Grummel V, Hemmer B, Hoshi M, Knier B, Korn T, Lill CM, Luessi F, Muhlau M, Zipp F, Dardiotis E, Agliardi C, Amoroso A, Barizzone N, Benedetti MD, Bernardinelli L, Cavalla P, Clarelli F, Comi G, Cusi D, Esposito F, Ferre L, Galimberti D, Guaschino C, Leone MA, Martinelli V, Moiola L, Salvetti M, Sorosina M, Vecchio D, Zauli A, Santoro S, Zuccala M, Mescheriakova J, van Duijn C, Bos SD, Celius EG, Spurkland A, Comabella M, Montalban X, Alfredsson L, Bomfim IL, Gomez-Cabrero D, Hillert J, Jagodic M, Linden M, Piehl F, Jelcic I, Martin R, Sospedra M, Baker A, Ban M, Hawkins C, Hysi P, Kalra S, Karpe F, Khadake J, Lachance G, Molyneux P, Neville M, Thorpe J, Bradshaw E, Caillier SJ, Calabresi P, BAC C, Cross A, Davis MF, de Bakker P, Delgado S, Dembele M, Edwards K, Fitzgerald K, Frohlich IY, Gourraud P-A, Haines JL, Hakonarson H, Kimbrough D, Isobe N, Konidari I, Lathi E, Lee MH, Li T, An D, Zimmer A, Lo A, Madireddy L, Manrique CP, Mitrovic M, Olah M, Patrick E, Pericak-Vance MA, Piccio L, Schaefer C, Weiner H, Lage K, Compston A, Hafler D,

Harbo HF, Hauser SL, Stewart G, D'Alfonso S, Hadjigeorgiou G, Taylor B, Barcellos LF, Booth D, Hintzen R, Kockum I, Martinelli-Boneschi F, JL MC, Oksenberg JR, Oturai A, Sawcer S, Ivinson AJ, Olsson T, De Jager PL. The multiple sclerosis genomic map: role of peripheral immune cells and resident microglia in susceptibility. bioRxiv 2017;.

[31] Moutsianas L, Jostins L, Beecham AH, Dilthey AT, Xifara DK, Ban M, Shah TS, Patsopoulos NA, Alfredsson L, Anderson CA, Attfield KE, Baranzini SE, Barrett J, Binder TM, Booth D, Buck D, Celius EG, Cotsapas C, D'Alfonso S, Dendrou CA, Donnelly P, Dubois B, Fontaine B, Lar Fugger L, Goris A, Gourraud PA, Graetz C, Hemmer B, Hillert J, International IBDGC, Kockum I, Leslie S, Lill CM, Martinelli-Boneschi F, Oksenberg JR, Olsson T, Oturai A, Saarela J, Sondergaard HB, Spurkland A, Taylor B, Winkelmann J, Zipp F, Haines JL, Pericak-Vance MA, Spencer CC, Stewart G, Hafler DA, Ivinson AJ, Harbo HF, Hauser SL, De Jager PL, Compston A, McCauley JL, Sawcer S, McVean G, International Multiple Sclerosis Genetics C. Class II HLA interactions modulate genetic risk for multiple sclerosis. Nat Genet. 2015;47(10):1107–13. https://doi.org/10.1038/ng.3395. PubMed PMID: 26343388; PMCID: PMC4874245.

[32] Patsopoulos NA, Barcellos LF, Hintzen RQ, Schaefer C, van Duijn CM, Noble JA, Raj T, Imsgc, Anzgene, Gourraud PA, Stranger BE, Oksenberg J, Olsson T, Taylor BV, Sawcer S, Hafler DA, Carrington M, De Jager PL, de Bakker PI. Fine-mapping the genetic association of the major histocompatibility complex in multiple sclerosis: HLA and non-HLA effects. PLoS Genet. 2013;9(11):e1003926. https://doi.org/10.1371/journal.pgen.1003926. PubMed PMID: 24278027; PMCID: PMC3836799.

[33] International Multiple Sclerosis Genetics Consortium. Electronic address ccye, International multiple sclerosis genetics C. low-frequency and rare-coding variation contributes to multiple sclerosis risk. Cell. 2018;175(6):1679–87.e7. Epub 2018/10/23. https://doi.org/10.1016/j.cell.2018.09.049. PubMed PMID: 30343897; PMCID: PMC6269166.

[34] Gourraud PA, Sdika M, Khankhanian P, Henry RG, Beheshtian A, Matthews PM, Hauser SL, Oksenberg JR, Pelletier D, Baranzini SE. A genome-wide association study of brain lesion distribution in multiple sclerosis. Brain. 2013;136(Pt. 4):1012–24. https://doi.org/10.1093/brain/aws363. PubMed PMID: 23412934; PMCID: PMC3613709.

[35] Matsushita T, Madireddy L, Sprenger T, Khankhanian P, Magon S, Naegelin Y, Caverzasi E, Lindberg RL, Kappos L, Hauser SL, Oksenberg JR, Henry R, Pelletier D, Baranzini SE. Genetic associations with brain cortical thickness in multiple sclerosis. Genes Brain Behav. 2015;14(2):217–27. https://doi.org/10.1111/gbb.12190. PubMed PMID: 25684059; PMCID: PMC4857705.

[36] Buck D, Albrecht E, Aslam M, Goris A, Hauenstein N, Jochim A, International Multiple Sclerosis Genetics C, Wellcome Trust Case Control C, Cepok S, Grummel V, Dubois B, Berthele A, Lichtner P, Gieger C, Winkelmann J, Hemmer B. Genetic variants in the immunoglobulin heavy chain locus are associated with the IgG index in multiple sclerosis. Ann Neurol. 2013;73(1):86–94. https://doi.org/10.1002/ana.23749. PubMed PMID: 23225573; PMCID: PMC3661208.

[37] Delgado-Garcia M, Matesanz F, Alcina A, Fedetz M, Garcia-Sanchez MI, Ruiz-Pena JL, Fernandez O, Pinto Medel MJ, Leyva L, Arnal C, Delgado C, Lopez Guerrero JA, Gonzalez-Perez A, Saez ME, Villar LM, Alvarez-Cermeno JC, Picon C, Arroyo R, Varade J, Urcelay E, Izquierdo G, Lucas M. A new risk variant for multiple sclerosis at the immunoglobulin heavy chain locus associates with intrathecal IgG, IgM index and oligoclonal bands. Mult Scler. 2015;21(9):1104–11. https://doi.org/10.1177/1352458514556302. PubMed PMID: 25392328.

[38] Coyle PK. Pharmacogenetic biomarkers to predict treatment response in multiple sclerosis: current and future perspectives. Mult Scler Int. 2017;2017:6198530. Epub 2017/08/15. https://doi.org/10.1155/2017/6198530. PubMed PMID: 28804651; PMCID: PMC5540248.

[39] Chitnis T, Krupp L, Yeh A, Rubin J, Kuntz N, Strober JB, Chabas D, Weinstock-Guttmann B, Ness J, Rodriguez M, Waubant E. Pediatric multiple sclerosis. Neurol Clin. 2011;29(2):481–505. Epub 2011/03/29. https://doi.org/10.1016/j.ncl.2011.01.004. PubMed PMID: 21439455.

[40] Yeh EA, Chitnis T, Krupp L, Ness J, Chabas D, Kuntz N, Waubant E, Excellence USNoPMSCo. Pediatric multiple sclerosis. Nat Rev Neurol. 2009;5(11):621–31. Epub 2009/10/15. https://doi.org/10.1038/nrneurol.2009.158. PubMed PMID: 19826402.

[41] Gianfrancesco MA, Stridh P, Shao X, Rhead B, Graves JS, Chitnis T, Waldman A, Lotze T, Schreiner T, Belman A, Greenberg B, Weinstock-Guttman B, Aaen G, Tillema JM, Hart J, Caillier S, Ness J, Harris Y, Rubin J, Candee M, Krupp L, Gorman M, Benson L, Rodriguez M, Mar S, Kahn I, Rose J, Roalstad S, Casper TC, Shen L, Quach H, Quach D, Hillert J, Hedstrom A, Olsson T, Kockum I, Alfredsson L, Schaefer C, Barcellos LF, Waubant E. Network of pediatric multiple sclerosis C. Genetic risk factors for pediatric-onset multiple sclerosis. Mult Scler 2017;24(14):1825–34. Epub 2017/10/06. https://doi.org/10.1177/1352458517733551. PubMed PMID: 28980494; PMCID: PMC5878964.

[42] De Jager PL, Chibnik LB, Cui J, Reischl J, Lehr S, Simon KC, Aubin C, Bauer D, Heubach JF, Sandbrink R, Tyblova M, Lelkova P, Steering committee of the Bs, Steering committee of the Bs, Steering committee of the LTFs, Steering committee of the CCRs, Havrdova E, Pohl C, Horakova D, Ascherio A, Hafler DA, Karlson EW. Integration of genetic risk factors into a clinical algorithm for multiple sclerosis susceptibility: a weighted genetic risk score. Lancet Neurol. 2009;8(12):1111–9. https://doi.org/10.1016/S1474-4422(09)70275-3. PubMed PMID: 19879194; PMCID: PMC3099419.

[43] Gourraud PA, McElroy JP, Caillier SJ, Johnson BA, Santaniello A, Hauser SL, Oksenberg JR. Aggregation of multiple sclerosis genetic risk variants in multiple and single case families. Ann Neurol. 2011;69(1):65–74. https://doi.org/10.1002/ana.22323. PubMed PMID: 21280076; PMCID: PMC3511846.

[44] Sondergaard HB, Petersen ER, Magyari M, Sellebjerg F, Oturai AB. Genetic burden of MS risk variants distinguish patients from healthy individuals but are not associated with disease activity. Mult Scler Relat Disord. 2017;13:25–7. Epub 2017/04/22. https://doi.org/10.1016/j.msard.2017.01.015. PubMed PMID: 28427696.

[45] Isobe N, Keshavan A, Gourraud PA, Zhu AH, Datta E, Schlaeger R, Caillier SJ, Santaniello A, Lizee A, Himmelstein DS, Baranzini SE, Hollenbach J, Cree BA, Hauser SL, Oksenberg JR, Henry RG. Association of HLA genetic risk burden with disease phenotypes in multiple sclerosis. JAMA Neurol. 2016;73(7):795–802. Epub 2016/06/01. https://doi.org/10.1001/jamaneurol.2016.0980. PubMed PMID: 27244296; PMCID: PMC5081075.

[46] Hilven K, Patsopoulos NA, Dubois B, Goris A. Burden of risk variants correlates with phenotype of multiple sclerosis. Mult Scler. 2015;21(13):1670–80. https://doi.org/10.1177/1352458514568174. PubMed PMID: 25948629.

[47] Lindberg RL, De Groot CJ, Certa U, Ravid R, Hoffmann F, Kappos L, Leppert D. Multiple sclerosis as a generalized CNS disease—comparative microarray analysis of normal appearing white matter and lesions in secondary progressive MS. J Neuroimmunol. 2004;152(1–2):154–67. Epub 2004/06/30. https://doi.org/10.1016/j.jneuroim.2004.03.011. PubMed PMID: 15223248.

[48] Dutta R, McDonough J, Chang A, Swamy L, Siu A, Kidd GJ, Rudick R, Mirnics K, Trapp BD. Activation of the ciliary neurotrophic factor (CNTF) signalling pathway in cortical neurons of multiple sclerosis patients. Brain. 2007;130(Pt. 10):2566–76. Epub 2007/09/28. https://doi.org/10.1093/brain/awm206. PubMed PMID: 17898009.

[49] Dutta R, McDonough J, Yin X, Peterson J, Chang A, Torres T, Gudz T, Macklin WB, Lewis DA, Fox RJ, Rudick R, Mirnics K, Trapp BD. Mitochondrial dysfunction as a cause of axonal degeneration in multiple sclerosis patients. Ann Neurol. 2006;59(3):478–89. Epub 2006/01/05. https://doi.org/10.1002/ana.20736. PubMed PMID: 16392116.

[50] Mycko MP, Brosnan CF, Raine CS, Fendler W, Selmaj KW. Transcriptional profiling of microdissected areas of active multiple sclerosis lesions reveals activation of heat shock protein genes. J Neurosci Res. 2012;90(10):1941–8. Epub 2012/06/21. https://doi.org/10.1002/jnr.23079. PubMed PMID: 22715030.

[51] Lock C, Hermans G, Pedotti R, Brendolan A, Schadt E, Garren H, Langer-Gould A, Strober S, Cannella B, Allard J, Klonowski P, Austin A, Lad N, Kaminski N, Galli SJ, Oksenberg JR, Raine CS, Heller R, Steinman L. Gene-microarray analysis of multiple sclerosis lesions yields new targets validated in autoimmune encephalomyelitis. Nat Med. 2002;8(5):500–8. Epub 2002/05/02. https://doi.org/10.1038/nm0502-500. PubMed PMID: 11984595.

[52] Mycko MP, Papoian R, Boschert U, Raine CS, Selmaj KW. Microarray gene expression profiling of chronic active and inactive lesions in multiple sclerosis. Clin Neurol Neurosurg. 2004;106(3):223–9. Epub 2004/06/05. https://doi.org/10.1016/j.clineuro.2004.02.019. PubMed PMID: 15177772.

[53] Chabas D, Baranzini SE, Mitchell D, Bernard CC, Rittling SR, Denhardt DT, Sobel RA, Lock C, Karpuj M, Pedotti R, Heller R, Oksenberg JR, Steinman L. The influence of the proinflammatory cytokine, osteopontin, on autoimmune demyelinating disease. Science. 2001;294(5547):1731–5. Epub 2001/11/27. https://doi.org/10.1126/science.1062960. PubMed PMID: 11721059.

[54] Clemente N, Comi C, Raineri D, Cappellano G, Vecchio D, Orilieri E, Gigliotti CL, Boggio E, Dianzani C, Sorosina M, Martinelli-Boneschi F, Caldano M, Bertolotto A, Ambrogio L, Sblattero D, Cena T, Leone M, Dianzani U, Chiocchetti A. Role of anti-osteopontin antibodies in multiple sclerosis and experimental autoimmune encephalomyelitis. Front Immunol. 2017;8:321. Epub 2017/04/08. https://doi.org/10.3389/fimmu.2017.00321. PubMed PMID: 28386258; PMCID: PMC5362623.

[55] Junker A, Ivanidze J, Malotka J, Eiglmeier I, Lassmann H, Wekerle H, Meinl E, Hohlfeld R, Dornmair K. Multiple sclerosis: T-cell receptor expression in distinct brain regions. Brain. 2007;130(Pt. 11):2789–99. Epub 2007/09/25. https://doi.org/10.1093/brain/awm214. PubMed PMID: 17890278.

[56] Hendrickx DAE, van Scheppingen J, van der Poel M, Bossers K, Schuurman KG, van Eden CG, Hol EM, Hamann J, Huitinga I. Gene expression profiling of multiple sclerosis pathology identifies early patterns of demyelination surrounding chronic active lesions. Front Immunol. 2017;8:1810. Epub 2018/01/10. https://doi.org/10.3389/fimmu.2017.01810. PubMed PMID: 29312322; PMCID: PMC5742619.

[57] Tajouri L, Mellick AS, Ashton KJ, Tannenberg AE, Nagra RM, Tourtellotte WW, Griffiths LR. Quantitative and qualitative changes in gene expression patterns characterize the activity of plaques in multiple sclerosis. Brain Res Mol Brain Res. 2003;119(2):170–83. Epub 2003/11/20. PubMed PMID: 14625084.

[58] Itoh N, Itoh Y, Tassoni A, Ren E, Kaito M, Ohno A, Ao Y, Farkhondeh V, Johnsonbaugh H, Burda J, Sofroniew MV, Voskuhl RR. Cell-specific and region-specific transcriptomics in the multiple sclerosis model: focus on astrocytes. Proc Natl Acad Sci USA. 2018;115(2):E302-E9. Epub 2017/12/28. https://doi.org/10.1073/pnas.1716032115. PubMed PMID: 29279367; PMCID: PMC5777065.

[59] Marchetti G, Ziliotto N, Meneghetti S, Baroni M, Lunghi B, Menegatti E, Pedriali M, Salvi F, Bartolomei I, Straudi S, Manfredini F, Voltan R, Basaglia N, Mascoli F, Zamboni P, Bernardi F. Changes in expression profiles of internal jugular vein wall and plasma protein levels

in multiple sclerosis. Mol Med. 2018;24(1):42. Epub 2018/08/24. https://doi.org/10.1186/s10020-018-0043-4. PubMed PMID: 30134823; PMCID: PMC6085618.

[60] Falcao AM, van Bruggen D, Marques S, Meijer M, Jakel S, Agirre E, Samudyata, Floriddia EM, Vanichkina DP, Ffrench-Constant C, Williams A, Guerreiro-Cacais AO, Castelo-Branco G. Disease-specific oligodendrocyte lineage cells arise in multiple sclerosis. Nat Med. 2018;24(12):1837–44. Epub 2018/11/14. https://doi.org/10.1038/s41591-018-0236-y. PubMed PMID: 30420755.

[61] Corvol JC, Pelletier D, Henry RG, Caillier SJ, Wang J, Pappas D, Casazza S, Okuda DT, Hauser SL, Oksenberg JR, Baranzini SE. Abrogation of T cell quiescence characterizes patients at high risk for multiple sclerosis after the initial neurological event. Proc Natl Acad Sci USA. 2008;105(33):11839–44. Epub 2008/08/12. https://doi.org/10.1073/pnas.0805065105. PubMed PMID: 18689680; PMCID: PMC2504481.

[62] Frisullo G, Nociti V, Iorio R, Patanella AK, Marti A, Mirabella M, Tonali PA, Batocchi AP. The persistency of high levels of pSTAT3 expression in circulating CD4+ T cells from CIS patients favors the early conversion to clinically defined multiple sclerosis. J Neuroimmunol. 2008;205(1–2):126–34. Epub 2008/10/18. https://doi.org/10.1016/j.jneuroim.2008.09.003. PubMed PMID: 18926576.

[63] Achiron A, Gurevich M, Friedman N, Kaminski N, Mandel M. Blood transcriptional signatures of multiple sclerosis: unique gene expression of disease activity. Ann Neurol. 2004;55(3):410–7. Epub 2004/03/03. https://doi.org/10.1002/ana.20008. PubMed PMID: 14991819.

[64] Brynedal B, Khademi M, Wallstrom E, Hillert J, Olsson T, Duvefelt K. Gene expression profiling in multiple sclerosis: a disease of the central nervous system, but with relapses triggered in the periphery? Neurobiol Dis. 2010;37(3):613–21. Epub 2009/12/01. https://doi.org/10.1016/j.nbd.2009.11.014. PubMed PMID: 19944761.

[65] Weinstock-Guttman B, Badgett D, Patrick K, Hartrich L, Santos R, Hall D, Baier M, Feichter J, Ramanathan M. Genomic effects of IFN-beta in multiple sclerosis patients. J Immunol. 2003;171(5):2694–702. Epub 2003/08/21. PubMed PMID: 12928423.

[66] Reder AT, Velichko S, Yamaguchi KD, Hamamcioglu K, Ku K, Beekman J, Wagner TC, Perez HD, Salamon H, Croze E. IFN-beta1b induces transient and variable gene expression in relapsing-remitting multiple sclerosis patients independent of neutralizing antibodies or changes in IFN receptor RNA expression. J Interferon Cytokine Res. 2008;28(5):317–31. Epub 2008/06/13. https://doi.org/10.1089/jir.2007.0131. PubMed PMID: 18547162.

[67] De Felice B, Mondola P, Sasso A, Orefice G, Bresciamorra V, Vacca G, Biffali E, Borra M, Pannone R. Small non-coding RNA signature in multiple sclerosis patients after treatment with interferon-beta. BMC Med Genomics. 2014;7:26. Epub 2014/06/03. https://doi.org/10.1186/1755-8794-7-26. PubMed PMID: 24885345; PMCID: PMC4060096.

[68] Harari D, Orr I, Rotkopf R, Baranzini SE, Schreiber G. A robust type I interferon gene signature from blood RNA defines quantitative but not qualitative differences between three major IFNbeta drugs in the treatment of multiple sclerosis. Hum Mol Genet. 2015;24(11):3192–205. Epub 2015/02/28. https://doi.org/10.1093/hmg/ddv071. PubMed PMID: 25721402.

[69] Cox MB, Bowden NA, Scott RJ, Lechner-Scott J. Altered expression of the plasminogen activation pathway in peripheral blood mononuclear cells in multiple sclerosis: possible pathomechanism of matrix metalloproteinase activation. Mult Scler. 2013;19(10):1268–74. Epub 2013/02/13. https://doi.org/10.1177/1352458513475493. PubMed PMID: 23401127.

[70] Nickles D, Chen HP, Li MM, Khankhanian P, Madireddy L, Caillier SJ, Santaniello A, Cree BA, Pelletier D, Hauser SL, Oksenberg JR, Baranzini SE. Blood RNA profiling in a large cohort of multiple sclerosis patients and healthy controls. Hum Mol Genet.

2013;22(20):4194–205. Epub 2013/06/12. https://doi.org/10.1093/hmg/ddt267. PubMed PMID: 23748426; PMCID: PMC3781642.

[71] Friess J, Hecker M, Roch L, Koczan D, Fitzner B, Angerer IC, Schroder I, Flechtner K, Thiesen HJ, Winkelmann A, Zettl UK. Fingolimod alters the transcriptome profile of circulating CD4+ cells in multiple sclerosis. Sci Rep. 2017;7:42087. Epub 2017/02/06. https://doi.org/10.1038/srep42087. PubMed PMID: 28155899; PMCID: PMC5290459 Novartis and Teva. A.W. received speaking fees and travel funds from Bayer HealthCare, Genzyme, Merck Serono and Novartis. U.K.Z. received research support as well as speaking fees and travel funds from Almirall, Bayer HealthCare, Biogen, Merck Serono, Novartis, Sanofi and Teva. J.F., L.R., D.K., B.F., I.C.A., I.S., K.F. and H.J.T. declare that they have no competing interests.

[72] Roch L, Hecker M, Friess J, Angerer IC, Koczan D, Fitzner B, Schroder I, Flechtner K, Thiesen HJ, Meister S, Winkelmann A, Zettl UK. High-resolution expression profiling of peripheral blood CD8(+) cells in patients with multiple sclerosis displays fingolimod-induced immune cell redistribution. Mol Neurobiol. 2017;54(7):5511–25. Epub 2016/09/16. https://doi.org/10.1007/s12035-016-0075-0. PubMed PMID: 27631876.

[73] Gafson AR, Kim K, Cencioni MT, van Hecke W, Nicholas R, Baranzini SE, Matthews PM. Mononuclear cell transcriptome changes associated with dimethyl fumarate in MS. Neurol Neuroimmunol Neuroinflamm. 2018;5(4):e470. Epub 2018/10/05. https://doi.org/10.1212/NXI.0000000000000470. PubMed PMID: 30283812; PMCID: PMC6168332.

[74] Starossom SC, Imitola J, Wang Y, Cao L, Khoury SJ. Subventricular zone microglia transcriptional networks. Brain Behav Immun. 2011;25(5):991–9. Epub 2010/11/16. https://doi.org/10.1016/j.bbi.2010.11.002. PubMed PMID: 21074605; PMCID: PMC3109092.

[75] Lewis ND, Hill JD, Juchem KW, Stefanopoulos DE, Modis LK. RNA sequencing of microglia and monocyte-derived macrophages from mice with experimental autoimmune encephalomyelitis illustrates a changing phenotype with disease course. J Neuroimmunol. 2014;277(1–2):26–38. Epub 2014/10/02. https://doi.org/10.1016/j.jneuroim.2014.09.014. PubMed PMID: 25270668.

[76] Peferoen LA, Vogel DY, Ummenthum K, Breur M, Heijnen PD, Gerritsen WH, Peferoen-Baert RM, van der Valk P, Dijkstra CD, Amor S. Activation status of human microglia is dependent on lesion formation stage and remyelination in multiple sclerosis. J Neuropathol Exp Neurol. 2015;74(1):48–63. Epub 2014/12/04. https://doi.org/10.1097/NEN.0000000000000149. PubMed PMID: 25470347.

[77] Gustafsson M, Gawel DR, Alfredsson L, Baranzini S, Bjorkander J, Blomgran R, Hellberg S, Eklund D, Ernerudh J, Kockum I, Konstantinell A, Lahesmaa R, Lentini A, Liljenstrom HR, Mattson L, Matussek A, Mellergard J, Mendez M, Olsson T, Pujana MA, Rasool O, Serra-Musach J, Stenmarker M, Tripathi S, Viitala M, Wang H, Zhang H, Nestor CE, Benson M. A validated gene regulatory network and GWAS identifies early regulators of T cell-associated diseases. Sci Transl Med. 2015;7(313):313ra178. Epub 2015/11/13. https://doi.org/10.1126/scitranslmed.aad2722. PubMed PMID: 26560356.

[78] Annibali V, Umeton R, Palermo A, Severa M, Etna MP, Giglio S, Romano S, Ferraldeschi M, Buscarinu MC, Vecchione A, Annese A, Policano C, Mechelli R, Pizzolato Umeton R, Fornasiero A, Angelini DF, Guerrera G, Battistini L, Coccia EM, Salvetti M, Ristori G. Analysis of coding and non-coding transcriptome of peripheral B cells reveals an altered interferon response factor (IRF)-1 pathway in multiple sclerosis patients. J Neuroimmunol. 2018;324:165–71. Epub 2018/10/03. https://doi.org/10.1016/j.jneuroim.2018.09.005. PubMed PMID: 30270021.

[79] Loda E, Balabanov R. Interferon regulatory factor 1 regulation of oligodendrocyte injury and inflammatory demyelination. Rev Neurosci. 2012;23(2):145–52. Epub 2012/04/14. https://doi.org/10.1515/revneuro-2011-068. PubMed PMID: 22499673.

[80] Iwanowski P, Losy J, Kramer L, Wojcicka M, Kaufman E. CXCL10 and CXCL13 chemokines in patients with relapsing remitting and primary progressive multiple sclerosis. J Neurol Sci. 2017;380:22–6. Epub 2017/09/06. https://doi.org/10.1016/j.jns.2017.06.048. PubMed PMID: 28870573.

[81] Ottoboni L, Keenan BT, Tamayo P, Kuchroo M, Mesirov JP, Buckle GJ, Khoury SJ, Hafler DA, Weiner HL, De Jager PL. An RNA profile identifies two subsets of multiple sclerosis patients differing in disease activity. Sci Transl Med. 2012;4(153):153ra31. Epub 2012/09/29. https://doi.org/10.1126/scitranslmed.3004186. PubMed PMID: 23019656; PMCID: PMC3753678.

[82] Bustamante MF, Nurtdinov RN, Rio J, Montalban X, Comabella M. Baseline gene expression signatures in monocytes from multiple sclerosis patients treated with interferon-beta. PLoS One. 2013;8(4):e60994. Epub 2013/05/03. https://doi.org/10.1371/journal.pone.0060994. PubMed PMID: 23637780; PMCID: PMC3630153.

[83] Fraga MF, Ballestar E, Paz MF, Ropero S, Setien F, Ballestar ML, Heine-Suner D, Cigudosa JC, Urioste M, Benitez J, Boix-Chornet M, Sanchez-Aguilera A, Ling C, Carlsson E, Poulsen P, Vaag A, Stephan Z, Spector TD, Wu YZ, Plass C, Esteller M. Epigenetic differences arise during the lifetime of monozygotic twins. Proc Natl Acad Sci USA. 2005;102(30):10604–9. Epub 2005/07/13. https://doi.org/10.1073/pnas.0500398102. PubMed PMID: 16009939; PMCID: PMC1174919.

[84] Baranzini SE, Mudge J, van Velkinburgh JC, Khankhanian P, Khrebtukova I, Miller NA, Zhang L, Farmer AD, Bell CJ, Kim RW, May GD, Woodward JE, Caillier SJ, McElroy JP, Gomez R, Pando MJ, Clendenen LE, Ganusova EE, Schilkey FD, Ramaraj T, Khan OA, Huntley JJ, Luo S, Kwok PY, Wu TD, Schroth GP, Oksenberg JR, Hauser SL, Kingsmore SF. Genome, epigenome and RNA sequences of monozygotic twins discordant for multiple sclerosis. Nature. 2010;464(7293):1351–6. Epub 2010/04/30. https://doi.org/10.1038/nature08990. PubMed PMID: 20428171; PMCID: PMC2862593.

[85] Kumagai C, Kalman B, Middleton FA, Vyshkina T, Massa PT. Increased promoter methylation of the immune regulatory gene SHP-1 in leukocytes of multiple sclerosis subjects. J Neuroimmunol. 2012;246(1–2):51–7. Epub 2012/03/31. https://doi.org/10.1016/j.jneuroim.2012.03.003. PubMed PMID: 22458980; PMCID: PMC3335962.

[86] Mastronardi FG, Wood DD, Mei J, Raijmakers R, Tseveleki V, Dosch HM, Probert L, Casaccia-Bonnefil P, Moscarello MA. Increased citrullination of histone H3 in multiple sclerosis brain and animal models of demyelination: a role for tumor necrosis factor-induced peptidylarginine deiminase 4 translocation. J Neurosci. 2006;26(44):11387–96. Epub 2006/11/03. https://doi.org/10.1523/JNEUROSCI.3349-06.2006. PubMed PMID: 17079667.

[87] Ma J, Wang R, Fang X, Ding Y, Sun Z. Critical role of TCF-1 in repression of the IL-17 gene. PLoS One. 2011;6(9):e24768. Epub 2011/09/22. https://doi.org/10.1371/journal.pone.0024768. PubMed PMID: 21935461; PMCID: PMC3173465.

[88] Yang H, Lee SM, Gao B, Zhang J, Fang D. Histone deacetylase sirtuin 1 deacetylates IRF1 protein and programs dendritic cells to control Th17 protein differentiation during autoimmune inflammation. J Biol Chem. 2013;288(52):37256–66. Epub 2013/11/12. https://doi.org/10.1074/jbc.M113.527531. PubMed PMID: 24214980; PMCID: PMC3873578.

[89] Coquet JM, Middendorp S, van der Horst G, Kind J, Veraar EA, Xiao Y, Jacobs H, Borst J. The CD27 and CD70 costimulatory pathway inhibits effector function of T helper 17 cells and attenuates associated autoimmunity. Immunity. 2013;38(1):53–65. Epub 2012/11/20. https://doi.org/10.1016/j.immuni.2012.09.009. PubMed PMID: 23159439.

[90] Du C, Liu C, Kang J, Zhao G, Ye Z, Huang S, Li Z, Wu Z, Pei G. MicroRNA miR-326 regulates TH-17 differentiation and is associated with the pathogenesis of multiple sclerosis. Nat Immunol. 2009;10(12):1252–9. Epub 2009/10/20. https://doi.org/10.1038/ni.1798. PubMed PMID: 19838199.

[91] Escobar T, Yu CR, Muljo SA, Egwuagu CE. STAT3 activates miR-155 in Th17 cells and acts in concert to promote experimental autoimmune uveitis. Invest Ophthalmol Vis Sci. 2013;54(6):4017–25. Epub 2013/05/16. https://doi.org/10.1167/iovs.13-11937. PubMed PMID: 23674757; PMCID: PMC3680004.

[92] Murugaiyan G, Beynon V, Mittal A, Joller N, Weiner HL. Silencing microRNA-155 ameliorates experimental autoimmune encephalomyelitis. J Immunol. 2011;187(5):2213–21. Epub 2011/07/27. https://doi.org/10.4049/jimmunol.1003952. PubMed PMID: 21788439; PMCID: PMC3167080.

[93] Murugaiyan G, da Cunha AP, Ajay AK, Joller N, Garo LP, Kumaradevan S, Yosef N, Vaidya VS, Weiner HL. MicroRNA-21 promotes Th17 differentiation and mediates experimental autoimmune encephalomyelitis. J Clin Invest. 2015;125(3):1069–80. Epub 2015/02/03. https://doi.org/10.1172/JCI74347. PubMed PMID: 25642768; PMCID: PMC4362225.

[94] Ponomarev ED, Veremeyko T, Barteneva N, Krichevsky AM, Weiner HL. MicroRNA-124 promotes microglia quiescence and suppresses EAE by deactivating macrophages via the C/EBP-alpha-PU.1 pathway. Nat Med. 2011;17(1):64–70. Epub 2010/12/07. https://doi.org/10.1038/nm.2266. PubMed PMID: 21131957; PMCID: PMC3044940.

[95] Gandhi R, Healy B, Gholipour T, Egorova S, Musallam A, Hussain MS, Nejad P, Patel B, Hei H, Khoury S, Quintana F, Kivisakk P, Chitnis T, Weiner HL. Circulating microRNAs as biomarkers for disease staging in multiple sclerosis. Ann Neurol. 2013;73(6):729–40. Epub 2013/03/16. https://doi.org/10.1002/ana.23880. PubMed PMID: 23494648.

[96] Kimura K, Hohjoh H, Fukuoka M, Sato W, Oki S, Tomi C, Yamaguchi H, Kondo T, Takahashi R, Yamamura T. Circulating exosomes suppress the induction of regulatory T cells via let-7i in multiple sclerosis. Nat Commun. 2018;9(1):17. Epub 2018/01/04. https://doi.org/10.1038/s41467-017-02406-2. PubMed PMID: 29295981; PMCID: PMC5750223.

[97] Hemond CC, Healy BC, Tauhid S, Mazzola MA, Quintana FJ, Gandhi R, Weiner HL, Bakshi R. MRI phenotypes in MS: longitudinal changes and miRNA signatures. Neurol Neuroimmunol Neuroinflamm. 2019;6(2):e530. Epub 2019/02/26. https://doi.org/10.1212/NXI.0000000000000530. PubMed PMID: 30800720; PMCID: PMC6384020.

[98] Comabella M, Fernandez M, Martin R, Rivera-Vallve S, Borras E, Chiva C, Julia E, Rovira A, Canto E, Alvarez-Cermeno JC, Villar LM, Tintore M, Montalban X. Cerebrospinal fluid chitinase 3-like 1 levels are associated with conversion to multiple sclerosis. Brain. 2010;133(Pt. 4):1082–93. Epub 2010/03/20. https://doi.org/10.1093/brain/awq035. PubMed PMID: 20237129.

[99] Ottervald J, Franzen B, Nilsson K, Andersson LI, Khademi M, Eriksson B, Kjellstrom S, Marko-Varga G, Vegvari A, Harris RA, Laurell T, Miliotis T, Matusevicius D, Salter H, Ferm M, Olsson T. Multiple sclerosis: identification and clinical evaluation of novel CSF biomarkers. J Proteomics. 2010;73(6):1117–32. Epub 2010/01/23. https://doi.org/10.1016/j.jprot.2010.01.004. PubMed PMID: 20093204.

[100] Stoop MP, Dekker LJ, Titulaer MK, Burgers PC, Sillevis Smitt PA, Luider TM, Hintzen RQ. Multiple sclerosis-related proteins identified in cerebrospinal fluid by advanced mass spectrometry. Proteomics. 2008;8(8):1576–85. Epub 2008/03/21. https://doi.org/10.1002/pmic.200700446. PubMed PMID: 18351689.

[101] Liu S, Bai S, Qin Z, Yang Y, Cui Y, Qin Y. Quantitative proteomic analysis of the cerebrospinal fluid of patients with multiple sclerosis. J Cell Mol Med. 2009;13(8A):1586–603. Epub 2009/07/16. https://doi.org/10.1111/j.1582-4934.2009.00850.x. PubMed PMID: 19602050; PMCID: PMC3828869.

[102] Hammack BN, Fung KY, Hunsucker SW, Duncan MW, Burgoon MP, Owens GP, Gilden DH. Proteomic analysis of multiple sclerosis cerebrospinal fluid. Mult Scler. 2004;10(3):245–60. Epub 2004/06/30. https://doi.org/10.1191/1352458504ms1023oa. PubMed PMID: 15222687.

[103] Ly L, Barnett MH, Zheng YZ, Gulati T, Prineas JW, Crossett B. Comprehensive tissue processing strategy for quantitative proteomics of formalin-fixed multiple sclerosis lesions. J Proteome Res. 2011;10(10):4855–68. Epub 2011/08/30. https://doi.org/10.1021/pr200672n. PubMed PMID: 21870854.

[104] Noben JP, Dumont D, Kwasnikowska N, Verhaert P, Somers V, Hupperts R, Stinissen P, Robben J. Lumbar cerebrospinal fluid proteome in multiple sclerosis: characterization by ultrafiltration, liquid chromatography, and mass spectrometry. J Proteome Res. 2006;5(7):1647–57. Epub 2006/07/11. https://doi.org/10.1021/pr0504788. PubMed PMID: 16823972.

[105] De Masi R, Vergara D, Pasca S, Acierno R, Greco M, Spagnolo L, Blasi E, Sanapo F, Trianni G, Maffia M. PBMCs protein expression profile in relapsing IFN-treated multiple sclerosis: a pilot study on relation to clinical findings and brain atrophy. J Neuroimmunol. 2009;210(1–2):80–6. Epub 2009/03/31. https://doi.org/10.1016/j.jneuroim.2009.03.002. PubMed PMID: 19329191.

[106] Comabella M, Montalban X. Body fluid biomarkers in multiple sclerosis. Lancet Neurol. 2014;13(1):113–26. Epub 2013/12/18. https://doi.org/10.1016/S1474-4422(13)70233-3. PubMed PMID: 24331797.

[107] Brewis IA, Brennan P. Proteomics technologies for the global identification and quantification of proteins. Adv Protein Chem Struct Biol. 2010;80:1–44. Epub 2010/11/27. https://doi.org/10.1016/B978-0-12-381264-3.00001-1. PubMed PMID: 21109216.

[108] Querol L, Clark PL, Bailey MA, Cotsapas C, Cross AH, Hafler DA, Kleinstein SH, Lee JY, Yaari G, Willis SN, O'Connor KC. Protein array-based profiling of CSF identifies RBPJ as an autoantigen in multiple sclerosis. Neurology. 2013;81(11):956–63. Epub 2013/08/08. https://doi.org/10.1212/WNL.0b013e3182a43b48. PubMed PMID: 23921886; PMCID: PMC3888197.

[109] Broadwater L, Pandit A, Clements R, Azzam S, Vadnal J, Sulak M, Yong VW, Freeman EJ, Gregory RB, McDonough J. Analysis of the mitochondrial proteome in multiple sclerosis cortex. Biochim Biophys Acta. 2011;1812(5):630–41. Epub 2011/02/08. https://doi.org/10.1016/j.bbadis.2011.01.012. PubMed PMID: 21295140; PMCID: PMC3074931.

[110] Rosenling T, Stoop MP, Attali A, van Aken H, Suidgeest E, Christin C, Stingl C, Suits F, Horvatovich P, Hintzen RQ, Tuinstra T, Bischoff R, Luider TM. Profiling and identification of cerebrospinal fluid proteins in a rat EAE model of multiple sclerosis. J Proteome Res. 2012;11(4):2048–60. Epub 2012/02/11. https://doi.org/10.1021/pr201244t. PubMed PMID: 22320401.

[111] Liu T, Donahue KC, Hu J, Kurnellas MP, Grant JE, Li H, Elkabes S. Identification of differentially expressed proteins in experimental autoimmune encephalomyelitis (EAE) by proteomic analysis of the spinal cord. J Proteome Res. 2007;6(7):2565–75. Epub 2007/06/19. https://doi.org/10.1021/pr070012k. PubMed PMID: 17571869; PMCID: PMC2430926.

[112] Hakobyan S, Luppe S, Evans DR, Harding K, Loveless S, Robertson NP, Morgan BP. Plasma complement biomarkers distinguish multiple sclerosis and neuromyelitis optica spectrum disorder. Mult Scler. 2017;23(7):946–55. Epub 2016/09/11. https://doi.org/10.1177/1352458516669002. PubMed PMID: 27613120.

[113] Boziki M, Grigoriadis N. An update on the role of matrix metalloproteinases in the pathogenesis of multiple sclerosis. Med Chem. 2018;14(2):155–69. Epub 2017/09/07. https://doi.org/10.2174/1573406413666170906122803. PubMed PMID: 28875862.

[114] Gray E, Thomas TL, Betmouni S, Scolding N, Love S. Elevated myeloperoxidase activity in white matter in multiple sclerosis. Neurosci Lett. 2008;444(2):195–8. Epub 2008/08/30. https://doi.org/10.1016/j.neulet.2008.08.035. PubMed PMID: 18723077.

[115] Ohl K, Tenbrock K, Kipp M. Oxidative stress in multiple sclerosis: central and peripheral mode of action. Exp Neurol. 2016;277:58–67. Epub 2015/12/03. https://doi.org/10.1016/j.expneurol.2015.11.010. PubMed PMID: 26626971.

[116] Dagley LF, Emili A, Purcell AW. Application of quantitative proteomics technologies to the biomarker discovery pipeline for multiple sclerosis. Proteomics Clin Appl. 2013;7(1–2):91–108. Epub 2012/11/01. https://doi.org/10.1002/prca.201200104. PubMed PMID: 23112123.

[117] Jaros JA, Guest PC, Bahn S, Martins-de-Souza D. Affinity depletion of plasma and serum for mass spectrometry-based proteome analysis. Methods Mol Biol. 2013;1002:1–11. Epub 2013/04/30. https://doi.org/10.1007/978-1-62703-360-2_1. PubMed PMID: 23625390.

[118] Farias AS, Pradella F, Schmitt A, Santos LM, Martins-de-Souza D. Ten years of proteomics in multiple sclerosis. Proteomics. 2014;14(4–5):467–80. Epub 2013/12/18. https://doi.org/10.1002/pmic.201300268. PubMed PMID: 24339438.

[119] Calais G, Forzy G, Crinquette C, Mackowiak A, de Seze J, Blanc F, Lebrun C, Heinzlef O, Clavelou P, Moreau T, Hennache B, Zephir H, Verier A, Neuville V, Confavreux C, Vermersch P, Hautecoeur P. Tear analysis in clinically isolated syndrome as new multiple sclerosis criterion. Mult Scler. 2010;16(1):87–92. Epub 2009/12/24. https://doi.org/10.1177/1352458509352195. PubMed PMID: 20028709.

[120] Lebrun C, Forzy G, Collongues N, Cohen M, de Seze J, Hautecoeur P, Club francophone de la SEP, Risconsortium. Tear analysis as a tool to detect oligoclonal bands in radiologically isolated syndrome. Rev Neurol (Paris). 2015;171(4):390–3. Epub 2015/01/24. https://doi.org/10.1016/j.neurol.2014.11.007. PubMed PMID: 25613196.

[121] Manconi B, Liori B, Cabras T, Vincenzoni F, Iavarone F, Lorefice L, Cocco E, Castagnola M, Messana I, Olianas A. Top-down proteomic profiling of human saliva in multiple sclerosis patients. J Proteomics. 2018;187:212–22. Epub 2018/08/08. https://doi.org/10.1016/j.jprot.2018.07.019. PubMed PMID: 30086402.

[122] Robinson WH, Fontoura P, Lee BJ, de Vegvar HE, Tom J, Pedotti R, DiGennaro CD, Mitchell DJ, Fong D, Ho PP, Ruiz PJ, Maverakis E, Stevens DB, Bernard CC, Martin R, Kuchroo VK, van Noort JM, Genain CP, Amor S, Olsson T, Utz PJ, Garren H, Steinman L. Protein microarrays guide tolerizing DNA vaccine treatment of autoimmune encephalomyelitis. Nat Biotechnol. 2003;21(9):1033–9. Epub 2003/08/12. https://doi.org/10.1038/nbt859. PubMed PMID: 12910246.

[123] Lucchinetti C, Bruck W, Parisi J, Scheithauer B, Rodriguez M, Lassmann H. Heterogeneity of multiple sclerosis lesions: implications for the pathogenesis of demyelination. Ann Neurol. 2000;47(6):707–17. Epub 2000/06/14. PubMed PMID: 10852536.

[124] Quintana FJ, Farez MF, Viglietta V, Iglesias AH, Merbl Y, Izquierdo G, Lucas M, Basso AS, Khoury SJ, Lucchinetti CF, Cohen IR, Weiner HL. Antigen microarrays identify unique serum autoantibody signatures in clinical and pathologic subtypes of multiple sclerosis. Proc Natl Acad Sci USA. 2008;105(48):18889–94. Epub 2008/11/26. https://doi.org/10.1073/pnas.0806310105. PubMed PMID: 19028871; PMCID: PMC2596207.

[125] Jakel S, Agirre E, Mendanha Falcao A, van Bruggen D, Lee KW, Knuesel I, Malhotra D, Ffrench-Constant C, Williams A, Castelo-Branco G. Altered human oligodendrocyte heterogeneity in multiple sclerosis. Nature. 2019;566(7745):543–7. Epub 2019/02/13. https://doi.org/10.1038/s41586-019-0903-2. PubMed PMID: 30747918.

Chapter 17

Systemic sclerosis

Sevdalina Lambova[a], Ulf Müller-Ladner[b,c]
[a]*Department of Propaedeutics of Internal Diseases, Medical University, Plovdiv, Bulgaria,* [b]*Department of Internal Medicine and Rheumatology, Justus-Liebig University Giessen, Giessen, Germany,* [c]*Department of Rheumatology and Clinical Immunology, Campus Kerckhoff, Bad Nauheim, Germany*

Definition

Systemic sclerosis (SSc) is a chronic, multisystem connective tissue disease characterized by microangiopathy, fibrosis of the skin and internal organs, and activation of humoral as well as cellular immune responses.

Subsets of systemic sclerosis

SSc is divided into two major subsets defined by the extension of skin involvement: limited SSc, characterized by skin involvement at the face, neck and the skin distal to the elbows and knees, and diffuse SSc, with skin thickening distal and proximal to the knees and elbows and thickening involving the trunk. CREST syndrome (calcinosis, Raynaud phenomenon (RP), esophageal dysmotility, sclerodactyly, and telangiectasia) is a form of limited SSc.

Epidemiology

The prevalence and incidence of SSc vary in different populations, suggesting the involvement of genetic and/or environmental factors. The prevalence of SSc is 276 cases per million adults in the United States [1] and 8 cases per 100,000 adults in Europe [2]. The annual incidence of SSc is 18–23 cases per million. SSc usually begins between 30 and 50 years of age and is three times more common in women than in men [1, 3, 4].

Etiology and pathogenesis

The role of the causative agents: Environmental and infectious agents

Environmental factors that have been proposed to be SSc-causative agents are organic solvents, vinyl chloride, silica, metal dust, certain pesticides, hair dyes

Genomic and Precision Medicine. https://doi.org/10.1016/B978-0-12-801496-7.00017-4

and toxins [3, 5]. Several infectious agents (e.g., herpes viruses, retroviruses and human cytomegalovirus) triggering molecular mimicry have also been hypothesized to be causative agents in SSc. This hypothesis has been supported by the presence of IgG anti-human cytomegalovirus antibodies and sequence homologies between retroviral proteins and the target of anti-Scl-70 antibody—a DNA topoisomerase I antigen [3, 6].

Microchimerism

Microchimeric cells, which can be transferred from one person to another during pregnancy, blood transfusion, bone marrow and solid organ transplantation, have been detected in peripheral blood and tissue of SSc patients and may support a graft-vs-host disease triggering SSc [3] but the evidence for this confounding factor is sparse [7].

Pathogenic pathways

The pathogenesis of SSc is complex and is based on (a) microvascular damage, (b) activation of the immune system, and (c) progressive fibrosis. Owing to this multifactorial pathophysiology, the search for reliable biomarkers is still ongoing [8].

Vascular damage

Vascular dysfunction can be found very early in SSc and consists of apoptosis of microvascular endothelial cells induced by anti-endothelial antibodies, entailing the subsequent inflammatory and finally fibrotic complications [9]. Increased levels of free oxygen radicals, anti-endothelial antibodies and cytokines produced by activated lymphocytes contribute to this activation [3, 10, 11]. The endothelial injury in SSc results in decreased production of vasodilators such as nitric oxide (NO), prostacyclin (which also inhibits platelet aggregation) and increased levels of vasoconstrictors such as endothelin-1 (ET). The exposition of the subendothelium to the blood stream may induce platelet adhesion and intravascular thrombus formation [12–14]. Pathomorphologic alterations of the blood vessels in SSc include intimal proliferation, narrowing of the vessel lumen and reduced blood flow [3, 15]. More recently, it could be also shown that stem cells mobilize from the bone marrow into the peripheral blood, differentiate in circulating endothelial progenitors (EPCs), and home to site of ischemia to contribute to de novo vessel formation resulting in the therapeutic idea to use autologous EPCs as a tool for therapeutic vascularization and vascular repair [16].

Immune activation

Mononuclear cell infiltrates are found in affected skin and visceral organ in SSc. CD4+ T cells are the predominant cell type [3]. The activation of humoral immunity and the role of B-cells is demonstrated by the numerous autoantibodies detected in the serum of SSc patients [17]. Antinuclear autoantibodies (ANA) are present in >90–95% of SSc patients and facilitate early diagnosis of the disease [18–20]. Due to their high specificity and diagnostic potential some of them, i.e., anti-centromere antibodies (ACA), anti-DNA topoisomerase I (ATA I, topo I, anti-Scl-70) and anti-RNA polymerase III were recently included in the updated 2013 American College of Rheumatology/European League against Rheumatism (ACR/EULAR) SSc classification criteria [21, 22].

ATA I was initially named anti-Scl-70 as an antibody reacting with a 70 kDa protein and was later recognized that the full length autoantigen was DNA topoisomerase I (topo I) and the 70 kDa protein is a breakdown product of a full-length 100 kDa protein. Although the terminology of anti-Scl-70 is still used, the more accurate nomenclature is ATA I antibody [23–25]. Their frequency in SSc patients is 20–30% [26, 27]. ATAs have been associated with diffuse cutaneous subset of the disease and presence of interstitial pulmonary fibrosis [28, 29]. Disappearance of ATAs in the disease course has been suggested to predict a more favorable outcome [30]. They are suggested to be a marker of disease activity [31].

ACA were initially described when HEp-2 cells were used as the substrate for ANA detection via indirect immunofluorescence, resulting in the characteristic speckled staining pattern [18]. Centromeres are differentiated chromosomal domains that specify the mitotic behavior of the chromosomes [32]. A number of centromere proteins (CENP) are recognized, the major targets being CENP-A, CENP-B, CENP-C, etc. [33]. It has been found that ELISA test using CENP-B autoantigen shows 100% correlation with anti-centromere staining by indirect immunofluorescence. Thus, antibody against CENP-B antigen is regarded diagnostic for the presence of ACA [34]. ATAs have been associated with diffuse cutaneous involvement and ACA with limited cutaneous form of SSc. However, autoantibody profiles do not completely predict the clinical form of the disease [29]. Thus, ACA could be found in diffuse cutaneous SSc and vice versa [2, 24, 29]. ACA and ATAs are almost always mutually exclusive and are present simultaneously in <1% of SSc patients [35, 36]. The frequency of ACA is 20–30% [28, 29, 37]. ACA positive SSc patients show lower frequency of pulmonary fibrosis, musculoskeletal and myocardial involvement vs ATAs positive scleroderma patients but higher frequency of pulmonary arterial hypertension without fibrosis due to vascular involvement [29].

Anti-RNA polymerase III antibodies are detected in approximately 20% of patients with SSc [24, 38]. Their presence is associated with diffuse cutaneous involvement [38, 39]. Of note, a negative association between the presence of anti-RNA polymerase III antibodies and ACA and ATAs has been observed [38, 40]. Their presence has been suggested to be a predictive risk factor for the development of renal crisis [39, 41], but a negative association with the development of interstitial fibrosis has also been observed. Survival in SSc patients positive for anti-polymerase III antibodies is better than those with ATAs as renal crisis is now more easily treated than pulmonary fibrosis [28]. Autoantibodies to RNA polymerase I and III often co-exist and are highly specific for SSc [24, 42]. RNA polymerase

I is localized in nucleoli, while RNA polymerase III is localized in nuclei [43]. It has been found that indirect immunofluorescence is not a sensitive method for anti-RNA polymerase III antibodies, as the punctate nucleolar staining associated with anti-RNA polymerase I antibodies may be covered by the coexisting nuclear speckled staining of anti-RNA polymerase II and III antibodies. Thus, ELISA testing for anti-RNA polymerase III is suggested to be performed, regardless of the ANA staining pattern [43a].

Other SSc-specific autoantibodies that are found in a lower proportion of SSc patients are anti-Th/To and anti-U3-RNP/fibrillarin, both producing a nucleolar staining at indirect immunofluorescence [24, 44]. Anti-Th/To autoantibodies are directed against subunits of the ribonuclease mitochondrial RNA processing and ribonuclease P complexes [45]. They are positive in approximately 5% of scleroderma patients and their presence is associated with limited cutaneous involvement [28]. The major difference between SSc patients with limited cutaneous involvement who are ACA or Th/To positive is that, in cases with Th/To antibody positivity the patients develop both interstitial fibrotic and vascular pulmonary disease [28, 46]. Anti-U3-RNP antibody is directed against fibrillarin that is a 34-kDa basic protein. It is detected in <7% of cases [24].

Other autoantibodies that are less specific for SSc and could be detected both in SSc and in other systemic autoimmune diseases include anti-U1-RNP, anti-PM-Scl, anti- Ku, anti-Ro/SS-A, La/SS-B. These autoantibodies are associated with SSc-overlap syndromes [24, 33].

Of note, some autoantibodies against non-nuclear autoantigens, i.e., anti-platelet-derived growth factor (PDGF) receptor antibodies, anti-endothelial cells antibodies, anti-fibroblast, antibodies against matrix metalloproteinases (MMPs), anti-angiotensin type 1 receptor, and anti-endothelin-1 type A receptor have been suggested to inherit pathogenic properties [24, 33, 47, 48].

More recently, autoantibodies against distinct signaling receptors have been detected. Of these, G protein-coupled receptors (GPCRs) comprise the largest and most diverse family of integral membrane proteins that participate in different physiological processes such as the regulation of distinct functions of the immune systems. As experimental data show that SSc-associated functional autoantibodies are able to bind GPCRs to trigger or block intracellular signaling pathways, they could result in agonistic or antagonistic SSc modulating effects [49]. Taken together, aside the known "anti"bodies, SSc is one of the few entities in rheumatology, in which agonistic antibodies play a significant role [50].

Fibrosis

In SSc, transforming growth factor β (TGF β) is one of the key cytokines involved in tissue fibrosis [48, 51, 52]. Human B lymphocytes, among other cells, can be a source of TGF-β and express receptors for this cytokine [48, 53]. TGF-β promotes collagen deposition, and inhibits collagen degradation by decreasing matrix metalloproteinases and increasing the expression of tissue inhibitors of metalloproteinases [51, 54]. In addition, a cross-talk between TGF-β and PDGF has been found with positive up-regulation of PDGF α receptors after stimulation of scleroderma fibroblasts by TGF-β [55]. Connective tissue growth factor (CTDF), which is produced by fibroblasts, vascular smooth muscle cells and endothelial cells in response to TGF-β stimulation, is another potentially important mediator of fibrosis [3]. A body of researchers have continuously added significant data around the dysregulation and aberrant regulation of growth factors [56] including factors of the innate immune system such as Toll like receptors [57], distinct proinflammatory cytokines including IL-17 [58], intracellular signaling molecules [210] and novel cellular aspects addressing platelets and myofibroblasts [59, 60].

Genetic markers and epigenetics

It has been shown that risk of SSc is increased among first-degree relatives of patients compared to the general population. In a study of 703 families in the United States, including 11 multiplex SSc families, the relative risk for first-degree relatives to develop SSc was approximately 13 with a 1.6% recurrence rate, compared to 0.026% in the general population [61]. The risk ratio for siblings was approximately 15 (ranging from 10 to 27 across cohorts). However, twin studies revealed that fewer than 5% of mono- and dizygotic twins are concordant for SSc. On the other hand, when gene expression was compared by high-throughput microarray techniques, the pattern of dermal fibroblasts in unaffected monozygotic twins was not significantly different from that of SSc patients. In this study, additional functional experiments revealed that healthy fibroblasts that were incubated with serum from an SSc-affected patient or with serum of the unaffected monozygotic twin developed a typical SSc pattern with an increased expression of collagen1 A2, SPARC (secreted protein, acidic and rich in cysteine, osteonectin) and CTGF [62].

At present, no whole genome or proteome analysis has revealed an unique pattern in SSc patients, tissues or cells [63, 64], although several approaches have found distinct polymorphisms and disease associations [65–67].

Knowledge from genetics addressing target genes and the respective protein groups such as the topoisomerase I complex will likely provide novel ideas in this field [68]. In addition, cDNA array techniques were successfully used to identify the gene expression of disease-related cell types such as endothelial cells. When compared to normal skin endothelial cells, it has been demonstrated that microvascular endothelial cells of patients with SSc show abnormalities in a variety of genes that are able to account for defective angiogenesis [69]. In another approach, cDNA arrays were used to compare the gene expression profiles of peripheral blood mononuclear cells of patients with early SSc, which revealed a distinct up-regulation of 18 interferon-inducible genes, selectins and integrins, supporting the idea of an infectious trigger in the early phases of the disease [70]. CD8-positive lung T cells, on the other hand, resulted in two distinct gene cluster groups, with one showing a type II T cell activation in combination with profibrotic factors and MMPs [71].

Regarding genetic markers, a limited but increasing number of studies have examined the presence and pattern of single nucleotide polymorphisms (SNPs) in SSc [72]. Also, as SSc is not inherited in a Mendelian manner, experimental and clinical research has focused on genetic alterations in numerous genes known to be operative in SSc path physiology. This has revealed interesting aspects, especially with regard to growth factors, matrix-related molecules and inflammation markers such as TGF, fibrillin-1, TNF-α,TNF-α receptor type II, TNF-β, and IL-1 (Table 1).

The initial analysis of the TGF gene revealed no strong genetic abnormalities, which were also not found for PDGF [73]. However, a detailed analysis at codon 10 showed that SSc patients are prone to high TGF synthesis, irrespective of limited or diffuse disease [74]. Interestingly, adenoviral gene transfer of TGF receptor type I (TGF-RI) into fibroblasts in combination with cDNA array revealed a distinct TGF-RI-induced profibrotic phenotype with up-regulation of collagen type I and CTGF [75]. CTGF is another promising candidate for detailed gene and polymorphism analysis [76].

Further research on matrix metabolism-regulating genes showed an association of the stromelysin promoter with SSc [77], but no association of the MMP-1 promoter with the disease [78].

Tumor necrosis factor (TNF) is one of the driving proinflammatory molecules in several autoimmune diseases. Based on knowledge of the inflamed initial stages in SSc pathophysiology, numerous groups have examined the presence of TNF gene polymorphisms in SSc patients. In the first intron at locus 252,

TABLE 1 Loci of interest of genetic abnormalities of different molecules involved in SSc pathophysiology

Molecule/gene	Loci
TGF β	Codon 10, −1133 bp in promoter
Fibrillin-1	SNP in 5′-untranslated region, CT insertion in exon A
TNF α	TNF α 13 microsatellite, −863A allele
TNIP1	SNP at the 5q33 locus
IL-1	CTG/CTG diplotype SNP, – 889 allele polymorphism
CCL2	SNP
ACE	Insertions/deletions in chromosome 17
eNOS	Polymorphisms

the two homozygous genotypes of TNF were significantly associated in Japanese SSc patients. In the TNF gene itself, however, statistical power was not sufficient to prove a similar association [79]. Notably, the rare GG genotype in exon 6 of the TNF receptor type II was found to be more frequent in another European diffuse SSc cohort [80].

The three strongly associated SNPs at the 5q33 locus are located within the TNFAIP3 interacting protein 1 (TNIP1) gene. TNIP1 is a very interesting new candidate gene for SSc. The protein encoded by this gene exerts a negative regulation of NF-kappaB (NF-κB). The reduced inhibition of NF-κB favors inflammatory/immune responses and potentially contributes to the overproduction of extra-cellular matrix. Subsequently, in a recent large genome-wide association study of SSc, two new SSc-risk loci were identified, psoriasis susceptibility 1 candidate 1 (PSORS1C1) and TNIP1 [66].

Genomic evaluation of the interleukin-1 (IL-1) gene revealed distinct genetic aberrations in Japanese SSc patients, and SNP analyses showed that a distinct CTG/CTG diplotype associated strongly with the development of interstitial lung disease in those patients [81]. A study in Czech patients revealed also a polymorphism in the IL-1A gene at position 889 [82].

A SNP of the gene encoding monocyte chemoattractant protein-1 (CCL2) has previously been suggested to be involved in susceptibility to SSc, but in a recent study that gene was not implicated in susceptibility nor in the SSc phenotype [83].

With regard to cellular immunology, CD19-positive B cells appear to bear a 499G-T polymorphism in the CD19 coding region in SSc patients, which was also associated with susceptibility to the disease [84]. In Korean patients, transporters associated with antigen processing-1 and -2 polymorphisms were found to be independent of other HLADR associations [85], suggesting different roles of genomic alterations in antigen-presenting cells in SSc [86]. However, distinct HLA alleles appear to be linked directly to SSc subtypes, as it could be shown that in male SSc patients, HLA class II allele DQA1*0501 was associated with diffuse but not limited disease [87]. In that population, maternal HLA compatibility was not a significant risk factor for development of the disease [88]. A recent study found that DRB1*0407 and *1304 are independent risk factors for the development of scleroderma renal crisis [89].

Another study has confirmed the presence of an association between the TNFSF 4 gene promoter polymorphism and SSc genetic susceptibility, especially in patients with the limited form of the disease who are also positive for ACA. The TNFSF4 gene encodes OX40L, which is expressed on activated antigen-presenting cells and endothelial cells in acute inflammation. Furthermore, it enhances B cell proliferation and differentiation and promotes proliferation and survival of T cells [90].

Alterations in genes regulating microvasculature development, intravascular thrombosis, dysregulated fibrinolysis and perivascular fibrosis have also been addressed by genomic analyses. A study in an Italian population showed that patients with SSc appear to have a higher prevalence of angiotensin-converting enzyme (ACE) insertion/deletions on chromosome 17 and polymorphisms within the endothelial NO synthase gene [91]. In contrast, other groups showed that in a French Caucasian population eNOS polymorphisms neither influenced the course of SSc nor did they enhance susceptibility [92, 93]. A recent investigation performed in Korea also could not find a difference in the frequencies of all ACE insertion/deletion genotypes between patients and controls, nor between diffuse and limited SSc patients [94]. However, identifying distinct disease-specific genetic abnormalities remains difficult [95].

As in other autoimmune diseases, an increasing body of data illustrates the potential role of epigenetic factors such as DNA methylation, histone modifications, and microRNAs, in the initiation and perpetuation of the disease [96]. For example, it could be shown that profibrotic cytokines can trigger epigenetic changes which contribute to the persistently activated fibroblast phenotype. In addition, several epigenetic alterations have been described in SSc and associated with various pathogenic aspects of the disease, including particularly aberrant fibroblast activation, tissuc fibrosis, vascular dysfunction and inflammation.

Clinical presentation

Initial symptoms

RP is the initial symptom in the majority of patients with the limited form of the disease, while in those with the diffuse form, the classic initial complaints are swelling of the hands, skin thickening, and sometimes arthritis. Occasionally, the earliest symptom is visceral involvement, such as gastroesophageal involvement (dysphagia, heartburn) or dyspnea [15].

Raynaud's phenomenon

RP is one of the most frequent SSc symptoms. It occurs in the course of SSc in >95% of patients. Usually it is the initial symptom, which can precede the other features of the disease by years. RP affects hands, feet and, less frequently—the tip of the nose, the earlobes, and the tongue. RP in SSc is severe, and often presents with digital ulcers [15, 97]. Small areas of ischemic necrosis or ulceration of the fingertip are a frequent finding, often leaving pitting scars. Digital necrosis of the terminal portions also develops in SSc patients (Fig. 1). The capillaroscopic pattern in SSc is the most specific among rheumatic diseases, and is characterized by the presence of dilated and giant capillaries, hemorrhages, avascular areas, and neoangiogenesis. There are three phases of capillaroscopic changes in SSc: an early phase (appearance of few dilated and/or giant capillaries and few hemorrhages), an active phase (a high number of giant capillaries and hemorrhages, moderate loss of capillaries, slight derangement and diffuse pericapillary oedema), and a late phase (extensive avascular areas, appearance of ramified and bushy capillaries) [98–106] (Fig. 2).

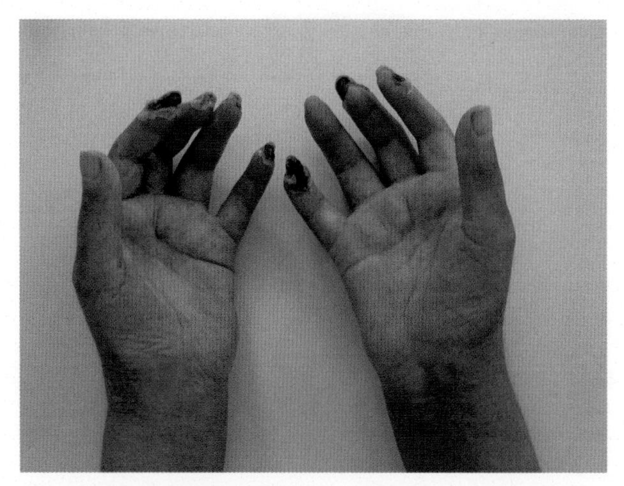

FIG. 1 Female patient with limited SSc and digital necrosis.

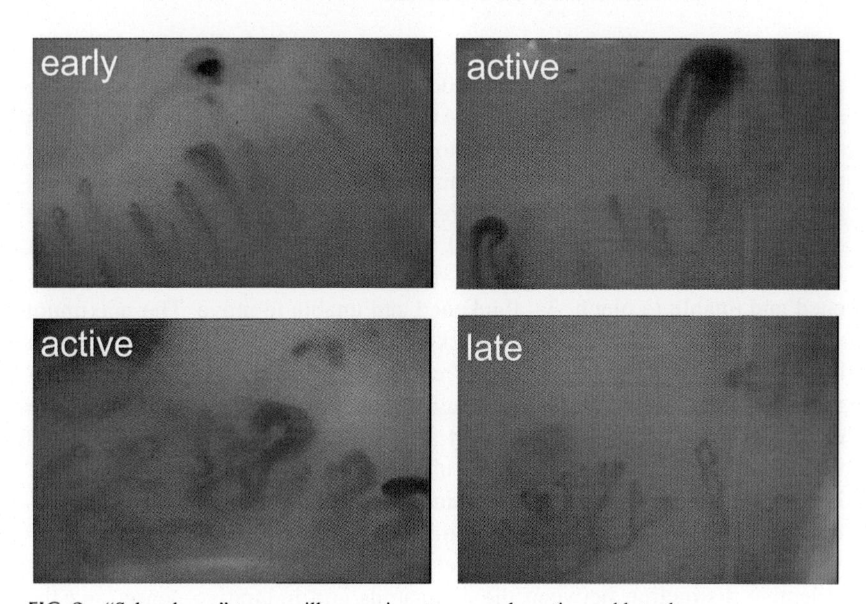

FIG. 2 "Scleroderma"-type capillaroscopic pattern—early, active and late phase.

Skin

The earliest skin changes in SSc are tight, puffy fingers, especially in the morning (the edematous phase). Edema may last indefinitely (in patients with limited SSc) or may be replaced gradually by thickening and tightening of the skin (the indurative phase) after several weeks or months. According to the extent of skin changes, SSc is divided into two distinct subsets—limited and diffuse. Facial skin changes may result in the characteristic expressionless

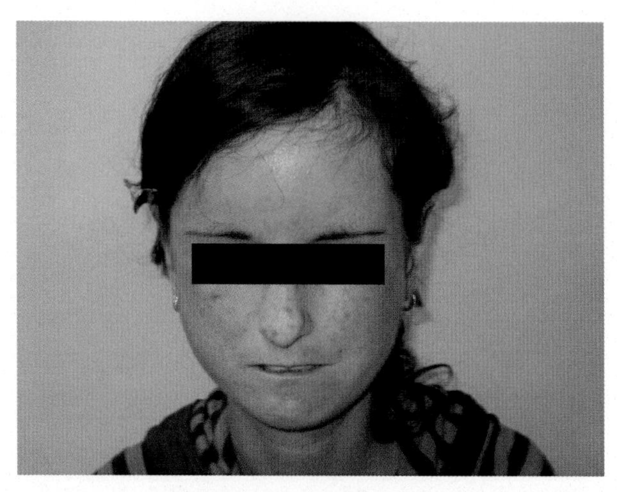

FIG. 3 Face of a 24-year-old female patient with diffuse SSc with long duration of the disease, since childhood. Note skin thickening and microstomia.

appearance, with thin, tightly pursed lips, vertical folds around the mouth and reduced oral aperture (Fig. 3). After several years, the dermis tends to soften, and in many cases it reverts to normal and become thinner than normal (the atrophic phase) [15]. The modified Rodnan skin score (mRSS) is a simple, inexpensive, reliable and reproducible method for the assessment of skin thickening in SSc being still the gold standard for skin assessment in SSc. It includes four grades: 0—normal, 1—thickened skin, 2—thickened and unable to pinch, 3—thickened and unable to move. The maximum skin score is 51 [107, 108]. Skin thickness score correlates closely with skin biopsy thickness [107, 109]. High-frequency ultrasound also inherits a potential for objective and quantitative assessment of skin thickness and skin echogenicity in SSc [110]. A recent systematic literature review evaluated the current evidence for the role of ultrasound and ultrasound elastography in SSc. Based on the available literature data, it has been concluded that ultrasound may be an additional option for quantitative assessment of skin involvement in SSc including earlier detection of subclinical skin involvement, objective distinction between the edematous and indurative phase, and the progression of disease. However, future research is necessary to clarify the existing uncertainties including definition and standardization of the number and exact location of skin sites to be examined; definition of outcomes of interest: thickness, echogenicity, stiffness, and vascularity [111].

Calcinosis

Patients with limited cutaneous SSc or late-stage diffuse cutaneous form of the disease commonly develop intra- or subcutaneous calcifications composed of hydroxyapatite. They are located mainly in the digital pads and periarticular tissues, along the extensor surface of the forearms, in the olecranon bursae, prepatellar areas, and buttocks. Calcinosis may be complicated by ulceration of the overlying skin and secondary bacterial infection [15, 112]. Calcinosis is also one of the unsolved and hitherto irreversible problems in management of SSc [113].

Lung involvement

Pulmonary involvement occurs in >70% of patients and is the leading cause of morbidity and mortality in patients with SSc [4]. Alveolitis, membrane thickening and/or modification of microvascular function are the characteristic features of lung involvement [3, 114]. During progression of the disease, these changes may lead to interstitial lung disease and to pulmonary arterial hypertension (PAH) [115] and pulmonary fibrosis. Chest radiography has low sensitivity for detection of early lung disease, and may be negative in >10% of symptomatic patients in the early stage of disease [116, 117] (Fig. 4). At present, chest high-resolution computed tomography is the preferred noninvasive standard technique for diagnosis of interstitial lung disease in SSc [118, 119]. The usefulness of bronchoalveolar lavage to define alveolitis in SSc has been questioned [120]. Neutrophilia has been found to be associated with early mortality [121].

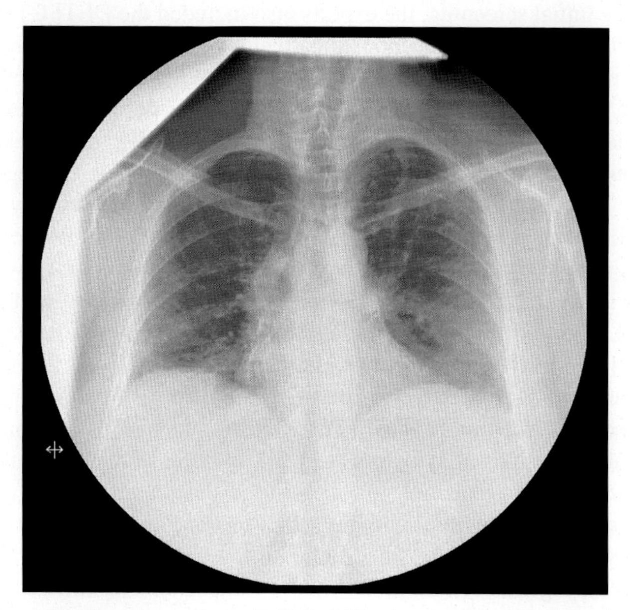

FIG. 4 X-ray of the lungs in a patient with limited SSc, demonstrating bilateral pulmonary fibrosis. Note interstitial thickening in a reticular pattern.

Pulmonary arterial hypertension

The prevalence of PAH in SSc is 10–12% [122], varying in the different reports between 4.9% and 26.7% [123]. In SSc, two pathogenic pathways for development of the syndrome are recognized: PAH associated with pulmonary fibrosis and PAH without pulmonary fibrosis. Pulmonary fibrosis can be found in more than one-third of SSc patients with either the diffuse or limited cutaneous form of the disease. PAH in SSc patients without pulmonary fibrosis is a severe complication due to narrowing or occlusion of small pulmonary arteries caused by smooth muscle hypertrophy, intimal hyperplasia, inflammation, thrombosis in situ. Isolated PAH, in the absence of lung fibrosis, was found to be more frequent in the limited form of SSc (45%) than in the diffuse form of the disease (26%) [29]. In these cases, the rate of progression of dyspnea from normal exercise tolerance to oxygen dependency is about 6–12 months with a subsequent mean duration of survival of 2 years. In contrast, SSc patients with PAH in the context of interstitial lung disease develop a similar degree of disability, but progress more slowly for a period of more than two and up to 10 years [15, 115, 124, 125]. PAH is still a leading cause for increased mortality in SSc, and due to its high prevalence, recommendations for screening and early detection have been developed. Initial screening in all SSc patients includes transthoracic echocardiography (TTE), pulmonary function tests (PFT) with single breath diffusion carbon monoxide (DLCO) and measurement of NT-pro-BNP. All three investigations should be also performed if new signs and symptoms appear. TTE and PFT with diffusing capacity (DLCO) are repeated on annual basis. In the initial screening, the experts also included the DETECT algorithm (DETECTion of PAH in SSc) in SSc patients with disease duration >3 years from the time of the first non-Raynaud's symptom and a DLCO <60% [126]. According to the echocardiographic findings, PAH is defined as mean pulmonary arterial pressure (PAP) > 25 mmHg at rest, >30 mmHg during exercise, or systolic pulmonary pressure >40 mmHg. Echocardiographic findings suggestive of the complication are elevated tricuspid regurgitation velocity jet above 2.8 m/s, dilated right ventricule or dilated atrium. As the mean PAP cannot easily be determined by echocardiography, an estimated systolic PAP > 35 mmHg and/or an increased tricuspid velocity are used as indicators of probable PAH [115, 123]. Right heart catheterization (RHC) is mandatory for the diagnosis of PAH [126]. On resting RHC, PAH is defined a mean PAP of ≥25 mmHg with a pulmonary capillary wedge pressure of ≤15 mmHg with pulmonary vascular resistance of >3 Wood units [115, 127]. Referring to RHC is recommended in SSc patients with signs or symptoms of PAH and tricuspid regurgitation velocity jet of 2.5–2.8 m/s (equating to a trans-tricuspid gradient 25–32 mmHg) as well as in all patients (with or without symptoms of PAH) with tricuspid regurgitation velocity jet >2.8 m/s. In addition, all patients with right atrial or right ventricular enlargement (right atrium major dimension >53 mm and right ventricle mid cavity dimension >35 mm), irrespective of tricuspid regurgitation jet (including non-measurable or <2.5 m/s) should also be referred to RHC. RHC is also recommended in patients symptomatic for PAH and a forced vital capacity %/DLCO % ratio > 1.6 and/or a DLCO <60% if TTE does not reveal overt systolic dysfunction, higher than grade I diastolic dysfunction or mild mitral or aortic valve disease [126].

The DETECT algorithm consists of 2-step approach. "Step 1" assessment determines referral to echocardiography and "step 2" assessment guides referral to RHC. "Step 1" includes six variables, i.e., forced vital capacity/DLCO (% predicted), presence of current/past telangiectasias, serum ACA, serum NT-proBNP, serum urate and right-axis deviation on electrocardiogram. In "step 2" assessment, tricuspid regurgitant jet velocity and right atrium area are included. It has been postulated that using tricuspid regurgitant jet velocity as single marker, 20% of PAH patients would fail to be diagnosed when using a PAH cutoff threshold of ≥2.5 m/s, 36% when using a threshold of >2.8 m/s, and 63% when using a threshold of >3.4 m/s [128].

Cardiac involvement

Subclinical fibrosis of both the myocardium and the conducting system with "patchy" distribution has been described in SSc [129, 130]. Unfortunately, cardiac involvement in SSc is most commonly subclinical especially in the early stages of the disease. The knowledge about its pathophysiologic basis and the diagnostic methods for its detection, including high-end functional MRI [131] are being constantly developed and improved [132]. Malignant ventricular arrhythmias are dreadful complications, frequently requiring a pacemaker.

Gastrointestinal involvement

Gastrointestinal abnormalities in SSc involve motility dysfunction and mucosal damage in different segments of the alimentary tract. Myogenic and neurogenic factors are involved in the pathogenesis of these abnormalities, and in the past years, data on abnormalities of the microbiome have been added [133].

Although any part of the gastrointestinal tract can be involved, esophageal disease occurs in nearly all patients with SSc. Common esophageal manifestations in SSc include esophageal dysmotility, which occurs in 75–90% of SSc patients, gastroesophageal reflux, Barrett's esophagus, adenocarcinoma, and infectious esophagitis [134]. Stomach involvement occurs in 10–75% of SSc patients, small bowel involvement in 30–70%, colon involvement—in 10–50% and ano-rectal involvement in 50–70% [134, 135]. Watermelon stomach is a rare finding, which presents with gastric antral venous ectasia. It may result in persistent bleeding in patients with SSc [136]. Esophageal manometry, pH-monitoring tests and endoscopy are the technical methods for evaluation of the upper gastrointestinal tract disorders (those of the esophagus and the stomach). The upper intestine can be evaluated with deep duodenal biopsy, transit time, H2-breath tests, D-xylose tests, lactose tolerance tests, beta-carotin, and endomysium- and transglutaminase autoantibodies. The lower intestine can be evaluated with colonoscopy, computed tomography scans, and rectoscopy [137–142].

Musculoskeletal involvement

Joint symptoms have been noted in 12–66% of patients at the time of diagnosis and in 24–97% of patients at a given time during the course of their illness. Histological evidence of synovitis has been found in up to 66% of synovial biopsies from patients with SSc, but, clinically, arthralgias have been considered to occur more frequently than true arthritis. Radiographic abnormalities, which are detected in 46% of SSc cases, are juxta-articular demineralization, terminal phalangeal tuft resorption, articular calcification, joint erosions, and joint space narrowing [143, 144] (Fig. 5). Approximately 20% of SSc patients have a subtle myopathic process with weakness, noted by the examining physician, mild or no serum muscle enzyme elevation, perimysial and endomysial fibrosis on histologic examination. A minority of patients develop more pronounced proximal muscle weakness classified as SSc overlap with polymyositis [15, 145].

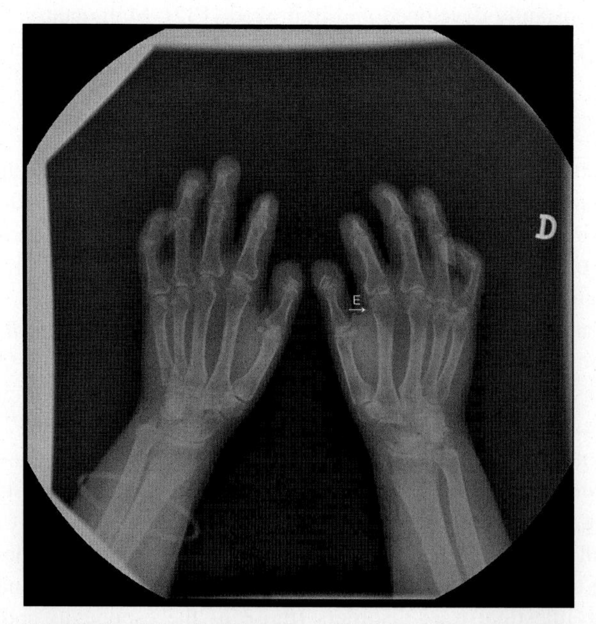

FIG. 5 X-ray of the hands of a female SSc patient with limited SSc, demonstrating acroosteolysis of all of the fingers bilaterally and erosion of the distal surface of the second right metacarpal bone.

Kidney involvement

Scleroderma renal crisis (SRC) is the most acute and life-threatening manifestation of SSc. It occurs most often in patients with SSc with diffuse cutaneous involvement. Until the availability of ACE inhibitors in 1979, it was almost always associated with malignant hypertension, rapidly progressive renal failure and early mortality [146].

Later, having learnt that especially high doses of steroids in the early stages of SSc contributed to SRC-dependent mortality, its occurrence has decreased significantly. However, kidney fibrosis is still an unresolved problem [147].

Development of classification criteria for SSc

ACR criteria (1980) have been used for years for the diagnosis of SSc [148]. Presence of the major criterion (skin thickening, proximal of metacarpophalangeal joints) or two minor criteria (sclerodactyly, fingertip pitting scars, bibasilar pulmonary fibrosis) were required for the diagnosis. In the further course, Nagy et al. observed a high specificity of capillaroscopy for early diagnosis of SSc in 447 patients with connective tissue diseases. Abnormal "scleroderma-type" capillaroscopic pattern was found in patients who did not fulfill the ACR criteria—those patients presented with sclerodactyly, telangiectasia, subcutaneous calcinosis, esophageal dysmotility, and other symptoms [149]. Of note, the ACR criteria (1980) have been established prior to the discovery of SSc-specific autoantibodies and capillaroscopic changes, which both have been shown to improve the early diagnosis of the disease [150, 151]. Thus, Le Roy and Medsger proposed patients with objectively documented RP and abnormal nailfold capillaroscopic changes or positive, SSc-selective autoantibodies (ACA, anti-topoisomerase I, anti-fibrillarin, anti-PM-Scl, anti-fibrillin or anti-RNA polymerase I or III in a titer of 1:100 or higher) to be diagnosed as early SSc even in the absence of other features of the disease. If RP is only subjectively reported both "scleroderma" type capillaroscopic pattern and SSc-selective autoantibodies are required [152]. More recently, the multicenter project VEDOSS (Very Early Diagnosis of Systemic Sclerosis) added the key data that resulted in the new classification criteria for SSc (EULAR/ACR, 2013), which are based on a number of validated criteria, e.g., disease-specific autoantibodies and capillaroscopic changes, RP, telangiectasias, PAH, and interstitial lung disease. These new EULAR/ACR classification criteria for SSc specifically aim at the detection of SSc in the early stages [21, 22] (Table 2).

Prognostic markers and systemic sclerosis activity score

The difficulty of measuring disease activity is an important barrier in the study of SSc. The *Valentini Disease Activity Index* appears to be simple and easy to use, and is now increasingly being used in research settings (Table 3). The disease is considered active if the sum of the scores of detected items is ≥ 3 [153, 154].

TABLE 2 EULAR/ACR classification criteria for systemic sclerosis (2013), [21, 22].

Criteria	Score
1. Skin thickening of the fingers of both hands extending proximal to the metacarpophalangeal joints	9
2. Skin thickening of the fingers *(only count the higher score)*	
– Puffy fingers	2
– Sclerodactyly of the fingers (distal to the metacarpophalangeal joints but proximal to the proximal interphalangeal joints)	4
3. Fingertip lesions *(only count the higher score)*	
– Digital tip ulcers	2
– Fingertip pitting scars	3
4. Telangiectasia	2
5. Abnormal nailfold capillaries	2
6. Pulmonary arterial hypertension and/or interstitial lung disease *(maximum score is 2)*	
– Pulmonary arterial hypertension	2
– Interstitial lung disease	2
7. Raynaud's phenomenon	3
8. SSc-related autoantibodies *(maximum score is 3)*	3
– anticentromere	
– anti-topoisomerase I (anti-Scl-70)	
– anti-RNA polymerase III	

The total score is determined by adding the maximum weight (score) in each category. Patients with a total score of ≥9 are classified as having definite SSc.

Treatment

There have been substantial advances in evidence-based treatment of SSc as an entity, and specifically in the disease complications, which are summarized in the updated EULAR recommendations [155].

Immunosuppressive and antifibrotic therapies

Counteracting one of the key features of SSc, fibrosis, is one of the most challenging goals when treating the affected patients and at present, still no reliable antifibrotic agent exists [156].

The first approaches included D-penicillamine, which has been introduced decades ago. Owing to the unfavorable outcome of controlled trials [157–159], D-penicillamine is not recommended for treatment of SSc by EULAR [155]. Treatment with methotrexate in patients with early (<3 years) diffuse SSc for 1 year led to significantly improved mRSS compared with placebo [160] and can be used in early active disease.

TABLE 3 Valentini disease activity index for SSc [153].

Parameter	Score
1. Modified Rodnan skin score > 14	1
2. Scleredema	0.5
3. Change in skin symptoms in the last month[a]	2
4. Digital necrosis	0.5
5. Change in vascular symptoms in the last month[a]	0.5
6. Arthritis	0.5
7. DLco <80% of the predicted value	0.5
8. Change in cardiopulmonary symptoms[a]	2
9. ESR >30 mm for the first hour (Westergren method)	1.5
10.Hypocomplementemia (C3 and/or C4)	1
Total disease activity index score	10

[a] *Assessed by the patient.*

Although IL-2-dependent pathways are active in SSc, the nephrotoxicity of cyclosporin limits its use in SSc patients [161, 162]. The production of fibrogenic cytokines, such as IL-4, IL-6, IL-17, and TGF-β1, was attenuated by rapamycin, but convincing clinical studies are lacking [163]. A randomized controlled trial (RCT) comparing azathioprine and placebo after cyclophosphamide induction therapy found only marginal benefits [164, 165].

Treatment with mycophenolate mofetil has demonstrated similar efficacy to oral cyclophosphamide on lung and skin fibrosis [166, 167]

Specific management of the different disease manifestations

Raynaud's phenomenon

Patient education in avoiding exposure to cold and emotional stress, in wearing warm clothes, and in smoking cessation is very important [12, 168, 169]. Dihydropyridine-type calcium-channel blockers (CCBs) such as nifedipine (3 × 1–2 tablets 5–10 mg, or twice daily slow-release tablets 20 mg), felodipine (2.5–10 mg bid), amlodipine (5–10 mg daily) are usually the first choice in RP [170]. The dihydropyridines are vasodilators with direct effect on vascular smooth muscles that are indicated for the treatment of arterial hypertension and some forms of ischemic heart disease (stable angina, vasospastic angina) [171]. However, their administration in RP is off-label. Side effects of CCBs include hypotension, tachycardia, flushing, dizziness, headache, ankle edema, constipation, etc. Slow-release forms are better tolerated and most commonly used in everyday practice [169]. Diltiazem (30–120 mg tid) also shows a good therapeutic effect in RP (Leighton, 2001) [172].

ACE inhibitors are potent vasodilators, but the potential benefit of renin-angiotensin blockade for vascular disease in organs other than the kidney is unknown [168, 173]. The direct vasodilator—prazosin and the oral serotonin antagonist—ketanserin have also demonstrated therapeutic efficacy in RP [172].

NO is a potent vasodilator and an inhibitor of platelet activation and vascular smooth muscle proliferation. Synthesis of NO is regulated by the family of NO synthases, and its effect is mediated via cyclic guanosine monophosphate (cGMP). The intracellular concentration of cGMP is regulated by phosphodiesterases (PDE), which rapidly degrade cGMP in vivo. Sildenafil, tadalafil, vardenafil are selective inhibitors of the PDE-5 isoform that is abundant in the vascular smooth muscle cells and lung tissue. They lead to improved blood flow in patients with severe secondary RP [174–179]. A meta-analysis of six RCTs (two with sildenafil at a dose of 50 mg bid and modified released sildenafil 200 mg daily; three with tadalafil at a dose of 20 mg daily or 20 mg on alternate days as add-on therapy and one with vardenafil at a dose of 10 mg bid) demonstrated that PDE-5 inhibitors improve Raynaud's Condition Score, frequency and duration of RP attacks [180]. Side effects during treatment with PDE-5 inhibitors include headache, back pain, fluid retention, vasomotor changes, nasal stuffiness, flushing, dizziness, palpitations [181–183].

Pentoxifylline inhibits phosphodiesterase and elevates cyclic adenosine monophosphate levels. It improves blood flow mainly via increased red blood cell deformability [184, 185]. It is administered orally or intravenously at a maximal daily dose of 1200 mg (400 mg tid orally, 600 mg bid orally, 100–600 mg intravenously once or twice daily or combined oral and intravenous treatment) [169].

Intravenous iloprost is recommended for the treatment of severe SSc-related RP after failure of oral therapy as outlined in the EULAR recommendations [155].

Fluoxetine (a serotonin-specific reuptake inhibitor and antidepressant) is also suggested as an option in SSc-related RP despite limited evidence for its efficacy [155]. Fluoxetine (at a dose of 20 mg daily) has been studied in a small trial that included 26 patients with primary RP and 27 with SSc-related RP and demonstrated better efficacy in comparison with nifedipine (at a dose of 40 mg daily). Side effects of fluoxetine include apathy, lethargy, and impaired concentration [186]. Fluoxetine is suggested by EULAR experts as an alternative for treatment of SSc-related RP, especially in cases who do not tolerate or fail to respond to vasodilators [155].

Taken together, the current EULAR recommendations (2017) for the management of RP in SSc include dihydropyridine-type CCBs, PDE-5 inhibitors, intravenous iloprost and fluoxetine [155].

Improvement of RP, with significant reductions in the Raynaud's Condition Score and patient assessment by visual analog scale, has been observed during atorvastatin treatment of SSc patients at the dose 10 mg daily for 24 months. Thus, statins may be beneficial in treating vascular manifestations of SSc [187].

Digital ulcers

Intravenous iloprost (a prostacyclin analogue) has demonstrated efficacy in SSc patients with severe RP and digital ulcers. It is administered as intravenous infusion for 6–8 h, at the dose of 0.5–3 ng/kg/min. The mean duration of the therapeutic course is 5–10 days. These effects may be maintained by a 1-day infusion at intervals of several weeks according to the therapeutic effect [188–190].

A meta-analysis of RCTs that evaluated several selective PDE-5 inhibitors (sildenafil 50 mg bid, modified-release sildenafil 100 mg/daily increased up to 200 mg/daily and tadalafil 20 mg on alternate days) in SSc patients demonstrated that PDE-5 inhibitors can heal digital ulcers in SSc patients [191]. In one small RCT, it has been demonstrated that tadalafil may prevent development of new digital ulcers in SSc [183]. Thus, PDE-5 inhibitors are recommended by EULAR experts for treatment of digital ulcers in SSc patients [155].

The dual endothelin-receptor antagonist (ERA)—bosentan (ET-A and ET-B receptors) has no proven efficacy in the treatment of active digital ulcers in SSc patients. However, in two high quality RCTs (the RAPIDS-1 and -2 studies) it has been observed that bosentan prevents development of digital ulcers in SSc patients, in particularly in those with multiple digital ulcers [192]. The effect of bosentan on the prevention of new digital ulcers in SSc has not been proven for other ERA (DUAL-1 and DUAL-2 study for macitentan) [155, 193].

Pulmonary arterial hypertension

It has been accepted that the therapeutic approach in severe PAH in the context of autoimmune disease should follow the same treatment strategy as in primary PAH patients. As a general recommendation, low-level aerobic exercise such as walking is recommended for PAH patients. Restriction of sodium consumption to 2400 mg per day in patients with right ventricular failure is also a common measure in these patients. Routine immunizations against influenza and pneumococcal pneumonia are advised to prevent severe infections. The 2009 American College of Cardiology guidelines recommend oxygen therapy in PAH patients with pulse oximetry saturation of <90% at rest or with exercise [115]. Abnormalities of the activated clotting system, impaired fibrinolysis, abnormal platelet function and histological evidence of microvascular thrombosis provide the rationale for anticoagulation therapy in these patients [115, 194, 195]. The 2009 ACC treatment guidelines recommend warfarin anticoagulation in patients with idiopathic PAH titrated to an international normalized ratio of 1.5–2.5. Treatment of right heart failure with digoxin, diuretics and oxygen therapy is thought to offer little more than palliation [123]. Epoprostenol improves exercise capacity and survival in SSc-related PAH [115, 196]. Both intravenous and subcutaneous treprostinil are possible options and produce similar hemodynamic effects compared to those of epoprostenol in patients with PAH. Prostaglandins for inhalation are also used in PAH and inhaled iloprost has been found to be effective therapy for patients with severe PAH. When administered via inhalation, its pulmonary vasodilative potency was similar to that of prostacyclin, but its effects lasted for 30–90 min, as compared with 15 min for prostracyclin [197]. Beraprost is the first oral prostacyclin analogue with vasodilative and antiplatelet action and a half-life of approximately 1 h [198, 199].

Bosentan is approved for treatment of PAH associated with SSc based on high quality RCTs in patients with WHO functional classes III and IV. Therapeutic benefit has been also demonstrated in mild PAH (WHO class II), (EARLY study) [200]. It is used at a dose of 62.5 mg twice daily for 4 weeks before titration up to 125 mg twice daily [115, 192, 201]. *Macitentan* is a novel dual ERA (ET-A and ET-B receptors) approved for the treatment of SSc-related PAH. It is administered at a dose of 10 mg daily [202]. *Ambrisentan* is a potent and selective inhibitor of ET-A receptors (ET-A:ET-B—4000:1). The half-life of the medication is 9–15 h, which allows once-daily dosing (5–10 mg per day). Effectiveness of ambrisentan has been evaluated in two phase III RCTs (ARIES-1, -2) in patients with PAH. Ambrisentan improves exercise capacity with a dose-dependent effect [203]. Side effects during treatment with ERA are headache, elevation of liver enzymes, anemia, nasal congestion, palpitations, chest pain and peripheral edema [192, 200, 201, 203, 204]. Potential liver injury and teratogenicity are the two major concerns related to the use of ERA [155]. Increase in hepatic aminotransferases occurred in approximately 10% of patients during treatment with bosentan and were found to be dose dependent and reversible after dose reduction or discontinuation [205]. During bosentan treatment, it is recommended liver function tests to be checked monthly and hematocrit every 3 months [115]. Ambrisentan has been associated with lower liver toxicity as compared with bosentan [203, 206].

The isoform 5 of the PDE enzyme is present in large amounts in the lung [176, 207]. All three PDE-5 inhibitors (e.g., sildenafil—50 mg daily dose), tadalafil—20, 40 and 60 mg daily dose, vardenafil—10 and 20 mg daily dose) have been found to cause significant pulmonary vasorelaxation. Significant improvement in arterial oxygenation (equally to NO inhalation), however, was only noted with sildenafil.

The combination therapy of sildenafil with intravenous epoprostenol has led to improvement in patients with idiopathic PAH, while in SSc-related PAH, efficacy remains to be determined. The combination of sildenafil and bosentan improved the functional class in idiopathic PAH, while patients with PAH in SSc did not experience significant improvement [195].

Riociguat is a drug from a novel therapeutic class acting as soluble guanylate cyclase stimulators. It possesses a dual mode of action as it stimulates soluble guanylate cyclase directly, independent of NO, and also sensitizes soluble guanylate cyclase to endogenous NO by stabilizing its binding and increasing in this way the level of cGMP [208].

NO-guanylate cyclase-cGMP pathway plays a crucial role in the pathogenesis of PAH. NO is synthesized in endothelial cells primarily through oxidation of the amino group of L-arginine, catalyzed by NO synthases. NO exhibits its effects via cGMP—a key second messenger that controls dilation and cellular proliferation within the vascular wall.

NO diffuses to smooth muscle cells, where it activates the intracellular enzyme—soluble guanylate cyclase, which in turn catalyzes the conversion of guanosine-5′-triphosphate to cGMP. Decreased level of NO and subsequently of cGMP leads to increased pressure within the pulmonary circulation [208, 209]. Of note, soluble guanylate cyclase stimulators may also exert antifibrotic effects by interfering with non-canonical profibrotic TGFβ signaling [210].

In RCT, it has been shown that riociguat improves significantly exercise capacity (assessed by 6-min walk distance), time to clinical worsening and hemodynamic parameters [211]. It is administered at a dose of 1 mg tid for 2 weeks with dose elevation with 0.5 mg tid every 2 weeks to a maximal dose 2.5 mg tid. Drug-related serious adverse events included syncope, elevated hepatic enzymes, dizziness, presyncope, acute renal failure, and hypotension [211]. Riociguat is recommended by EULAR experts for the treatment of PAH in SSc [155].

Imatinib is a small molecule tyrosine kinase inhibitor that binds competitively to the ATP-binding pocket of Abelson-kinase (c-Abl) and thereby efficiently blocks its tyrosine kinase activity. C-Abl is an important downstream signaling molecule of TGF-β and PDGF. Thus, imatinib targets simultaneously and selectively two major pro-fibrotic pathways activated in SSc [212, 213]. Based on evidence that PDGF signaling is an important process in the pathophysiology of PAH, imatinib has been tested and shown to be effective in experimental models of PAH. Of the available surgical options, atrial septostomy creates a right to left interatrial shunt, decreasing right heart loading pressures and improving right heart function and left heart load in patients with worsened right heart function in severe PAH. Single and double lung transplantation and combined heart and lung transplantation are ultimate therapeutic options in patients with end-stage disease [115].

Skin

Two RCTs have demonstrated that methotrexate (at a dose of 15 weekly intramuscularly and 10 mg weekly, respectively) improves skin score in early diffuse SSc. Thus, EULAR recommendations suggest methotrexate as a therapeutic option for skin manifestations in early diffuse SSc [160, 214].

Lung

There is limited evidence based therapeutic strategies to counteract interstitial lung disease and progressive fibrosis, although immunosuppressive agents are being used off-label in a high number of patients [215].

With respect to controlled trials, in patients with active alveolitis, oral cyclophosphamide at a dose of 1–2 mg/kg daily for a period of 1 year has led to significant therapeutic benefit [216, 217]. Combination therapy of low-dose prednisolone and six infusions of intravenous cyclophosphamide (600 mg/m^2/monthly) at 4-week intervals followed by oral azathioprine (2.5 mg/kg/daily) for 6 months was compared with placebo. A trend toward statistically significant improvement of forced vital capacity was registered ($P=0.08$), while diffusing capacity and secondary outcome measures (high-resolution computed tomography and dyspnea scores) were not improved. Thus, it has been suggested that treatment with low-dose prednisolone and intravenous cyclophosphamide followed by azathioprine stabilizes lung function in a subset of SSc with interstitial pulmonary involvement [164]. Based on the results from two high-quality RCTs [164, 217] and despite its toxicity, cyclophosphamide is recommended for treatment of SSc-related interstitial lung disease, especially for the progressive forms of lung involvement [155].

Hematopoietic stem cells transplantation (HSCT) has emerged as a new therapeutic procedure for patients affected by severe SSc refractory to conventional treatments. In the ASSIST trial that included 19 SSc patients, it has been observed that HSCT (with conditioning regimen that included 200 mg/kg cyclophosphamide in four equal fractions from day 3 to day 2 before stem cell infusion and rabbit antithymocyte globulin administered from day 5 to day 1 prior stem cell infusion at total dose of 6.5 mg/kg) was superior to intravenous cyclophosphamide (1 g/m^2/monthly for 6 months) for improvement of skin score and lung volumes. Peripheral blood stem cells were mobilized at least 2 weeks prior the conditioning regimen preceded by intravenous cyclophosphamide (2 g/m^2) and 10 μg/kg subcutaneous filgrastim from day 5 after cyclophosphamide administration [218].

However, the key trial in this field has been the ASTIS trial (Autologous Stem Cell Transplantation International Scleroderma) [219]. This phase 3, multicenter, randomized (1:1), open-label, parallel-group, clinical trial was conducted in 10 countries at 29 centers with access to a European Group for Blood and Marrow Transplantation-registered transplant facility. From March 2001 to October 2009, 156 patients with early diffuse cutaneous systemic sclerosis were recruited and followed up until October 31, 2013. The group could show that among patients with early diffuse cutaneous systemic sclerosis, HSCT conferred a significant long-term event-free survival benefit but was associated with increased treatment-related mortality in the first year after treatment [220].

Kidney involvement

All patients with SSc who have new-onset hypertension, with or without evidence of renal involvement, should be treated and maintained on a maximum tolerable dose of an ACE inhibitor, even if renal failure develops [146]. Patients with scleroderma renal crisis, who received ACE inhibitors had an impressive 1-year survival of 76% and a 5-year survival of 65% vs 15%—1-year survival and 10%—5-year survival of patients not receiving ACE inhibitors, despite other aggressive antihypertensive treatment [146]. In a small group of patients with "normotensive" scleroderma renal crisis, ACE inhibitors also have been shown to improve survival [221]. A higher risk of scleroderma renal crisis was observed in SSc patients on corticosteroid treatment. Thus, blood pressure and renal function should be carefully monitored in patients receiving steroids [147, 155] (Table 4).

TABLE 4 EULAR recommendations for the treatment of SSc [155].

Recommendations
I. SSc-related RP
1. A meta-analysis on dihydropyridine-type CCBs and PDE-5 inhibitors indicate that nifedipineand PDE-5 inhibitors reduce the frequency and severity of SSc-related RP attacks. Considering long-term experience and good safety profile, dihydropyridine-type CCBs, usually oral nifedipine, are recommended as first-line therapy for SSc-related RP. PDE-5 inhibitors should also be considered for treatment in patients with SSc with severe RP and/or those who do not satisfactorily respond to CCBs
2. A meta-analysis of RCTs on prostanoids indicates that intravenous iloprost reduces frequency and severity of SSc-related RP attacks. Efficacious in healing digital ulcers in patients with SSc. Intravenous iloprost should be considered for severe SSc-related RP
3. Intravenous iloprost is considered for treatment of SSc-RP after failure of oral therapy
4. In one small study, it has been observed that fluoxetine might improve SSc-RP attacks. Thus, it has been recommended that Fluoxetine might be considered in treatment of SSc-RP attacks
II. Digital ulcers in SSc
5. Intravenous iloprost is efficacious in healing digital ulcers SSc that has been demonstrated in two RCTs. Intravenous iloprost should be considered in the treatment of digital ulcers SSc
6. PDE-5 inhibitors have proven efficacy in RCTs in healing of digital ulcers in SSc patients. In one small RCT, it has been demonstrated that tadalafil may prevent development of new digital ulcers in SSc. Thus, PDE-5 inhibitors are recommended for treatment of digital ulcers in SSc patients
7. Bosentan has no proven efficacy in the treatment of active digital ulcers in SSc patients. In two high-quality RCTs, bosentan has proven efficacy to reduce the number of new digital ulcers in SSc patients. Bosentan should be considered in SSc patients, particularly those with multiple digital ulcers despite treatment with CCBs, PDE-5 inhibitors and iloprost to prevent the development of new digital ulcers

Continued

TABLE 4 EULAR recommendations for the treatment of SSc [155].—cont'd

Recommendations
III.SSc-related PAH 8. Based on the results of high-quality RCTs that include also PAH associated with connective tissue diseases, several ERA (ambrisentan, bosentan and macitentan), PDE-5 inhibitors (sildenafil, tadalafil) and riociguat have been approved for treatment of PAH associated with connective tissue diseases. ERA, PDE-5 inhibitors or riociguat should be considered to treat SSc-related PAH 9. One high-quality RCT indicates that continuous intravenous epoprostenol improves exercise capacity, functional class and hemodynamic measures in SSc-PAH. Intravenous epoprostenol should be considered for the treatment of patients with severe SSc-PAH (class III and IV) 10. Based on the results of high-quality RCTs that include PAH associated with CTD, other prostacyclin analogues (iloprost, treprostinil) have also been registered for treatment of PAH associated with connective tissue diseases. Prostacyclin analogues should also be considered for the treatment of patients with SSc-PAH
IV.Skin and lung disease 11. Two RCTs have shown that methotrexate improves skin score in early diffuse SSc. Positive effects on other organ manifestations have not been established. Methotrexate may be considered for treatment of skin manifestations of early diffuse SSc 12. In view of the results from two high-quality RCTs and despite its known toxicity, cyclophosphamide should be considered for treatment of SSc-related interstitial lung disease, especially in patients with progressive interstitial lung disease 13. Two RCTs have shown that hematopoietic stem cell transplantation (HSCT) leads to improvement of skin involvement and stabilization of lung function in patients with SSc and one large RCT reports improvement in event-free survival in patients with SSc as compared with cyclophosphamide in both trials. HSCT should be considered for treatment of selected patients with rapidly progressive SSc at risk of organ failure
V. Scleroderma renal crisis 14. Several cohort studies have demonstrated benefit in survival with use of ACE inhibitors in patients with SRC. Immediate administration of ACE inhibitors in the treatment of SRC is recommended 15. Several retrospective studies suggest that glucocorticoids are associated with a higher risk of scleroderma renal crisis. Blood pressure and renal function should be carefully monitored in SSc patients on steroid therapy
VI.SSc-related gastrointestinal disease 16. Despite the lack of specific RCTs, experts recommend that proton pump inhibitors should be used for the treatment of SSc-related GERD and prevention of esophageal ulcers and strictures 17. Despite the lack of RCTs, experts recommend that prokinetic drugs should be used for the management of SSc-related symptomatic motility disturbances (dysphagia, GERD, early satiety response, bloating, pseudo-obstruction, etc.) 18. Despite the lack of RCTs, experts recommend the use of intermittent or rotating antibiotics to treat symptomatic small intestine bacterial overgrowth in SSc patients

Gastrointestinal involvement

Adequate differential treatment of manifestations of SSc at the gastrointestinal tract is still challenging [222]. An upright posture and avoidance of exertion after feeding are common measures in cases with esophageal disease to minimize the possibility of heartburn [3]. For symptoms of heart burn and dyspepsia, proton pump inhibitors should be used. In refractory cases, doubling of the dose could be considered as well as the addition of a H2 blocker at bedtime because of different mechanisms of action to prevent nocturnal breakthrough [223–226]. Despite the lack of specific RCTs, proton pump inhibitors are suggested in the EULAR recommendations for the treatment of SSc-related gastroesophageal reflux disease (GERD) and prevention of esophageal ulcers and strictures [155].

In cases of gastroparesis, pro-motility agents such as erythromycin (100–150 mg qid), azithromycin (400 mg daily), metoclopramide (10–15 mg qid), prucalopride (5–15 mg/day), and domperidon (10–20 mg qid), may be helpful [135, 223, 227]. Watermelon stomach is treated by several transendoscopic procedures—laser photocoagulation, bipolar electrocoagulation, heater probe coagulation, and injection sclerotherapy. Surgical intervention (partial gastrectomy, antrectomy) is indicated in some cases [136].

Malabsorption due to suspected bacterial overgrowth irrespective of the results of breath testing is treated with rotating broad spectrum antibiotics. A 10-day course of a selective antibiotic could be tried, with subsequent treatment with a selective antibiotic for the first 10 days of 4 consecutive months [223, 228]. Therapy with octreotide may be considered in cases with refractory symptoms from small bowel involvement [223, 229]. Pseudo-obstruction of the small and large intestines occurs with patients with either the limited or diffuse form of SSc and is frequently lethal regardless of the therapeutic efforts as outlined above [15]. Of interest, recent studies could show beneficial effects of repeated application of intravenous immunoglobulins on several gastrointestinal dysfunctions [230, 231].

Musculoskeletal involvement

Articular complaints can be treated with non-steroidal anti-inflammatory drugs. Corticosteroids are rarely necessary, but patients with true synovitis may benefit from prednisone—5 to 7.5 mg daily. Patients with diffuse cutaneous involvement often develop digital contractures, with deformity and severe impairment of hand function. A vigorous exercise program should be recommended with use of analgesics before each exercise session to relieve the pain and maximize the results [232]. In cases of severe finger contractures, the hand is vulnerable to trauma. Repeated skin breakdown and infection, including septic arthritis, are potential complications. In these cases joint fusion is a treatment option, while joint replacement will not be effective. Active polymyositis should be treated with moderate doses of corticosteroids (prednisone, 15–20 mg/day) and other drugs used in myositis. In cases of incomplete response, methotrexate, azathioprine or another immunosuppressive agent can be added [15].

New therapies

Biologic agents in SSc

Rituximab has been examined in SSc patients. It has shown to be well-tolerated and may have efficacy for skin disease, but clinical trials in larger patient groups are warranted [233, 234]. A recent double-blind, RCT, phase II was the first to demonstrate evidence of a potential disease-modifying effect of subcutaneous *tocilizumab* in SSc patients. In patients receiving tocilizumab, a clinically meaningful but not statistically significant decline in mRSS vs placebo after week 48 was observed. In the open-label period of the study, both tocilizumab and placebo group were switched on subcutaneous tocilizumab until week 96. In this period, an absolute decline >10% of the predicted forced vital capacity was not observed in either treatment group who completed week 96 or withdrew. Patients from placebo-tocilizumab group experienced a similar decrease in mRSS by week 96 to patients who received tocilizumab throughout the study. Patients from the continuous tocilizumab group maintained and continued the decrease in mRSS observed during the first 48 weeks of treatment. Although the change in mRSS appeared to decrease from week 72 onward, there were still improvements for individual patients between weeks 72 and 96. Amelioration of the skin score and stabilization of the forced vital capacity in the double-blind period were observed in placebo-treated patients who switched to tocilizumab and were maintained in the open-label period. Infections were the most commonly reported adverse events and serious adverse events [235]. Trials using TNF-inhibitors do not provide optimism for TNF as a specific therapeutic target in SSc [165, 236–239].

Prognosis

Historic factors negatively affecting survival include male sex and older age at diagnosis. In contrast, survival of patients with SRC after initiation of treatment with ACE-inhibitors has been improved. Thus, currently cardiopulmonal complications are the leading cause of death. Of these, specifically PAH is one of the complications of SSc that determines poor prognosis. Historically, a 1-year survival rate in SSc-associated PAH was 45% [122, 240]. A subsequent 6-year follow-up (2001–2006) of 315 SSc patients with PAH who have been documented in the UK National registry has revealed 1-, 2- and 3-year survival rates of 78%, 58% and 47%, respectively. Although survival in patients with SSc-associated PAH is better in the modern treatment era than in historical cohorts, it remains unacceptably poor [122]. A large-scale analysis [241] could show that over a median follow-up of 2.3 years, 1072 (9.6%) of 11,193 patients from the EUSTAR sample died, from cardiac disease in 27% and respiratory causes in 17%.

References

[1] Mayes MD, Lacey JVJ, Beebe-Dimmer J, et al. Prevalence, incidence, survival, and disease characteristics of systemic sclerosis in a large US population. Arthritis Rheum 2003;48:2246–55.

[2] Allcock RJ, Forrest I, Corris PA, et al. A study of the prevalence of systemic sclerosis in northeast England. Rheumatology (Oxford) 2004;43:596–602.

[3] Matucci-Cerinic M, Miniati I, Denton CP. Systemic sclerosis. In: JWJ B, et al., editors. EULAR compendium on rheumatic diseases. 1st ed. London: BMJ Publishing group; 2009. p. 290–6.

[4] Meier FM, Frommer KW, Dinser R. Update on the profile of the EUSTAR cohort: an analysis of the EULAR Scleroderma Trials and Research group database. Ann Rheum Dis 2012;71:1355–60.

[5] Mayes MD. Epidemiologic studies of environmental agents and systemic autoimmune diseases. Environ Health Perspect 1999;107(Suppl. 5):743–8.

[6] Namboodiri AM, Rocca KM, Pandey JP. IgG antibodies to human cytomegalovirus late protein UL94 in patients with systemic sclerosis. Autoimmunity 2004;37(3):241–4.

[7] Sahin A, Ozkan T, Türkçapar N, et al. Peripheral blood mononuclear cell microchimerism in Turkish female patients with systemic sclerosis. Mod Rheumatol 2014;24(1):97–105.

[8] Wermuth PJ, Piera-Velazquez S, Rosenbloom J, Jimenez SA. Existing and novel biomarkers for precision medicine in systemic sclerosis. Nat Rev Rheumatol 2018;14(7):421–32.

[9] Sgonc R, Gruschwitz M, Dietrich H, et al. Endothelial cell apoptosis is a primary pathogenetic event underlying skin lesions in avian and human scleroderma. J Clin Invest 1996;98:785–92.

[10] Herrick AL, Matucci-Cerinic M. The emerging problem of oxidative stress and the role of antioxidants in systemic sclerosis. Clin Exp Rheumatol 2001;19:4–8.

[11] Kahaleh B, Meyer O, Scorza R. Assessment of vascular involvement. Clin Exp Rhematol 2003;21(Suppl. 29):S9–14.

[12] Block JA, Sequeira W. Raynaud's phenomenon. Lancet 2001;357:2042–8.

[13] Ho M, Belch JJF. Raynaud's phenomenon: state of the art 1998. Scand J Rheumatol 1998;27:319–22.

[14] Konttinen YT, Mackiewicz Z, Ruuttila P, et al. Vascular damage and lack of angiogenesis in systemic sclerosis skin. Clin Rheumatol 2003;22:196–202.

[15] Silver RM, Medsger Jr TA, Bolster MB. Systemic sclerosis and scleroderma variants: clinical aspects. In: Koopman WJ, Moreland LW, editors. Arthritis and allied conditions a textbook in rheumatology. 15th ed. Philadelphia: Lippincot Williams&Wilkins; 2005. p. 1633–80.

[16] Del Papa N, Pignataro F. The role of endothelial progenitors in the repair of vascular damage in systemic sclerosis. Front Immunol 2018;18(9):1383.

[17] Sato S, Fujimoto M, Hasegawa M, et al. Altered blood B lymphocyte homeostasis in systemic sclerosis: expanded naive B cells and diminished but activated memory B cells. Arthritis Rheum 2004;50:1918–27.

[18] Ho KT, Reveille JD. The clinical relevance of autoantibodies in scleroderma. Arthritis Res Ther 2003;5:80–93.

[19] Kuwana M, Okano Y, Kaburaki J, Tojo T, Medsger Jr TA. Racial differences in the distribution of systemic sclerosis-related antinuclear antibodies. Arthritis Rheum 1994;37(6):902–6.

[20] Meyer O. Prognostic markers for systemic sclerosis. Joint Bone Spine 2006;73:490–4.

[21] van den Hoogen F, Khanna D, Fransen J, et al. 2013 classification criteria for systemic sclerosis: an American college of rheumatology/European league against rheumatism collaborative initiative. Ann Rheum Dis 2013;72(11):1747–55.

[22] van den Hoogen F, Khanna D, Fransen J, et al. 2013 classification criteria for systemic sclerosis: an American college of rheumatology/European league against rheumatism collaborative initiative. Arthritis Rheum 2013;65(11):2737–47.
[23] Douvas AS, Achten M, Tan EM. Identification of a nuclear protein (Scl-70) as a unique target of human antinuclear antibodies in scleroderma. J Biol Chem 1979;254:10514–22.
[24] Mehra S, Walker J, Patterson K, Fritzler MJ. Autoantibodies in systemic sclerosis. Autoimmun Rev 2013;12:340–54.
[25] Shero JH, Bordwell B, Rothfield NF, Earnshaw WC. (1986) High titers of autoantibodies to topoisomerase I (Scl-70) in sera from scleroderma patients. Science 1986;231:737–40.
[26] Koenig M, Dieudé M, Senécal JL. Predictive value of antinuclear autoantibodies: the lessons of the systemic sclerosis autoantibodies. Autoimmun Rev 2008;7(8):588–93.
[27] Reveille JD, Solomon DH, American College of Rheumatology Ad Hoc Committee of Immunologic Testing Guidelines. Evidence-based guidelines for the use of immunologic tests: anticentromere, Scl-70, and nucleolar antibodies. Arthritis Rheum 2003;49(3):399–412.
[28] Steen VD. Autoantibodies in systemic sclerosis. Semin Arthritis Rheum 2005;35:35–42.
[29] Walker UA, Tyndall A, Czirják LO, et al. Clinical risk assessment of organ manifestations in systemic sclerosis: a report from the EULAR Scleroderma trials and research group database. Ann Rheum Dis 2007;66:754–63.
[30] Kuwana M, Kaburaki J, Mimori T, Kawakami Y, Tojo T. Longitudinal analysis of autoantibody response to topoisomerase I in systemic sclerosis. Arthritis Rheum 2000;43:1074–84.
[31] Hu PQ, Fertig N, Medsger Jr TA, Wright TA. Correlation of serum anti-DNA topo-isomerase I antibody levels with disease severity and activity in systemic sclerosis. Arthritis Rheum 2003;48:1363–73.
[32] Sullivan KF, Hechenberger M, Masri K. Human CENP-A contains a histone H3 related histone fold domain that is required for targeting to the centromere. J Cell Biol 1994;127:581–92.
[33] Kayser C, Fritzler MJ. Autoantibodies in systemic sclerosis: unanswered questions. Front Immunol 2015;6:167.
[34] Earnshaw W, Bordwell B, Marino C, Rothfield N. Three human chromosomal autoantigens are recognized by sera from patients with anti-centromere antibodies. J Clin Invest 1986;77:426–30.
[35] Dick T, Mierau R, Bartz-Bazzanella P, et al. Coexistence of antitopoisomerase I and anticentromere antibodies in patients with systemic sclerosis. Ann Rheum Dis 2002;61(2):121–7.
[36] Ferri C, Valentini G, Cozzi F, Sebastiani M. Systemic sclerosis: demographic, clinical, and serologic features and survival in 1,012 Italian patients. Medicine 2002;81:139–53.
[37] Walker JG, Fritzler MJ. Update on autoantibodies in systemic sclerosis. Curr Opin Rheumatol 2007;19:580–91.
[38] Santiago M, Baron M, Hudson M, Burlingame RW. The Canadian scleroderma research group, Fritzler MJ. Antibodies to RNA polymerase III in systemic sclerosis as detected by an ELISA. J Rheumatol 2007;34:1528–34.
[39] Nikpour M, Hissaria P, Byron J, et al. Prevalence, correlates and clinical usefulness of antibodies to RNA polymerase III in systemic sclerosis: a cross-sectional analysis of data from an Australian cohort. Arthritis Res Ther 2011;13:R211.
[40] Satoh T, Ishikawa O, Ihn H, et al. Clinical usefulness of anti-RNA polymerase III antibody measurement by enzyme-linked immunosorbent assay. Rheumatology (Oxford) 2009;48:1570–4.
[41] Hesselstrand R, Scheja A, Wuttge DM. Scleroderma renal crisis in a Swedish systemic sclerosis cohort: survival, renal outcome, and RNA polymerase III anti-bodies as a risk factor. Scand J Rheumatol 2012;41:39–43.

[42] Kuwana M, Kaburaki J, Mimori T, Tojo T, Homma M. Autoantibody reactive with three classes of RNA polymerases in sera from patients with systemic sclerosis. J Clin Invest 1993;91:1399–404.

[43] Jones E, Kimura H, Vigneron M, et al. Isolation and characterization of monoclonal antibodies directed against subunits of human RNA polymerases I, II, and III. Exp Cell Res 2000;254:163–72.

[43a] Yamasaki Y, Honkanen-Scott M, Hernandez L, et al. Nucleolar staining cannot be used as a screening test for the scleroderma marker anti-RNA polymerase I/III antibodies. Arthritis Rheum 2006;54(9):3051–6.

[44] Okano Y, Medsger Jr TA. Autoantibody to Th ribonucleoprotein (nucleolar 7-2 RNA protein particle) in patients with systemic sclerosis. Arthritis Rheum 1990;33:1822–8.

[45] van Eenennaam H, Vogelzangs JHP, Lugtenberg D, et al. Identity of the RNase MRP- and RNase P-associated Th/To autoantigen. Arthritis Rheum 2002;46:3266–72.

[46] Mitri GM, Lucas M, Fertig N, Steen VD, Medsger Jr TA. A comparison between anti-Th/To- and anticentromere antibody-positive systemic sclerosis patients with limited cutaneous involvement. Arthritis Rheum 2003;48(1):203–9.

[47] Gabrielli A, Svegliati S, Moroncini G, Avvedimento EV. Pathogenic autoantibodies in systemic sclerosis. Curr Opin Immunol 2007;19:640–5.

[48] Kraaij MD, van Laar JM. The role of B cells in systemic sclerosis. Biologics 2008;2(3):389–95.

[49] Cabral-Marques O, Riemekasten G. Functional autoantibodies targeting G protein-coupled receptors in rheumatic diseases. Nat Rev Rheumatol 2017;13(11):648–56.

[50] Moroncini G, Svegliati Baroni S, Gabrielli A. Agonistic antibodies in systemic sclerosis. Immunol Lett 2018;195:83–7.

[51] Sgonc R, Wick G. Pro- and anti-fibrotic effects of TGF-β in scleroderma. Rheumatology (Oxford) 2008;47:v5–7.

[52] Varga J, Abraham D. Systemic sclerosis: a prototypic multisystem fibrotic disorder. J Clin Invest 2007;117(3):557–67.

[53] Kehrl JH, Roberts AB, Wakefield LM, et al. Transforming growth factor beta is an important immunomodulatory protein for human B lymphocytes. J Immunol 1986;137(12):3855–60.

[54] Verrecchia F, Mauviel A, Farge D. Transforming growth factor-beta signaling through the Smad proteins: role in systemic sclerosis. Autoimmun Rev 2006;5(8):563–9.

[55] Trojanowska M. Role of PDGF in fibrotic disease and systemic sclerosis. Rheumatology (Oxford) 2008;47:v2–4.

[56] Antic M, Distler JH, Distler O. Treating skin and lung fibrosis in systemic sclerosis: a future filled with promise? Curr Opin Pharmacol 2013;13(3):455–62.

[57] O'Reilly S. Toll like receptors in systemic sclerosis: An emerging target. Immunol Lett 2018;195:2–8.

[58] Chizzolini C, Dufour AM, Brembilla NC. Is there a role for IL-17 in the pathogenesis of systemic sclerosis? Immunol Lett 2018;195:61–7.

[59] Scherlinger M, Guillotin V, Truchetet ME, et al. Systemic lupus erythematosus and systemic sclerosis: All roads lead to platelets. Autoimmun Rev 2018;17(6):625–35.

[60] Schulz JN, Plomann M, Sengle G, et al. New developments on skin fibrosis—essential signals emanating from the extracellular matrix for the control of myofibroblasts. Matrix Biol 2018;68-69:522–32.

[61] Arnett FC, Cho M, Chatterjee S, et al. Familial occurrence frequencies and relative risks for systemic sclerosis (scleroderma) in three United States cohorts. Arthritis Rheum 2001;44:1359–62.

[62] Zhou X, Tan FK, Xiong M, et al. Monozygotic twins clinically discordant for scleroderma show concordance from fibroblast gene expression profiles. Arthritis Rheum 2005;52(10):3305–14.

[63] Ahmed SS, Tan FK. Identification of novel targets in scleroderma, update on population studies, cDNA arrays, SNP analysis, and mutations. Curr Opin Rheumatol 2003;15:766–71.

[64] Feghali-Bostwick CA. Genetics and proteomics in scleroderma. Curr Rheumatol Rep 2005;7:129–34.

[65] Allanore Y, Dieude P, Boileau C. Updating the genetics of systemic sclerosis. Curr Opin Rheumatol 2010;22:665–70.

[66] Allanore Y, Saad M, Dieude P, et al. Genome-wide scan identifies TNIP1, PSORS1C1, and RHOB as novel risk loci for systemic sclerosis. PLoS Genet 2011;7(7):e1002091.

[67] Arnett FC, Gourh P, Shete S, et al. Major histocompatibility complex (MHC) class II alleles, haplotypes, and epitopes which confer susceptibility or protection in the fibrosing autoimmune disease systemic sclerosis: analyses in 1300 Caucasian, African-American and Hispanic cases and 1000 controls. Ann Rheum Dis 2010;69:822–7.

[68] Czubaty A, Girstun A, Kowalska-Loth B, et al. Proteomic analysis of complexes formed by human topoisomerase I. Biochim Biophys Acta 2005;1749:133–41.

[69] Giusti B, Fibbi G, Margheri F, et al. A model of anti-angiogenesis, differential transcriptosome profiling of microvascular endothelial cells from diffuse systemic sclerosis patients. Arthritis Res Ther 2006;8:R115.

[70] Tan FK, Hildebrand BA, Lester MS, et al. Classification analysis of the transcriptosome of nonlesional cultured dermal fibroblasts from systemic sclerosis patients with early disease. Arthritis Rheum 2005;52:865–76.

[71] Luzina IG, Atamas SP, Wise R, et al. Occurrence of an activated, profibrotic pattern of gene expression in lung CD8 T cells from scleroderma patients. Arthritis Rheum 2003;48:2262–74.

[72] Assassi S, Mayes MD, McNearney T, et al. Polymorphisms of endothelial nitric oxide synthase and angiotensin-converting enzyme in systemic sclerosis. Am J Med 2005;118:907–11.

[73] Zhou X, Tan FK, Stivers DN, Arnett FC. Microsatellites and intragenic polymorphisms of transforming growth factor beta and platelet-derived growth factor and their receptor genes in Native Americans with systemic sclerosis (scleroderma), a preliminary analysis showing no genetic associations. Arthritis Rheum 2000;43(5):1068–73.

[74] Crilly A, Hamilton J, Clark CJ, et al. Analysis of transforming growth factor beta1 gene polymorphisms in patients with systemic sclerosis. Ann Rheum Dis 2002;61:678–81.

[75] Pannu J, et al. Increased levels of transforming growth factor beta receptor type I and up-regulation of matrix gene program: a model of scleroderma. Arthritis Rheum 2006;54:3011–21.

[76] Zhu H, Bona C, McGaha TL. Polymorphisms of the TGF-beta1 promoter in tight skin (TSK) mice. Autoimmunity 2004;37(1):51–5.

[77] Marasini B, Casari S, Zeni S, et al. Stromelysin promoter polymorphism is associated with systemic sclerosis. Rheumatology (Oxford) 2001;40:475–6.

[78] Johnson KL, Nelson JL, Furst DE, et al. Fetal cell microchimerism in tissue from multiple sites in women with systemic sclerosis. Arthritis Rheum 2001;44:1848–54.

[79] Pandey JP, Takeuchi F. TNF-alpha and TNF-beta gene polymorphisms in systemic sclerosis. Hum Immunol 1999;60:1128–30.

[80] Tolusso B, Fabris M, Caporali R, et al. 238 and 489 TNF-alpha along with TNFRII gene polymorphisms associate with the diffuse phenotype in patients with systemic sclerosis. Immunol Lett 2005;96:103–8.

[81] Kawaguchi Y, Tochimoto A, Ichikawa N, et al. Association of IL1A gene polymorphisms with susceptibility to and severity of systemic sclerosis in the Japanese population. Arthritis Rheum 2003;48:186–92.

[82] Hutyrova B, Lukác J, Bosák V, et al. Interleukin 1alpha single-nucleotide polymorphism associated with systemic sclerosis. J Rheumatol 2004;31:81–4.

[83] Radstake TR, Vonk MC, Dekkers M. The -2518A>G promoter polymorphism in the CCL2 gene is not associated with systemic sclerosis susceptibility or phenotype: results from a multicenter study of European Caucasian patients. Hum Immunol 2009;70:130–3.

[84] Tsuchiya N, Kuroki K, Fujimoto M, et al. Association of a functional CD19 polymorphism with susceptibility to systemic sclerosis. Arthritis Rheum 2004;50:4002–7.

[85] Takeuchi F, Nakano K, Yamada H, et al. Association of HLA-DR with progressive systemic sclerosis in Japanese. J Rheumatol 1994;21:857–63.

[86] Song YW, Lee EB, Whang DH, et al. Association of TAP1 and TAP2 gene polymorphisms with systemic sclerosis in Korean patients. Hum Immunol 2005;66:810–7.

[87] Lambert NC, Evans PC, Hashizumi TL, et al. Cutting edge, persistent fetal microchimerism in T lymphocytes is associated with HLA-DQA1 * 0501, implications in autoimmunity. J Immunol 2000;164:5545–8.

[88] Lambert NC, Distler O, Müller-Ladner U, et al. HLA-DQA1 * 0501 is associated with diffuse systemic sclerosis in Caucasian men. Arthritis Rheum 2000;43:2005–10.

[89] Nguyen B, Mayes MD, Arnett FC. HLA-DRB1*0407 and *1304 are risk factors for scleroderma renal crisis. Arthritis Rheum 2011;63:530–4.

[90] Bossini-Castillo L, Broen JC, Simeon CP, et al. A replication study confirms the association of TNFSF4 (OX40L) polymorphisms with systemic sclerosis in a large European cohort. Ann Rheum Dis 2011;70:638–41.

[91] Fatini C, Guiducci S, Abbate R, Matucci-Cerinic M. Vascular injury in systemic sclerosis, angiotensin-converting enzyme insertion/deletion polymorphism. Curr Rheumatol Rep 2004;6:149–55.

[92] Allanore Y, Borderie D, Lemaréchal H, et al. Lack of association of eNOS (G894T) and p22hox NAPDH oxidase subunit (C242T) polymorphism with systemic sclerosis in a cohort of French Caucasian patients. Clin Chim Acta 2004;350(1–2):51–5.

[93] Tikly M, Gulumian M, Marshall S. Lack of association of eNOS(G849T) and p22hox NADPH oxidase submit (C242T) polymorphisms with systemic sclerosis in a cohort of French Caucasian patients. Clin Chim Acta 2005;358:196–7.

[94] Joung CI, Park YW, Kim SK, et al. Angiotensin-converting enzyme gene insertion/deletion polymorphism in Korean patients with systemic sclerosis. J Korean Med Sci 2006;21:329–32.

[95] Terao C, Kawaguchi T, Dieude P, et al. Transethnic meta-analysis identifies GSDMA and PRDM1 as susceptibility genes to systemic sclerosis. Ann Rheum Dis 2017;76:1150–8.

[96] Bergmann C, Distler JH. Epigenetic factors as drivers of fibrosis in systemic sclerosis. Epigenomics 2017;9(4):463–77.

[97] Seibold JR, Steen VD, Klippel JH, Dieppe PA. Systemic sclerosis. In: Rheumatology. 1st ed. London: Mosby; 1994. p. 6.8–6.11.

[98] Bollinger A, Fagrell B. Clinical capillaroscopy—a guide to its use in clinical research and practice. Toronto: Hogrete&Huber Publishers; 1990.p.1–158.

[99] Cutolo M, Sulli A, Pizzorni C, Accardo S. Nailfold videocapillaroscopy assessment of micro vascular damage in systemic sclerosis. J Rheumatol 2000;27:155–60.

[100] Cutolo M, Grassi W, Matucci-Cerinic M. Raynaud's phenomenon and the role of capillaroscopy. Arthritis Rheum 2003;48(11):3023–30.

[101] Cutolo M, Pizzorni C, Sulli A. Capillaroscopy. Best Pract Res Clin Rheumatol 2005;19(3):437–52.

[102] Lambova S, Hermann W, Müller-Ladner U. Nailfold capillaroscopy—its role in diagnosis and differential diagnosis of microvascular damage in systemic sclerosis. Curr Rheumatol Rev 2013;9(4):254–60.

[103] Lambova S, Müller-Ladner U. Nailfold capillaroscopy in rheumatology. Curr Rheumatol Rev 2018;14:2–4.
[104] Maricq HR, LeRoy EC, D'Angelo WA, et al. Diagnostic potential of in vivo capillary microscopy in scleroderma and related disorders. Arthritis Rheum 1980;23(2):183–9.
[105] Maricq HR, Harper FE, Khan MM, et al. Microvascular abnormalities as possible predictors of disease subsets in Raynaud phenomenon and early connective tissue disease. Clin Exp Rheumatol 1983;1(3):195–205.
[106] Mihai C, Smith V, Dobrota R, et al. The emerging application of semi-quantitative and quantitative capillaroscopy in systemic sclerosis. Microvasc Res 2018;118:113–20.
[107] Clements P, Lachenbruch P, Seibold J, et al. Inter- and intraobserver variability of total skin thickness score (modified Rodnan TSS) in systemic sclerosis. J Rheumatol 1995;22:1281–5.
[108] Kaheleh MB, Sultany GL, Smith EA, et al. A modified scleroderma skin scoring method. Clin Exp Rheumatol 1986;4:367–9.
[109] Verrecchia F, Laboureau J, Verola O, et al. Skin involvement in scleroderma—where histological and clinical scores meet. Rheumatology (Oxford) 2007;46:833–41.
[110] Moore TL, Lunt M, McManus B, Anderson ME, Herrick AL. Seventeen-point dermal ultrasound scoring system—a reliable measure of skin thickness in patients with systemic sclerosis. Rheumatology (Oxford) 2003;42(12):1559–63.
[111] Santiago T, Santiago M, Ruaro B, et al. Ultrasonography for the assessment of skin in systemic sclerosis: a systematic review. Arthritis Care Res (Hoboken) 2019;71(4):563–74.
[112] Scheja A, Akkeson A. Comparison of high frequency (20 MHz) ultrasound and palpation for the assessment of skin involvement in systemic sclerosis (scleroderma). Clin Exp Rheumatol 1997;15:283–8.
[113] Pai S, Hsu V. Are there risk factors for scleroderma-related calcinosis? Mod Rheumatol 2018;28(3):518–22.
[114] Kim DS, Yoo B, Lee JS. The major histopathologic pattern of pulmonary fibrosis in scleroderma is non specific interstitial pneumonia. Sarcoidosis Vasc Diffuse Lung Dis 2002;19:121–7.
[115] McLaughlin VV, Archer SL, Badesch DB, et al. ACCF/AHA 2009 expert consensus document on pulmonary hypertension A report of the American college of cardiology foundation task force on expert consensus documents and the American heart association. J Am Coll Cardiol 2009;53:1573–619.
[116] Peters-Golden M, Wise RA, Hochberg MC, et al. Carbon monoxide diffusing capacity as predictor of outcome in systemic sclerosis. Am J Med 1984;77:1027–34.
[117] Wells AU, Hansell DM, Rubens MB, et al. Fibrosing alveolitis in systemic sclerosis: indices of lung function in relation to extend of disease in computed tomography. Arthritis Rheum 1997;40:1229–36.
[118] Remy-Jardin M, Remy J, Wallaert B. Pulmonary involvement in progressive systemic sclerosis: sequential evaluation with CT, pulmonary tests and broncho-alveolar lavage. Radiology 1993;188:499–506.
[119] Remy-Jardin M, Giraud F, Remy J, Copin MC, Gosselin B, Duhamel A. Importance of ground-glass attenuation in chronic diffuse infiltrative lung disease: pathologic-CT correlation. Radiology 1993;189(3):693–8.
[120] Witt C, Borges AC, John M, et al. Pulmonary involvement in diffuse cutaneous systemic sclerosis: broncheoalveolar fluid granulocytosis predicts progression of fibrosing alveolitis. Ann Rheum Dis 1999;58(10):635–40.
[121] Goh NS, Veeraraghavan S, Desai SR, et al. Bronchoalveolar lavage cellular profiles in patients with systemic sclerosis-associated interstitial lung disease are not predictive of disease progression. Arthritis Rheum 2007;56:2005–12.

[122] Condiffe R, Kiely DG, Peacock AJ, et al. Connective tissue disease-associated pulmonary arterial hypertension in the modern treatment era. Am J Respir Crit Care Med 2009;179(2):151–67.
[123] Proudman SM, Stevens WM, Sahhar J, Celermajer D. Pulmonary arterial hypertension in systemic sclerosis: the need for early detection and treatment. Intern Med J 2007;37:485–94.
[124] Humbert M, Morrell NW, Archer AL, et al. Cellular and molecular pathobiology of pulmonary arterial hypertension. J Am Coll Cardiol 2004;43(12):13S–24S.
[125] Jeffery TK, Morrell NW. Molecular and cellular basis of pulmonary vascular remodeling in pulmonary hypertension. Prog Cardiovasc Dis 2002;45:173–202.
[126] Khanna D, Gladue H, Channick R, et al. Recommendations for screening and detection of connective-tissue disease associated pulmonary arterial hypertension. Arthritis Rheum 2013;65(12):3194–201.
[127] Hoeper MM. Definition, classification, and epidemiology of pulmonary arterial hypertension. Semin Respir Crit Care Med 2009;30(4):369–75.
[128] Coghlan JG, Denton CP, Grünig E, et al. Evidence-based detection of pulmonary arterial hypertension in systemic sclerosis: the DETECT study. Ann Rheum Dis 2014;73:1340–9.
[129] Bulkley B, Ridolfi R, Salyer W, et al. Myocardial lesions of progressive systemic sclerosis: a cause of cardiac dysfunction. Circulation 1976;53:483–90.
[130] Kahan A, Nitenberg A, Foult JM, et al. Decreased coronary reserve in primary scleroderma myocardial disease. Arthritis Rheum 1985;28:637–46.
[131] Lee DC, Hinchcliff ME, Sarnari R, et al. Diffuse cardiac fibrosis quantification in early systemic sclerosis by magnetic resonance imaging and correlation with skin fibrosis. J Scleroderma Relat Disord 2018;3(2):159–69.
[132] Smolenska Z, Barraclough R, Dorniak K, Szarmach A, Zdrojewski Z. Cardiac involvement in systemic sclerosis: diagnostic tools and evaluation methods. Cardiol Rev 2019;27(2):73–9.
[133] Bellocchi C, Volkmann ER. Update on the gastrointestinal microbiome in systemic sclerosis. Curr Rheumatol Rep 2018;20(8):49.
[134] Wielosz E, Borys O, Żychowska I, Majdan M. Gastrointestinal involvement in patients with systemic sclerosis. Pol Arch Med Wewn 2010;120(4):132–5.
[135] Ntoumazios SK, Voulgari PV, Potsis K, et al. Esophageal involvement in scleroderma: gastroesophageal reflux, the common problem. Semin Arthritis Rheum 2006;36:173–81.
[136] Elkayam O, Oumanski M, Yaron M, Caspi D. Watermelon stomach following and preceding systemic sclerosis. Semin Arthritis Rheum 2000;30:127–31.
[137] Davidson A, Russell C, Littlejohn GO. Assessment of esophageal abnormalities in progressive systemic sclerosis using radionuclide transit. J Rheumatol 1985;12(3):472–7.
[138] Goldblatt F, Gordon TP, Waterman SA. Antibody-mediated gastrointestinal dysmotility in scleroderma. Gastroenterology 2002;123(4):1144–50.
[139] Greydanus MP, Camilleri M. Abnormal postcibal antral and small bowel motility due to neuropathy or myopathy in systemic sclerosis. Gastroenterology 1989;96:110–5.
[140] Segel MC, Campbell WL, Medsger Jr TA, Roumm AD. Systemic sclerosis (scleroderma) and esophageal adenocarcinoma: is increased patient screening necessary? Gastroenterology 1985;89(3):485–8.
[141] Weber P, Ganser G, Frosch M, et al. Twenty-four hour intraesophageal pH monitoring in children and adolescents with scleroderma and mixed connective tissue disease. J Rheumatol 2000;27(11):2692–5.
[142] Wegener M, Adamek RJ, Wedmann J, et al. Gastrointestinal transit through oesophagus, stomach, small and large intestine in patients with progressive systemic sclerosis. Dig Dis Sci 1994;39:2209–15.

[143] Allali F, Tahiri L, Senjari A, et al. Erosive arthropathy in systemic sclerosis. BMC Public Health 2007;7:260.
[144] Baron M, Lee P, Keystone EC. The articular manifestations of progressive systemic sclerosis (scleroderma). Ann Rheum Dis 1982;41:147–52.
[145] Clements PJ, Furst DE, Campion DS, et al. Muscle disease in progressive systemic sclerosis: diagnostic and therapeutic considerations. Arthritis Rheum 1978;21(1):62–71.
[146] Steen VD, Costantino JP, Shapiro AP, Medsger Jr. TA. Outcome of renal crisis in systemic sclerosis: relation to availability of angiotensin converting enzyme (ACE) inhibitors. Ann Intern Med 1990;113:352–7.
[147] Galluccio F, Müller-Ladner U, Furst DE, Khanna D, Matucci-Cerinic M. Points to consider in renal involvement in systemic sclerosis. Rheumatology 2017;56(Suppl_5):v49–52.
[148] Masi AT, Rodnan GP, Medsger Jr TA, et al. Preliminary criteria for the classification of systemic sclerosis (scleroderma). Subcommittee for scleroderma criteria of the American rheumatism association diagnostic and therapeutic criteria committee. Arthritis Rheum 1980;23(5):581–90. 1980.
[149] Nagy Z, Czirjac L. Nailfold digital capillaroscopy in 447 patients with connective tissue disease and Raynaud disease. J Europ Acad Dermatol Venerol 2004;18:62–8.
[150] Hudson M, Taillefer S, Steele R, et al. Improving the sensitivity of the American College of Rheumatology classification criteria for systemic sclerosis. Clin Exp Rheumatol 2007;25(5):754–7.
[151] Lonzetti LS, Joyal F, Raynauld JP, et al. Updating the American college of rheumatology preliminary classification criteria for systemic sclerosis: addition of severe nailfold capillaroscopy abnormalities markedly increases the sensitivity for limited scleroderma. Arthritis Rheum 2001;44(3):735–6.
[152] Le Roy EC, Medsger Jr TA. Criteria for the classification of early systemic sclerosis. J Rheumatol 2001;28(7):1573–6.
[153] Valentini G, Silman AJ, Veale D. Assessment of disease activity. Clin Exp Rheumatol 2003;21(3 Suppl. 29):S39–41.
[154] Hudson M, Steele R, Canadian Scleroderma Research Group (CSRG), Baron M. Update on indices of disease activity in systemic sclerosis. Semin Arthritis Rheum 2007;37:93–8.
[155] Kowal-Bielecka O, Fransen J, Avouac J, et al. Update of EULAR recommendations for the treatment of systemic sclerosis. Ann Rheum Dis 2017;76(8):1327–39.
[156] Distler JH, Feghali-Bostwick C, Soare A, et al. Review: frontiers of antifibrotic therapy in systemic sclerosis. Arthritis Rheumatol 2017;69(2):257–67.
[157] Clements PJ, Furst DE, Wong WK, et al. High-dose versus low-dose D-penicillamine in early diffuse systemic sclerosis—analysis of a two-year, double-blind, randomized, controlled clinical trial. Arthritis Rheum 1999;42(6):1194–203.
[158] Clements PJ, Seibold JR, Furst DE, et al. High-dose versus low-dose D-penicillamine in early diffuse systemic sclerosis trial: lessons learned. Semin Arthritis Rheum 2004;33(4):249–63.
[159] Steen VD, Medsger Jr TA, Rodnan GP. D-penicillamine therapy in progressive systemic sclerosis (scleroderma). A retrospective analysis. Ann Intern Med 1982;97:652–9.
[160] Pope JE, Bellamy N, Seibold JR, et al. A randomized, controlled trial of methotrexate versus placebo in early diffuse scleroderma. Arthritis Rheum 2001;44(6):1351–8.
[161] Clements PJ, Lachenbruch PA, Sterz M, et al. Cyclosporine in systemic sclerosis—results of a forty-eight-week open safety study in ten patients. Arthritis Rheum 1993;36(1):75–83.
[162] Denton CP, Black CM. Targeted therapy comes in age of scleroderma. Trends Immunol 2005;26(11):596–602.

[163] Yoshizaki A, Yanaba K, Yoshizaki A, et al. Treatment with rapamycin prevents fibrosis in tight-skin and bleomycin-induced mouse models of systemic sclerosis. Arthritis Rheum 2010;62(8):2476–87.

[164] Hoyles RK, Ellis RW, Wellsbury J, et al. A multicenter, prospective, randomized, double-blind, placebo-controlled trial of corticosteroids and intravenous cyclophosphamide followed by oral azathioprine for the treatment of pulmonary fibrosis in scleroderma. Arthritis Rheum 2006;54:3962–70.

[165] Ramos-Casals M, Fonollosa-Pla V, Brito-Zerón P, Sisó-Almirall A. Targeted therapy for systemic sclerosis: how close are we? Nat Rev Rheumatol 2010;6(5):269–78.

[166] Namas R, Tashkin DP, Furst DE, et al. Efficacy of mycophenolate mofetil and oral cyclophosphamide on skin thickness: post-hoc analyses from the Scleroderma Lung Study I and II. Arthritis Care Res (Hoboken) 2018;70(3):439–44.

[167] Tashkin DP, Roth MD, Clements PJ, et al. Mycophenolate mofetil versus oral cyclophosphamide in scleroderma-related interstitial lung disease (SLS II): a randomised controlled, double-blind, parallel group trial. Lancet Respir Med 2016;4:708–19.

[168] Herrick AL. Treatment of Raynaud's phenomenon: new insights and developments. Curr Rheumatol Rep 2003;5:168–74.

[169] Wigley FM. Raynaud's phenomenon. N Engl J Med 2002;347(13):1001–7.

[170] Thompson AE, Shea B, Welch V, et al. Calcium-channel blockers for Raynaud's phenomenon in systemic sclerosis. Arthritis Rheum 2001;44(8):1841–7.

[171] Godfraind T. Discovery and development of calcium channel blockers. Front Pharmacol 2017;8:286.

[172] Leighton C. Drug treatment in scleroderma. Drugs 2001;61(3):419–27.

[173] Dziadzio M, Denton CP, Smith R, et al. Losartan therapy for Raynaud's phenomenon and scleroderma. Arthritis Rheum 1999;42(12):2646–55.

[174] Galie N, Ghofrani HA, Torbicki A, et al. Sildenafil citrate therapy for pulmonary arterial hypertension. N Engl J Med 2005;353:2148–57.

[175] Hoeper MM, Welte T. Sildenafil citrate therapy for pulmonary arterial hypertension. N Engl J Med 2006;354:1091–3.

[176] Kamata Y, Kamimura T, Iwamoto M, Minota S. Comparable effects of sildenafil citrate and alprostadil on severe Raynaud's phenomenon in a patient with systemic sclerosis. Clin Exp Dermatol 2005;30:451.

[177] Kamata Y, Minota S. Effects of phosphodiesterase type 5 inhibitors on Raynaud's phenomenon. Rheumatol Int 2014;34(11):1623–6.

[178] Kumana CR, Cheung GTY, Lau CS. Severe digital ischaemia treated with phosphodiesterase inhibitors. Ann Rheum Dis 2004;63:1522–4.

[179] Rosenkranz S, Diet F, Karasch T, et al. Sildenafil improved pulmonary hypertension and peripheral blood flow in a patient with scleroderma-associated lung fibrosis and Raynaud's phenomenon. Ann Intern Med 2003;139(10):871–3.

[180] Roustit M, Blaise S, Allanore Y, et al. Phosphodiesterase-5 inhibitors for the treatment of secondary Raynaud's phenomenon: systematic review and meta-analysis of randomised trials. Ann Rheum Dis 2013;72(10):1696–9.

[181] Fries R, Shariat K, von Wilmowsky H, Böhm M. Sildenafil in the treatment of Raynaud's phenomenon resistant to vasodilatory therapy. Circulation 2005;112(19):2980–5.

[182] Schiopu E, Hsu VM, Impens AJ, et al. Randomized placebo-controlled crossover trial of tadalafil in Raynaud's phenomenon secondary to systemic sclerosis. J Rheumatol 2009;36:2264–8.

[183] Shenoy PD, Kumar S, Jha LK, et al. Efficacy of tadalafil in secondary Raynaud's phenomenon resistant to vasodilator therapy: a double-blind randomized cross-over trial. Rheumatology 2010;49(12):2420–8.

[184] Maderazo EG, Breaux S, Woronick CL, Krause PJ. Efficacy, toxicity, and pharmacokinetics of pentoxifylline and its analogs in experimental Staphylococcus aureus infections. Antimicrob Agents Chemother 1990;34(6):1100–6.

[185] Neirotti M, Longo F, Molaschi M, et alPernigotti L. Functional vascular disorders: treatment with pentoxifylline. Angiology 1987;38(8):575–80.

[186] Coleiro B, Marshall SE, Denton CP, et al. Treatment of Raynaud's phenomenon with the selective serotonin reuptake inhibitor fluoxetine. Rheumatology 2001;40:1038–43.

[187] Kuwana M, Okazaki Y, Kaburaki J. Long-term beneficial effects of statins on vascular manifestations in patients with systemic sclerosis. Mod Rheumatol 2009;19(5):530–5.

[188] Della Bella S, Molteni M, Mocellin C, et al. Novel mode of action of iloprost: in vitro downregulation of endothelial cell adhesion molecules. Prostaglandins Other Lipid Mediat 2001;65:73–83.

[189] Mittag MP, Beckheinrich UF. Systemic sclerosis-related Raynaud's phenomenon: effects of iloprost infusion therapy on serum cytokine, growth factor and soluble adhesion molecule levels. Acta Derm Venereol 2001;81:294–7.

[190] Scorza R, Caronni M, Mascagni B, et al. Effects of long-term cyclic iloprost therapy in systemic sclerosis with Raynaud's phenomenon. A randomized controlled study. Clin Exp Rheumatol 2001;19:503–8.

[191] Tingey T, Shu J, Smuczek J, Pope J. Meta-analysis of healing and prevention of digital ulcers in systemic sclerosis. Arthritis Care Res 2013;65:1460–71.

[192] Korn JH, Mayes M, Matucci-Cerinic M, et al. Digital ulcers in systemic sclerosis: prevention by treatment with bosentan, an oral endothelin receptor antagonist. Arthritis Rheum 2004;50(12):3985–93.

[193] Khanna D, Denton CP, Merkel PA, et al. Effect of macitentan on the development of new ischemic digital ulcers in patients with systemic sclerosis: DUAL-1 and DUAL-2 randomized clinical trials. JAMA 2016;315:1975–88.

[194] Johnson SR, Mehta S, Granton JT. Anticoagulation in pulmonary arterial hypertension: a qualitative systematic review. Eur Respir J 2006;28:999–1004.

[195] Mathai SC, Hassoun PM. Therapy for pulmonary arterial hypertension associated with systemic sclerosis. Curr Opin Rheumatol 2009;21:642–8.

[196] Badesch DB, Tapson VF, McGoon MD, et al. Continuous intravenous epoprostenol for pulmonary hypertension due to the scleroderma spectrum disease. Ann Intern Med 2000;132:425–34.

[197] Hoeper MM, Spiekerkoetter E, Westerkamp V, Gatzke R, Fabel H. Intravenous iloprost for treatment failure of aerosolised iloprost in pulmonary arterial hypertension. Eur Respir J 2002;20:339–43.

[198] Galie N, Humbert M, Vachiery JL, et al. Effects of beraprost sodium, an oral prostacyclin analogue, in patients with pulmonary arterial hypertension: a randomized, double-blind, placebocontrolled trial. J Am Coll Cardiol 2002;39:1496–502.

[199] Saji T, Ozawa Y, Ishikita T, et al. Short term hemodynamic effect of a new oral PgI2 analogue, Beraprost, in pulmonary and secondary pulmonary hypertension. Am J Cardiol 1996;78:244–7.

[200] Galie N, Rubin LJ, Koeper MM, et al. Treatment of patients with mildly symptomatic pulmonary arterial hypertension with bosentan (EARLY study): a double-blind, randomized controlled trial. Lancet 2008;371:2093–100.

[201] Rubin LJ, Badesch DB, Barst RJ, et al. Bosentan therapy for pulmonary hypertension. N Engl J Med 2002;346:896–903.
[202] Patel T, McKeage K. Macitentan: first global approval. Drugs 2014;74:127–33.
[203] Galie N, Olschewski H, Oudiz RJ, et al. Ambrisentan for the treatment of pulmonary arterial hypertension. Results of the ambrisentan in pulmonary arterial hypertension, randomized, doubleblind, placebo-controlled, multicenter, efficacy (ARIES) study 1 and 2. Circulation 2008;117:3010–9.
[204] Denton CP, Humbert M, Rubin L, Black CM. Bosentan treatment for pulmonary arterial hypertension related to connective tissue disease: a subgroup analysis of the pivotal clinical trials and their open-label extensions. Ann Rheum Dis 2006;65:1336–40.
[205] Galiè N, Humbert M, Vachiery JL, et al. 2015 ESC/ERS Guidelines for the diagnosis and treatment of pulmonary hypertension: the joint task force for the diagnosis and treatment of pulmonary hypertension of the European society of cardiology (ESC) and the European respiratory society (ERS): endorsed by: association for European paediatric and congenital cardiology (AEPC), international society for heart and lung transplantation (ISHLT). Eur Respir J 2015;46(4):903–75.
[206] McGoon M, Frost A, Oudiz R, et al. Ambrisentan therapy in patients with pulmonary arterial hypertension who discontinued bosentan or sitaxsentan due to liver function test abnormalities. Chest 2009;135:122–9.
[207] Giaid A, Saleh D. Reduced expression of endothelial nitric oxide synthase in the lungs of patients with pulmonary hypertension. N Engl J Med 1995;333(4):214–21.
[208] Lian TY, Jiang X, Jing ZC. Riociguat: a soluble guanylate cyclase stimulator for the treatment of pulmonary hypertension. Drug Des Devel Ther 2017;11:1195–207.
[209] Klinger JR, Abman SH, Gladwin MT. Nitric oxide deficiency and endothelial dysfunction in pulmonary arterial hypertension. Am J Respir Crit Care Med 2013;188(6):639–46.
[210] Matei AE, Beyer C, Györfi AH, et al. Protein kinases G are essential downstream mediators of the antifibrotic effects of sGC stimulators. Ann Rheum Dis 2018;77(3):459.
[211] Ghofrani HA, Galiè N, Grimminger F, et al. Riociguat for the treatment of pulmonary arterial hypertension. N Engl J Med 2013;369(4):330–40.
[212] Distler JHW, Distler O. Intracellular tyrosine kinase as a novel targets for anti-fibrotic therapy in systemic sclerosis. Rheumatology 2008;47:v10–1.
[213] Hassoun PM. Therapies for scleroderma-related pulmonary arterial hypertension. Expert Rev Respir Med 2009;3(2):187–96.
[214] van den Hoogen FH, Boerbooms AM, Swaak AJ, et al. Comparison of methotrexate with placebo in the treatment of systemic sclerosis: a 24 week randomized double-blind trial, followed by a 24 week observational trial. Br J Rheumatol 1996;35(4):364–72.
[215] Adler S, Huscher D, Siegert E, et al. Systemic sclerosis associated interstitial lung disease—individualized immunosuppressive therapy and course of lung function: results of the EUSTAR group. Arthritis Res Ther 2018;20(1):17.
[216] Akesson A, Schema A, Lundin A, Wollheim FA. Improved pulmonary function in systemic sclerosis after treatment with cyclophosphamide. Arthritis Rheum 1994;37(5):729–35.
[217] Tashkin DP, Elashoff R, Clements PJ, et al. Cyclophosphamide versus placebo in scleroderma lung disease. N Engl J Med 2006;354:2655–66.
[218] Burt RK, Shah SJ, Dill K, et al. Autologous non-myeloablative haemopoietic stem-cell transplantation compared with pulse cyclophosphamide once per month for systemic sclerosis (ASSIST): an open-label, randomised phase 2 trial. Lancet 2011;378:498–506.

[219] van Laar JM, Farg D, Sont JK, et al. Autologous hematopoietic stem cell transplantation vs intravenous pulse cyclophosphamide in diffuse cutaneous systemic sclerosis: a randomized clinical trial. JAMA 2014;311:2490–8.

[220] Walker UA, Saketkoo LA, Distler O. Haematopoietic stem cell transplantation in systemic sclerosis. RMD Open 2018;4(1):e000533.

[221] Helfrich DJ, Banner B, Steen VD, Medsger Jr TA. Normotensive renal failure in systemic sclerosis. Arthritis Rheum 1989;32:1128–34.

[222] Emmanuel A. Current management of the gastrointestinal complications of systemic sclerosis. Nat Rev Gastroenterol Hepatol 2016;13(8):461–72.

[223] Baron M, Bernier P, Cote LF, et al. Screening and management for malnutrition and related gastro-intestinal disorders in systemic sclerosis: recommendations of a North American expert panel. Clin Exp Rheumatol 2010;28(2 Suppl. 58):S42–6.

[224] Khan M, Santana J, Donnelan C, et al. Medical treatments in the short term management of reflux esophagitis. Cochrane Database Syst Rev 2007;18(2):CD003244.

[225] Mainie I, Tutuian R, Castell DO. Addition of a H2 receptor antagonist to PPI improves acid control and decreases nocturnal acid breakthrough. J Clin Gastroenterol 2008;42:676–9.

[226] Williams C, McColl KE. Review article: proton pump inhibitors and bacterial overgrowth. Aliment Pharmacol Ther 2006;23:3–10.

[227] Ramirez-Mata M, Ibañez G, Alarcon-Segovia D. Stimulatory effect of metoclopramide on the esophagus and lower esophageal sphincter of patients of patients with PSS. Arthritis Rheum 1977;20(1):30–4.

[228] Marie I, Ducrotte P, Denis P, Menard JF, Levesque H. Small intestine bacterial overgrowth in systemic sclerosis. Rheumatology (Oxford) 2009;48:1314–9.

[229] Perlemuter G, Cacoub P, Chaussade S, et al. Octreotide treatment of chronic intestinal pseudoobstruction secondary to connective tissued diseases. Arthritis Rheum 1999;42:1545–9.

[230] Raja J, Nihtyanova SI, Murray CD, et al. Sustained benefit from intravenous immunoglobulin therapy for gastrointestinal involvement in systemic sclerosis. Rheumatology 2016;55(1):115–9.

[231] Sanges S, Rivière S, Mekinian A, et al. Intravenous immunoglobulins in systemic sclerosis: data from a French nationwide cohort of 46 patients and review of the literature. Autoimmun Rev 2017;16(4):377–84.

[232] Pettersson H, Boström C, Bringby F, et al. Muscle endurance, strength, and active range of motion in patients with different subphenotypes in systemic sclerosis: a cross-sectional cohort study. Scand J Rheumatol 2019;48(2):141–8.

[233] Smith V, Van Praet JT, Vandooren B, et al. Rituximab in diffuse cutaneous systemic sclerosis: an open-label clinical and histopathological study. Ann Rheum Dis 2010;69:193–7.

[234] van Laar JM. B-cell depletion with rituximab: a promising treatment for diffuse cutaneous systemic sclerosis. Arthritis Res Ther 2010;12:112.

[235] Khanna D, Denton CP, Lin CJF, et al. Safety and efficacy of subcutaneous tocilizumab in systemic sclerosis: results from the open-label period of a phase II randomised controlled trial (faSScinate). Ann Rheum Dis 2018;77(2):212–20.

[236] Allanore Y, Devos-François G, Caramella C, et al. Fatal exacerbation of fibrosing alveolitis associated with systemic sclerosis in a patient treated with adalimumab. Ann Rheum Dis 2006;65(6):834–5.

[237] Bruni C, Praino E, Allanore Y, et al. Use of biologics and other novel therapies for the treatment of systemic sclerosis. Expert Rev Clin Immunol 2017;13(5):469–82.

[238] Denton CP, Engelhart M, Tvede N, et al. An open-label pilot study of infliximab therapy in diffuse cutaneous systemic sclerosis. Ann Rheum Dis 2009;68(9):1433–9.

[239] Ostor AJ, Crisp AJ, Somerville MF, Scott DG. Fatal exacerbation of rheumatoid arthritis associated fibrosing alveolitis in patients given infliximab. BMJ 2004;329:1266.
[240] Koh ET, Lee P, Gladman D, Abu-Shakram M. Pulmonary hypertension in systemic sclerosis: an analysis of 17 patients. Br J Rheumatol 1996;35:989–93.
[241] Elhai M, Meune C, Boubaya M, et al. Mapping and predicting mortality from systemic sclerosis. Ann Rheum Dis 2017;76(11):1897–905.

Chapter 18

Leveraging genomics to uncover the genetic, environmental and age-related factors leading to asthma

Brian D. Modena, Ali Doroudchi[a], Parth Patel[a], Varshini Sathish[a]
Division of Allergy, National Jewish Health, Denver, CO, United States

Introduction

An estimated 9% of children and 6% of adults in the United States have asthma [1]. The total number of asthma sufferers worldwide is estimated to be over 300 million, with an additional 100 million expected to develop asthma by 2025 [2–5]. Developed countries are the most affected, with some of the highest rates found in the United Kingdom, Australia, New Zealand and the Republic of Ireland [3]. Asthma prevalence is rising significantly in developing countries in transition to a more Western lifestyle [3]. In 2007, the cost of disease in the United States was estimated to be $56 billion in relation to medical expenses, missed days of work, and early deaths [1]. The rate of asthma deaths has likely plateaued, but is still as high as 250,000 per year worldwide [6]. Morbidity and mortality are particularly high in ethnic minorities living below or near the poverty line, and African American children had a death rate 10 times that of non-Hispanic white children in 2015 [7]. Thus, asthma is a costly, growing health problem associated with high morbidity and mortality.

Clinically, asthma is characterized by episodes of coughing, chest tightness, wheezing, dyspnea, or sputum production. Often, asthma sufferers experience a combination of these symptoms, or some symptoms more than others. Pulmonary breathing tests typically demonstrate variable airway obstruction and hyperreactivity, but may be normal, even in patients with severe and uncontrolled disease [8]. Thus, the diagnosis of asthma, which is based on general clinical symptoms and variable lung function testing, is non-specific and heavily dependent on clinical history. Within the "umbrella" diagnosis of asthma there

a With contributions.

Genomic and Precision Medicine. https://doi.org/10.1016/B978-0-12-801496-7.00018-6

exists a diverse array of differing clinical phenotypes [9]. For example, childhood asthma is often associated with personal and parental atopic diseases (i.e., atopic dermatitis, food allergy, eosinophilic esophagitis, allergic rhinitis), viral infections, and tobacco smoke exposure [10]. Alternatively, adult-onset asthma is less associated with atopic disease [11, 12], but more associated with female sex [13], sinus disease [14], and preceding respiratory infections such as pneumonia [15]. In addition, adult-onset disease is often of higher severity [12, 16] with a faster and more persistent decline in lung function [17]. Moreover, although severe patients are found in every demographic and age group, the most common phenotype is an adult female that is older and obese [18].

The underlying molecular and inflammatory mechanisms of asthma are complex and remain incompletely understood. The structural changes include epithelial alterations, subepithelial fibrosis, mucous gland enlargement, smooth muscle hypertrophy (increased size) and hyperplasia (increased number), and neovascularization [19]. Within the epithelial layer, there is an increased density of goblet cells creating a tendency for mucous plugging [20–24], replacement of ciliated epithelial cells with columnar cells having damaged or poor-functioning cilia [25–27], shedding of the epithelium, increased growth factor release and up-regulation of epidermal growth factor receptors [28–32]. As the first line of defense against inhaled microbes and pollutants, airway epithelial cells (AECs) secrete potent chemokines and cytokines when damaged or activated by the presence of pathogens. These chemical messengers include the innate cytokines IL-1, IL-6, IL-8, interferons-α and β (i.e. Type I interferons) and TNF-α [33, 34], in addition to potent eosinophil chemokines known as eotaxins [35, 36] and several pro-Type 2 inflammatory chemokines, including thymic stromal lymphopoietin (TSLP) [37–39], IL-25 [39, 40], and IL-33 [39, 41]. AECs in the asthmatic airway appear perpetually and inappropriately activated. AEC secretion of these pro-inflammatory messengers, particularly TSLP [42, 43], IL-33 [41] and eotaxins [44, 45], are upstream drivers of inflammation. Similarly, while epidermal growth factor receptor (EGFR) signaling is required for airway repair, its perpetual up-regulation in asthma [28–31] may increase fibrosis and induce the innate immune response [28, 46]. EGFR activation in AECs has also demonstrated an ability to induce mucin synthesis in response to bacterial products, cigarette smoke, mechanical damage, and inflammatory cytokines [47].

The subepithelial fibrosis seen in asthma occurs just below the epithelial basement membrane in the lamina reticularis [19], and is strongly associated with asthma severity [48, 49]. Moreover, subepithelial deposition of collagen types I and III has been positively correlated with both severity and airway hyperreactivity [49–51]. The cause of epithelial fibrosis is not well understood, although mechanistically occurs via an increased deposition and decreased degradation of extra-cellular matrix (ECM) proteins. This profibrotic balance may relate to a deficit in matrix metalloproteinase (MMP) activity, which are a family of proteases produced by interstitial cells, macrophages and neutrophils, and implicated in collage degradation. Several MMPs (MMP-2, -3, -8 and -9) have been implicated in asthma, although many studies show an up-regulation of these proteases in relation to healthy controls [19, 52–54]. Taken together, it suggests that although these proteases are increased in asthma vs. healthy airways, a net deficit of fibrosis relative to collagen breakdown by MMPs results in an overall profibrotic state. Beneath the subepithelial layer, airway smooth muscle mass has similarly correlated to asthma severity [55]. Although wall thickness may limit reactivity, as it has also been inversely correlated with airway reactivity to methacholine [56]. Platelet-derived epidermal growth factor (PDEGF), and epithelial-secreted epidermal growth factor (EGF) and TGF-β all likely play a role in airway smooth muscle proliferation [57–61]. TGF-β also plays an important role in the development of subepithelial fibrosis [57].

Asthma airways have been characterized by the Type 2 (T2) inflammatory profile, deriving its name from $CD4^+$ T (Th2) lymphocytes that secrete T2 putative cytokines IL-4, IL-5 and IL-13. As mentioned, the release of epithelial cytokines TSLP, IL-25, IL-33 and chemokine eotaxin-3 are important proximal steps to the initiation of T2 inflammation [62]. Specifically, TSLP upregulates OX40 on dendritic cells, promoting Th2 cell differentiation in naïve CD^+ T cells [63]. Interestingly, a new monoclonal antibody targeting TSLP has recently demonstrated in a Phase II trial the ability to suppress both T2 and non-T2 inflammation in moderate-to-severe asthmatic adults, indicating TSLP activation is important to both T2 and non-T2 inflammatory pathways [64]. IL-33 binds ST2 receptors (*IL1RL1*) and activates $CD4^+$ T cells, mast cells and a newly identified population of cells known as innate lymphoid cells (ILCs) [65]. ILCs were initially described as a small population of non-delineated T cells that can produce T2 cytokines when stimulated with IL-33 and IL-25 in combination with IL-2 [66]. Since this time, there has been increasing evidence that ILCs, which have been shown to produce relatively large amounts of T2 cytokines when activated, play a major role in pathogenesis of asthma [67]. In parallel with these recent advances, i.e., the elucidation by which epithelial cells regulate airway T2 inflammation and the discovery of ILCs, several large-scale genome wide association studies (GWAS) found that many of these newly-discovered cytokines or their cognate receptor (i.e., IL-33, TSLP, IL-33 receptor ST2 (*IL1RL1*)) are located at highly significance and reproductible asthma susceptibility loci.

Although the numerous pathways of Type 2 inflammation in asthma is complex and beyond the scope of this chapter, worth mentioning are the seminal T2 inflammatory mechanisms that pertain to asthma. Under the influence of Type 2 cytokines, antigen-presenting cells, such as dendritic cells and macrophages, promote the creation of allergen-reactive Th2 memory cells. These "sensitized" Th2 cells then promote B cells to produce IgE antibodies (i.e. isotype class switching) that target allergen protein epitopes. The Fc portion of allergen-specific IgE antibodies bind the high affinity IgE receptor (FcεR1) on the surface of mast cells, causing mast cell degranulation and release of potent pre-formed mediators on subsequent allergen exposure. Together, this process creates the mechanism for the acute allergic response following aeroallergen exposure, while mast cell degranulation and release of cytokines also leads to the delayed inflammatory response in asthmatic airways. T2 cytokine IL-5 recruits and activates eosinophils, augmenting inflammation through the release leukotrienes, platelet activation factor, major basic protein, and other cytokines [23–26]. As mentioned, IL-4 and IL-13 induce airway epithelial alterations, and activate epithelial cells to produce nitric oxide, a reactive oxidative species, eotaxin-3, a strong chemotaxis mediator for eosinophils, anti-microbial products and another pro-inflammatory cytokines.

Interestingly, not all asthma patients demonstrate evidence of T2 inflammation on cross-sectional analysis. This discovery was made possible by mRNA global gene expression, which was first used to identify a T2-specific gene signature in bronchial and nasal airway epithelial cells [68, 69], and then used to show that only ~1/2 of asthma patients have evidence of T2 inflammation on cross-sectional analysis [68–70]. One possibility was that the T2 "low" subjects simply had a wane, or corticosteroid-induced suppression, of T2 inflammatory signaling at the time of sampling. Notwithstanding, recent data have confirmed the presence of T2 "high" vs. T2 "low" phenotypes that are stable over time, noting that those with persistently T2 high disease had higher severity and worse outcomes over time [71]. Determining the inflammatory mechanisms underlying or causative in T2 low is an area of high research interest. Some of the leading hypotheses are that these individuals have activation of the Type 1 inflammatory pathway (i.e., IFN-γ signaling), or innate immune pathways featuring cytokines IL-1, IL-6, IL-18 or TNF-α, or others. Supportive of numerous and different active inflammatory pathways, asthma patients are well known to demonstrate differences in the airway inflammatory milieu, where eosinophilic, neutrophilic and paucigranulocytic phenotypes are all readily observed [70, 72, 73]. Again, such findings highlight the heterogeneity of asthma, where multiple molecular roots with different inflammatory pathways are thought to lead to a common clinical phenotype.

Hereditability of asthma

The hereditability of asthma is estimated to be between 35% and 95% [74–86]. Evidence for the inheritability of asthma is best demonstrated by twin studies that show concordances rates in monozygotic twins are two times higher than dizygotic twins (Table 1). In some patients, asthma occurs comorbidly with atopic dermatitis (AD), allergic rhinitis (AR) and food allergy (FA), which typically occurs in childhood, and known as the "atopic march." The atopic march is also generally associated with positive skin prick tests (or positive allergen-specific IgE blood testing), blood eosinophilia, and maternal history of atopic disease. In children with atopic dermatitis (AD), 20–30% also develops asthma [87]. In patients with severe AD, as high as 70% will develop asthma [87]. In patients with allergic rhinitis, the reported prevalence of asthma varies considerably, but has been reported as high as 80% [88]. Thus, the case for asthma as an inherited disease is perhaps best made in those with childhood-onset asthma in association with comorbid atopic disease.

TABLE 1 Twin studies show that the heritability of liability to asthma

Study	Population	Age (yrs)	Method	Dx	MZf	MZm	DZf	DZm	Variance in liability due to genetic factors
Duffy [75]	3808 pairs Australian twins	18–88	Ques	Self	0.59	0.76	0.26	0.19	0.60–0.70
Nieminen [76]	13,888 pairs Finnish twins		Records						35.6
Harris [77]	5864 pairs Norwegian twins	0–25	Ques	Self	0.37	0.52	0.07	0.13	0.75
Skadhauge [79]	11,688 pairs Danish twins	12–41	Ques	Self	0.42 (12–26 yrs); 0.38 (27–41 yrs)	0.48 (12–26 yrs); 0.51 (27–41 yrs)	0.26 (12–26 yrs); 0.09 (27–41 yrs)	0.19 (12–26 yrs); 0.16 (27–41 yrs)	F: 0.50–0.61 M: 0.76–0.79

Continued

TABLE 1 Twin studies show that the heritability of liability to asthma–cont'd

Study	Population	Age (yrs)	Method	Dx	MZf	MZm	DZf	DZm	Variance in liability due to genetic factors
Laitinen [78]	2483 pairs Finnish twins	~16	Ques	Phys	0.25	0.55	0.11 (ss)	0.27 (ss)	0.79 (general) 0.87 (parent w/ asthma);
Koeppen-Schomerus [80]	4910 pairs English twins	4	Ques	Self	0.66	0.64	0.34 (ss), 0.34 (os)	0.45 (ss)	0.68
Hallstrand [81]	1384 pairs American twins	~ 32	Ques	Self	0.51 (combined)		0.34 (combined)		0.77
Fagnani [84]	392 pairs Italian twins	8–17	Ques	Self	0.67 (combined)		0.35 (combined)		0.92

Abbreviations: DZf: dizygotic female, DZm: dizygotic male, MZf: monozygotic female, MZm: monozygotic male, Ques: Questionnaires, os: opposite-sexed, ss: single-sexed, yrs.: years.

Types of genetic variation

Genetic variation can be single nucleotide polymorphisms (SNPs, pronounced "snips"), insertion deletion polymorphisms (indels), or gene copy number variations (CNVs). A SNP is a change in a single nucleotide base (i.e., A, T, C or G) from one base to another. SNPs occur once in every 1000 nucleotides in humans, meaning that ~ 4 to 5 million common SNPs, those with minor-allele frequency of at least 5%, are inherited across generations in blocks [89, 90]. Moreover, more than 100 million SNPs have been identified in the global population [91]. A haplotype is a term used for when an individual(s) has a combination of SNPs or alleles. Meanwhile, an insertion is the addition of one or more nucleotide base pairs into the DNA. Alternatively, a deletion is the loss of one or more nucleotide base pairs. Thus, the term "indel" is used to label either an insertion or deletion, and indels can range in size from 1 to 10,000 base pairs. While the total number of indels in the human genome are less than the number of SNPs founds, indels cause similar levels of variation in terms of base pair variation [92]. Gene copy number variations are a type of indel where fairly large stretches of DNA, typically >1 kb and spanning numerous genes, are repeated or deleted, acting to increase or decrease the numbers of gene copies [93].

Candidate gene association studies

Candidate gene association studies look at the genetic variation associated with disease within a limited number of pre-specified genes. Candidate gene studies are typically structured as case control studies. Cases of disease and controls are first identified, and then genetic differences determined, i.e., identifying variants (SNPs, haplotypes, indels, CNVs) found more commonly in one group or the other. Results are typically reported as odds ratios (ORs) or relative risks (RRs), depending on whether the study was structured as a case control or cross-sectional study, respectively.

Whereas genome wide association studies (GWAS) investigate genetic variants spanning the entire genome, candidate gene studies limit the analysis to a relatively few number of genes. As a result, candidate gene studies have increased statistical power to detect differences. They are also unavoidably biased toward genes and biological pathways of "interest," i.e., candidate genes are hand-picked by investigators based on scientific interest (i.e., functional candidate) or based on chromosomal position (i.e., positional candidate). As a result, the candidate gene variants discovered may be given undue attention.

Nonetheless, many candidate gene studies were performed and identified hundreds of genetic variants associated with asthma. The results were summarized in several review articles [94–96], and will not be repeated here. One review of 500 candidate gene studies showed that 25 genes have been associated with asthma in 6 or more populations, including T2 inflammatory response genes (*IL13, IL4, IL4RA, STAT6*), the IgE receptor (*FCER1B*), a coreceptor for bacterial lipopolysaccharide that is preferentially expressed on monocytes and macrophages (*CD14*), a prostanoid receptor to thromboxane (*TXAR2*), two MHC class II genes (*HLA-DRB1, HLA-DQB1*), glutathione-S-transferases (*LTC4S, GSTM1, GSTP1*), pro-inflammatory products (*TNF, CCL5, NOD1, LTA*) and proteins responsible for immune suppression (*IL10, TGFB1, CTLA4*). *ADAM33*, which encodes a metalloproteinase expressed in airway smooth muscles and lung fibroblasts and likely involved in airway remodeling [97], was one of the first gene variants discovered by a candidate gene study in a Caucasian population [98]. *ADAM33* polymorphisms were reproducibly associated with asthma [99–101] and a more rapid decline in postbronchodilator forced expiratory volume in 1 s (FEV_1) [102] in subsequent studies. As a result, *ADAM33* has garnered much attention over the years [103] Yet, the exact mechanism by which *ADAM33* variants increase susceptibility to asthma was never elucidated and its locus was not identified as an asthma susceptibility locus in any of the major subsequent genome wide association studies (GWAS).Thus, ADAM33 serves as cautionary tale where preliminary genctic findings may lead researchers down fruitless pathways.

Several additional weaknesses of gene candidate studies are worth mentioning. Another difficulty with candidate gene studies is that differences in population substructures may result in the detection of alleles that differ between case and control, but are unrelated to disease. Moreover, the number of genetic roots underlying asthma is unknown and could be numerous given its heterogeneity. Therefore, the likelihood of identifying a common causative variant among a small population of asthmatics, as was the case for most gene candidate studies, may be unpredictably low. Finally, gene candidate studies had often limited the analysis to the protein-coding regions of genes and neglected how neighboring genes and introns affect gene expression or epigenetics of candidate genes. For these reasons and others, the accuracy and importance of the early candidate gene association studies are now questioned.

Genome-wide approaches

With advancements in microarray and sequencing technologies, genome-wide approaches have largely replaced candidate gene studies. Unlike candidate gene studies, genome-wide approaches scale the entire genome. The most popular genome-wide study has been the genome-wide association study (GWAS). In a GWAS, a large number of genetic variants (i.e., SNPs) are associated to disease, and ranked according to their associated *P*-value significance. GWAS results are typically presented through a Manhattan plot. In a Manhattan plot, SNPs are positioned along the x-axis according to chromosomal position. Plotted on the y-axis is the negative log of the SNP's associated *P*-value. SNPs with the lowest *P*-value significance (i.e., highest association with disease) are thus positioned at the top of the graph, giving the plot the appearance of a Manhattan skyline.

Two major advantages of a genome-wide approach are the number of genes (and haplotypes) interrogated, and that the studies maintain "hypothesis free" testing. Thus, GWAS are much more objective than candidate gene studies.

The major disadvantage of a GWAS is the large patient populations required. A GWAS involves testing thousands to millions of SNPs for an association with disease. Performing such a large number of tests makes it difficult to detect statistical significance, as the current accepted practice is to correct *P*-values in accordance to the number of tests performed. Or another standard practice is to set a very low *P*-value cut-off in order to be considered a statistically significant result, e.g., *P*-value $<10^{-6}$. Both of these methods are accepted and have the same goal, which is to avoid false positive results. However, as mentioned, the human genome has ~4–5 million SNPs. Thus, a GWAS interrogating millions of SNPs would require two very large populations to overcome this enormous statistical burden. As will be discussed below, GWAS have tried to overcome this limitation by interrogating only a limited number of "marker" SNPs.

Genome-wide linkage analysis (GWLA)

Genetic linkage studies require families where more than one relative is affected by the disease. The underlying premise is that disease-causing genes will co-segregate with the family members affected by disease. Such loci are tracked by microsatellites or short tandem repeats (STRs) [104]. STRs are small units of nucleotides (2–13 nucleotides long) repeated hundreds of times in a row. STRs are highly variable between individuals (i.e., highly polymorphic), and can thereby be used to track the transfer of genetic material between generations. Using STRs to track genetic material also relies on an important concept known as linkage disequilibrium (LD). LD is the tendency of alleles at two loci to be inherited together. Typically, the closer two alleles are in chromosomal position, the more likely they are inherited together. Thus, unique STRs can be used to track of blocks of DNA as they are passed from parents to offspring.

One advantage of linkage studies is that a limited number of marker loci are required. Most linkage studies require interrogation of less than 10,000 highly polymorphic microsatellite markers. Another advantage (although often forgotten) is the ability help investigators localize gene regions inferring disease risk. As will be discussed, the risk of developing asthma may depend on the existence of multiple rare variants. Thus, linkage studies could potentially be used to help investigators narrow in on asthma susceptibility loci that contain rare variants responsible for disease. Although, once an asthma susceptibility locus is identified, the discovery of the multiple rare variants contributing to the linkage peak is often difficult. Thus, the major disadvantage of linkage studies is their poor resolution. Linkage analyses may point to a region containing hundreds of genes, but without an ability to narrow it down any further. Another weakness is that they require families to perform the studies. As a result, the scalability to large populations is much more difficult.

Since the 1990s, numerous genome-linkage studies have been performed, showing linkage on multiple chromosomal areas (e.g., 2q22-33, 5q31-33, 6p21, 12q14-24, 13q14, 14q11-13, 19q13), although reproducibility has been low [96, 104]. Following, there were two meta-analyses of genome-wide linkage studies in relation to asthma and related traits that have been performed [104, 105]. The first analysis by Denham et al. included 11 Caucasian studies taken over 10 years [104]. The authors found that genes in the regions **17q12-21** and **3p14.1-q22.1** influence allergen skin prick responses (SPT), while genes in **5q.11.2-q14.3** and **6pter-p21.1** may relate to IgE production. Finally, the **6p22.3-p21.1** region was identified as containing susceptibility genes related to bronchodilator response (a region including the HLA locus), but that no chromosomal region showed genome-wide significance using asthma as a phenotype. Alternatively, the second meta-analysis by Bouzigon et al. featured genome-wide linkage studies in 20 independent populations from all different ethnic origins (>3204 families with >10,000 subjects) [105]. In agreement with the Denham et al., the region **3p25.3-q24** showed genome-wide significance in relation to SPT for all families. In European families, **2q32-34** showed significance for blood eosinophil levels, **5q23-33** for quantitative scores of SPTs, and **17q12-24** again showed significance for SPT. Moreover, the authors found additional regions that reached genome-wide significance for asthma: 2p21-p14, 4q34.3-qter and 5q31.1-q31.2.

Genome-wide association studies (GWAS)

In a GWAS, the most commonly variant SNPs are interrogated. These common variant SNPs act as tags for inherited blocks of DNA, while other bases in the region can be imputed based on population data. Through the advancement of high throughput genotyping platforms, over a million SNPs can now be quickly and affordably genotyped. In comparison to a GWLA, a GWAS does not require families, has a higher resolution, and offers an ability to detect SNPs with modest effect sizes. In contrast to candidate gene studies, GWAS are hypothesis-free. Disease-incurring SNPs are identified and ranked based only on their association with the disease of interest. The major disadvantage of a GWAS is the statistical burden created through the multiple hypothesis testing, i.e., reaching statistical significance for an association between a SNP and the disease is difficult after correcting for the large number of tests performed (typically >1 million SNPs tested). In practice, a *P*-value significance of less than 10^{-8} or 10^{-7} is the typical threshold for a SNP to be considered significantly associated with a disease or trait. As a result, very large samples sizes (e.g., several thousand subjects) are required to obtain the necessary statistical power to detect associations. Another disadvantage, GWASs have a reduced power to detect SNPs with lower frequencies in the population, i.e., so-called rare variants, as they interrogate only common SNP variants. For asthma, it is likely that uncommon risk variants are shared poorly between individuals, and yet responsible for disease.

Thus, it's perhaps not surprising that only a small percentage (~3%) of the heritability of asthma can be explained by the asthma risk variants identified in all the GWAS performed to date [106].

Numerous GWASs have been performed in populations of asthma with variability in race, disease severity, age of onset, and comorbid conditions (e.g., presence of atopic disease such as allergic rhinitis). Table 2 provides a summary list of the major asthma susceptibility loci identified reproducibility by GWASs, organized by chromosomal position. In one of the first GWAS of asthma, Moffatt et al. interrogated 317,000 SNPs in 994 childhood-onset asthma patients compared to 1243 non-asthmatic, healthy controls [109]. The results were replicated in a second cohort of 2320 German children and 3301 healthy subjects from the British 1958 Birth cohort. The authors found multiple markers in the chromosomal region 17q21 were strongly and reproducibly associated with childhood onset asthma (P-value $<10^{-12}$). As mentioned above, the **17q12-21** location was the same region associated with allergen skin prick test (SPT) positivity in the GWLAs performed by Denham et al. and Bouzigon et al. Further, Moffatt et al. showed that a SNP located at the asthma susceptibility locus at 17q21 strongly associated with mRNA expression a nearby gene, *ORMDL3*, in Epstein-Barr virus-transformed lymphoblastoid cell lines derived from children with asthma.

Sleiman et al. investigated 317,000 SNPs in two children cohorts with physician-diagnosed asthma at a large children's hospital in the United States [110]. The first cohort included 807 Caucasian asthmatic subjects in comparison to 2583 healthy controls, and results replicated in a second cohort of 1456 African American (AA) asthmatic children vs. 1973 controls. In the Caucasian cohort, seven of the nine most significant SNPs identified by Moffatt et al. were reproduced, and again, a SNP in the 17q21 region significantly associated with the gene expression of *ORMDL3*. Yet, in the AA population, none of the SNPs associated with asthma were replicated. Halapi et al. also replicated a SNP in the **17q12-21** region (a SNP in the intron region of *ORMDL3*) in six European and one Asian cohort (4917 cases vs. 34,589 controls), but only for early childhood (ages 0–5) and adolescence (ages 1–17) subgroups [111]. This SNP also correlated positively with asthma severity in the early onset cases. In agreement, Bouzigon et al. had shown that the risk conferred by 17q21 genetic variants is restricted to early-onset disease, and increased by early-life exposure to tobacco smoke [112].

Ober et al. performed a GWAS in a founder population of European descent, the Hutterites [113]. A population of 652 Hutterites were first assessed for asthma based on the presence of characteristic clinical symptoms, which included both evidence of bronchial hyperresponsiveness and a physician's opinion. A total of 76 (11.7%) were diagnosed with asthma, although an additional 80 (12.3%) subjects demonstrated bronchial hyperresponsiveness without clinical symptoms. At the time, the authors were interested in a chitinase-like protein (YKL-40), encoded by the *CHI3L1* gene. Chitinases are evolutionary-conserved proteins implicated in the inflammatory response and

TABLE 2 Summary of asthma susceptibility loci identified by GWAS or meta-analysis of GWAS

Region	Gene(s)	Trait	Strongest SNP	Studies	Population	*P*-value	OR or beta-coefficient (95% CI)
1q23.1	*PYHIN1*	Asthma	rs158932907	[107]	Diverse N. American		
1q31.3	*DENND1B, CRB1*	Asthma	rs2786098			2×10–13	1.18 (1.08–1.30)
2q12.1	*IL18R1*	Asthma	rs3771166	[124]	European	3×10–9	1.15 (1.10–1.20)
2q12.1	*IL1RL1*	Asthma	rs102957348	[107]	Diverse N. American		
		Asthma		[107a][a]	Puerto Rican	4×10–3	0.84
2q12	*IL1RL1, IL1RL2, IL18R1*	Asthma	rs1420101	[142]	European and multi-ancestry	9.1×10–20 (Euro); 3.9×10–21 (Multi)	1.12 (both)
4q31	*GAB1*	Asthma		[120]	Japanese		
5q12.1	*PDE4D*	Asthma	rs1588265	[108]	European American	4×10–7	0.60 (0.49–0.73)
			rs1544791	[108]	European American	10×10–7	0.61 (0.54–0.78)

5q22.1	*TSLP*	Asthma	rs110401872	[107]	Diverse N American		
		Asthma		[107a][a]		3×10–4	0.82
		Asthma		[120]	Japanese		
		Asthma	rs10455025	[142]	European and multi-ancestry	2.0×10–25 (Euro); 9.4×10–26 (Multi)	1.15 (both)
5q31.1	*RAD50*	Asthma	rs2244012	[107b]	Diverse N American	3×10–7	1.64 (1.36–1.97)
5q31.1	*IL4*	Asthma/Atopy		Candidate Gene [96]			
5q31.1	*IL13*	Asthma	rs1295686	[124] [107b]	European Diverse N American	1×10–7	1.15 (1.09–1.20)
5q31	*IL13, RAD50, IL4*	Asthma	rs20541	[142]	European and multi-ancestry	1.4×10–14 (Euro); 5.0×10–16	0.89 (both)
5q31.1	*SLC22A5*	Asthma	rs2073643	[124]	European	2×10–7	1.11 (1.06–1.15)
5q31.3	*CD14*	Asthma/Atopy		Candidate Gene [96]			

Continued

TABLE 2 Summary of asthma susceptibility loci identified by GWAS or meta-analysis of GWAS—cont'd

Region	Gene(s)	Trait	Strongest SNP	Studies	Population	*P*-value	OR or beta-coefficient (95% CI)
5q31.3	*NDFIP1, GNDPA1, SPRY4*	Asthma	rs7705042	[142]	European and multi-ancestry	1.6×10–6 (Euro); 7.9×10–9 (Multi)	1.08 (Euro); 1.09 (Multi)
5q32	*ADRB2*	Asthma/Atopy		Candidate Gene [96]			
6p21.32	*HLA-DQ*	Asthma	rs9273349	[124]	European	7×10–14	1.18 (1.13–1.24)
	HLA-DR/DQ		rs1063355	Li et al. 2010	Diverse N Amer		
		Asthma	rs9272346	[142]	European and multi-ancestry	4.8×10–28 (Euro); 5.7×10–24 (Multi)	1.16 (both)
6p21.33	*MICB, HCP5, MCCD1*	Asthma	rs2855812	[142]	European and multi-ancestry	1.7×10–8 (Euro); 8.9×10–12 (Multi)	1.10 (both)
6p22.1	*GPX5, TRIM27*	Asthma	rs1233578	[142]	European and multi-ancestry	5.3×10–9 (Euro); 5.9×10–7 (Multi)	1.11 (Euro); 1.09 (Multi)
6q15	*BACH2, GJA10, MAP3K7*	Asthma	rs2325291	[142]	European and multi-ancestry	8.6×10–13 (Euro); 2.2×10–12	0.91 (both)

8q21.13	*TPD52, ZBTB10*	Asthma plus allergic rhinitis	rs12543811	[142]	European and multi-ancestry	3.4×10–8 (Euro); 1.1×10–10 (Multi)	0.93 (Euro); 0.92 (Multi)
9p24.1	*IL33*	Asthma	rs1342326	[124]	European	9×10–10	1.20 (1.13–1.28)
		Asthma	rs6193455	[107]	Diverse N Amer		
		Asthma	rs992969	[142]	European and multi-ancestry	1.1×10–17 (Euro); 7.2×10–20 (Multi)	0.85 (Euro); 0.86 (Multi)
10p14		Asthma		[120]	Japanese		
10p14	*GATA3, CELF2*	Asthma	rs2589561	[142]	European and multi-ancestry	1.4×10–8 (Euro); 3.5×10–9 (Multi)	0.90 (Euro); 0.91 (Multi)
11q13.5	*C11orf30, LRRC32*	Asthma	rs7927894	[142]	European and multi-ancestry	3.5×10–11 (Euro); 2.2×10–14 (Multi)	1.10 (both)
12q13		Asthma		[120]	Japanese		

Continued

TABLE 2 Summary of asthma susceptibility loci identified by GWAS or meta-analysis of GWAS—cont'd

Region	Gene(s)	Trait	Strongest SNP	Studies	Population	*P*-value	OR or beta-coefficient (95% CI)
12q13.3	*ZNF652, PHB*	Asthma	rs17637472	[142]	European and multi-ancestry	1.6×10–7 (Euro); 3.9×10–9 (Multi)	1.08 (both)
15q21.2	*SCG3*	Asthma	rs17525472			2×10–6	
15q22.2	*RORA*	Asthma	rs11071559	[124]		1×10–7	1.18 (1.11–1.25)
15q22	*RORA, NARG2, VPS13C*	Asthma	rs11071558	[142]	European and multi-ancestry	1.9×10–10 (Euro); 1.3×10–9 (Multi)	0.89 (both)
15q22.33	*SMAD3*	Asthma	rs744910	[124]		4×10–9	1.12 (1.09–1.16)
15q22.33	*SMAD3, SMAD6, AAGAB*	Asthma	rs2033784	[142]	European and multi-ancestry	2.5×10–14 (Euro); 7.4×10–15 (Multi)	1.11 (Euro), 1.10 (Multi)
16p13.13	*CLEC16A, DEX1, SOCS1*	Asthma plus allergic rhinitis	rs17806299	[142]	European and multi-ancestry	2.1×10–10 (Euro); 2.7×10–10	0.90 (Euro); 0.91 (Multi)

17q12	*ORMDL3*	Asthma	rs7216389			9×10–11	1.45 (1.17–1.81)
	GSDMB	Asthma	rs2305480	[124]		1×10–7	1.18 (1.11–1.23)
		Asthma	rs38064405	[107]	Diverse N Amer		
		Asthma		[107a][a]		8×10–12	0.72
17q21	*IKZF3*	Asthma	rs907092	[107a][a]		1×10–12	0.71
17q12–21	*ERBB2, PCAP3, C17orf37*	Asthma	rs2952156	[142]	European and multi-ancestry	7.6×10–29 (Euro); 2.2×10–30	0.86 (Euro); 0.87 (Multi)
17q21.1	*GSDMA*	Asthma	rs3894194	[124]		5×10–9	1.17 (1.11–1.23)
20p13	*KIAA1271*	Asthma	rs4815617			8×10–6	
22q12.3	*IL2RB*	Asthma	rs2284033	[124]		1×10–8	1.12 (1.08–1.16)

[a] *Denotes a meta-analysis.*
Abbreviations: CI: confidence interval, Meta: meta-analysis, N Amer: North American, OR: odds ratio, SNP: single nucleotide polymorphism.

tissue remodeling. Preliminary evidence had suggested YKL-40 was important to asthma pathogenesis [114]. The authors measured YKL-40 protein levels in frozen serum samples from 632 Hutterites and tested for gene variants that associated with YKL-40 levels. The authors also looked at gene variants associated with both pulmonary function and total serum IgE levels. The results were replicated in 344 asthmatic children vs. 294 healthy controls of European descent from Germany, and a second Caucasian population of 99 adult and pediatric asthma subjects vs. 197 healthy controls from Chicago. The authors showed that serum YKL-40 protein levels were a biomarker for both asthma susceptibility and reduced lung function. Moreover, genetic variation in *CHI3L1* influenced YKL-40 blood protein levels, and was associated with asthma risk, bronchial hyperresponsiveness, and reduced lung function.

Next, Weidinger et al. investigated SNPs in relation to total serum IgE levels [115]. High total serum IgE has been correlated with the presence of allergic inflammatory diseases, including the severity of asthma and allergic rhinitis [116, 117]. Moreover, pedigree- and twin-based studies has provided strong evidence that genetics contribute to the variability of total IgE levels [118, 119]. Here the authors tested 353,569 SNPs for association with serum IgE levels in 1530 European adults, with results replicated in 9769 samples from 4 independent study populations. In the primary cohort, no single SNP reached genome-wide significance, likely because the study was not sufficiently powered to overcome the high statistical burden of a GWAS. Nonetheless, in the larger replication cohort, functional variants in *FCER1A*, which encodes the alpha chain of the high affinity receptor for IgE (*FCER1A*) on chromosome 1q23, were significantly correlated to IgE levels (combined P-value$=1.85\times10^{-20}$). Correspondingly, the authors also found that levels of FCER1A proteins on basophils associated with the G allele of a SNP located at the 5′ region of *FCER1A*. In agreement with prior GWASs, a variant SNP in chromosomal regions **17q12-21** associated with decreased mRNA expression levels in both *GSDMB* and *ORMDL3* genes in peripheral blood mononuclear cells of asthma subjects. Of note, several variants in the 17q12-21 region were also shown to associate with early onset asthma, and were more strongly associated with *GSDMB* or *ORMDL3* gene expression [111]. Given the extensive linkage disequilibrium (LD) between SNPs in the area, in combination with the potential for gene expression of neighboring genes to be co-regulated by epigenetic modifications, it has been difficult to determine exactly which SNPs in the 17q12-21 region are most relevant to disease.

Hirota et al. performed a large GWAS in a Japanese population with 7171 asthma subjects and 27,912 controls [120]. Using a primary cohort of 1532 cases vs. 3304 controls with interrogation of 458,847 SNPs, the study identified 5 loci associated with adult asthma. Two loci were close to areas that had been previously identified. One was located at **5q22** near genes *TSLP/WDR36,* and of particular interest given the importance of thymic stromal lymphopoetin (TSLP) to asthma disease [121], and the recent clinical success of anti-TSLP antibodies in the treatment of severe asthma [64, 122]. Moreover, Hunninghake et al.

conducted sex-stratified associations between SNPs in TSLP and asthma in a family-based study of Costa Rican children. They showed that TSLP variants again associated with asthma, but in sex-specific manner [123]. Another susceptibility locus identified by Hirota et al. was at **6p21** near the major histocompatibility complex, but also equally close to other genes in this region, such as *NOTCH*, *C6orf10*, *BTNL2*, etc. The authors reported three additional loci: USP38/GAB1 at **4q31,** a locus on chromosome **10p14** and another gene-rich region on chromosome **12q13**.

GABRIEL Consortium

The GABRIEL (A Multidisciplinary Study to Identify the Genetic and Environmental Causes of Asthma) Consortium was a unique, large-scale effort supported by the European Commission to perform the largest GWAS of asthma to date [124]. GABRIEL included 23 different studies from 14 European countries with a total of 10,365 cases vs. 16,110 controls with 582,892 SNPs interrogated. Some of the same researchers had conducted the prior GWAS (Moffatt et al. [109, 112].) showing that the region 17q21 was most strongly associated with childhood-onset asthma. Yet, the GABRIEL study had a sample size 10 times that of the older one. The authors used a physician diagnosis to determine asthma, and childhood onset asthma was defined as asthma diagnosed before the age of 16 years.

Using a stringent cutoff (P-values $<7.2\times10^{-8}$), in this new study, Moffatt et al. reported significant associations in or near genes *HLA-DQ* (chr **6p21**), *ORMDL3/GSDML* (chr **17q21**), *IL33* (chr **9p24**), *IL18R1* (chr **2q12**), *IL2RB* (chr **22q12**) and *SMAD3* (chr **15q22**). The strongest overall association was a SNP near *HLA-DQ* of the major histocompatibility complex (MHC) (P-value$=7.0\times10^{-14}$; odds ratio, 1.14), which was more associated with adult than childhood-onset asthma. Alternatively, the *ORMDL3/GSDML* was markedly associated with childhood-onset asthma (6.0×10^{-23}), but not adult onset. For the ORMDL3/GSDML susceptibility locus, there was in fact a 380 kb span of variants associated with disease. Of interest to asthma disease pathogenesis, the *IL18R1* SNP was in high LD ($r^2=0.96$) to a SNP related to the *IL1RL1* (aka ST2). Both genes encode receptors for interleukin-1 cytokine family members IL-18 and IL-33, respectively, and share similar pro-inflammatory responses [125]. SNPs flanking the *IL33* gene (P-value$=9\times10^{-10}$) were also significantly associated with asthma. As mentioned, there is increasing evidence that IL-18 and IL-33 signaling axes play significant roles in the development of asthma [126–128]. An additional three genes with P-values close to the cutoff (P-values $<5.0\times10^{-7}$) were also discovered: genes *SLCA22A5* (chr **5q31.1**), *IL13* (chr **5q31.1**) and *RORA* (chr **15q22.2**). Although *SCL22A5* and IL13 are positionally near one another, *SCL22A5* is 273 kb upstream and independent of the *IL13* locus. Finally, in a sub analysis (7087 asthma subjects vs. 7667 controls), the authors identified another locus within the class II region of MHC

(P-value $= 8\times10^{-15}$) to be associated with total blood IgE, but independent of the *HLA-DQ* SNP mentioned above.

EVE Consortium

The EVE Consortium comprises nine U.S. institutions and GWAS results for more than 15,000 individuals of ethnically diverse populations, including European Americans, African Americans and Latino Americans. Unlike the GABRIEL Consortium that reported results as a large GWAS, the EVE Consortium has performed mostly meta-analyses and sub-analyses of prior GWASs. The first aim of the EVE Consortium was to replicate SNPs previously associated with asthma in one or more ethnic groups. The second aim was to conduct meta-analyses of samples that have undergone GWAS. Included in the second aim was the possible detection of copy number variations that may also associate with asthma. Also included in the second aim was the conduction of GWASs that take into consideration sex, environmental and gene-by-gene interactions.

Torgerson et al. performed a meta-analysis of 3246 cases vs. 3385 controls, and 1702 case-parent trios while including populations of European Americans, African Americans, and Latino American [107]. Of note, the study was least powered to detect SNPs in Latino Americans where there were only 606 cases. In comparison, there were 1486 and 1154 asthma cases in European Americans and African Americans, respectively. Different genotyping platforms were used (i.e., Illumina vs. Affymetrix arrays), and the number of SNPs interrogated varied from 550,000 (Illumina 550K arrays) to 1.8 million (Affymetrix 6.0). Using computation methods to impute SNPs, roughly 2 million HapMap SNPs were interrogated per individual. Using a stringent threshold P-value (P-value $<2\times10^{-8}$), SNPs were identified in each individual population before being interrogated in the combined population. Significant SNPs in the primary cohort were replicated in a second cohort of 12,649 individuals from the same ethnic populations. The authors found that SNPs near the **17q21** locus, and other SNPs near genes *IL1RL1*, *TSLP*, and *IL33* associated with asthma for all three ethnic groups. As mentioned, this replicated results of the GABRIEL Consortium for genes *ORMDL3/GSDML* (17q21) and *IL1RL1*, and the results of Hirota et al. for gene *TSLP* [123, 129].

A new association was found with gene *PYHIN1* (chr **1q23.1**), that was specific to the African American population. As mentioned, functional variants in *FCER1A*, also at **1q23** near *PYHIN1*, had been significantly correlated to IgE level by Weidinger et al. [115]. Interestingly, the *PYHIN1* SNPs identified occurred only in minor allele frequencies (0.26–0.29) in the African American controls, were not polymorphic in the European Americans, and occurred at low frequencies (<0.05) in the Latino Americans. Thus, the influence of this SNP to asthma may be limited to, and unique to, the African American population.

At *P*-values $<10^{-6}$ (i.e., just above the threshold cutoff used to call statistically significant), the authors also identified 34 SNPs associated with asthma in European Americans, 4 in African Americans, 32 in Latino Americans, and 75 in the combined population. In the European Americans, 33 (97%) were at **17q21** locus while the remaining locus was on chromosome 17, but 27 megabases (Mb) away from the 17q21 locus. In the African Americans, two SNPs were in the *PYHIN1* gene and the remaining two in the intergenic region between *NNMT* and *C11orf71* genes (chr **11q23**). In the Latino Americans, 13 SNPs were at **3q27** near the *RTP2* gene, 1 near gene *GALNT10* (chr **5q33**), 12 at **17q21**, and 2 between *CCNE1* and *C19orf2* loci (**19q12**). Only one of these SNPs, within the intron of *RTP2*, reached genome-wide significance for Latino Americans (*P*-Value $=4.4\times10^{-9}$). For the replication studies, the authors selected one SNP from each of the 15 regions that contained a SNP with a *P*-value $<10^{-6}$. Two SNPs, one near the *SRP9* gene on chromosome 1q and another near the *AURK* gene on 17p, failed genotyping. SNPs near the 17q21 locus and the genes *IL1RL1*, *TSLP*, and *IL33* genes replicated in all three ethnic groups. Again, the SNP in *PYHIN1* associated with asthma only in the African American replication samples.

Myers et al. performed an extended replication study of the results from the original EVE Consortium meta-analysis [130]. The authors selected 3186 candidate SNPs with *P*-values of the original meta-analysis of less than 10^{-4}. These SNPs were then genotyped in an ethnically diverse population of 7702 cases, 6426 controls and 507 case-parent trios. This replication study had roughly double the number of subjects in comparison to the original meta-analysis. Although 5560 SNPs were selected for replication based on the original meta-analysis, only 3186 were successfully genotyped. Like the original meta-analysis, these populations were of European ($N=3019$), African American ($N=2639$) or Latino American descent ($N=1544$). Latinos were again the group with the least number of cases, and thereby the least powered. In summary, two novel asthma-associated SNPs were identified in the European American population. The first SNP lies in the *KLK3* gene (chr **19q13**), a gene encoding prostate specific antigen (PSA), which can be pro-inflammatory through induction of IFN-γ [131]. The second SNP lies between two pseudogenes, RPP40P2 and POLR3KP1, located at chr 13q21. The region of **13q21** had been previously reported in association with asthma only in an older linkage analysis study [132].

Next, Levin et al. (EVE Consortium) performed a meta-analysis of GWASs for total serum IgE [133]. As mentioned, a GWAS by Weidinger et al. performed in European adults had implicated the alpha chain of the high affinity receptor for IgE (*FCER1A*) in strong association with total IgE level variability [115]. Levin et al. performed this follow-up meta-analysis in 4292 individuals (2469 African Americans, 1564 European Americans and 259 Latinos). With a *P*-value threshold of less than 5.0×10^{-6}, the authors identified 10 unique regions in association with total serum IgE levels. A SNP corresponding to *HLA-DQB2,* located between *HLA-DQB1* and *HLA*-DQA2, was the most associated with total IgE levels in both the discovery and replication sets (combined *P*-Values $=0.007$ and 2.45×10^{-7}, respectively) [133]. In comparison, Moffatt et al. had identified *HLA-DRB1* and *HLA-DQB1* in association with both IgE levels and asthma [124]. Levin et al. also demonstrated that variants in this lead SNP influenced the number of MHC class II molecules on lymphoblastoid cell lines [133]. The authors also identified SNPs near genes *PTBP2*, *SUCLG2*, *MAT2B*, *TBX19*, *SOBP*, *TLE4*, *CCDC82*, *WWP2* and *LINC00469*, all with *P*-values $<3.26 \times 10^{-6}$. Finally, they found evidence for replication in many genes previously identified, including *DARC*, *FCER1A*, *RAD50*, *IL13*, *HLA-DQA2*, *STAT6* and *IL4R/IL21R*.

Myers et al. next used the EVE Consortium cohorts to investigate for sex-specific asthma risk alleles [135]. Asthma is more prevalent in boys before puberty, but more prevalent and more severe in adult women compared to adult men [136]. Moreover, the risk of asthma decreases after menopause in women [137], while roughly 1/3 of women report worsening of asthma before or during their menstrual cycle [138]. Further, patterns of airway inflammation are likely different in men vs. women [138]. Serum IgE levels are higher in boys [139, 140], as are cytokine responses [138]. Thus, to investigate the possibility of sex-specific alleles, Myers et al. performed a meta-analysis of genome-wide genotype-by-sex associations in 2653 male cases and 2566 female cases vs. 3830 controls from the EVE Consortium. The authors found that none of the associations reached the threshold for genome-wide significance (*P*-value $<2.3 \times 10^{-8}$), possibly due to being underpowered. Yet, six independent loci had an association at a *P*-value $<1 \times 10^{-6}$. The most significant was a male-specific SNP in the European American population at the interferon regulatory factor (*IRF1*) gene on chromosome **5q31.1**. One SNP in the **5q31.1** region had a *P*-value of 8.90×10^{-7}, while an additional 20 SNPs at this locus had a *P*-Value $<1 \times 10^{-5}$. As mentioned, numerous genome-linkage studies performed in the 1990s had found an association between asthma and the 5q31 region [96, 104]. In addition, Moffatt et al. found independent SNPs in this region related to asthma for genes *SLCA22A5* (**5q31.1**) and *IL13* (**5q31.1**) in the GABRIEL GWAS [124]. In African Americans, the highest association (*P*-Value $=3.57 \times 10^{-7}$) was a male-specific SNP between genes *EMX2* and *RAB11FIP2* (chr **10q26.11**). In the female specific analysis, 4 variants had *P*-values $<1 \times 10^{-6}$. One SNP was in African Americans and three in Latino Americans. In the African Americans, the associated SNP was located in an uncharacterized gene. In the Latino Americans, one SNP was in gene *C6orf118* (chr **6p27**), another in an intron of *ERBB4* (chr **2q34**), and another in the 3′ untranslated region of gene *RAP1GAP2* (chr **17p13.3**).

Although the known hereditability of asthma is estimated between 35% and 95% [74–86], all the variants identified by GWAS have accounted for only a small percent (~3%) of the genetic risk [107, 120, 124, 141, 142]. There are a few proposed mechanisms that might explain this "missing heritability" of asthma [143]. One hypothesis is that environmental exposures (i.e., viruses, microbiome, allergens, smoke, air pollution) cause inherited epigenetic changes to the DNA [144, 145]. Another possibility is that the non-specific nature of the asthma diagnosis, where multiple phenotypes exist under the same umbrella term [72], has limited the effectiveness of asthma GWASs. Moreover, GWAS do not adequately model polygenic interactions [145], which is likely to exist in asthma. Yet, still another possible explanation is that asthma is caused by rare variants not identified by genotyping arrays. Igartua et al. investigated the role of rare variants in asthma in the EVE Consortium [146]. In general, GWASs are vastly underpowered to detect individual associations or rare variants [147–149]. As a result, it is not straightforward how to perform such studies. The authors investigated an association between rare variants (defined as a minor allele frequency (MAF) $<1\%$) and low-frequency variants (defined as a MAF between 1% and 5%) and asthma. Note: MAF is defined as the population frequency of the second most common allele. SNPs with a MAF $>5\%$ are targeted by the International HapMap project, and thereby included in commercial SNP arrays. In an innovative computational analysis, Igartua et al. limited the analysis space to only functional exonic variants, and then investigated the individual effects of low-frequency alleles a relatively large population of 11,225 study participants, i.e., the authors improved the study's power by limiting the number of tests and increasing the sample size. In addition, the authors used a gene-based statistical test (i.e., combining all variant associations within a gene) to improve statistical power [150]. In summary, the authors identified two novel associations: a missense mutation in the gene *GRASP* in Latino Americans (P-value $=4.3\times10^{-6}$) and a rare missense mutation in the gene *MTHFR* in African Americans (P-value $=1.7\times10^{-6}$). A missense mutation in gene *GSDMB* was also identified in Latino Americans and in the combined samples (P-values $=7.8\times10^{-8}$ and 4.1×10^{-8}, respectively).

Most recently, Demenais et al. performed a very large meta-analysis of worldwide asthma GWASs consisting of 23,948 cases vs. 118,538 controls from ethnically diverse populations [142]. The authors used GWASs with high-density genotyped and imputed SNP data (2.83 million SNPs) in populations of European (19,954 cases vs. 107,715 controls), African (2149 cases vs. 6055 controls), Japanese (1239 cases vs. 3976 controls) and Latino (606 cases vs. 792 controls) populations. Ancestry-specific meta-analyses were conducted, as well as a subgroup analysis using only those with childhood-onset asthma. In the ancestry-specific analysis, there were identified 673 genome-wide significant SNPs (P-Values $<5\times10^{-8}$) at 16 loci in populations of European ancestry. No risk loci were detected in African, Japanese or Latino populations. An additional 205 SNPs were identified in the multi-ancestry meta-analysis. Altogether, 878 SNPs at 18 loci reached genome-wide significance. All but one variant was found in non-coding sequences, with it being a missense variant in *IL13*. Of the 18 loci, 5 are on the q arm of chromosome 5, including 5q22.1, 5q22.2, 5q22.33, 5q31, 5q31.3. As mentioned, Hirota et al. [120] identified a locus associated with adult asthma in a Japanese population located at

5q22 near genes *TSLP/WDR36*. Likewise, variants in the **5q31.1** region had been associated with asthma in males (*IRF1* gene) [135] and identified in the GABRIEL GWAS (genes *SLCA22A5* and *IL13*) [124]. Five of the loci (identified by Demenais et al.) were found on chromosome six, including 6p13.13, 6p21.32, 6p21.33, 6p22.1, and 6q15. The region **6p21.32** is at the MHC Class II locus (HLA-DRB1, HLA-DQA1), and asthma susceptibility loci have been found at or near this area by both early linkage studies and several GWAS [96, 104, 120, 124]. Of the remaining eight loci, five were near genes of importance to asthma pathogenesis, including Th2 transcription factor *STAT6* (**12q13.3**), pro-inflammatory transcription factor *GATA3* (**10p14**), *SMAD3* (**15q22.33**), and IL-1 family members *IL1RL1/IL1RL2/IL18R1* (**2q12**) and *IL33* (**9p24.1**). Finally, the region of 17q12-21 was the most associated with asthma risk, in a region containing genes *ERBB2, ORMDL3/GSDMB* and *PGAP3.*

In the multi-ancestry analysis, the lead SNP was within gene *ERBB2* (P-value $= 2.2 \times 10^{-30}$), 180 kb from the *GSDMB/ORMDL3* locus (see GABRIEL and EVE meta-analyses above). Further, variants in this region were in strong LD ($r^2 > 0.9$) with missense variants of *ERBB2*, i.e., variants in this region were associated with point mutations that result in amino acid changes to the ERBB2 protein. Another lead SNP in this region, which was 3.6 kb proximal to *GSDMB*, was associated with asthma predominantly in the pediatric population. This was perhaps in agreement with Moffatt et al. whom initially showed that variants near *ORMDL3/GSDML* were associated with childhood-onset asthma [124]. Interestingly, variants located at the *STARD3/PGAP3* locus had independently been associated with atopic asthma in comparison to non-atopic forms of disease [151].

Messenger RNA global gene expression

As a first step to producing functional gene products from DNA sequences, protein-encoding genes are "transcribed" into messenger RNA (mRNA). mRNA transcripts are then "translated" into proteins or degraded. Typically, sections of DNA sequences are first transcribed into precursor mRNA sequences. Precursor sequences require the non-coding sections (called "introns") to be removed, and remaining sections (called "exons") to be rejoined. This process is known as "RNA splicing." All the exons in the DNA are together called the exome, which together comprise only 1% of the total genome, or ~30 Mbs. The number of protein-encoding genes in the exome is now estimated at roughly 21,000. The remaining 99% of the genome, comprising many Mbs of non-protein-coding RNA, was once controversially called "junk DNA." The National Human Genome Research Institute set out in 2003 to determine how much of the DNA was in fact non-useful. As a result, an international consortium known as the Encyclopedia of DNA Elements (ENCODE) was formed. ENCODE has since demonstrated that the majority of the human genome likely serves a purpose [152]. Noncoding DNA is transcribed into non-coding RNA molecules, such as transfer RNA, ribosomal RNA and regulatory RNAs. These non-coding RNA molecules were found to regulate both transcription and translation. MicroRNA (miRNA) are probably the most studied of the non-coding RNA molecules. miRNA are small molecules (~22 nucleotides), which through RNA silencing and post-translational modification, are estimated to modulate up to 60% of the protein-coding genes at the translational level [153].

Microarrays vs. RNAseq

Microarrays are solid surfaces where single-stranded fragments of DNA molecule, called "probes," are attached in an orderly arrangement of rows and columns. mRNA is extracted from cells and copied into double-stranded "copy DNA (cDNA)" to improve stability. cDNA is then fragmented, fluorescently labeled, and allowed to bind to the ordered array of complementary sequences or probes. The measure of fluorescent intensity at each probe indicates relative abundance of mRNA for each oligonucleotide, i.e., how strongly the small section of DNA is being expressed. Microarray technology has several limitations that are beyond the scope of this chapter. Briefly, their limitations include a susceptibility to batch effects and an inability to investigate alternative splicing, small non-coding RNAs or transcripts at low abundance [154, 155]. Thus, microarrays have been used mostly to determine differences in gene expression between two cohorts of samples (e.g., disease tissue/cells vs. healthy tissue/cells).

Alternatively, none of the limitations of microarrays apply to newer RNA sequencing technologies (RNAseq). RNAseq provides an ability to look at alternative splicing, small non-coding RNA, transcripts of low abundance, post-transcriptional modifications, gene fusions and mutations. As a result, microarrays are largely being replaced by the RNAseq. A discussion of the different technologies used for RNAseq is beyond the scope of this chapter. Briefly, RNA is isolated from tissue, converted to cDNA, fragmented into small strands of cDNA, amplified using sequencing adaptors added to one or more ends, sequenced using next generation sequencing, and then sequenced reads aligned to the genome using powerful computers. While many of the early asthma mRNA expression experiments were performed using microarrays, the most recent studies are being performed mostly using RNAseq.

Gene expression in airway epithelial cells

Global gene expression, measured in various lung and blood cells, has acted as a window into both the molecular mechanisms of inflammation [29, 70, 156–166] and the cellular responses to medications [159]. Woodruff et al. compared global gene expression by microarray of bronchial airway epithelial brushings from 42 mild-to-moderate asthma subjects in comparison to 28 non-smoking, healthy controls [166]. In a seminal finding, three of the airway epithelial cell (AEC) genes differentially up-regulated in asthma were found to be markedly up-regulated in AECs (grown ex vivo in an air liquid interface) when stimulated with Type 2 cytokine IL-13 (50 ng/mL for 4 days). These three genes (chloride channel, calcium activated family member 1 (*CLCA1*), periostin (*POSTN*) and serine peptidase inhibitor, clade B (ovalbumin), member 2 (*SERPINB2*), defined the T2 inflammatory gene signature. Further, treatment of these mild-to-moderate, untreated asthma subjects with inhaled corticosteroids lowered the expression of these genes to the level of healthy controls. Alternatively, gene FK506-binding protein 51 (*FKBP51*) was increased after corticosteroid treatment, identifying a marker for active steroid use.

The discovery of an AEC T2 gene signature then equipped scientists with an ability to classify asthma patients as either T2 inflammatory "high" vs. T2 "low." As later discussed, this ability is increasingly important in determining the likelihood of clinical response to new monoclonal antibody treatments that target T2 inflammation. For example, Dupilumab, a monoclonal antibody directed against interleukin-4 receptor subunit α (IL-4Rα) of IL-4 and IL-13 receptors, is more likely to be effective in T2 high asthma patients [167, 168]. In a follow-up study involving the same data, Woodruff et al. used the gene expression of *POSTN*, *CLCA1* and *SERPINB2* to cluster subjects into different subgroups using a hypothesis-free method known as hierarchical clustering [69]. The authors found that while all the subjects with increased expression of the T2 gene signature had asthma, not asthma subjects had these genes up-regulated. In fact, 1/2 of asthma subjects had expression levels of the T2 signature comparable to healthy controls. Subjects with T2 high asthma tended to have higher airway hyperresponsiveness, increased skin prick reactivity (especially to dog dander and house dust mites), higher total serum IgE, and higher eosinophil counts in both blood and bronchoalveolar lavage fluid. Further, T2 high subjects had increased thickness in reticular basement membranes, and differences in gel-forming mucin production. In comparison to T2 low airways, T2 high subjects had induction of *MUC5AC* and *MUC2* mucins, but repression of mucin *MUC5B*. Alveolar macrophages in T2 high subjects had increased expression of pro-inflammatory enzyme 15-lipoxygenase and cytokine TNF-α. Yet, no clear inflammatory mechanism for those with T2 low disease was identified. All asthma subjects were then treated for 4 weeks with inhaled corticosteroids. Lung function improved, as determined by a FEV1 increase of greater than 150 mL, only in the T2 high subjects. In summary, for patients with mild-to-moderate, untreated asthma, an elevation in the epithelial T2 gene signature associates with higher inflammatory and allergy markers including eosinophilia, characteristic airway changes, but stronger improvement following inhaled corticosteroid treatment.

These findings have set in motion a search for more easily obtained markers of the T2 inflammatory response. Poole et al. demonstrated that the T2 gene signature (*POSTN*, *CLCA1*, *SERPINB2*) could also be detected in nasal AECs of adolescent children with asthma. Further, expression profiles in the nose closely matched profiles in bronchial AECs among these children with atopic disease [68]. Advancing this discovery, Pandey et al. recently used machine learning techniques to show that nasal gene expression could accurately classify subjects into asthma (mild-to-moderate) vs. non-asthma categories [169].

Jia et al. showed that serum periostin (*POSTN*) protein levels were higher in asthma patients with airway eosinophilia [170], reinforcing the connection between the T2 inflammatory response and eosinophilia. T2 cytokine IL-5 is known to activate and act as a survival factor for eosinophils in the airway. Likewise, T2 cytokines IL-4 and IL-13 induce the secretion of eosinophil chemotaxis protein Eotaxin-3 (*CCL26*) by epithelial cells [70, 171]. Further, periostin is found in bronchial fibroblasts [172] and is secreted into the extracellular matrix from the basolateral surface of AECs when stimulated with IL-13 [173]. Thus, the use of serum periostin as a surrogate for airway eosinophilia was biologically plausible and of great interest. Notwithstanding, Jia et al. also showed that elevated serum periostin is sensitive, but not specific for airway eosinophilia. Moreover, in comparison of groups with high tissue eosinophilia (>22 cells/mm^2) vs. low tissue eosinophilia (<22 cells/mm^2), serum periostin level differences were relatively small (fold-change ~1.2, *P*-value=0.023). Nonetheless, Jia et al. concluded that serum periostin level acted as the single best predictor of airway eosinophilia when compared to sex, age, body mass index, IgE levels, blood eosinophils and fractional exhaled nitric oxide (FeNO) in a logistic regression model.

Gene expression as tool to help phenotype asthma subjects

In one of the first efforts to identify asthma subphenotypes, Moore et al. clustered 856 participants from the Severe Asthma Research Program (SARP) using 34 clinical characteristics, including gender, race, BMI, age of asthma onset, lung function, medication use, allergy testing, etc. [9]. Using a hypothesis-free method known as hierarchical, unsupervised clustering, the authors identified five subject clusters. Over 70% of the severe asthma subjects fell into two clusters, Clusters 4 and 5. Cluster 5 had mostly adult onset disease patterns without a strong association to atopy. In contrast, Cluster 4 was characterized by childhood-onset asthma associated with atopic disease and clinical indicators of high T2 inflammation. Following this seminal study, several groups began using inflammatory markers and global gene expression in cluster analyses aimed at improving disease phenotyping [70, 73].

Fractional exhaled nitric oxide (FeNO) is a biomarker for asthma that associates with atopy, eosinophilia and the T2 inflammatory response [9, 174]. Moreover, treatment of asthma patients with Dupilumab, a monoclonal antibody directed against interleukin-4 receptor subunit α (IL-4Rα) of IL-4 and IL-13 receptors, is known to suppress FeNO [167, 168, 175]. Using participants from the Severe Asthma Research Program (SARP), airway epithelial gene expression was directly correlated to FeNO levels [70]. The top correlated gene was inducible nitric oxide synthase, *NOS2* ($\rho = 0.72$, *P*-value $<10^{-7}$), supporting the biologic plausibility of the study. After *NOS2*, the T2 gene signature genes (*POSTN*, *CLCA1* and *SERPINB2*), eotaxin-3 (*CCL26*) and the IL-33 receptor (*IL1RL1*), were the top genes to correlate with FeNO, as illustrated in Fig. 1. In agreement with Woodruff et al., the mucin gene *MUC5B,* which is known to be strongly repressed by IL-13 [174], had the strongest negative correlation

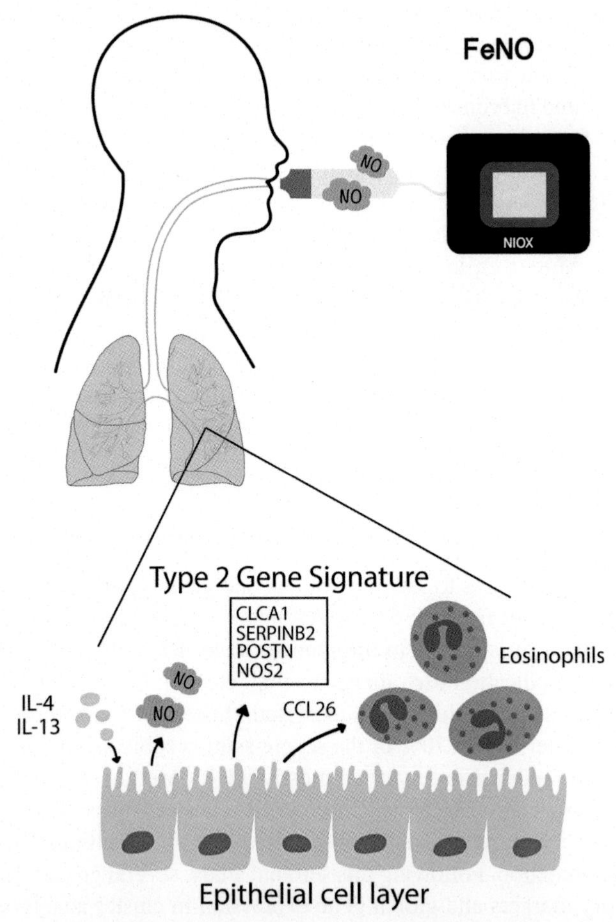

FIG. 1 Schematic detailing the Type 2 gene signature and its relationships to clinical traits. The Type 2 gene expression signature was identified by *in vitro* stimulation of epithelial cells with Type 2 cytokines (IL-4/IL-13), which demonstrated marked up-regulation in the expression of genes *CLCA1*, *POSTN*, *SERPINB2* (169). The gene encoding inducible nitric oxide synthase (iNOS), which is responsible for exhaled nitric oxide production in the airways was subsequently shown to correlate strongly or be co-expressed with the Type 2 gene signature, connecting fractional exhaled nitric oxide (FeNO) measurements to Type 2 inflammation (71). The Type 2 signature also includes CCL26 (Eotaxin-3), suggesting IL-4/IL-13 signaling in airway epithelial cells is upstream to eosinophilic inflammation.

with FeNO. A folate receptor (*FOLR1*) also had one of the strongest negative correlations, and recent systematic review and meta-analysis showed that a polymorphism of the methylene-tetrahydrofolate reductase (*MTHFR*) C677T, a mutation potentially impairing folic acid metabolism, is a risk factor for asthma [176].

Next, the genes that correlated with FeNO (n=549) were used to objectively cluster asthma subjects into subgroups. In agreement with Moore et al., most of the severe asthma patients clustered into 2 subject clusters (SCs) (SC2 and SC3). One severe asthma cluster (SC2) had high T2 inflammation, as evidence by a high FeNO, bronchoalveolar lavage and blood eosinophilia, and up-regulated expression of T2 signature and mast cell genes. The other severe asthma cluster (SC3) had lower levels of T2 clinical biomarkers and gene expression, in addition to a later age of onset, increased incidence of nasal polyposis and higher levels of neutrophilic inflammation. Roughly 1/2 of all asthma subjects had evidence of high T2 inflammatory response (by clinical biomarkers and gene expression), confirming the prior findings of Woodruff et al. in a more severe and steroid-treated patient population. In general, both severe asthma clusters (SC2 and SC3) were older and more obese than the other non-severe subclusters. Further, both of the severe SCs demonstrated suppression of genes associated with cilia function, neuronal function, cell adhesion and wound repair. These findings suggested that airway epithelial defense, repair, neuronal function are an integral part of a healthy epithelial layer and perhaps prevention of severe asthma.

In a re-analysis of this data, gene expression was next correlated directly to four clinical traits indicative of severe and uncontrolled disease: lung function (i.e., forced expiratory volume, 1 s (FEV_1)), rates of ED visit or hospitalization, medication use, and Juniper quality of life scores [177]. In agreement with Woodruff et al., the T2 signature was high in many asthma subjects with mild, corticosteroid naïve asthma. In further agreement, the T2 gene signature dropped to near normal levels in mild-to-moderate asthma subjects started on inhaled corticosteroids. Yet, the T2 signal became progressively higher in those on inhaled corticosteroids as severity worsened. This finding indicated either poor medication adherence or the emergence of poor corticosteroid responsiveness as severity increases. As expected, FeNO levels demonstrated the same pattern. This was in agreement with prior studies that showed FeNO elevation is observed across a range of disease states ranging from atopy without asthma to mild untreated asthma to corticosteroid-refractory asthma [178, 179]. It may also help explain some of the prior limitations and inconsistencies in using FeNO to guide clinical care [180–182]. In summary, elevation of FeNO and the T2 gene signature was a marker for uncontrolled and severe disease only in patients already on corticosteroid treatment.

Further, in a network analysis of this data, a T2 inflammatory network was found to positively correlate with FeNO and BMI, but did not track well with either disease control and severity in the overall population. Yet, a network of epithelial growth and repair genes (designated the "EGR" network) was found to be strongly suppressed in relation to worsening control and higher severity. Interestingly, *ERBB2* was the center hub gene in this identified network. As mentioned above, the meta-analysis of GWASs by Demenais et al. showed that the lead SNP in the multi-ancestry ancestry was within *ERBB2*, which is positioned within the asthma susceptibility locus at **17q12-21** [142]. Recently, Nicodemus-Johson et al. investigated airway epithelial cells transcriptomes in relation to the epigenome and genetic variations (i.e., high-density SNP arrays) in 74 asthmatic and 41 non-asthmatic adults [183]. The authors integrated genetic, epigenetic and transcriptomic data, and concluded that a lead SNP within the *ERBB2* had the most effect on gene expression. Thus, there is increasing evidence that genetic variants and epigenetic alterations of *ERBB2*, along with disrupted *ERBB2* gene expression, may play an important role in the development of asthma.

ERBB2 (HER2) is an orphan-receptor, a member of the epidermal growth factor receptor (EGFR) family of receptor tyrosine kinases that includes HER1/EGFR, HER2/ERBB2, HER3/ERBB3 and HER4/ERBB4, and likely required for normal epithelial repair [184–186]. ERBB2 dimerizes with ERBB3 in the lungs, which acts to stabilize ligand binding and enhance signaling [187]. Neuregulin 1 (NRG1) binds the ERBB2/3 complex, and this signaling axis is particularly important for epithelial recovery following injury [187]. In differentiated human airway epithelia, NRG1 is present exclusively in the apical membrane, while ERBB2-4 segregates to the basolateral membrane [188]. Following injury, NRG1 actives ERBB2 at the edge of the wound, hastening restoration of epithelial integrity [188]. Further, Vermeer et al. showed that blocking ERBB2 with an anti-ERBB2 antibody (trastuzumab) attenuated airway epithelial differentiation, and decreased the ciliated cell number by 49%, while at the same time increasing metaplastic, flat cell number by 640% [189]. Alternatively, treatment with neuregulin 1 increased epithelial cell height and decreased the number of non-ciliated columnar cells [188]. Thus, a deficit in the neuregulin 1—ERBB2/ERBB3 signaling pathway may relate to the transformation of the epithelial layer from ciliated epithelial cells to non-ciliated columnar cells typically observed in asthma. Kettle et al. showed that NRG1 induces expression of mucins MUC5AC and MUC5B through ERBB2/ERBB3 receptors [190], but it is unclear whether this is harmful or helpful to asthmatic airways, as MUC5AC is generally increased in asthma airways, but MUC5B is generally decreased. Although the exact role of ERBB2 in asthma pathogenesis has not yet been well elucidated, it serves as a good example in how genomics can help direct mechanistic research.

Bronchoalveolar lavage (BAL) genomic studies

BAL fluid, which is typically composed of ~90% monocyte/macrophages, is an important sampling of the lung cellular milieu. Woodruff et al. showed that alveolar macrophages from Type 2 high asthma subjects have increased expression levels of pro-inflammatory enzyme 15-lipoxygenase (*ALOX15*) and cytokine TNF-α [69]. Esnault et al. performed BAL before and 48h after a segmental bronchoprovocation with allergen in mild, atopic asthma subjects. The authors also performed the bronchoprovocation 1 month before and after administering a dose of mepolizumab, an anti-IL-5 monoclonal antibody. As expected, eosinophilic-specific transcripts (e.g., *IL5RA*, *SIGLEC8*, etc.) were up-regulated following allergen challenge prior to mepolizumab, but not after the administration of anti-IL-5 therapy. Yet, allergen challenge also up-regulated 36 genes related to immune defense, wound-healing and apoptosis. Among these 36 genes, 13 up-regulated genes related to lung remodeling (e.g., *MMP9*, *MMP12*, etc.), and were not impacted by the administration of mepolizumab.

Leroy et al. exposed 19 subjects (7 with asthma) to clean air (0ppb), medium (100ppb) and high (200ppb) ambient levels of ozone for 4h in climate chamber, followed by bronchoscopy with BAL 24h later. Increasing doses of inhaled ozone caused a dose-dependent up-regulation of genes related to cell trafficking, remodeling, inflammation and tissue repair. Interesting, gene expression patterns differed between asthma vs. non-asthma, and between those with and without lung function response (so-called “ozone responders”). Gene *SPP1* was most highly expressed in BAL cells after ozone exposure, and its encoded protein osteopontin (OPN) showed a dose-dependent increase with ozone exposure. Of interest, administration of OPN during pulmonary antigenic challenge decreased Type 2 response and protected mice from allergic disease [191]. Moreover, Arjomandi et al. showed that variants in *SPP1* (chr **4q22.1**) were associated with a diagnosis of asthma, severity of asthma, and post-bronchodilator FEV1 in Puerto Ricans [192]. In summary, although gene expression in BAL cells has indicated that these cells are important to remodeling, inflammation and likely significant to disease pathology, not enough studies have been performed to reach clear conclusions.

Sputum genomic studies

The characteristics of induced sputum can serve as a window into the inflammatory patterns of the lower airways. Using differential cells counts (cell proportions) of induced sputum cells from 93 subjects with asthma and 42 healthy controls, Simpson et al. characterized asthma subjects into eosinophilic (EA) ($n=38$ (41%)), neutrophilic (NA) ($n=19$ (20%)), paucigranulocytic (PA) ($n=29$ (31%)), and mixed granulocytic ($n=7$ (7.5%)) subtypes [193]. The definition of eosinophilic asthma was based on greater than 1.01% eosinophils in the sputum that was maintained over 4weeks. Some subjects changed subtype over the course of 4 weeks, with the majority changing from eosinophilic to non-eosinophilic. Advancing this study, Baines et al. first classified 47 asthma subjects according to the proportions of induced sputum granulocytes ($n=17$ subjects with EA, $n=12$ subjects with NA, and $n=18$ subjects with PA), and then measured gene expression differences between the three groups. They found that expression levels of six genes (*CLC*, *CPA3*, *DNASE1L3*, *IL1B*, *ALPL*, and *CXCR2*) reproducibly discriminated the inflammatory phenotypes of asthma, and could predict response to inhaled corticosteroids. Interestingly, *CLC* (an eosinophil-specific gene), *CPA3* (mast cell gene), *DNASE1L3* were higher in EA, whereas *IL1B*, *ALPL* and *CXCR2* were higher in NA. Although EA had previously associated with a positive response to inhaled corticosteroids [194], the six-gene signature performed better than the sputum eosinophil percentage in predicting ICS response. Clinically, EA had the highest level of FeNO and sputum eosinophils, in addition to a depressed $FEV_1\%$ predicted (~70%). NA had lower levels of FeNO and atopy, but the highest levels of sputum neutrophils. PA had higher proportions of sputum macrophages, low numbers of both eosinophils and neutrophils, and clinically had the mildest disease.

Hastie et al. stratified 242 Severe Asthma Research Program (SARP) subjects by sputum eosinophils (<2% or ≥2%) and neutrophils (<40% or ≥40%), and assessed for differences in inflammatory mediators by performing protein microarrays based on 120 inflammatory proteins [195]. In a four-way stratification, subjects with elevated levels of both sputum eosinophils and neutrophils had the worse asthma control, greatest symptoms, and lowest lung function in comparison to the other groups. Protein levels of Brain-derived neurotrophic factor (BDNF), IL-1β and MIP-3a/CCL20 were increased in sputa with >40% neutrophils, regardless of eosinophil %. Moreover, BDNF, CXCL13 and epidermal growth factor (*EGF*) were significantly increased in severe disease vs. non-severe disease. In agreement, Watanabe et al. recently showed that BDNF mRNA and protein levels are higher in severe asthma compared to healthy controls, and have positive correlation with both FeNO and sputum eosinophilia [196].

The Unbiased Biomarkers for the Prediction of Respiratory Diseases Outcomes (U-BIOPRED) project is 5-year European-wide project that is using clinical and genomic data to better understand severe asthma. Lefaudeux et al. clustered a large U-BIOPRED cohort of asthma and non-asthma subjects (which included current or previous smokers) in a hypothesis-free manner according to clinical traits, such as age of disease onset, pack years of cigarette smoking, etc. [158]. Four subject clusters were first identified in a training set ($N=266$), and then reproduced in a validation set ($N=152$). The four subject clusters were described as [1] moderate-to-severe, well controlled, [2] severe, late-onset with obstruction, high BMI, smoking and OCS use, [3] severe asthma with obstruction and OCS but no smoking, [4] severe asthma with female predominance, high BMI, frequent exacerbations and OCS, but not smoking or airway obstruction. Sputum was induced and used to further identify differences in transcriptomic and proteomic profiles, although only performed on a subset of the training ($n=86$) and validation ($n=94$) subjects. Ten of 1129 proteins measured were differentially abundant across the 4 phenotypes, including IL-16, CTAP-III, Trypsin 2, GM-CSF, HPLN1, Cathepsin G, ARSB, PSA2, LYN and FUT5. In the differential gene analysis, there were only eight differentially expressed genes (DEGs) between phenotypes 1 vs. 2, 3 DEGs between 2 vs. 4, and 14 DEGs between 3 vs. 4. Yet, 147 DEGs were found between 1 vs. 3, showing that the largest gene expression differences were observed between the non-smoking, severe asthma cluster and the moderate-to-severe, well controlled asthma subjects.

Following, Kuo et al. used sputum cell transcriptomics from 104 moderate-to-severe asthma vs. 14 non-asthma subjects of U-BIOPRED to again classify asthma phenotypes [197]. In a novel analysis, the authors first identified DEGs between eosinophilic asthma (EA) (defined by sputum eosinophil ≥1.5%) vs. non-EA (sputum eosinophils <1.5%) ($n=508$). Then, a clustering technique was used to identify three asthma phenotypes, named the "transcriptomic-associated clusters" or TACs. TAC1 was characterized by mostly severe patients with high eosinophilia, high FeNO, high serum periostin, high oral corticosteroid use, nasal polyps and severe airflow obstruction. TAC1 was also defined by up-regulation of genes related to mast cells (*CPA3*, *TPSB2*), eosinophils (*CLC*), and Type 2 inflammation (*IL1RL1*, *ALOX15*, etc.). In comparison to the clustering studies mentioned above, TAC1 was most similar to the eosinophilic asthma (EA) group in the Baines et al. study that used sputum cell proportions, Cluster 4 in the Moore et al. study that used clinical traits, and Subject Cluster 2 in the Modena et al. study that used epithelial cell gene expression. TAC2 was characterized by high sputum neutrophils, high serum C-reactive protein (CRP), a greater prevalence of eczema, and up-regulation of genes related to neutrophilic, interferon and TNF-driven inflammation. Again, this phenotype matched the neutrophilic asthma (NA) group and Subject Cluster 3 in the Baines and Modena studies, respectively. TAC3 was characterized by altered inflammasome-associated and metabolic pathways, and is perhaps similar to a pauci-granulocytic phenotype, although TAC3 also showed normal to moderately high sputum eosinophils.

Most recently, Jones et al. compared gene expression by RNAseq of sputum-derived cells in 4 groups: house dust mite (HDM)-sensitized asthma, non-HDM-sensitized asthma, HDM-sensitized without asthma, and nonatopic controls [29]. The subjects were relatively young (median ages ~22 years) with normal lung function, and roughly 1/4 of subjects were current smokers. As expected, the T2 gene signature was higher in both atopic asthma and atopic non-asthma subjects compared to non-atopic controls. Interestingly, eosinophil developmental transcription factors GATA1 and GATA2 were significantly higher in non-HDM-sensitized asthma subjects compared to non-atopic controls, suggesting that eosinophilic inflammation may be independent of atopic status. The authors only measured IgE to HDM, and so it remains possible that non-HDM sensitized asthma patients were still atopic.

A network analysis using Ingenuity pathway analysis showed that three major networks differentiated atopic asthma vs. atopic non-asthma subjects. At the center of these networks were epidermal growth factor receptor (*EGFR*), EGF receptor family member Her-2 (*ERBB2*) and E-cadherin (*CDH1*). In comparing

HDM-sensitized asthma vs. non-atopic controls, three differential networks were similarly identified with *IL13*, *ERBB2* and *CDHR3* as the dominant hubs. Interestingly, *CDHR3* is a known susceptibility gene for asthma exacerbations and receptor for rhinovirus C [198, 199]. CDHR3, ERBB2 and EGFR protein levels were higher in columnar epithelial cells of 8 atopic asthma patients as compared to non-atopic controls on staining. These results are actually in opposition to the expression results reported by Modena et al., and the observed methylation of this region observed by Nicodemus-Johnson et al. [183]. This discrepancy may have been due to differences in disease phenotypes. While the SARP patients in the Modena et al. study were older with mostly severe disease and reduced lung function, the subjects in the Jones et al. study were younger with preserved lung function. Thus, one explanation may be that epithelial repair mechanisms may be preserved in younger, healthy asthma subjects with normal lung functions, but reduced in older, more severe asthma related to aging or medication use. As indicated by Nicodemus-Johnson et al., perhaps such age-related changes may be due to epigenetic modifications of the DNA. Alternatively, it may all be related to the heterogeneity of the disease, as the protein analysis of Jones et al. only included eight asthma subjects.

Blood genomic studies

Peripheral blood is an easily-obtained source of inflammatory immune cells, some of which should traffic to the lungs during active inflammation. Some evidence suggests that gene expression can be used to help identify asthma and to classify asthma subjects into disease phenotypes. Hakonarson et al. first showed that gene expression profiles in isolated peripheral blood mononuclear cells (PBMCs) could separate out glucocorticoid responders vs. non-responders, and potentially be used to predict corticosteroid response [200]. George et al. implemented a novel, multi-step decision tree that used both blood global gene expression and clinical biomarkers to classify subjects into separate subgroups. The algorithm could differentiate asthma subjects from non-asthma, and classify the asthma subjects into distinct asthma subgroups, or endotypes [201].

Kong et al. compared global genome expression in PBMCs of asthmatic children vs. healthy controls, and found differentially expressed genes related to immune response, stress response and apoptosis [202], Further, a combination of *ADAM33*, *Smad7*, and LIGHT (*TNFSF14*) expression provided excellent discriminating power between the two groups. As mentioned above, *ADAM33*, was one of the first gene variants discovered in a Caucasian population [98], and polymorphisms associated with asthma in multiple studies [99–101]. In mice, allergen exposure induces LIGHT in immune cells (B and T cells), whereas inhibition of LIGHT reduces fibrosis, smooth muscle hyperplasia and airway hyperresponsiveness along [203]. Further, Herro et al. showed that LIGHT (*TNFSF14*) induces TSLP directly in bronchial epithelial cells, and synergizes with IL-13 and TGF-β in vivo to induce TSLP and drive fibrosis in mice [204].

Bjornsdottir and colleagues enrolled a relatively large cohort ($N=337$ subjects) to determine PBMC gene expression changes coincident with asthma exacerbations [205]. The authors collected 166 exacerbation samples from 118 subjects that had experienced at least one exacerbation. They used a covariate analysis to account for steroid use, age, sex and differences in cell proportions between samples. During an exacerbation, PBMC gene expression reflected activation of innate immune pathways, and a systemic type I interferon response. Interestingly, a large proportion of patients during exacerbation demonstrated a robust signature of innate immune activation despite reporting no symptoms of a respiratory viral infection. The authors speculated that other sources of immune activation, such as air pollution, allergic response, or resident viruses, were at play. Similarly, Croteau-Chonka et al. performed a gene set enrichment analysis (GSEA) using peripheral blood expression profiles in uncontrolled vs. controlled asthma subjects from the Asthma BRIDGE and CAMP cohorts [206]. Put simply, GSEA is a computational technique that determines if a predefined group of genes (e.g., the T2 signature genes) are differentially expressed between groups. Using GSEA, Croteau and colleagues found that TREM-1 signaling, which is a critical activator of inflammatory responses by innate immune receptors [207], was activated during asthma exacerbations.

Recently, Bigler et al. obtained transcriptomic profiles by microarray in 610 subjects from U-BIOPRED, including severe asthma and non-smoking (NSM) ($n=311$), severe asthma and smoking (SM) ($n=110$), mild/moderate asthma and non-smoking ($n=88$), and non-asthma and non-smoking ($n=101$). The author found that a total of 1693 genes were differentially expressed between severe asthma vs. subjects without asthma, but that most differential expression was due to differences in the proportions of circulating immune cell populations [157]. E.g., eosinophilic-specific genes were elevated in samples with higher proportions of eosinophils. Herein lies the major difficulty with using peripheral blood for either gene expression studies or to identify asthma markers—cell proportions in peripheral blood are highly variable. In fact, Bigler et al. found that the number of differentially expressed genes reduced by 90% when differential cell counts were accounted for in the analysis. Nonetheless, the authors showed that not all genes were explained by proportions of circulating immune cells, including genes related to immune cell trafficking and B-lymphocyte development.

Epigenetics and asthma: A new frontier

As mentioned, genetic variant studies have accounted only for a small percent of the known hereditability of asthma [124, 208]. One possibility is that asthma is a multifactorial disease caused by *combinations* of alleles that interact with environmental factors, and thereby not easily identified. Another possibility is that asthma is caused by rare polymorphisms with high penetrance that cannot be easily interrogated by current statistical methods [208]. Yet, gene variants are unable to easily explain the recent and sharp rise in childhood asthma rates that has occurred over the last 30 years [144]. Alternatively, the transgenerational inheritance of epigenetic changes (e.g., DNA methylation and histone modifications) is a fascinating concept that may explain asthma's increasing prevalence and missing heritability.

Similar to the many genetic variant studies discussed, a number of candidate gene and genome-wide association studies have interrogated epigenetic changes associated with asthma [209–221]. To date, the results from these studies have been difficult to interpret, and poorly reproduced [222]. A discussion into all the asthma epigenetic studies, which have been mostly candidate-gene studies, is beyond the scope of this chapter. Although, the results have been well summarized by Begin et al. [144], Yang et al. [223], and DeVries et al. [224]. Nonetheless, some epigenetic association studies have made important discoveries worth mentioning. Naumova et al. found that DNA methylation at loci within *ZPBP2* gene, located at the asthma susceptibility locus 17q12-21, was critical for both gene expression regulation and predisposition to asthma [225]. Liang et al. interrogated methylation status at CpG sites in peripheral blood leukocytes within the proximal promoter regions of 14,475 genes in 335 subjects from 95 nuclear pedigrees [226]. These subjects had previously been part of genome-wide SNP associations studies performed for IgE levels and asthma [109]. The authors identified 34 loci associated with IgE levels in the primary cohort, but also a total of 36 loci were identified by combining their data with 2 replication cohorts. The authors reported associations with methylation sites on many genes related to eosinophilic (e.g., *CLC*, *GATA1*, *IL5RA*, *ZNF22*) and T2 inflammation (e.g., *IL4*, *TFF1*). Yang et al. compared DNA methylation profiles and gene expression in PBMCs of inner-city children with atopic asthma vs. healthy controls [227]. Eighty-one differentially methylated regions (DMRs) were identified, and among the 73 hypomethylated in asthma subjects were T2 cytokine *IL13*, asthma-associated transcription factor *RUNX3*, and genes related to maturation and function of NK cells (*KIR2DL4*, *KIR2DL3*, *KIR3DL1*, and *KLRD1*). Recently, Nicodemus-Johnson et al. investigated airway epithelial cells transcriptomes in relation to the epigenome and genetic variations (i.e., high-density SNP arrays) in 74 asthmatic and 41 nonasthmatic adults [183] The authors integrated genetic, epigenetic and transcriptomic data, and found that the lead SNP affecting gene expression in the region was found within the *ERBB2* gene.

Targeted therapies and future directions

Technological advances in the development of antigen specific monoclonal antibodies has given researchers and physicians the ability to target virtually any immune or inflammatory pathway. Omalizumab, a humanized monoclonal antibody targeting IgE, was the first monoclonal antibody to be FDA-approved for the treatment of moderate-to-severe asthma. Omalizumab is believed to treat asthma by stabilizing mast cells and preventing degranulation. Yet, recent data suggests that omalizumab also reduces susceptibility to viral infections [228, 229], perhaps because IgE receptor activation decreases IFN-γ responses in dendritic cells [230, 231]. After omalizumab came to market in 2003, there was a dearth in the development of new asthma treatment until the past few years. Now, an array of new monoclonal antibodies have been developed that target IL-5 (mepolizumab, reslizumab), the IL-5 receptor (benralizumab), IL-13 (lebrikizumab, Tralokinumab), IL-4 (pitrakinra, anrukinzumab), IL-4 receptor alpha (dupilumab), IL-17 (secukinumab), TSLP (tezepelumab), IL-33, and several others. Of these drugs, only mepolizumab, reslizumab, benralizumab, and dupilumab have been FDA-approved for the treatment of severe asthma. Of note, dupilumab has also been approved for the treatment of atopic dermatitis/ eczema in adults.

The benefits of new monoclonals are the excellent safety profiles in combination with an ability to prevent exacerbations (benralizumab [233, 234], dupilumab [168, 232], mepolizumab [235], reslizumab [236]), improve lung function (benralizumab [233], dupilumab [168, 232], reslizumab [237]), and reduce oral steroid use (benralizumab [234], dupilumab [167], mepolizumab [238]). The choice of which asthma biologic to use in practice is a subject of on-going debate. Not unexpectedly, the anti-IL-5 and anti-IL-5 receptor alpha antibodies are more effective in patients with elevated eosinophils in the blood [233–238]. Thus, blood eosinophilia (which quickly plummets after initiation of therapy), has been the most effective and widely-used biomarker to indicate whether a patient will respond positively to these agents. Yet, dupilumab, which blocks the alpha subunit of the IL-4 and IL-13 receptors (i.e., blocking both IL-4 and IL-13 signaling), is also more effective when eosinophils are elevated in the blood [167, 168, 232]. Dupilumab therapy has shown to reduce FeNO [232] and serum Eotaxin-3 (CCL26) [232]. Interestingly, blood eosinophils levels remain relatively unchanged or become elevated in some patients during dupilumab therapy [167, 168, 232], possibly reflecting a loss in eosinophil chemotaxis into the lungs. Furthermore, a re-analysis of two phase III clinical trials showed that blood eosinophilia also predicts a positive response omalizumab [239], which is known to significantly reduce blood eosinophilia during therapy [240]. In contrast, anti-TSLP antibody Tezepelumab has recently shown the ability to reduce exacerbation rates, independent of blood eosinophils, possibly blocking both T2 and non-T2 inflammatory responses [64].

The drawbacks to new monoclonals are the high costs (~$35,000/patient/ year) and the requirement that these medications be given either intravenously or injected subcutaneously. In each of the Phase III trials mentioned above, a significant proportion of patients are present that do not clearly benefit from these medications. As demands on an already resource-constrained healthcare system intensify, and to avoid unnecessary side effects, biomarkers that identify responders vs. non-responders are greatly needed. As discussed, the most reproducible finding in gene expression has been the T 2 gene signature of epithelial cells, first discovered in bronchial epithelial cells [69, 166] but reproduced in the nasal epithelium in atopic adolescents not on steroids [68]. Thus, one possible approach would be to use nasal epithelial gene expression for the T2 gene signature (i.e., *POSTN*, *CLCA1*, *SERPINB2*) as predictor for response to Dupilumab. Notwithstanding, Wenzel et al. has shown that dupilumab was effective in patients with high and low serum periostin levels. Further, in the nasal RNA gene expression classifier study by Pandey et al., the top classifiers to discriminate asthma from healthy controls did not include *POSTN*, *CLCA1*, *SERPINB2*, although it did contain *CDHR3* (see discussion of Jones et al. above), mast cells genes (*TPSAB1*, *TPSB2*, *CPA3*), and general inflammatory genes such as complement (*C3*) and Arachidonate 15-Lipoxygenase (*ALOX15B*).

In summary, asthma is a heterogeneous disease and likely a common phenotype resulting from numerous combinations of genetic, environmental and age-related factors. Thus far, genetic variant studies have identified asthma susceptibility loci, advanced disease understanding and perhaps helped guide new monoclonal antibody therapies. Global gene expression has advanced disease phenotyping and will have an important future role in connecting gene variants to disease mechanisms. Similarly, epigenetics will likely play a similarly important role in connecting environmental influences to disease. The task lying in front of the field is at the present time overwhelming, which is the more granular identification of the likely vast number of molecular mechanisms responsible for asthma. Yet, herein lies the key to better and more personalized treatment. In the words of Nelson Mandela, "It always seems impossible until it is done." Excitingly, the field of asthma genomics is at the current time completely open for new exploration.

References

[1] Hayes SM, Howlin R, Johnston DA, Webb JS, Clarke SC, Stoodley P, et al. Intracellular residency of *Staphylococcus aureus* within mast cells in nasal polyps: a novel observation. J Allergy Clin Immunol 2015;135(6):1648–51.

[2] Behera D, Sehgal IS. Bronchial asthma—issues for the developing world. Indian J Med Res 2015;141(4):380–2.

[3] Masoli M, Fabian D, Holt S, Beasley R. The global burden of asthma: executive summary of the GINA dissemination committee report. Allergy 2004;59(5):469–78.

[4] Beasley R. Worldwide variation in prevalence of symptoms of asthma, allergic rhinoconjunctivitis, and atopic eczema: ISAAC. The international study of asthma and allergies in childhood (ISAAC) steering committee. Lancet (London, England) 1998;351(9111):1225–32.

[5] The Global Asthma Report 2014. Auckland, New Zealand: Global Asthma Network, 2014.

[6] D'Amato G, Vitale C, Molino A, Stanziola A, Sanduzzi A, Vatrella A, et al. Asthma-related deaths. Multidiscip Respir Med 2016;11:37.

[7] Asthma and African Americans [internet]. U.S.: Department of health & human services; 2015. [cited October 8, 2018]. Available from, https://minorityhealth.hhs.gov/omh/browse.aspx?lvl=4&lvlid=15.

[8] Cowie RL, Underwood MF, Field SK. Asthma symptoms do not predict spirometry. Can Respir J 2007;14(6):339–42.

[9] Moore WC, Meyers DA, Wenzel SE, Teague WG, Li H, Li X, et al. Identification of asthma phenotypes using cluster analysis in the severe asthma research program. Am J Respir Crit Care Med 2010;181(4):315–23.
[10] Litonjua AA, Carey VJ, Burge HA, Weiss ST, Gold DR. Parental history and the risk for childhood asthma. Am J Respir Crit Care Med 1998;158(1):176–81.
[11] Shaaban R, Zureik M, Soussan D, Neukirch C, Heinrich J, Sunyer J, et al. Rhinitis and onset of asthma: a longitudinal population-based study. Lancet (London, England) 2008;372(9643):1049–57.
[12] de Nijs SB, Venekamp LN, Bel EH. Adult-onset asthma: is it really different? Eur Respir Rev 2013;22(127):44.
[13] Rackemann FM. Intrinsic asthma. J Allergy 1940;11(2):147–62.
[14] Jarvis D, Newson R, Lotvall J, Hastan D, Tomassen P, Keil T, et al. Asthma in adults and its association with chronic rhinosinusitis: the GA2LEN survey in Europe. Allergy 2012;67(1):91–8.
[15] Rantala A, Jaakkola JJK, Jaakkola MS. Respiratory infections precede adult-onset asthma. PLoS One 2011;6(12):e27912.
[16] de Marco R, Locatelli F, Cerveri I, Bugiani M, Marinoni A, Giammanco G. Incidence and remission of asthma: a retrospective study on the natural history of asthma in Italy. J Allergy Clin Immunol 2002;110(2):228–35.
[17] Ulrik CS, Lange P. Decline of lung function in adults with bronchial asthma. Am J Respir Crit Care Med 1994;150(3):629–34.
[18] Zein JG, Erzurum SC. Asthma is different in women. Curr Allergy Asthma Rep 2015;15(6):28.
[19] Bergeron C, Tulic MK, Hamid Q. Airway remodelling in asthma: from benchside to clinical practice. Can Respir J 2010;17(4):e85–93.
[20] Diamond G, Legarda D, Ryan LK. The innate immune response of the respiratory epithelium. Immunol Rev 2000;173:27–38.
[21] Knowles MR, Boucher RC. Mucus clearance as a primary innate defense mechanism for mammalian airways. J Clin Invest 2002;109(5):571–7.
[22] Rogers DF. The airway goblet cell. Int J Biochem Cell Biol 2003;35(1):1–6.
[23] Rogers DF, Barnes PJ. Treatment of airway mucus hypersecretion. Ann Med 2006;38(2):116–25.
[24] Randell SH, Boucher RC. Effective mucus clearance is essential for respiratory health. Am J Respir Cell Mol Biol 2006;35(1):20–8.
[25] Cokugras H, Akcakaya N, Seckin I, Camcioglu Y, Sarimurat N, Aksoy F. Ultrastructural examination of bronchial biopsy specimens from children with moderate asthma. Thorax 2001;56(1):25–9.
[26] Dunnill MS. The pathology of asthma, with special reference to changes in the bronchial mucosa. J Clin Pathol 1960;13(1):27–33.
[27] Laitinen LA, Heino M, Laitinen A, Kava T, Haahtela T. Damage of the airway epithelium and bronchial reactivity in patients with asthma. Am Rev Respir Dis 1985;131(4):599–606.
[28] Burgel PR, Nadel JA. Epidermal growth factor receptor-mediated innate immune responses and their roles in airway diseases. Eur Respir J 2008;32(4):1068.
[29] Jones AC, Troy NM, White E, Hollams EM, Gout AM, Ling K-M, et al. Persistent activation of interlinked type 2 airway epithelial gene networks in sputum-derived cells from aeroallergen-sensitized symptomatic asthmatics. Sci Rep 2018;8(1):1511.
[30] Zhou X, Trudeau JB, Wenzel SE. Epithelial human epidermal growth factor receptor 2 (ErbB2) is increased in severe asthma. C37 mediators of asthma and allergic lung disease. In: American thoracic society international conference abstracts. American Thoracic Society; 2012. p. A4299. A.

[31] Polosa R, Puddicombe SM, Krishna MT, Tuck AB, Howarth PH, Holgate ST, et al. Expression of c-erbB receptors and ligands in the bronchial epithelium of asthmatic subjects. J Allergy Clin Immunol 2002;109(1):75–81.

[32] Puddicombe SM, Polosa R, Richter A, Krishna MT, Howarth PH, Holgate ST, et al. Involvement of the epidermal growth factor receptor in epithelial repair in asthma. FASEB J 2000;14(10):1362–74.

[33] Berry M, Brightling C, Pavord I, Wardlaw A. TNF-alpha in asthma. Curr Opin Pharmacol 2007;7(3):279–82.

[34] Brightling C, Berry M, Amrani Y. Targeting TNF-α: a novel therapeutic approach for asthma. J Allergy Clin Immunol 2008;121(1):5–12.

[35] Larose MC, Chakir J, Archambault AS, Joubert P, Provost V, Laviolette M, et al. Correlation between CCL26 production by human bronchial epithelial cells and airway eosinophils: involvement in patients with severe eosinophilic asthma. J Allergy Clin Immunol 2015;136(4):904–13.

[36] Rochman Y, Leonard WJ. Thymic stromal lymphopoietin: a new cytokine in asthma. Curr Opin Pharmacol 2008;8(3):249–54.

[37] Allakhverdi Z, Comeau MR, Jessup HK, Yoon B-RP, Brewer A, Chartier S, et al. Thymic stromal lymphopoietin is released by human epithelial cells in response to microbes, trauma, or inflammation and potently activates mast cells. J Exp Med 2007;204(2):253–8.

[38] Tsilingiri K, Fornasa G, Rescigno M. Thymic stromal lymphopoietin: to cut a long story short. Cell Mol Gastroenterol Hepatol 2017;3(2):174–82.

[39] Divekar R, Kita H. Recent advances in epithelium-derived cytokines (IL-33, IL-25 and TSLP) and allergic inflammation. Curr Opin Allergy Clin Immunol 2015;15(1):98–103.

[40] Kouzaki H, Tojima I, Kita H, Shimizu T. Transcription of interleukin-25 and extracellular release of the protein is regulated by allergen proteases in airway epithelial cells. Am J Respir Cell Mol Biol 2013;49(5):741–50.

[41] Prefontaine D, Nadigel J, Chouiali F, Audusseau S, Semlali A, Chakir J, et al. Increased IL-33 expression by epithelial cells in bronchial asthma. J Allergy Clin Immunol 2010;125(3):752–4.

[42] Soumelis V, Reche PA, Kanzler H, Yuan W, Edward G, Homey B, et al. Human epithelial cells trigger dendritic cell–mediated allergic inflammation by producing TSLP. Nat Immunol 2002;3:673.

[43] Gauvreau GM, O'Byrne PM, Boulet L-P, Wang Y, Cockcroft D, Bigler J, et al. Effects of an anti-TSLP antibody on allergen-induced asthmatic responses. N Engl J Med 2014;370(22):2102–10.

[44] Matsukura S, Stellato C, Georas SN, Casolaro V, Plitt JR, Miura K, et al. Interleukin-13 upregulates eotaxin expression in airway epithelial cells by a STAT6-dependent mechanism. Am J Respir Cell Mol Biol 2001;24(6):755–61.

[45] Coleman JM, Naik C, Holguin F, Ray A, Ray P, Trudeau JB, et al. Epithelial eotaxin-2 and eotaxin-3 expression: relation to asthma severity, luminal eosinophilia and age at onset. Thorax 2012;67:1061–6.

[46] Hamilton LM, Torres-Lozano C, Puddicombe SM, Richter A, Kimber I, Dearman RJ, et al. The role of the epidermal growth factor receptor in sustaining neutrophil inflammation in severe asthma. Clin Exp Allergy 2003;33(2):233–40.

[47] Burgel P, Nadel J. Roles of epidermal growth factor receptor activation in epithelial cell repair and mucin production in airway epithelium. Thorax 2004;59(11):992–6.

[48] Little S, Sproule M, Cowan M, Macleod K, Robertson M, Love J, et al. High resolution computed tomographic assessment of airway wall thickness in chronic asthma: reproducibility and relationship with lung function and severity. Thorax 2002;57(3):247–53.

[49] Chetta A, Foresi A, Del Donno M, Bertorelli G, Pesci A, Olivieri D. Airways remodeling is a distinctive feature of asthma and is related to severity of disease. Chest 1997;111(4):852–7.

[50] Minshall E, Chakir J, Laviolette M, Molet S, Zhu Z, Olivenstein R, et al. IL-11 expression is increased in severe asthma: association with epithelial cells and eosinophils. J Allergy Clin Immunol 2000;105(2 Pt 1):232–8.

[51] Hoshino M, Nakamura Y, Sim JJ. Expression of growth factors and remodelling of the airway wall in bronchial asthma. Thorax 1998;53(1):21–7.

[52] Lemjabbar H, Gosset P, Lamblin C, Tillie I, Hartmann D, Wallaert B, et al. Contribution of 92 kDa gelatinase/type IV collagenase in bronchial inflammation during status asthmaticus. Am J Respir Crit Care Med 1999;159(4 Pt 1):1298–307.

[53] Prikk K, Maisi P, Pirila E, Reintam MA, Salo T, Sorsa T, et al. Airway obstruction correlates with collagenase-2 (MMP-8) expression and activation in bronchial asthma. Lab Invest 2002;82(11):1535–45.

[54] Vignola AM, Riccobono L, Mirabella A, Profita M, Chanez P, Bellia V, et al. Sputum metalloproteinase-9/tissue inhibitor of metalloproteinase-1 ratio correlates with airflow obstruction in asthma and chronic bronchitis. Am J Respir Crit Care Med 1998;158(6):1945–50.

[55] Benayoun L, Druilhe A, Dombret MC, Aubier M, Pretolani M. Airway structural alterations selectively associated with severe asthma. Am J Respir Crit Care Med 2003;167(10):1360–8.

[56] Niimi A, Matsumoto H, Takemura M, Ueda T, Chin K, Mishima M. Relationship of airway wall thickness to airway sensitivity and airway reactivity in asthma. Am J Respir Crit Care Med 2003;168(8):983–8.

[57] Bentley JK, Hershenson MB. Airway smooth muscle growth in asthma: proliferation, hypertrophy, and migration. Proc Am Thorac Soc 2008;5(1):89–96.

[58] Kelleher MD, Abe MK, Chao TS, Jain M, Green JM, Solway J, et al. Role of MAP kinase activation in bovine tracheal smooth muscle mitogenesis. Am J Physiol 1995;268(6 Pt 1):L894–901.

[59] Hirst SJ, Barnes PJ, Twort CH. PDGF isoform-induced proliferation and receptor expression in human cultured airway smooth muscle cells. Am J Physiol 1996;270(3 Pt 1):L415–28.

[60] Hirst SJ, Barnes PJ, Twort CH. Quantifying proliferation of cultured human and rabbit airway smooth muscle cells in response to serum and platelet-derived growth factor. Am J Respir Cell Mol Biol 1992;7(6):574–81.

[61] Goldsmith AM, Bentley JK, Zhou L, Jia Y, Bitar KN, Fingar DC, et al. Transforming growth factor-β induces airway smooth muscle hypertrophy. Am J Respir Cell Mol Biol 2006;34(2):247–54.

[62] Locksley RM. Asthma and allergic inflammation. Cell 2010;140(6):777–83.

[63] Ito T, Wang Y-H, Duramad O, Hori T, Delespesse GJ, Watanabe N, et al. TSLP-activated dendritic cells induce an inflammatory T helper type 2 cell response through OX40 ligand. J Exp Med 2005;202(9):1213–23.

[64] Corren J, Parnes JR, Wang L, Mo M, Roseti SL, Griffiths JM, et al. Tezepelumab in adults with uncontrolled asthma. N Engl J Med 2017;377(10):936–46.

[65] Lund S, Walford HH, Doherty TA. Type 2 innate lymphoid cells in allergic disease. Curr Immunol Rev 2013;9(4):214–21.

[66] Mjösberg JM, Trifari S, Crellin NK, Peters CP, van Drunen CM, Piet B, et al. Human IL-25- and IL-33-responsive type 2 innate lymphoid cells are defined by expression of CRTH2 and CD161. Nat Immunol 2011;12:1055.

[67] Kim HY, Umetsu DT, Dekruyff RH. Innate lymphoid cells in asthma: will they take your breath away? Eur J Immunol 2016;46(4):795–806.

[68] Poole A, Urbanek C, Eng C, Schageman J, Jacobson S, O'Connor BP, et al. Dissecting childhood asthma with nasal transcriptomics distinguishes subphenotypes of disease. J Allergy Clin Immunol 2014;133(3):670–678.e12.
[69] Woodruff PG, Modrek B, Choy DF, Jia G, Abbas AR, Ellwanger A, et al. T-helper type 2–driven inflammation defines major subphenotypes of asthma. Am J Respir Crit Care Med 2009;180(5):388–95.
[70] Modena BD, Tedrow JR, Milosevic J, Bleecker ER, Meyers DA, Wu W, et al. Gene expression in relation to exhaled nitric oxide identifies novel asthma phenotypes with unique biomolecular pathways. Am J Respir Crit Care Med 2014;190(12):1363–72.
[71] Peters MC, Kerr S, Dunican EM, Woodruff PG, Fajt ML, Levy BD, et al. Refractory airway type 2 inflammation in a large subgroup of asthmatic patients treated with inhaled corticosteroids. J Allergy Clin Immunol 2018;143:104–13.
[72] Wenzel SE. Asthma phenotypes: the evolution from clinical to molecular approaches. Nat Med 2012;18(5):716–25.
[73] Wu W, Bleecker E, Moore W, Busse WW, Castro M, Chung KF, et al. Unsupervised phenotyping of severe asthma research program participants using expanded lung data. J Allergy Clin Immunol 2014;133(5):1280–8.
[74] Ober C, Yao TC. The genetics of asthma and allergic disease: a 21st century perspective. Immunol Rev 2011;242(1):10–30.
[75] Duffy DL, Martin NG, Battistutta D, Hopper JL, Mathews JD. Genetics of asthma and hay fever in Australian twins. Am Rev Respir Dis 1990;142(6 Pt 1):1351–8.
[76] Nieminen MM, Kaprio J, Koskenvuo M. A population-based study of bronchial asthma in adult twin pairs. Chest 1991;100(1):70–5.
[77] Harris JR, Magnus P, Samuelsen SO, Tambs K. No evidence for effects of family environment on asthma. A retrospective study of Norwegian twins. Am J Respir Crit Care Med 1997;156(1):43–9.
[78] Laitinen T, Rasanen M, Kaprio J, Koskenvuo M, Laitinen LA. Importance of genetic factors in adolescent asthma: a population-based twin-family study. Am J Respir Crit Care Med 1998;157(4 Pt 1):1073–8.
[79] Skadhauge LR, Christensen K, Kyvik KO, Sigsgaard T. Genetic and environmental influence on asthma: a population-based study of 11,688 Danish twin pairs. Eur Respir J 1999;13(1):8.
[80] Koeppen-Schomerus G, Stevenson J, Plomin R. Genes and environment in asthma: a study of 4 year old twins. Arch Dis Child 2001;85(5):398–400.
[81] Hallstrand TS, Fischer ME, Wurfel MM, Afari N, Buchwald D, Goldberg J. Genetic pleiotropy between asthma and obesity in a community-based sample of twins. J Allergy Clin Immunol 2005;116(6):1235–41.
[82] Thomsen SF, van der Sluis S, Kyvik KO, Skytthe A, Backer V. Estimates of asthma heritability in a large twin sample. Clin Exp Allergy 2010;40(7):1054–61.
[83] Lichtenstein P, Svartengren M. Genes, environments, and sex: factors of importance in atopic diseases in 7–9-year-old Swedish twins. Allergy 1997;52(11):1079–86.
[84] Fagnani C, Annesi-Maesano I, Brescianini S, D'Ippolito C, Medda E, Nisticò L, et al. Heritability and shared genetic effects of asthma and hay fever: an Italian study of young twins. Twin Res Hum Genet 2008;11(2):121–31.
[85] Nystad W, Roysamb E, Magnus P, Tambs K, Harris JR. A comparison of genetic and environmental variance structures for asthma, hay fever and eczema with symptoms of the same diseases: a study of Norwegian twins. Int J Epidemiol 2005;34(6):1302–9.
[86] van Beijsterveldt CE, Boomsma DI. Genetics of parentally reported asthma, eczema and rhinitis in 5-yr-old twins. Eur Respir J 2007;29(3):516–21.

[87] Zheng T, Yu J, Oh MH, Zhu Z. The atopic March: progression from atopic dermatitis to allergic rhinitis and asthma. Allergy, Asthma Immunol Res 2011;3(2):67–73.
[88] Bousquet J, Khaltaev N, Cruz AA, Denburg J, Fokkens WJ, Togias A, et al. Allergic rhinitis and its impact on asthma (ARIA) 2008 update (in collaboration with the world health organization, GA(2)LEN and AllerGen). Allergy 2008;63(Suppl 86):8–160.
[89] Gabriel SB, Schaffner SF, Nguyen H, Moore JM, Roy J, Blumenstiel B, et al. The structure of haplotype blocks in the human genome. Science (New York, NY) 2002;296(5576):2225–9.
[90] Fareed M, Afzal M. Single nucleotide polymorphism in genome-wide association of human population: a tool for broad spectrum service. Egyptian J Med Hum Genet 2013;14(2):-123–34.
[91] The Genomes Project C, Auton A, Abecasis GR, Altshuler DM, Durbin RM, Abecasis GR, et al. A global reference for human genetic variation. Nature. 2015;526:68.
[92] Mullaney JM, Mills RE, Pittard WS, Devine SE. Small insertions and deletions (INDELs) in human genomes. Hum Mol Genet 2010;19(R2):R131–6.
[93] Thapar A, Cooper M. Copy number variation: what is it and what has it told us about child psychiatric disorders? J Am Acad Child Adolesc Psychiatry 2013;52(8):772–4.
[94] Vercelli D. Discovering susceptibility genes for asthma and allergy. Nat Rev Immunol 2008;8(3):169–82.
[95] Bossé Y, Hudson TJ. Toward a comprehensive set of asthma susceptibility genes. Annu Rev Med 2007;58(1):171–84.
[96] Ober C, Hoffjan S. Asthma genetics 2006: the long and winding road to gene discovery. Genes Immun 2006;7:95.
[97] Samitas K, Carter A, Kariyawasam HH, Xanthou G. Upper and lower airway remodelling mechanisms in asthma, allergic rhinitis and chronic rhinosinusitis: the one airway concept revisited. Allergy 2017;73(5):993–1002.
[98] Van Eerdewegh P, Little RD, Dupuis J, Del Mastro RG, Falls K, Simon J, et al. Association of the ADAM33 gene with asthma and bronchial hyperresponsiveness. Nature 2002;418(6896):426–30.
[99] Bijanzadeh M, Ramachandra NB, Mahesh PA, Mysore RS, Kumar P, Manjunath BS, et al. Association of IL-4 and ADAM33 gene polymorphisms with asthma in an Indian population. Lung 2010;188(5):415–22.
[100] Hirota T, Hasegawa K, Obara K, Matsuda A, Akahoshi M, Nakashima K, et al. Association between ADAM33 polymorphisms and adult asthma in the Japanese population. Clin Exp Allergy 2006;36(7):884–91.
[101] Blakey J, Halapi E, Bjornsdottir US, Wheatley A, Kristinsson S, Upmanyu R, et al. Contribution of ADAM33 polymorphisms to the population risk of asthma. Thorax 2005;60(4):274–6.
[102] Jongepier H, Boezen HM, Dijkstra A, Howard TD, Vonk JM, Koppelman GH, et al. Polymorphisms of the ADAM33 gene are associated with accelerated lung function decline in asthma. Clin Exp Allergy 2004;34(5):757–60.
[103] Mahesh PA. Unravelling the role of ADAM 33 in asthma. Indian J Med Res 2013;137(3):447–50.
[104] Denham S, Koppelman GH, Blakey J, Wjst M, Ferreira MA, Hall IP, et al. Meta-analysis of genome-wide linkage studies of asthma and related traits. Respir Res 2008;9(1):38.
[105] Bouzigon E, Forabosco P, Koppelman GH, Cookson WOCM, Dizier M-H, Duffy DL, et al. Meta-analysis of 20 genome-wide linkage studies evidenced new regions linked to asthma and atopy. Eur J Hum Genet 2010;18:700.
[106] Vicente CT, Revez JA, Ferreira MAR. Lessons from ten years of genome-wide association studies of asthma. Clin Transl Immunology 2017;6(12):e165.

[107] Torgerson DG, Ampleford EJ, Chiu GY, Gauderman WJ, Gignoux CR, Graves PE, et al. Meta-analysis of genome-wide association studies of asthma in ethnically diverse north American populations. Nat Genet 2011;43(9):887–92.

[108] Himes BE, Hunninghake GM, Baurley JW, Rafaels NM, Sleiman P, Strachan DP, et al. Genome-wide association analysis identifies PDE4D as an asthma-susceptibility gene. Am J Hum Genet 2009;84(5):581–93.

[109] Moffatt MF, Kabesch M, Liang L, Dixon AL, Strachan D, Heath S, et al. Genetic variants regulating ORMDL3 expression contribute to the risk of childhood asthma. Nature 2007;448(7152):470–3.

[110] Sleiman PMA, Annaiah K, Imielinski M, Bradfield JP, Kim CE, Frackelton EC, et al. ORMDL3 variants associated with asthma susceptibility in North Americans of European ancestry. J Allergy Clin Immunol 2008;122(6):1225–7.

[111] Halapi E, Gudbjartsson DF, Jonsdottir GM, Bjornsdottir US, Thorleifsson G, Helgadottir H, et al. A sequence variant on 17q21 is associated with age at onset and severity of asthma. Eur J Hum Genet 2010;18(8):902–8.

[112] Bouzigon E, Corda E, Aschard H, Dizier M-H, Boland A, Bousquet J, et al. Effect of 17q21 variants and smoking exposure in early-onset asthma. N Engl J Med 2008;359(19):1985–94.

[113] Ober C, Tan Z, Sun Y, Possick JD, Pan L, Nicolae R, et al. Effect of variation in CHI3L1 on serum YKL-40 level, risk of asthma, and lung function. N Engl J Med 2008;358(16):1682–91.

[114] Johansen JS. Studies on serum YKL-40 as a biomarker in diseases with inflammation, tissue remodelling, fibroses and cancer. Dan Med Bull 2006;53(2):172–209.

[115] Weidinger S, Gieger C, Rodriguez E, Baurecht H, Mempel M, Klopp N, et al. Genome-wide scan on total serum IgE levels identifies FCER1A as novel susceptibility locus. PLoS Genet 2008;4(8):e1000166.

[116] Limb SL, Brown KC, Wood RA, Wise RA, Eggleston PA, Tonascia J, et al. Adult asthma severity in individuals with a history of childhood asthma. J Allergy Clin Immunol 2005;115(1):61–6.

[117] Burrows B, Martinez FD, Halonen M, Barbee RA, Cline MG. Association of asthma with serum IgE levels and skin-test reactivity to allergens. N Engl J Med 1989;320(5):271–7.

[118] Jacobsen HP, Herskind AM, Nielsen BW, Husby S. IgE in unselected like-sexed monozygotic and dizygotic twins at birth and at 6 to 9 years of age: high but dissimilar genetic influence on IgE levels. J Allergy Clin Immunol 2001;107(4):659–63.

[119] Strachan DP, Wong HJ, Spector TD. Concordance and interrelationship of atopic diseases and markers of allergic sensitization among adult female twins. J Allergy Clin Immunol 2001;108(6):901–7.

[120] Hirota T, Takahashi A, Kubo M, Tsunoda T, Tomita K, Doi S, et al. Genome-wide association study identifies three new susceptibility loci for adult asthma in the Japanese population. Nat Genet 2011;43(9):893–6.

[121] Cianferoni A, Spergel J. The importance of TSLP in allergic disease and its role as a potential therapeutic target. Expert Rev Clin Immunol 2014;10(11):1463–74.

[122] Corren J, Parnes JR, Wang L, Mo M, Roseti SL, Griffiths JM, et al. Tezepelumab demonstrates clinically meaningful improvements in asthma control (ACQ-6) in patients with uncontrolled asthma: results from a phase 2b clinical trial. J Allergy Clin Immunol 2018;141(2):AB80.

[123] Hunninghake GM, Soto-Quirós ME, Avila L, Kim HP, Lasky-Su J, Rafaels N, et al. TSLP polymorphisms are associated with asthma in a sex-specific fashion. Allergy 2010;65(12):1566–75.

[124] Moffatt MF, Gut IG, Demenais F, Strachan DP, Bouzigon E, Heath S, et al. A large-scale, consortium-based genomewide association study of asthma. N Engl J Med 2010;363(13):1211–21.

[125] Garlanda C, Dinarello CA, Mantovani A. The interleukin-1 family: back to the future. Immunity 2013;39(6):1003–18.

[126] Borish L, Steinke JW. Interleukin-33 in asthma: how big of a role does it play? Curr Allergy Asthma Rep 2011;11(1):7–11.

[127] Sawada M, Kawayama T, Imaoka H, Sakazaki Y, Oda H, Takenaka S, et al. IL-18 induces airway hyperresponsiveness and pulmonary inflammation via CD4+ T cell and IL-13. PLoS One 2013;8(1):e54623.

[128] Sjöberg LC, Nilsson AZ, Lei Y, Gregory JA, Adner M, Nilsson GP. Interleukin 33 exacerbates antigen driven airway hyperresponsiveness, inflammation and remodeling in a mouse model of asthma. Sci Rep 2017;7(1):4219.

[129] He JQ, Hallstrand TS, Knight D, Chan-Yeung M, Sandford A, Tripp B, et al. A thymic stromal lymphopoietin gene variant is associated with asthma and airway hyperresponsiveness. J Allergy Clin Immunol 2009;124(2):222–9.

[130] Myers RA, Himes BE, Gignoux CR, Yang JJ, Gauderman WJ, Rebordosa C, et al. Further replication studies of the EVE Consortium meta-analysis identifies 2 asthma risk loci in European Americans. J Allergy Clin Immunol 2012;130(6):1294–301.

[131] Kodak JA, Mann DL, Klyushnenkova EN, Alexander RB. Activation of innate immunity by prostate specific antigen (PSA). Prostate 2006;66(15):1592–9.

[132] Ober C. A genome-wide search for asthma susceptibility loci in ethnically diverse populations. The collaborative study on the genetics of asthma (CSGA). Nat Genet 1997;15(4):389–92.

[133] Levin AM, Mathias RA, Huang L, Roth LA, Daley D, Myers RA, et al. A meta-analysis of genome-wide association studies for serum total IgE in diverse study populations. J Allergy Clin Immunol 2013;131(4):1176–84.

[134] Deleted in review.

[135] Myers RA, Scott NM, Gauderman WJ, Qiu W, Mathias RA, Romieu I, et al. Genome-wide interaction studies reveal sex-specific asthma risk alleles. Hum Mol Genet 2014;23(19):5251–9.

[136] Osman M. Therapeutic implications of sex differences in asthma and atopy. Arch Dis Child 2003;88(7):587–90.

[137] Troisi RJ, Speizer FE, Willett WC, Trichopoulos D, Rosner B. Menopause, postmenopausal estrogen preparations, and the risk of adult-onset asthma. A prospective cohort study. Am J Respir Crit Care Med 1995;152(4 Pt 1):1183–8.

[138] Fuseini H, Newcomb DC. Mechanisms driving gender differences in asthma. Curr Allergy and Asthma Rep 2017;17(3):19.

[139] Genuneit J. Sex-specific development of asthma differs between farm and nonfarm children: a cohort study. Am J Respir Crit Care Med 2014;190(5):588–90.

[140] Borish L, Chipps B, Deniz Y, Gujrathi S, Zheng B, Dolan CM. Total serum IgE levels in a large cohort of patients with severe or difficult-to-treat asthma. Ann Allergy Asthma Immunol 2005;95(3):247–53.

[141] Bonnelykke K, Sleiman P, Nielsen K, Kreiner-Moller E, Mercader JM, Belgrave D, et al. A genome-wide association study identifies CDHR3 as a susceptibility locus for early childhood asthma with severe exacerbations. Nat Genet 2014;46(1):51–5.

[142] Demenais F, Margaritte-Jeannin P, Barnes KC, Cookson WOC, Altmüller J, Ang W, et al. Multiancestry association study identifies new asthma risk loci that colocalize with immune cell enhancer marks. Nat Genet 2018;50(1):42–53.

[143] Manolio TA, Collins FS, Cox NJ, Goldstein DB, Hindorff LA, Hunter DJ, et al. Finding the missing heritability of complex diseases. Nature 2009;461(7265):747–53.

[144] Bégin P, Nadeau KC. Epigenetic regulation of asthma and allergic disease. Allergy Asthma Clin Immunol 2014;10(1):27.

[145] Ober C, Vercelli D. Gene-environment interactions in human disease: nuisance or opportunity? Trends Genet 2011;27(3):107–15.

[146] Igartua C, Myers RA, Mathias RA, Pino-Yanes M, Eng C, Graves PE, et al. Ethnic-specific associations of rare and low frequency DNA sequence variants with asthma. Nat Commun 2015;6:5965.

[147] Zuk O, Schaffner SF, Samocha K, Do R, Hechter E, Kathiresan S, et al. Searching for missing heritability: designing rare variant association studies. Proc Natl Acad Sci U S A 2014;111(4):E455–64.

[148] Torgerson DG, Capurso D, Mathias RA, Graves PE, Hernandez RD, Beaty TH, et al. Resequencing candidate genes implicates rare variants in asthma susceptibility. Am J Hum Genet 2012;90(2):273–81.

[149] Cirulli ET, Goldstein DB. Uncovering the roles of rare variants in common disease through whole-genome sequencing. Nat Rev Genet 2010;11(6):415–25.

[150] Lee S, Teslovich TM, Boehnke M, Lin X. General framework for meta-analysis of rare variants in sequencing association studies. Am J Hum Genet 2013;93(1):42–53.

[151] Lavoie-Charland E, Berube JC, Boulet LP, Bosse Y. Asthma susceptibility variants are more strongly associated with clinically similar subgroups. J Asthma 2016;53(9):907–13.

[152] Pennisi E. ENCODE project writes eulogy for junk DNA. Science (New York, NY) 2012;337(6099):1159.

[153] Friedman RC, Farh KK, Burge CB, Bartel DP. Most mammalian mRNAs are conserved targets of microRNAs. Genome Res 2009;19(1):92–105.

[154] Zhao S, Fung-Leung W-P, Bittner A, Ngo K, Liu X. Comparison of RNA-seq and microarray in transcriptome profiling of activated T cells. PLoS One 2014;9(1):e78644.

[155] Tan PK, Downey TJ, Spitznagel Jr EL, Xu P, Fu D, Dimitrov DS, et al. Evaluation of gene expression measurements from commercial microarray platforms. Nucleic Acids Res 2003;31(19):5676–84.

[156] Kuo C-HS, Pavlidis S, Loza M, Baribaud F, Rowe A, Pandis I, et al. A transcriptome-driven analysis of epithelial brushings and bronchial biopsies to define asthma phenotypes in U-BIOPRED. Am J Respir Crit Care Med 2016;195(4):443–55.

[157] Bigler J, Boedigheimer M, Schofield JPR, Skipp PJ, Corfield J, Rowe A, et al. A severe asthma disease signature from gene expression profiling of peripheral Blood from U-BIOPRED cohorts. Am J Respir Crit Care Med 2017;195(10):1311–20.

[158] Lefaudeux D, De Meulder B, Loza MJ, Peffer N, Rowe A, Baribaud F, et al. U-BIOPRED clinical adult asthma clusters linked to a subset of sputum omics. J Allergy Clin Immunol 2017;139(6):1797–807.

[159] Howrylak JA, Fuhlbrigge AL, Strunk RC, Zeiger RS, Weiss ST, Raby BA. Classification of childhood asthma phenotypes and long-term clinical responses to inhaled anti-inflammatory medications. J Allergy Clin Immunol 2014;133(5):1289–130112.

[160] Bosco A, Ehteshami S, Stern DA, Martinez FD. Decreased activation of inflammatory networks during acute asthma exacerbations is associated with chronic airflow obstruction. Mucosal Immunol 2010;3(4):399–409.

[161] Bosco A, Wiehler S, Proud D. Interferon regulatory factor 7 regulates airway epithelial cell responses to human rhinovirus infection. BMC Genomics 2016;17(1):76.

[162] Gomez JL, Crisafi GM, Holm CT, Meyers DA, Hawkins GA, Bleecker ER, et al. Genetic variation in chitinase 3-like 1 (<em>CHI3L1</em>) contributes to asthma severity and airway expression of YKL-40. J Allergy Clin Immunol 2015;136(1):51–58.e10.

[163] Bochkov YA, Hanson KM, Keles S, Brockman-Schneider RA, Jarjour NN, Gern JE. Rhinovirus-induced modulation of gene expression in bronchial epithelial cells from subjects with asthma. Mucosal Immunol 2009;3:69.

[164] Altman MC, Babineau D, Whalen E, Gill MA, Shao B, Liu AH, et al. Identification of inflammatory gene expression patterns associated with viral upper respiratory tract infections (URI) that cause asthma exacerbations. D92 genetics and genomics of obstructive lung disease. In: American thoracic society international conference abstracts. American Thoracic Society; 2018. p. A7365. A.

[165] Singhania A, Rupani H, Jayasekera N, Lumb S, Hales P, Gozzard N, et al. Altered epithelial gene expression in peripheral airways of severe asthma. PLoS One 2017;12(1):e0168680.

[166] Woodruff PG, Boushey HA, Dolganov GM, Barker CS, Yang YH, Donnelly S, et al. Genome-wide profiling identifies epithelial cell genes associated with asthma and with treatment response to corticosteroids. Proc Natl Acad Sci 2007;104(40):15858.

[167] Rabe KF, Nair P, Brusselle G, Maspero JF, Castro M, Sher L, et al. Efficacy and safety of dupilumab in glucocorticoid-dependent severe asthma. N Engl J Med 2018;378(26):2475–85.

[168] Castro M, Corren J, Pavord ID, Maspero J, Wenzel S, Rabe KF, et al. Dupilumab efficacy and safety in moderate-to-severe uncontrolled asthma. N Engl J Med 2018;378(26):2486–96.

[169] Pandey G, Pandey OP, Rogers AJ, Ahsen ME, Hoffman GE, Raby BA, et al. A nasal brush-based classifier of asthma identified by machine learning analysis of nasal RNA sequence data. Sci Rep 2018;8(1):8826.

[170] Jia G, Erickson RW, Choy DF, Mosesova S, Wu LC, Solberg OD, et al. Periostin is a systemic biomarker of eosinophilic airway inflammation in asthmatic patients. J Allergy Clin Immunol 2012;130(3):647–654.e10.

[171] Ito Y, Al Mubarak R, Roberts N, Correll K, Janssen W, Finigan J, et al. IL-13 induces periostin and eotaxin expression in human primary alveolar epithelial cells: comparison with paired airway epithelial cells. PLoS One 2018;13(4):e0196256.

[172] Takayama G, Arima K, Kanaji T, Toda S, Tanaka H, Shoji S, et al. Periostin: a novel component of subepithelial fibrosis of bronchial asthma downstream of IL-4 and IL-13 signals. J Allergy Clin Immunol 2006;118(1):98–104.

[173] Sidhu SS, Yuan S, Innes AL, Kerr S, Woodruff PG, Hou L, et al. Roles of epithelial cell-derived periostin in TGF-β activation, collagen production, and collagen gel elasticity in asthma. Proc Natl Acad Sci U S A 2010;107(32):14170–5.

[174] Yamamoto M, Tochino Y, Chibana K, Trudeau JB, Holguin F, Wenzel SE. Nitric oxide and related enzymes in asthma: relation to severity, enzyme function and inflammation. Clin Exp Allergy 2012;42(5):760–8.

[175] Corren J, Lemanske RF, Hanania NA, Korenblat PE, Parsey MV, Arron JR, et al. Lebrikizumab treatment in adults with asthma. N Engl J Med 2011;365(12):1088–98.

[176] Wang T, Zhang H-P, Zhang X, Liang Z-A, Ji Y-L, Wang G. Is folate status a risk factor for asthma or other allergic diseases? Allergy, Asthma Immunol Res 2015;7(6):538–46.

[177] Modena BD, Bleecker ER, Busse WW, Erzurum SC, Gaston BM, Jarjour NN, et al. Gene expression correlated with severe asthma characteristics reveals heterogeneous mechanisms of severe disease. Am J Respir Crit Care Med 2017;195(11):1449–63.

[178] Dweik RA, Sorkness RL, Wenzel S, Hammel J, Curran-Everett D, Comhair SA, et al. Use of exhaled nitric oxide measurement to identify a reactive, at-risk phenotype among patients with asthma. Am J Respir Crit Care Med 2010;181(10):1033–41.

[179] Fitzpatrick AM, Gaston BM, Erzurum SC, Teague WG. Features of severe asthma in school-age children: atopy and increased exhaled nitric oxide. J Allergy Clin Immunol 2006;118(6):1218–25.

[180] Dweik RA, Boggs PB, Erzurum SC, Irvin CG, Leigh MW, Lundberg JO, et al. An official ATS clinical practice guideline: interpretation of exhaled nitric oxide levels (FENO) for clinical applications. Am J Respir Crit Care Med 2011;184(5):602–15.

[181] Shaw DE, Berry MA, Thomas M, Green RH, Brightling CE, Wardlaw AJ, et al. The use of exhaled nitric oxide to guide asthma management: a randomized controlled trial. Am J Respir Crit Care Med 2007;176(3):231–7.

[182] Wysocki K, Park SY, Bleecker E, Busse W, Castro M, Chung KF, et al. Characterization of factors associated with systemic corticosteroid use in severe asthma: data from the severe asthma research program. J Allergy Clin Immunol 2014;133(3):915–8.

[183] Nicodemus-Johnson J, Myers RA, Sakabe NJ, Sobreira DR, Hogarth DK, Naureckas ET, et al. DNA methylation in lung cells is associated with asthma endotypes and genetic risk. JCI Insight 2016;1(20):e90151.

[184] Mitri Z, Constantine T, O'Regan R. The HER2 receptor in breast cancer: pathophysiology, clinical use, and new advances in therapy. Chemother Res Pract 2012;2012:743193.

[185] Klezovitch O, Chevillet J, Mirosevich J, Roberts RL, Matusik RJ, Vasioukhin V. Hepsin promotes prostate cancer progression and metastasis. Cancer Cell 2004;6(2):185–95.

[186] Fischer BM, Cuellar JG, Byrd AS, Rice AB, Bonner JC, Martin LD, et al. ErbB2 activity is required for airway epithelial repair following neutrophil elastase exposure. FASEB J 2005;19(10):1374–6.

[187] Faress JA, Nethery DE, Kern EFO, Eisenberg R, Jacono FJ, Allen CL, et al. Bleomycin-induced pulmonary fibrosis is attenuated by a monoclonal antibody targeting HER2. J Appl Physiol 2007;103(6):2077–83.

[188] Vermeer PD, Einwalter LA, Moninger TO, Rokhlina T, Kern JA, Zabner J, et al. Segregation of receptor and ligand regulates activation of epithelial growth factor receptor. Nature 2003;422:322.

[189] Vermeer PD, Panko L, Karp P, Lee JH, Zabner J. Differentiation of human airway epithelia is dependent on erbB2. Am J Physiol Lung Cellular Mol Physiol 2006;291(2):L175–80.

[190] Kettle R, Simmons J, Schindler F, Jones P, Dicker T, Dubois G, et al. Regulation of neuregulin 1β1–induced MUC5AC and MUC5B expression in human airway epithelium. Am J Respir Cell Mol Biol 2010;42(4):472–81.

[191] Xanthou G, Alissafi T, Semitekolou M, Simoes DCM, Economidou E, Gaga M, et al. Osteopontin has a crucial role in allergic airway disease through regulation of dendritic cell subsets. Nat Med 2007;13(5):570–8.

[192] Arjomandi M, Galanter JM, Choudhry S, Eng C, Hu D, Beckman K, et al. Polymorphism in osteopontin gene (SPP1) is associated with asthma and related phenotypes in a Puerto Rican population. Pediatr Allergy Immunol Pulmonol 2011;24(4):207–14.

[193] Simpson JL, Scott R, Boyle MJ, Gibson PG. Inflammatory subtypes in asthma: assessment and identification using induced sputum. Respirology (Carlton, Vic) 2006;11(1):54–61.

[194] Berry M, Morgan A, Shaw DE, Parker D, Green R, Brightling C, et al. Pathological features and inhaled corticosteroid response of eosinophilic and non-eosinophilic asthma. Thorax 2007;62(12):1043–9.

[195] Hastie AT, Moore WC, Meyers DA, Vestal PL, Li H, Peters SP, et al. Analyses of asthma severity phenotypes and inflammatory proteins in subjects stratified by sputum granulocytes. J Allergy Clin Immunol 2010;125(5):1028–1036.e13.

[196] Watanabe T, Fajt ML, Trudeau JB, Voraphani N, Hu H, Zhou X, et al. Brain-derived neurotrophic factor expression in asthma. Association with severity and type 2 inflammatory processes. Am J Respir Cell Mol Biol 2015;53(6):844–52.

[197] Kuo C-HS, Pavlidis S, Loza M, Baribaud F, Rowe A, Pandis I, et al. T-helper cell type 2 (Th2) and non-Th2 molecular phenotypes of asthma using sputum transcriptomics in U-BIOPRED. Eur Respir J 2017;49(2):1602135.

[198] Bønnelykke K, Sleiman P, Nielsen K, Kreiner-Møller E, Mercader JM, Belgrave D, et al. A genome-wide association study identifies CDHR3 as a susceptibility locus for early childhood asthma with severe exacerbations. Nat Genet 2013;46:51.

[199] Bochkov YA, Watters K, Ashraf S, Griggs TF, Devries MK, Jackson DJ, et al. Cadherin-related family member 3, a childhood asthma susceptibility gene product, mediates rhinovirus C binding and replication. Proc Natl Acad Sci 2015;112(17):5485.

[200] Hakonarson H, Bjornsdottir US, Halapi E, Bradfield J, Zink F, Mouy M, et al. Profiling of genes expressed in peripheral blood mononuclear cells predicts glucocorticoid sensitivity in asthma patients. Proc Natl Acad Sci U S A 2005;102(41):14789–94.

[201] George BJ, Reif DM, Gallagher JE, Williams-DeVane CR, Heidenfelder BL, Hudgens EE, et al. Data-driven asthma endotypes defined from blood biomarker and gene expression data. PLoS One 2015;10(2):e0117445.

[202] Kong Q, Li W-J, Huang H-R, Zhong Y-Q, Fang J-P. Differential gene expression profiles of peripheral blood mononuclear cells in childhood asthma. J Asthma 2015;52(4):343–52.

[203] Doherty TA, Soroosh P, Khorram N, Fukuyama S, Rosenthal P, Cho JY, et al. The tumor necrosis factor family member LIGHT is a target for asthmatic airway remodeling. Nat Med 2011;17(5):596–603.

[204] Herro R, Da Silva AR, Aguilera AR, Tamada K, Croft M. Tumor necrosis factor superfamily 14 (LIGHT) controls thymic stromal lymphopoietin to drive pulmonary fibrosis. J Allergy Clin Immunol 2015;136(3):757–68.

[205] Bjornsdottir US, Holgate ST, Reddy PS, Hill AA, McKee CM, Csimma CI, et al. Pathways activated during human asthma exacerbation as revealed by gene expression patterns in blood. PLoS One 2011;6(7):e21902.

[206] Croteau-Chonka DC, Qiu W, Martinez FD, Strunk RC, Lemanske RF, Liu AH, et al. Gene expression profiling in blood provides reproducible molecular insights into asthma control. Am J Respir Crit Care Med 2016;195(2):179–88.

[207] Colonna M, Facchetti F. TREM-1 (triggering receptor expressed on myeloid cells): a new player in acute inflammatory responses. J Infect Dis 2003;187(Suppl 2):S397–401.

[208] Cookson W, Moffatt M, Strachan DP. Genetic risks and childhood-onset asthma. J Allergy Clin Immunol 2011;128(2):266–70. quiz 71-2.

[209] Runyon RS, Cachola LM, Rajeshuni N, Hunter T, Garcia M, Ahn R, et al. Asthma discordance in twins is linked to epigenetic modifications of T cells. PLoS One 2012;7(11):e48796.

[210] Stefanowicz D, Hackett T-L, Garmaroudi FS, Günther OP, Neumann S, Sutanto EN, et al. DNA methylation profiles of airway epithelial cells and PBMCs from healthy, atopic and asthmatic children. PLoS ONE 2012;7(9):e44213.

[211] Soto-Ramirez N, Arshad SH, Holloway JW, Zhang H, Schauberger E, Ewart S, et al. The interaction of genetic variants and DNA methylation of the interleukin-4 receptor gene increase the risk of asthma at age 18 years. Clin Epigenetics 2013;5(1):1.

[212] Breton CV, Byun HM, Wang X, Salam MT, Siegmund K, Gilliland FD. DNA methylation in the arginase-nitric oxide synthase pathway is associated with exhaled nitric oxide in children with asthma. Am J Respir Crit Care Med 2011;184(2):191–7.

[213] Morales E, Bustamante M, Vilahur N, Escaramis G, Montfort M, de Cid R, et al. DNA hypomethylation at ALOX12 is associated with persistent wheezing in childhood. Am J Respir Crit Care Med 2012;185(9):937–43.

[214] Perera F, Tang WY, Herbstman J, Tang D, Levin L, Miller R, et al. Relation of DNA methylation of 5'-CpG island of ACSL3 to transplacental exposure to airborne polycyclic aromatic hydrocarbons and childhood asthma. PLoS One 2009;4(2):e4488.

[215] Fu A, Leaderer BP, Gent JF, Leaderer D, Zhu Y. An environmental epigenetic study of ADRB2 5'-UTR methylation and childhood asthma severity. Clin Exp Allergy 2012;42(11):1575–81.

[216] Isidoro-Garcia M, Sanz C, Garcia-Solaesa V, Pascual M, Pescador DB, Lorente F, et al. PTGDR gene in asthma: a functional, genetic, and epigenetic study. Allergy 2011;66(12):1553–62.

[217] Pascual M, Suzuki M, Isidoro-Garcia M, Padron J, Turner T, Lorente F, et al. Epigenetic changes in B lymphocytes associated with house dust mite allergic asthma. Epigenetics 2011;6(9):1131–7.

[218] Kim YJ, Park SW, Kim TH, Park JS, Cheong HS, Shin HD, et al. Genome-wide methylation profiling of the bronchial mucosa of asthmatics: relationship to atopy. BMC Med Genet 2013;14:39.

[219] Reinius LE, Gref A, Saaf A, Acevedo N, Joerink M, Kupczyk M, et al. DNA methylation in the neuropeptide S receptor 1 (NPSR1) promoter in relation to asthma and environmental factors. PLoS One 2013;8(1):e53877.

[220] Michel S, Busato F, Genuneit J, Pekkanen J, Dalphin JC, Riedler J, et al. Farm exposure and time trends in early childhood may influence DNA methylation in genes related to asthma and allergy. Allergy 2013;68(3):355–64.

[221] Rossnerova A, Tulupova E, Tabashidze N, Schmuczerova J, Dostal M, Rossner Jr P, et al. Factors affecting the 27K DNA methylation pattern in asthmatic and healthy children from locations with various environments. Mutat Res 2013;741–742:18–26.

[222] De Vries A, Vercelli D. Epigenetic mechanisms in asthma. Ann Am Thorac Soc 2016;13(Supplement 1):S48–50.

[223] Yang IV, Schwartz DA. Epigenetic mechanisms and the development of asthma. J Allergy Clin Immunol 2012;130(6):1243–55.

[224] DeVries A, Vercelli D. Early predictors of asthma and allergy in children: the role of epigenetics. Curr Opin Allergy Clin Immunol 2015;15(5):435–9.

[225] Naumova AK, Al Tuwaijri A, Morin A, Vaillancourt VT, Madore AM, Berlivet S, et al. Sex- and age-dependent DNA methylation at the 17q12-q21 locus associated with childhood asthma. Hum Genet 2013;132(7):811–22.

[226] Liang L, Willis-Owen SAG, Laprise C, Wong KCC, Davies GA, Hudson TJ, et al. An epigenome-wide association study of total serum immunoglobulin E concentration. Nature 2015;520(7549):670–4.

[227] Yang IV, Pedersen BS, Liu A, O'Connor GT, Teach SJ, Kattan M, et al. DNA methylation and childhood asthma in the inner city. J Allergy Clin Immunol 2015;136(1):69–80.

[228] Teach SJ, Gill MA, Togias A, Sorkness CA, Arbes SJ, Calatroni A, et al. Preseasonal treatment with either omalizumab or an inhaled corticosteroid boost to prevent fall asthma exacerbations. J Allergy Clin Immunol 2015;136(6):1476–85.

[229] Esquivel A, Busse WW, Calatroni A, Togias AG, Grindle KG, Bochkov YA, et al. Effects of omalizumab on rhinovirus infections, illnesses, and exacerbations of asthma. Am J Respir Crit Care Med 2017;196(8):985–92.

[230] Gill MA, Bajwa G, George TA, Dong CC, Dougherty II, Jiang N, et al. Counterregulation between the FcepsilonRI pathway and antiviral responses in human plasmacytoid dendritic cells. J Immunol 2010;184(11):5999–6006.

[231] Durrani SR, Montville DJ, Pratt AS, Sahu S, DeVries MK, Rajamanickam V, et al. Innate immune responses to rhinovirus are reduced by the high-affinity IgE receptor in allergic asthmatic children. J Allergy Clin Immunol 2012;130(2):489–95.

[232] Wenzel S, Ford L, Pearlman D, Spector S, Sher L, Skobieranda F, et al. Dupilumab in persistent asthma with elevated eosinophil levels. N Engl J Med 2013;368(26):2455–66.

[233] Bleecker ER, FitzGerald JM, Chanez P, Papi A, Weinstein SF, Barker P, et al. Efficacy and safety of benralizumab for patients with severe asthma uncontrolled with high-dosage inhaled corticosteroids and long-acting beta2-agonists (SIROCCO): a randomised, multicentre, placebo-controlled phase 3 trial. Lancet (London, England) 2016;388(10056):2115–27.

[234] Nair P, Wenzel S, Rabe KF, Bourdin A, Lugogo NL, Kuna P, et al. Oral glucocorticoid–sparing effect of benralizumab in severe asthma. N Engl J Med 2017;376(25):2448–58.

[235] Ortega HG, Liu MC, Pavord ID, Brusselle GG, FitzGerald JM, Chetta A, et al. Mepolizumab treatment in patients with severe eosinophilic asthma. N Engl J Med 2014;371(13):1198–207.

[236] Castro M, Zangrilli J, Wechsler ME, Bateman ED, Brusselle GG, Bardin P, et al. Reslizumab for inadequately controlled asthma with elevated blood eosinophil counts: results from two multicentre, parallel, double-blind, randomised, placebo-controlled, phase 3 trials. Lancet Respir Med 2015;3(5):355–66.

[237] Bjermer L, Lemiere C, Maspero J, Weiss S, Zangrilli J, Germinaro M. Reslizumab for inadequately controlled asthma with elevated blood eosinophil levels: a randomized phase 3 study. Chest 2016;150(4):789–98.

[238] Bel EH, Wenzel SE, Thompson PJ, Prazma CM, Keene ON, Yancey SW, et al. Oral glucocorticoid-sparing effect of mepolizumab in eosinophilic asthma. N Engl J Med 2014;371(13):1189–97.

[239] Casale TB, Chipps BE, Rosén K, Trzaskoma B, Haselkorn T, Omachi TA, et al. Response to omalizumab using patient enrichment criteria from trials of novel biologics in asthma. Allergy 2018;73(2):490–7.

[240] Massanari M, Holgate ST, Busse WW, Jimenez P, Kianifard F, Zeldin R. Effect of omalizumab on peripheral blood eosinophilia in allergic asthma. Respir Med 2010;104(2):188–96.

Chapter 19

Chronic obstructive pulmonary disease

Peter J Barnes
National Heart and Lung Institute, Imperial College, London, United Kingdom

Introduction

Chronic obstructive pulmonary disease (COPD) is characterized by progressive airflow limitation that is not fully reversible [1]. The term COPD encompasses small airway disease and emphysema with lung parenchymal destruction [2]. Chronic bronchitis, by contrast, is defined by a productive cough of more than three months duration for more than two successive years; this reflects mucous hypersecretion and is not necessarily associated with airflow limitation.

COPD is common and is increasing globally as populations age. It is the 3rd leading cause of death in the developed countries and the 5th commonest cause of chronic disability [3]. COPD affects more approximately 10% of the population over 40 years and is equally common in females as in males. It has been estimated that there are about 400 million patients with COPD in the world, with 3 million deaths annually [4].

There is persuasive evidence that the susceptibility to develop COPD is genetically determined as not all chronic smokers develop the disease, although the exact genes have not yet been identified. Genomics and proteomics are now being used to understand the abnormal protein expression in this disease as a way of understanding its complex pathophysiology. There are clearly many different phenotypes of COPD that need to be better defined and this will be increasingly important in genetic studies. Finally, pharmacogenomics may have an impact on COPD therapy in the future [5]. This chapter gives an overview of COPD and highlights where genomic medicine is relevant to the future understanding and management of this common disease.

Predisposition

Several environmental and endogenous factors, including genes, increase the risk of developing COPD (Table 1).

Genomic and Precision Medicine. https://doi.org/10.1016/B978-0-12-801496-7.00019-8

TABLE 1 Risk factors for COPD

Environmental factors	Endogenous (host) factors
Cigarette smoking	α_1-Antitrypsin deficiency
Active	Telomerase polymorphism
Passive	Other genetic polymorphisms
Maternal	Ethnic factors
Air pollution	Airway hyperresponsiveness?
Outdoor (traffic)	Asthma?
Indoor: biomass smoke	Low birth rate
Occupational exposure	
Dietary factors	
Low antioxidant vitamins	
Low unsaturated fatty acids	
Infections (especially tuberculosis)	

Environmental factors

In industrialized countries cigarette smoking accounts for most cases of COPD, but in developing countries other environmental pollutants, such as wood smoke and other biomass smoke associated with cooking are important [6, 7]. It is likely that there are important interactions between environmental factors and a genetic predisposition to develop the disease. Outdoor air pollution (particularly sulfur dioxide and diesel particulates), exposure to certain occupational chemicals such as cadmium and passive smoking may all be additional risk factors. The role of airway hyperresponsiveness and allergy as risk factors for COPD is still uncertain. Atopy, serum IgE and blood eosinophilia are not important risk factors. COPD is also seen following tuberculosis in developing countries [8]. Low birth weight is another risk factor for COPD, probably because poor nutrition in fetal life results in small lungs, so that decline in lung function with age starts from a lower peak value [9]. Perhaps the greatest risk factor for COPD is low socioeconomic status which incorporates multiple risk factors [10].

Genetic factors

Longitudinal monitoring of lung function in cigarette smokers reveals that only a minority (15–40% depending on definition) develop significant airflow obstruction due to an accelerated decline in lung function (2- to 5-fold higher than the normal decline of 15 to 30 mL FEV_1/year compared to the normal population and the remainder of smokers who have consumed an equivalent number of cigarettes (Fig. 1). The rate of decline in lung function in the general population (Framingham Study) has a heritability of about 50%. This strongly suggests that genetic factors may determine which smokers susceptible and develop airflow limitation. Further evidence that genetic factors are important is the familial clustering of patients with early onset COPD and the differences in COPD prevalence between different ethnic groups. Patients with α_1-antitrypsin deficiency (AATD) proteinase inhibitor (Pi)ZZ phenotype with α_1-antitrypsin levels <10% of normal values) develop early emphysema, which is exacerbated by smoking, indicating a clear genetic predisposition to COPD [11]. However, AATD accounts for <1% of patients with COPD and many other genetic variants of α_1-antitrypsin that are associated with lower than normal serum levels of this proteinase inhibitor have not been clearly associated with an increased risk of COPD, although heterozygous PiMZ has a small risk of developing emphysema. Another genotype linked to early development of emphysema is polymorphisms of the telomerase gene that result in greater telomere shortening and cellular senescence and these polymorphisms have also been linked to development of idiopathic pulmonary fibrosis [12].

This has lead to a search for associations between COPD and single nuclear polymorphisms (SNP) of other candidate genes that may be involved in

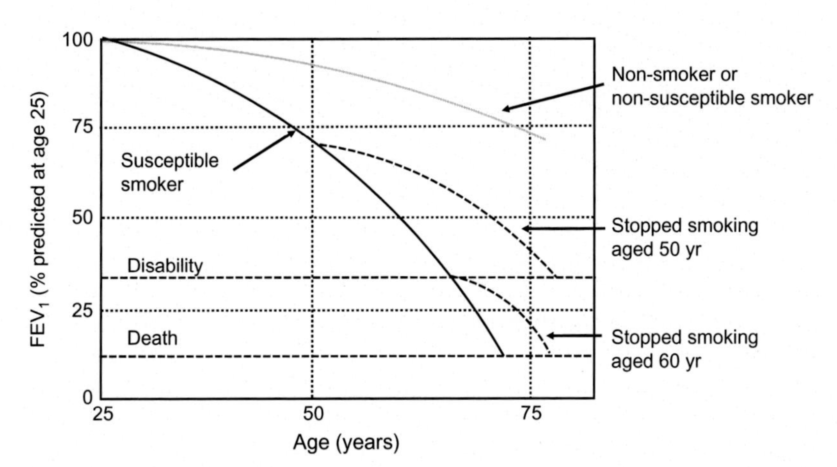

FIG. 1 Natural history of COPD. Annual decline in airway function showing accelerated decline in susceptible smokers and effects of smoking cessation. Patients with COPD usually show an accelerated annual decline in forced expiratory volume in 1 s (FEV_1), often >50 mL/year, compared to the normal decline of approximately 20 mL/year, although this is variable between patients. The classic studies of Fletcher and Peto established that 10–20% of cigarette smokers are susceptible to this rapid decline. However, with longer follow-up more smokers may develop COPD. The propensity to develop COPD amongst smokers is only weakly related to the amount of cigarettes smoked and this suggests that other factors play an important role in determining susceptibility. Most evidence points towards genetic factors, although the genes determining susceptibility have not yet been determined.

TABLE 2 Some of the genes associated with COPD susceptibility

Candidate genes	Risk
α_1-antitrypsin	ZZ genotype high risk MZ, SZ genotypes small risk
α_1-chymptrypsin	Associated in some populations
Telomerase polymorphisms	High risk
Matrix metalloproteinase-1, −2, −9, −12	Associated in some studies
Microsomal epoxide hydrolase	Increased risk
Glutathione S-transferase	Increased risk
Heme oxygenase-1	Small risk but consistent
Interleukin-13	Small risk
Vitamin D binding protein	Inconsistent
TNF-α promoter	Inconsistent
TGF-β promoter	Inconsistent
α-Nicotinic receptor	From GWAS: marker of nicotine addiction?
Hedgehog-interacting protein	From GWAS: involved in lung development

its pathophysiology (Table 2). Many genome-wide association studies (GWAS) have identified potential causative genes, but there is often poor replication between different studies. In the largest studies, genes involved in lung development, elastic fibers, epigenetics and nicotine addiction show an increased risk for COPD development [13, 14] and there appears to be an overlap with some genes linked to pulmonary fibrosis but not with asthma [15]. Although various SNPs have been associated with COPD, as defined by a reduced FEV_1, but there is emerging evidence that different aspects of COPD may relate to different genotypes. For example, a reduction in gas diffusion is correlated in SNPs of the microsomal epoxide hydrolase gene, reduced exercise performance is correlated with SNPs of the latent transforming growth factor-beta binding protein-4 (LTBP4), whereas dyspnea was linked to three SNPs in transforming growth factor β1 (TGF-β1). This suggests that more careful phenotyping is required in the future to sort out susceptibility genes.

One of the problems in the genetics of COPD is the clinical heterogeneity of the disease. There is now an effort to link different aspects of the disease to specific genetic associations in order to classify disease subtypes that might be useful in selecting patients for different types of therapy in the future.

Several SNPs were reported to be associated with severe COPD in almost 400 patients who participated in the National Emphysema Treatment Trial (NETT). Two of these genes microsomal epoxide hydrolase (EPHX1) and SERPINE2 were found to be significant associated with hypoxema and in a separate population of early onset COPD patients the same SNPs were associated with the requirement of domiciliary oxygen [16]. Another SNP in the surfactant protein B (SFTPB) gene was associated with pulmonary hypertension. Bronchodilator responsiveness to the inhaled β_2-agonist albuterol has been linked to SNPs in the SERPINE2 and EPHX1 genes, as well as in the β_2-adrenergic receptor (ADRB2) gene. Extracellular superoxide dismutase (SOD3) is an important antioxidant enzyme in the lungs and 2 novel SNPs of SOD3 have been linked to reduced FEV_1 values in a Danish population study and these polymorphisms were associated with hospitalization and mortality from COPD [17]. An SNP in the promoter region of MMP12 has been shown to be protective against the development of COPD in smokers [18].

Whole genome sequencing provides comprehensive screening of the whole genome, in contrast to GWAS and exome sequencing and in COPD has confirmed the risk of hedgehog interacting protein variant, which is involved in lung development, SERPINA1 [19]. Studies looking at gene=by-smoking interactions are also underway and new SNPs are being identified [20]. However, to date genetic mechanisms appear to account for very little susceptibility to COPD. Epigenetic mechanisms may account for the effects of cigarette smoking and other environmental risk factors [21]. Abnormal DNA methylation patterns have been described in COPD lung and airway epithelial cells that may affect gene expression [22–24].

Pathophysiology

COPD includes chronic obstructive bronchiolitis with fibrosis and obstruction of small airways, and emphysema with enlargement of airspaces and destruction of lung parenchyma, loss of lung elasticity and early closure of small airways. Chronic bronchitis, by contrast, is defined by a productive cough of more than three months duration for more than two successive years; this reflects mucous hypersecretion and is not necessarily associated with airflow limitation. Most patients with COPD have all three pathological mechanisms (chronic obstructive bronchitis, emphysema and mucus hypersecretion) as all are induced by smoking, but may differ in the proportion of emphysema and obstructive bronchitis (Fig. 2).

Small airway obstruction

There has been debate about the predominant mechanism of progressive airflow limitation and recent pathological studies suggest that is closely related to the degree of inflammation, narrowing and fibrosis in small airways [2].

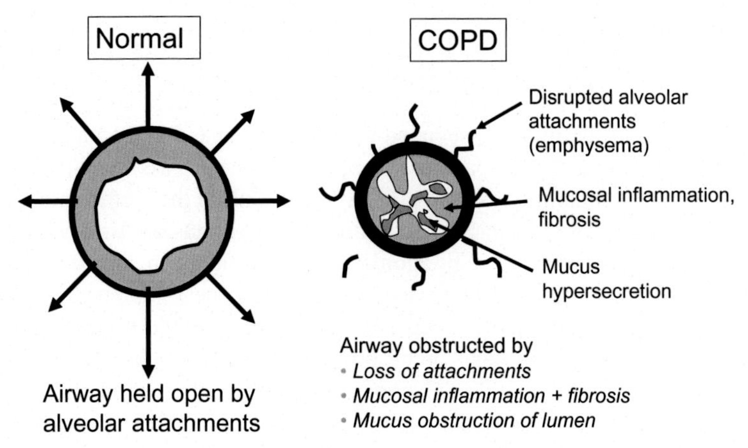

FIG. 2 Mechanisms of airflow limitation in COPD. The airway in normal subjects is distended by alveolar attachments during expiration, allowing alveolar emptying and lung deflation. In COPD these attachments are disrupted because of emphysema thus contributing to airway closure during expiration, trapping gas in the alveoli and resulting in hyperinflation. Peripheral airways are also obstructed and distorted by airways inflammation and fibrosis (chronic obstructive bronchiolitis) and by occlusion of the airway lumen by mucous secretions which may be trapped in the airways because of poor mucociliary clearance.

Emphysema may contribute to the small airway narrowing in the more advanced stages of the disease, with disruption of alveolar attachments facilitating small airway closure and gas trapping. This combined effect of small airway disease and early closure on expiration results in lung hyperinflation, which results in progressive exertional dyspnea, the predominant symptom of COPD. Increasing evidence suggests that small airway disease may be the initial lesion in the development of COPD [25].

Emphysema

Emphysema describes loss of alveolar walls due to destruction of matrix proteins (predominantly elastin) and loss of alveolar type 1 epithelial cells as a result of apoptosis. Several patterns of emphysema are recognized: centriacinar emphysema radiates from the terminal bronchiole; panacinar emphysema involves more widespread destruction and bullae are large airspaces. Emphysema results in airway obstruction by loss of elastic recoil and destruction of alveolar attachments of small airways so that intrapulmonary airways close more readily during expiration. Emphysema with loss of gas-exchanging surface also leads to progressive hypoxia and eventually to respiratory failure.

Pulmonary hypertension

Chronic hypoxia may lead to hypoxic vasoconstriction, with structural changes in pulmonary vessels that eventually lead to secondary pulmonary hypertension.

Inflammatory changes similar to those seen in small airways are also seen in pulmonary arterioles. Only a small proportion of COPD patients develop severe pulmonary hypertension and it is likely that genetic susceptibility plays a role.

Comobidities

COPD is usually associated with other chronic diseases, including cardiovascular and metabolic diseases [26]. The incidence of these comorbidities is greater than can be explained by common mechanisms such as cigarette smoking and has been attributed to the systemic inflammation that occurs in patients with COPD. A more likely explanation is that these comorbidities share common mechanisms such as accelerated aging [27]. There may be common genetic predispositions to COPD and comorbidities; for example here is evidence for genetic determinants of plasma C-reactive protein (CRP), which may be increased in the circulation of patients with COPD and cardiovascular disease [28].

Exacerbations

An important feature of COPD are exacerbations, with worsening of dyspnea and in increase is sputum production, which may lead to hospitalization [29]. Exacerbations are usually due to infections, either due to bacteria (especially *Haemophilus influenzae* or *Steptococcus pneumoniae*) or to upper respiratory tract virus infections (especially rhinovirus or respiratory syncytial virus). Exacerbations of COPD tend to increase as the disease progresses but some patients appear to have more frequent exacerbations than others, which may suggest genetic predisposing factors.

Cellular and molecular mechanisms

Inflammation

There is chronic inflammation predominantly in small airways and the lung parenchyma, with an increase in numbers of macrophages and neutrophils in early stages of the disease indicating an enhanced innate immune response, but in more advanced stages of the disease there is an increase in lymphocytes (particularly cytotoxic $CD8^+$ T cells), including lymphoid follicles that contain B- and T-lymphocytes, indicating acquired [2, 30] (Fig. 3). Alveolar macrophages play a critical role in the orchestration of this pulmonary inflammation, since they are activated by inhaled irritants such as cigarette smoke and release chemokines which attract inflammatory cells, such as monocytes, neutrophils and T cells, into the lungs.

Inflammatory mediators

Prominent mediators are those that amplify inflammation, such as tumor necrosis factor-α (TNF-α), interleukin(IL)-1β (IL-1β) and IL-6, and chemokines which

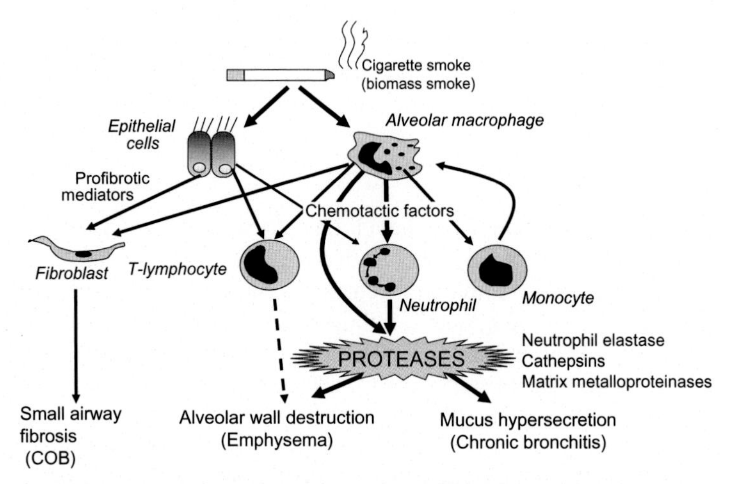

FIG. 3 Inflammatory mechanisms in COPD. Cigarette smoke (and other irritants such as biomass smoke) activate macrophages and epithelial cells in the respiratory tract to release multiple chemotactic factors that attract neutrophils, monocytes and T cells into the lung from the circulation. These cells release proteases that break down connective tissue in the lung parenchyma, resulting in emphysema, and also stimulate mucus hypersecretion. Profibrotic factors in the airway lead to small airway fibrosis that results in chronic obstructive bronchiolitis (COB).

attract inflammatory cells such as CXCL8 (IL-8), CXCL1 (GROα), CXCL10 (IP-10), CCL1 (MCP) and CCL5 (RANTES) [30]. Elastolytic enzymes account for the tissue destruction of emphysema and include neutrophil elastase and matrix metalloproteinase-9 (MMP-9). There is an imbalance between increased production of elastases and a deficiency of endogenous antiproteases, such as α_1-antritrypin, secretory leukoprotease inhibitor and tissue inhibitors of MMPs. MMP-9 may be the predominantly elastolytic enzyme causing emphysema and also activates transforming growth factor-β a cytokine that is expressed particularly in small airways that may result in the characteristic peribronchiolar fibrosis.

Oxidative stress is a prominent feature of COPD and is due to exogenous oxidants in cigarette smoke and endogenous oxidants release from activated inflammatory cells, such as neutrophils and macrophages [31]. Endogenous antioxidants may also be defective and there is some evidence for genetic polymorphisms in antioxidant genes, such as SOD3 and EPHX1 [16]. Oxidative stress enhances inflammation and may lead to corticosteroid resistance.

Differences from asthma

Although both COPD and asthma involve chronic inflammation of the respiratory tract, there are marked differences between the inflammatory mechanisms between these diseases [32]. While there are difference between the inflammation in mild asthma and COPD, patients with severe asthma become much more similar to patients with COPD, with involvement of neutrophils, macrophages, TNF-α, CXCL8, oxidative stress and a poor response to corticosteroids. This has suggested that there may be similarities in genetic predisposion to COPD in smokers and severe asthma. Several novel susceptibility genes, including DPP10, GPRA, PHF11 and ADAM33, that have been identified in severe asthma have now also been in COPD patients [33].

Gene regulation

Proinflammatory mediators, such as TNF-α and IL-1β, activate the transcription factor nuclear factor-κB (NF-κB), which is activated in the airways and lung parenchyma of COPD patients, particularly in epithelial cells and macrophages and there is further activation during exacerbations. NF-κB switches on many of the inflammatory genes that are activated in COPD lungs, including chemokines, adhesion molecules such as ICAM-1 and *E*-selectin, inflammatory enzymes such as cyclo-oxygenase-2 and inducible nitric oxide synthase, elastolytic enzymes such as MMP9 and proinflammatory mediators such as TNF-α and IL-1β which themselves activate NF-κB. NF-κB-activated genes result in acetylation of core histones (particularly histone-4), which is necessary for activation of inflammatory genes and this is reversed by histone deacetylase-2 (HDAC2). There is a marked reduction in HDAC2 activity and expression in lung parenchyma, airways and alveolar macrophages of COPD patients [34]. This is a mechanism that can account for the amplified pulmonary inflammation seen in COPD compared to smokers with normal lung function and also explains why COPD patients are not responsive to corticosteroids, since HDAC2 is the mechanism whereby corticosteroids switch off activated inflammatory genes [35].

Diagnosis and screening

Diagnosis

Diagnosis is commonly made from the history of progressive dyspnea in a chronic smoker and is confirmed by spirometry which shows an FEV_1/VC ratio of <70% and $FEV_1 < 80\%$ predicted normal. Staging of severity is made on the basis of FEV_1, but exercise capacity and the presence of systemic features may be more important determinants of clinical outcomes. Measurement of lung volumes by body plethysmography shows an increase in total lung capacity, residual volume and functional residual capacity, with consequent reduction in inspiratory capacity, representing hyperinflation as a result of small airway closure. This results in dyspnea which may be measured by dyspnea scales and reduced exercise tolerance. Carbon monoxide diffusion is reduced in proportion to the extent of emphysema (Fig. 4). Although COPD is currently defined by spirometric abnormalities, it is evident that smokes may develop respiratory symptoms before spirometry becomes abnormal and this may be reflected in abnormalities in small airway function [36].

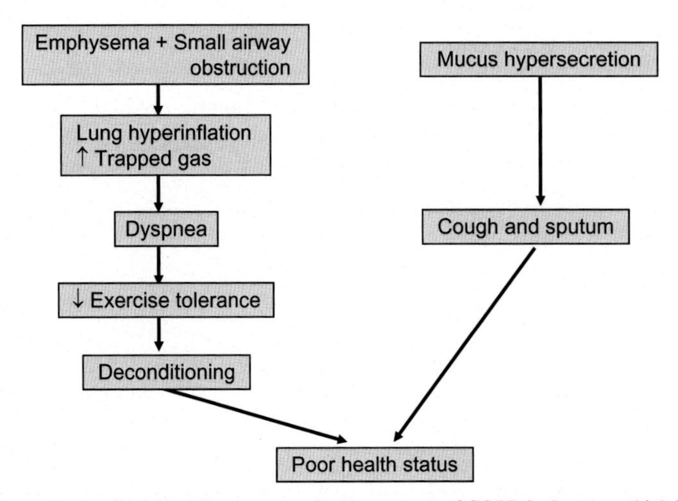

FIG. 4 Symptoms of COPD. The most prominent symptom of COPD is dyspnea which is largely due to hyperinflation of the lungs as a result of small airways collapse due to emphysema and narrowing due to fibrosis, so that the alveoli are not able to empty. Hyperinflation induces an uncomfortable sensation and reduces exercise tolerance. This leads to immobility and deconditioning and results in poor health status. Other common symptoms of COPD are cough and sputum production as a result of mucus hypersecretion, but not all patients have these symptoms and many smokers with these symptoms do not have airflow obstruction (simple chronic bronchitis).

A chest X-ray is rarely useful but may show hyperinflation of the lungs and the presence of bullae. High resolution computerized tomography demonstrates emphysema but is not used as a routing diagnostic test. Blood tests are rarely useful; a normocytic normochromic anemia is more commonly seen in patients with severe disease than polycythemia due to chronic hypoxia. Blood eosinophils counts may be useful in determining which patients show a useful clinical response to inhaled corticosteroids. Arterial blood gases are used to demonstrate hypoxia and in some patients hypercapnia.

Screening

COPD is grossly underdiagnosed as symptoms present late in the progression of disease [37]. Population screening with measurement of FEV_1 and FEV_1/VC ratio would pick up early COPD in smokers over 40 years. However, this would not identify COPD due to non-smoking causes which constitute 10–20% of the total. This would require screening spirometry of all individuals over 40 years in addition to the 20–40% of the population who smoke. It is hoped that gene expression studies might indicate a biomarker that is correlated with COPD susceptibility, but this is unlikely as so many genes are likely to be involved.

Prognosis

COPD is usually slowly progressive with an accelerated decline in FEV_1, leading to slowly increasing symptoms, fall in lung function and eventually to respiratory failure (Fig. 1). The only strategy to reduce disease progression is smoking cessation, although this is relatively ineffective once FEV_1 has fallen below 50% predicted and the patient is symptomatic. Factors other than FEV_1 are important in the prognosis of the disease, including exercise performance and muscle weakness, which signal a greater mortality. Patients who develop right heart failure (*cor pulmonale*) also have poor survival, although this may be improved by long-term oxygen therapy. It is hoped in the future that genetic approaches will help to improve the prediction of prognosis and identify the most effective management strategies.

Management

COPD is managed according to the severity of symptoms and the frequency of acute exacerbations [4].

Anti-smoking measures

Smoking cessation is the only measure so far shown to slow the progression of COPD, but in advanced disease stopping smoking has little effect and the chronic inflammation persists. Nicotine replacement therapy (gum, transdermal patch, inhaler) helps in quitting smoking, but a partial nicotinic agonist varenecline is currently the most effective drug for nicotine addiction. As discussed above, there may be genetic determinants of smoking cessation involving SNPs of nicotinic receptors and dopamine uptake which may make it difficult for some patients to quit.

Bronchodilators

Long-acting bronchodilators are currently the mainstay of drug therapy for COPD. The bronchodilator response measured by an increase in FEV_1 is limited in COPD, but bronchodilators may improve symptoms by reducing hyperinflation and therefore dyspnea, and may improve exercise tolerance, despite the fact that there is little improvement in spirometric measurements. Short-acting bronchodilators, including β_2-agonists and anticholinergics, have now largely been replaced by long-acting drugs, including inhaled long-acting β_2-agonists (LABA, such as salmeterol, formoterol, indacaterol) and once daily inhaled muscarinic antagonists (LAMA, such as tiotropium, glycopyrrolate, umeclidinium). In patients with more severe disease LABA and LAMA may have additive bronchodilator effects and may be given as fixed dose dual inhalers. Theophylline is also used as an add-on bronchodilator in patients with very severe disease, but systemic side effects may limit its value.

Antibiotics

Acute exacerbations of COPD are commonly assumed to be due to bacterial infections, since they may be associated with increased volume and purulence of the sputum. However, it is increasingly recognized that exacerbations may be triggered by upper respiratory tract viral infections or may be non-infective, questioning the place of antibiotic treatment in many patients. Controlled trials of antibiotics in COPD show a relatively minor benefit of antibiotics in terms of clinical outcomes and lung function. Macrolide antibiotics may reduce exacerbations in patients with frequent exacerbations who have stopped smoking [38].

Oxygen

Home oxygen accounts for a large proportion (over 30% in USA) of health care spending on COPD. Long-term oxygen therapy reduces mortality and improves quality of life in patients with severe COPD and chronic hypoxemia ($PaO_2 < 55$ mmHg) but has no benefit in other patients [39].

Corticosteroids

Inhaled corticosteroids (ICS) are the mainstay of chronic asthma therapy and the recognition that chronic inflammation is also present in COPD provided a rationale for their use in COPD. Indeed, ICS are now widely prescribed in the treatment of COPD. However, in most patients the inflammation in COPD shows little or no suppression even by high doses of inhaled or oral corticosteroids [35]. This may reflect the fact that neutrophilic inflammation in is not suppressible by corticosteroids as neutrophil survival is prolonged by steroids. There is also evidence for an active cellular resistance to corticosteroids, with no evidence that even high doses of corticosteroids suppress the synthesis of inflammatory mediators or enzymes. This is related to a decreased activity and expression of HDAC2. Approximately 10–20% of patients with stable COPD show some response to ICS with reduced exacerbations. These responders may have concomitant asthma and may be recognized by increased blood eosinophils over >300 cells/μl [40]. Long-term ICS fail to reduce mortality or disease progression and have a risk of long-term systemic side effects. High dose ICS may be safely withdrawn from COPD patients even with sever disease unless they have ≥2 exacerbations/year and increased blood eosinophils [41]. COPD patients on high doses of ICS have a risk of developing pneumonia, particularly when blood eosinophils are low (<100/μl) [42].

Other drug therapies

Systematic reviews show that mucolytic therapies reduce exacerbation by about 20%, but most of the benefit appears to derive from *N*-acetyl cysteine which is also an antioxidant. A controlled trial however did not show any overall benefit in reducing exacerbations or disease progression [43].

Pulmonary rehabilitation

Pulmonary rehabilitation consists of a structured program of education, exercises and physiotherapy and has been shown in controlled trials to improve the exercise capacity and quality of life of patients with severe COPD, with a reduction in health care utilization [44]. Pulmonary rehabilitation is now an important part of the management plan in patients with severe COPD. Most of the benefits appear to relate to exercise so that modified simplified programs are now often used.

Lung volume reduction

Surgical removal of emphysematous lung improves ventilatory function in carefully selected patients. Patients with localized upper lobe emphysema with poor exercise capacity do best, but there is a relatively high operative mortality, particularly with patients who have a low diffusing capacity. Surgoical long volume reducing is now being replaced by bronchoscopic lung volume reduction using various techniques, such as implanted valves and coils and thermal vapor ablation [45].

Management of acute exacerbations

Acute exacerbations of COPD should be managed by supplementary oxygen therapy, initially 24% oxygen and checking that there is no depression of ventilation. Antibiotics are given if the sputum is purulent or there are other signs of bacterial infection. High doses of systemic corticosteroids reduce hospital stay and are routinely given. Chest physiotherapy is usually given, but there is little objective evidence for benefit. Non-invasive ventilation is indicated for incipient respiratory failure and reduces the need for intubation [46].

Pharmacogenomics

There is little or no information about the impact of pharmacogenomics on COPD therapy [47]. The best studied area is polymorphisms of the β_2-receptor (*ADR2B*), but identified SNPs of this receptor have no clinical effect on the bronchodilator response to short- and long-acting β_2-agonists in COPD [48].

New treatments

New bronchodilators

Several LABAs and LAMAs and their combinations have now been approved for COPD and the preferred therapy for most patients. Triple inhalers with fixed doses of ICS LABA and LAMA have also been introduced but relatively few COPD patients benefit from ICS [49].

New anti-inflammatory treatments

It has proved very difficult to develop safe and effective anti-inflammatory treatments for COPD and most treatments are either ineffective or have dose-limiting side effects [50]. The phosphodiesterase-4 inhibitor roflumilast has anti-inflammatory side effects that markedly limit the dose that can be tolerated so that clinical benefits are minimal. Kinase inhibitors have also proved to be disappointing [51]. For those patients who have eosinophilic COPD anti-interleukin-5 therapies may be indicated, although even in carefully selected patients, an anti-IL-5 therapy (mepolizumab) provides little benefit in reducing exacerbation [52]. Cellular senescence is a key feature of COPD and several drugs are in development to target senescence [53].

Impacts of genomics on therapy

As indicated above, molecular genetics and genomics may identify novel targets for the development of new therapies. Pharmacogenomics might have an impact on choice of therapy in the future. It is unlikely that gene-based therapies will be useful, but silencing of inflammatory genes by inhaled small interference RNAs and antisense oligonucleotides might be feasible approaches in the future. MicroRNAs (miRNA) are small regulatory RNAs that inhibit the translation of several activated genes and may have application to the suppression of multiple inflammatory genes in the future.

Conclusions

COPD is a major global disease that is increasing. Although the major causes are cigarette smoking and biomass smoke exposure, relatively little is understood about the underlying inflammatory and immune mechanisms. The accelerated decline in lung function in the 10–20% of smokers who develop COPD is likely to be genetically determined, but there is no agreement on which genes are involved. More studies, particularly high density genome wide approaches and whole gene sequencing, are needed to identify susceptibility genes an to link these to different patient phenotypes. Identification of susceptibility genes may facilitate screening and early identification of disease. Microarray analysis is now revealing up-regulation of various inflammatory, immune and repair genes and down-regulation of other genes that may be protective in peripheral lung, macrophages and epithelial cells. More studies are needed in relevant cells from carefully phenotyped patients. At the moment it is not possible to give any guidance to family members of COPD patients, apart from those with PiZ α1-antitrypsin deficiency, about genetic risks since the genes determining risk have not yet been identified. Pharmacogenetics needs to be applied to COPD therapies in order to select patients that would benefit more from one treatment compared to another. Current therapy for COPD is unsatisfactory and there are no effective anti-inflammatory treatments. Identification of novel genes involved in COPD may also lead to the discovery of new drug targets.

References

[1] Barnes PJ, Burney PGJ, Silverman EK, Celli BR, Vestbo J, Wedzicha JA, et al. Chronic obstructive pulmonary disease. Nature Rev Primers 2015;1:1–21.

[2] Hogg JC, Timens W. The pathology of chronic obstructive pulmonary disease. Annu Rev Pathol 2009;4:435–59.

[3] Vogelmeier CF, Criner GJ, Martinez FJ, Anzueto A, Barnes PJ, Bourbeau J, et al. Global strategy for the diagnosis, management, and prevention of chronic obstructive lung disease 2017 report: GOLD executive summary. Am J Respir Crit Care Med 2017;195:557–82.

[4] GOLD. Global Initiative for Chronic Obstructive Lung Disease (GOLD). Global strategy for the diagnosis, management and prevention of COPD, www.goldcopd.com/goldreports; 2018.

[5] Hersh CP. Pharmacogenetics of chronic obstructive pulmonary disease: challenges and opportunities. Pharmacogenomics 2010;11:237–47.

[6] Salvi SS, Barnes PJ. Chronic obstructive pulmonary disease in non-smokers. Lancet 2009;374:733–43.

[7] Sood A, Assad NA, Barnes PJ, Churg A, Gordon SB, Harrod KS, et al. ERS/ATS workshop report on respiratory health effects of household air pollution. Eur Respir J 2018;51:1700698.

[8] Allwood BW, Myer L, Bateman ED. A systematic review of the association between pulmonary tuberculosis and the development of chronic airflow obstruction in adults. Respiration 2013;86:76–85.

[9] Lange P, Celli B, Agusti A, Boje Jensen G, Divo M, Faner R, et al. Lung-function trajectories leading to chronic obstructive pulmonary disease. N Engl J Med 2015;373:111–22.

[10] Burney P, Jithoo A, Kato B, Janson C, Mannino D, Nizankowska-Mogilnicka E, et al. Chronic obstructive pulmonary disease mortality and prevalence: The associations with smoking and poverty--a BOLD analysis. Thorax 2014;69:465–73.

[11] Stockley RA. Alpha1-antitrypsin review. Clin Chest Med 2014;35:39–50.

[12] Stanley SE, Chen JJ, Podlevsky JD, Alder JK, Hansel NN, Mathias RA, et al. Telomerase mutations in smokers with severe emphysema. J Clin Invest 2015;125:563–70.

[13] Wain LV, Shrine N, Artigas MS, Erzurumluoglu AM, Noyvert B, Bossini-Castillo L, et al. Genome-wide association analyses for lung function and chronic obstructive pulmonary disease identify new loci and potential druggable targets. Nature Genet 2017;49:416–25.

[14] Castaldi PJ, Cho MH, Zhou X, Qiu W, McGeachie M, Celli B, et al. Genetic control of gene expression at novel and established chronic obstructive pulmonary disease loci. Hum Mol Genet 2015;24:1200–10.

[15] Hobbs BD, de Jong K, Lamontagne M, Bosse Y, Shrine N, Artigas MS, et al. Genetic loci associated with chronic obstructive pulmonary disease overlap with loci for lung function and pulmonary fibrosis. Nature Genet 2017;49:426–32.

[16] Castaldi PJ, Cho MH, Cohn M, Langerman F, Moran S, Tarragona N, et al. The COPD genetic association compendium: a comprehensive online database of COPD genetic associations. Hum Mol Genet 2010;19:526–34.

[17] Dahl M, Bowler RP, Juul K, Crapo JD, Levy S, Nordestgaard BG. Superoxide dismutase 3 polymorphism associated with reduced lung function in two large populations. Am J Respir Crit Care Med 2008;178:906–12.

[18] Hunninghake GM, Cho MH, Tesfaigzi Y, Soto-Quiros ME, Avila L, Lasky-Su J, et al. MMP12, lung function, and COPD in high-risk populations. N Engl J Med 2009;361:2599–608.

[19] Prokopenko D, Sakornsakolpat P, Loehlein Fier H, Qiao D, Parker MM, McDonald MN, et al. Whole genome sequencing in severe chronic obstructive pulmonary disease. Am J Respir Cell Mol Biol 2018;59:614–22.

[20] Park B, Koo SM, An J, Lee M, Kang HY, Qiao D, et al. Genome-wide assessment of gene-by-smoking interactions in COPD. Sci Rep 2018;8:9319.
[21] Zong DD, Ouyang RY, Chen P. Epigenetic mechanisms in chronic obstructive pulmonary disease. Eur Rev Med Pharmacol Sci 2015;19:844–56.
[22] Song J, Heijink IH, Kistemaker LEM, Reinders-Luinge M, Kooistra W, Noordhoek JA, et al. Aberrant DNA methylation and expression of SPDEF and FOXA2 in airway epithelium of patients with COPD. Clin Epigenetics 2017;9:42.
[23] Sood A, Petersen H, Blanchette CM, Meek P, Picchi MA, Belinsky SA, et al. Wood smoke exposure and gene promoter methylation are associated with increased risk for COPD in smokers. Am J Respir Crit Care Med 2010;182:1098–104.
[24] Morrow JD, Cho MH, Hersh CP, Pinto-Plata V, Celli B, Marchetti N, et al. DNA methylation profiling in human lung tissue identifies genes associated with COPD. Epigenetics 2016;11:730–9.
[25] Koo H, Vasilescu DM, Booth S, Hsieh A, Katsamenis OL, Fishbane N, et al. Small airways disease in mild and moderate chronic obstructive pulmonary disease: a cross-sectional study. Lancet Resp Med 2018;6:591–602.
[26] Vanfleteren LE, Spruit MA, Groenen M, Gaffron S, van Empel VP, Bruijnzeel PL, et al. Clusters of comorbidities based on validated objective measurements and systemic inflammation in patients with chronic obstructive pulmonary disease. Am J Respir Crit Care Med 2013;187:728–35.
[27] Barnes PJ. Mechanisms of development of multimorbidity in the elderly. Eur Respir J 2015;45:790–806.
[28] Hersh CP, Miller DT, Kwiatkowski DJ, Silverman EK. Genetic determinants of C-reactive protein in chronic obstructive pulmonary disease. Eur Respir J 2006;28:1156–62.
[29] Wedzicha JA, Singh R, Mackay AJ. Acute COPD exacerbations. Clin Chest Med 2014;35:157–63.
[30] Barnes PJ. Inflammatory mechanisms in COPD. J Allergy Clin Immunol 2016;138:16–27.
[31] Kirkham PA, Barnes PJ. Oxidative stress in COPD. Chest 2013;144:266–73.
[32] Barnes PJ. Cellular and molecular mechanisms of asthma and COPD. Clin Sci (Lond) 2017;131:1541–58.
[33] Postma DS, Rabe KF. The asthma-COPD overlap syndrome. N Engl J Med 2015;373:1241–9.
[34] Barnes PJ. Role of HDAC2 in the pathophysiology of COPD. Annu Rev Physiol 2009;71:451–64.
[35] Barnes PJ. Corticosteroid resistance in patients with asthma and chronic obstructive pulmonary disease. J Allergy Clin Immunol 2013;131:636–45.
[36] Woodruff PG, Barr RG, Bleecker E, Christenson SA, Couper D, Curtis JL, et al. Clinical significance of symptoms in smokers with preserved pulmonary function. N Engl J Med 2016;374:1811–21.
[37] Diab N, Gershon AS, Sin DD, Tan WC, Bourbeau J, Boulet LP, et al. Under-diagnosis and over-diagnosis of chronic obstructive pulmonary disease. Am J Respir Crit Care Med 2018;198:1130–9.
[38] Uzun S, Djamin RS, Kluytmans JA, Mulder PG, van't Veer NE, Ermens AA, et al. Azithromycin maintenance treatment in patients with frequent exacerbations of chronic obstructive pulmonary disease (COLUMBUS): a randomised, double-blind, placebo-controlled trial. Lancet Resp Med 2014;2:361–8.
[39] Branson RD. Oxygen therapy in COPD. Respir Care 2018;63:734–48.
[40] Barnes NC, Sharma R, Lettis S, Calverley PM. Blood eosinophils as a marker of response to inhaled corticosteroids in COPD. Eur Respir J 2016;47:1374–82.

[41] Calverley PMA, Tetzlaff K, Vogelmeier C, Fabbri LM, Magnussen H, Wouters EFM, et al. Eosinophilia, frequent exacerbations, and steroid response in chronic obstructive pulmonary disease. Am J Respir Crit Care Med 2017;196:1219–21.

[42] Contoli M, Pauletti A, Rossi MR, Spanevello A, Casolari P, Marcellini A, et al. Long-term effects of inhaled corticosteroids on sputum bacterial and viral loads in COPD. Eur Respir J 2017;50:1700451.

[43] Decramer M, Rutten-van Molken M, Dekhuijzen PN, Troosters T, van Herwaarden C, Pellegrino R, et al. Effects of N-acetylcysteine on outcomes in chronic obstructive pulmonary disease (bronchitis randomized on NAC cost-utility study, BRONCUS): a randomised placebo-controlled trial. Lancet 2005;365:1552–60.

[44] Casaburi R, ZuWallack R. Pulmonary rehabilitation for management of chronic obstructive pulmonary disease. N Engl J Med 2009;360:1329–35.

[45] Kontogianni K, Eberhardt R. Endoscopic approaches for treating emphysema. Expert Rev Respir Med 2018;12:641–50.

[46] Osadnik CR, Tee VS, Carson-Chahhoud KV, Picot J, Wedzicha JA, Smith BJ. Non-invasive ventilation for the management of acute hypercapnic respiratory failure due to exacerbation of chronic obstructive pulmonary disease. Cochrane Database Syst Rev 2017;7 [Cd004104].

[47] Hizawa N. Pharmacogenetics of chronic obstructive pulmonary disease. Pharmacogenomics 2013;14:1215–25.

[48] Nielsen AO, Jensen CS, Arredouani MS, Dahl R, Dahl M. Variants of the ADRB2 gene in COPD: systematic review and meta-analyses of disease risk and treatment response. COPD 2017;14:451–60.

[49] Calverley PMA, Magnussen H, Miravitlles M, Wedzicha JA. Triple therapy in COPD: what we know and what we don't. COPD 2017;14:648–62.

[50] Barnes PJ. New anti-inflammatory treatments for chronic obstructive pulmonary disease. Nat Rev Drug Discov 2013;12:543–59.

[51] Barnes PJ. Kinases as novel therapeutic targets in asthma and COPD. Pharmacol Rev 2016;68:788–815.

[52] Pavord ID, Chanez P, Criner GJ, Kerstjens HAM, Korn S, Lugogo N, et al. Mepolizumab for eosinophilic chronic obstructive pulmonary disease. N Engl J Med 2017;377:1613–29.

[53] Barnes PJ. Senescence in COPD and its comorbidities. Annu Rev Physiol 2017;79:517–39.

Chapter 20

Precision medicine in solid organ transplantation

Brian I. Shaw[a], **Eileen Tsai Chambers**[b]
[a]*Department of Surgery, Duke University, Durham, NC, United States,* [b]*Department of Pediatrics, Duke University, Durham, NC, United States*

Introduction

The unique challenge of addressing dual immune systems, that of donor and recipient, makes solid organ transplantation an ideal field for precision medicine. Although transplantation is the treatment of choice for end stage organ disease, its therapy is constrained by the existing organ shortage. Currently, there are approximately 120,000 patients awaiting transplantation with over 7000 waitlist deaths in 2016 [1]. The problem is amplified by the fact that long-term outcomes have remained stagnant, further contributing to the organ shortage. The promise of precision medicine is to harness molecular knowledge of donors and recipients, refine management of allografts, and improve both individual and population outcomes.

Biomarkers remain the foundation of precision medicine in transplantation and generally fall into three categories- risk/susceptibility, diagnostic, and prognostic biomarkers (Fig. 1). Within this context, biomarkers from multiple "omics" disciplines have been utilized to advance specific priorities in transplantation: (1) identification of patients undergoing acute events which threaten immediate allograft survival such as acute rejection or infection (2) identification of patients at risk for recurrent underlying disease or malignancy (3) prediction and improvement of long term allograft outcomes. We will, therefore, focus our review on these categories.

To achieve improved outcomes in transplantation, there are a myriad of technologies available in the laboratory and clinical practice. We will concentrate on *modalities* that are applicable across transplantation types and the various *compartments* that may be examined (Fig. 2). Additionally, we will address the future of precision medicine including the incorporation of machine learning.

Genomic and Precision Medicine. https://doi.org/10.1016/B978-0-12-801496-7.00020-4

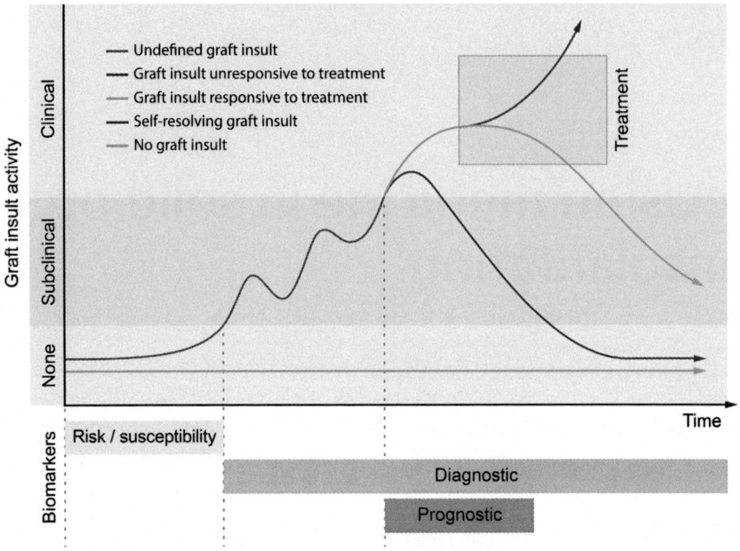

Graft insult activity

None Subclinical Clinical

— Undefined graft insult
— Graft insult unresponsive to treatment
— Graft insult responsive to treatment
— Self-resolving graft insult
— No graft insult

Treatment

Time

Biomarkers

Risk / susceptibility

Diagnostic

Prognostic

	Biomarker:	Outcome:	Graft type:
Risk / Susceptibility	serum Gd-IgA1-specific IgG antibodies[29]	IgA nephropathy recurrence	kidney
Risk / Susceptibility	urine-soluble urokinase receptor (SuPAR) and antibody panel[27,28]	FSGS recurrence	kidney
Risk / Susceptibility	NGS panel[26]	FSGS recurrence	kidney
Risk / Susceptibility	*APOL1 genotype*[36–38]	risk of graft failure or renal failure in donor	kidney
Risk / Susceptibility	HLA mismatching[40,41,44]	*de novo* DSA, ABMR, graft survival	kidney
Risk / Susceptibility	RETREAT and HALT HCC scores[33,44]	HCC recurrence	liver
Diagnostic	gene expression profiles (tissue[7], blood[15–17] and urine[18])	acute rejection	kidney, heart
Diagnostic	microRNA profiles (plasma[23] and urine[22])	acute rejection or infection	kidney
Diagnostic	cell-free DNA signature[10–14]	acute rejection	kidney, heart, liver, lung
Prognostic	GoCAR mRNA set[45]	graft fibrosis	kidney
Prognostic	molecular histology/ Molecular Microscope[47]	ABMR	kidney

FIG. 1 The utility of clinically relevant biomarkers to monitor the allograft is highlighted. Risk/susceptibility biomarkers define the risk of subsequent clinical events prior to the event. Diagnostic biomarkers assist in the identification of concurrent events, such as rejection. Prognostic biomarkers are used to identify the likelihood of a transplant outcome given a certain event, such as fibrosis after rejection. Abbreviations: FSGS-Focal Segmental Glomerulosclerosis; NGS-Next Generation Sequencing; HLA-Human Leukocyte Antigen; DSA-Donor Specific Antibody; ABMR-Antibody Mediated Rejection; HCC-Hepatocellular Carcinoma

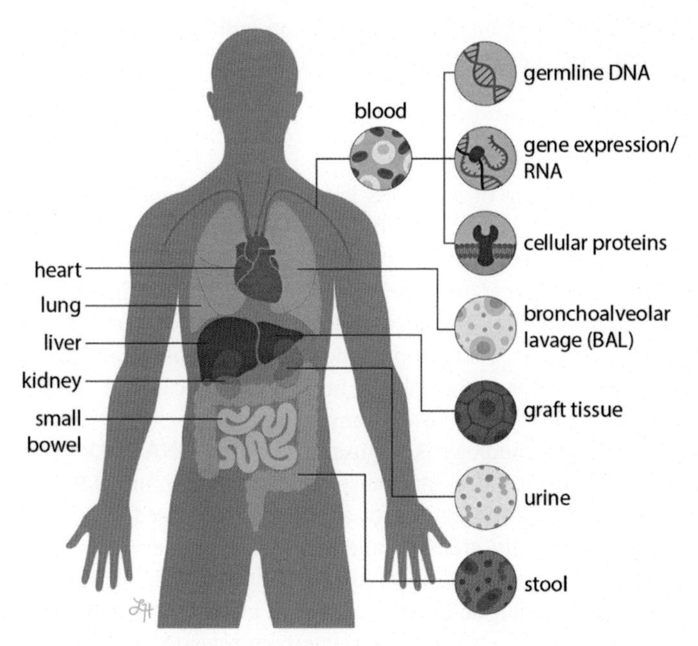

FIG. 2 Leveraging the potential of precision medicine requires utilizing samples from multiple compartments in the transplant recipient. Current solid organs that are commonly transplanted and their respective "compartments" for organ surveillance are featured.

Diagnosis of acute events

Gene expression

Molecular adjuncts to classical histology

The diagnosis of rejection remains an area of uncertainty in transplantation. In kidney transplantation, this has led to the canonical Banff histopathology classification system for kidney biopsies that is updated every 2 years [2]. However, there are intrinsic limitations to this system including the loss of consistency due to changes in the guidelines over time and low inter-rater reliability due to the subjective nature of histopathology itself [3,4]. This has led to the use of microarray technology to optimize and standardize the characterization of biopsy samples on a molecular level.

Early studies detected differences between acute rejection, chronic changes in the allograft, and "normal" kidneys [5]. Other studies have defined subtypes of rejection including Antibody Mediated Rejection (ABMR) [6], assisted in the diagnosis of T-Cell Mediated Rejection (TCMR) [4], and refined histologic definitions of TCMR in both heart and kidney transplantation [7,8]. Furthermore, the use of a Molecular Microscope Diagnostic System was incorporated into clinical practice and was shown to be concordant with treating physician judgment more often than traditional histology [7]. Unfortunately, the use of molecular diagnostics in conjunction with traditional pathology has been limited in practice as it still requires an invasive procedure and the "Gold Standard" is subjective, leaving no true comparator group.

Peripheral monitoring of allografts

Non-invasive monitoring of allografts is a goal for transplant precision medicine to decrease both the technical difficulty of biopsies, such as sampling error, and morbidities related to tissue biopsies [8].

Multiple peripheral markers have been developed across different transplantation modalities (see Fig. 2). When considering the peripheral compartment, non-invasive sources to consider include: peripheral blood in all transplant modalities, urine in kidney transplants, bronchoalveolar lavage (BAL) in lung transplants, and stool in small bowel transplants.

One developing technology is the use of cell free DNA (cfDNA) to detect acute rejection across solid organ transplantation. "Germline" DNA in a free circulating form is used to monitor for acute rejection based on DNA released from the allograft. Initially, cfDNA could only be used if the donor was male and recipient female. However, more sophisticated technologies have allowed for methods of distinguishing donor from recipient cfDNA regardless of sex pairing [9]. cfDNA has been shown to correlate with rejection in kidney and liver transplantation in a male-to-female transplant paradigm [10,11] and in a gender independent model in heart and kidney transplant utilizing the new technology of digital droplet polymerase chain reaction (PCR) [12,13]. In lung transplantation, a cfDNA signature that discriminates infection and acute rejection has also been developed, which is an improvement over previous biomarkers [14].

Gene expression panels have made the most strides in terms of incorporation into clinical practice. In kidney transplantation, two peripheral blood assays have been able to detect acute rejection, *kSort* and *TruGraf* [15,16]. Both utilize a gene expression profile of multiple genes to identify rejection and both are in various stages of clinical development. In heart transplantation, the *AlloMap* system also uses peripheral gene expression to anticipate rejection with the goal of decreasing the need for invasive cardiac biopsies [17]. Urine has been utilized as a sample medium for a urinary mRNA based assay in kidney transplantation to identify acute rejection using a three gene signature [18]. Despite concerns regarding mRNA degradation, it has been validated in a multi-center clinical trial [19]. Additionally, both a plasma and a urinary microRNA signature have been shown to discriminate rejection and infection from stable allografts in kidney transplantation [20,21].

Beyond molecular assays

Increasingly, more sophisticated methods of analyzing the proteome, metabolome, and microbiome, have been undertaken. One such study of the urine proteome has distinguished chronic rejection, acute rejection, and BK nephritis [22]. Additionally, the metabolome of urine has been shown to correlate with acute rejection in kidney transplantation [23]. In liver, a high throughput, mass spectrometry based proteomics platform has identified a signature of acute rejection in peripheral blood [24]. Finally, in small bowel transplant, the fecal microbiome, as surveyed by 16s rRNA, has been shown to correlate with acute rejection episodes [25]. These studies highlight the diverse methods that may be utilized to monitor grafts outside of histology.

Risk stratification for disease recurrence

Identifying high risk patients, especially those susceptible to autoimmune disease or cancer recurrence, is a major goal in transplantation precision medicine. As these pathologies are intrinsic to the patient, it is important to stratify patients by the likelihood that the primary disease may return.

Autoimmune diseases as an etiology of organ failure present the challenge of recurrence as the host immune system is not changed. In renal transplantation, focal segmental glomerulosclerosis (FSGS) has been well studied. FSGS can be categorized as genetic with a 1–5% risk of recurrence and differentiated from idiopathic FSGS with a 30–70% risk of recurrence using next generation DNA sequencing of multiple genes [26]. Additionally, recurrent FSGS can be predicted both by urine soluble urokinase receptor (SuPAR), a soluble urine protein, as well as a panel of circulating antibodies which are related to SuPAR expression [27,28]. In IgA Nephropathy, serum Gd-IgA1-specific IgG and IgA antibodies have been shown to predict recurrence [29]. In liver transplantation, there has been less success in determining risk factors for recurrence of autoimmune disorders such as primary sclerosing cholangitis, primary biliary cirrhosis, and autoimmune hepatitis, although there is some suggestion that greater differences in HLA haplotypes between donor and recipient may be important [30].

One area which has been successful is the prediction of recurrent hepatocellular carcinoma (HCC) in liver transplantation. As one of the few malignancy-associated indications for transplantation, it is imperative to ensure that patients will benefit before pursuing transplantation. The classic Milan criteria were the first set of rules to determine the risk of recurrence in HCC [31]. These criteria are based on imaging alone and there have been multiple adjustments over the years [32]. More recently, two large studies have sought to incorporate aspects of tumor biology and immune status to determine recurrence in a more robust way [33,34]. These studies are impressive in their translation of biologically relevant measures into biomarkers that are assay agnostic (i.e., they incorporate all relevant assays, regardless of specific study type) in order to create a more thorough risk score.

Prediction of long term outcomes

Donor factors

In examining long term transplant outcomes, one potential analytic method is to look for genotypes which predict outcome. Although many Genome Wide Association Studies (GWAS) have been performed in transplantation, few have held up to strict scrutiny [28,29]. One donor polymorphism that has been shown to be relevant to graft survival in renal transplantation is the gene coding for apolipoprotein L1, *APOL1*. Interestingly, this gene, which is represented in people of African but not European descent, has been shown to increase the risk of kidney failure in the general population [35]. Most importantly, it has been shown to decrease graft survival in recipients who receive kidneys from individuals with specific *APOL1* variants in a multicenter study [36]. Finally, there is some evidence that living donors with specific *APOL1* have greater decline in estimated glomular filtration rate after donation and are more likely to develop chronic kidney disease [37,38]. Therefore, this gene may be used to both risk stratify individuals considering living donor renal donation for later renal failure and assess the probability of durable survival of the donated kidney.

Interplay between donor and recipient factors

The role of human leukocyte antigen

One of the first ways in which molecular medicine improved transplantation outcomes was in the utilization of Human Leukocyte Antigen (HLA) typing to ensure compatibility between donor and recipient to prevent alloimmune responses and rejection [39].

Recently, studies have moved away from simply looking at HLA type, as defined by either serology or sequencing, to attempt to define differences in HLA that are antigenically meaningful. Some studies have compared "eplet mismatches"—mismatches in highly polymorphic amino acids in HLA molecules—to attempt to predict outcomes in kidney transplant. One early study found an increase in the formation *de novo* donor specific antibodies (*dn*DSA) with greater eplet mismatches [40]. Other studies found similar results with eplet mismatches, amino acid mismatches, and the physiochemical electrostatic mismatch score [41,42]. These *dn*DSA are correlated with shortened graft survival [43,44]. A more recent study built upon this concept and correlated predicted donor HLA antigen/recipient HLA pair recognizable epitopes with long term outcomes in kidney transplantation. As the number of these recognizable epitopes increased, the risk of graft failure increased twofold, with most of this excess risk occurring in the first 6 months of transplant [45]. This score could potentially be utilized in a risk calculator to identify the most ideal recipient for a given graft.

Prognostic biomarkers after transplantation

Gene expression assays

Though gene expression is dynamic, there have been efforts to utilize gene expression at certain time points to determine long term outcomes of allografts. In the GoCAR study, a 13 gene signature was identified which accurately predicted fibrosis in renal allografts at 1 year based on RNA extracted from graft tissue at 3 months. [46]. In another similar study, biopsy transcriptome data was analyzed using unsupervised machine learning methods to recapitulate histologic diagnoses (i.e., ABMR vs. TCMR vs. no rejection) and was able to identify six independent clusters, which were more predictive of overall graft failure than traditional histologic diagnoses [47]. Additionally, this technique has been used to stratify the risk of subsequent renal allograft failure after early ABMR [48].

Proteomics, microbiomics, and cellular assays

The most storied proteins that have been evaluated in regards to long term transplant outcomes are donor specific antibodies (DSA). DSA may be present prior to transplantation either because of exposure to blood products, pregnancy, previous transplantation or due to "heterologous immunity" where antibodies to pathogens cross react with alloantigens [49]. In studies spanning all of solid organ transplantation, DSA are shown to correlate with poor long-term clinical outcomes [50–52]. However, there are many continuing studies to determine the exact type of DSA necessary to be clinically relevant [53].

The urinary microbiome has also been interrogated to determine if it can predict long-term transplant outcomes. In a seminal paper, changes in the urinary microbiome were seen to correlate with the kidney graft outcome of Interstitial Fibrosis and Tubular Atrophy (IFTA) [54]. These data suggest that modulation in the microbiome may either contribute to or be a marker of long term renal transplant outcomes. In the realm of cellular assays, a study of the BAL of lung transplant patients showed that certain cellular phenotypes were associated with primary graft dysfunction in the newly realized setting of ex-vivo lung perfusion [55]. Cellular immunophenotyping has also been used in BAL samples from lung transplant recipients to predict risk of rejection and bronchiolitis obliterans [56]. Finally, a flow cytometric assay that examined the phenotype of B cells has been shown to correlate with immune tolerance in kidney transplantation [57]. These studies highlight the way in which alterations in the immune system ultimately affect allograft outcomes.

Opportunities in machine learning and multi-omics

The application of machine learning methodologies for precision transplantation holds promise for a variety of reasons. First, given the difficulty of classifying clinical entities whose final common pathway is "graft damage," machine learning offers an opportunity to catalog the specific type of "graft damage" in an unbiased way. This line of inquiry has been most developed by Halloran and colleagues in the development of their Molecular Microscope for the detection and classification of rejection events [47] and also holds promise in distinguishing rejection from infectious events. Additionally, machine learning methodologies have been applied to biomarker discovery, such as the development of the *kSort* assay for peripheral renal allograft monitoring [16,58]. Machine learning may even be applied to traditional histology in the form of image recognition paradigms as has recently been proposed for cardiac biopsies [59]. Finally, a new study using time-dependent analysis of both biomarkers and clinical events has shown some proof of concept in helping to create dynamic decision support system to guide clinical practice [60].

Efforts to take a "multi-omics" approach are also increasingly common, as evidenced by the recent attempts to combine transcriptional data with cellular assays for the prediction of tolerance in kidney transplantation [57]. In the prediction of HCC recurrence, we have seen a steady move towards this paradigm as well [33,34]. Increasingly, "assay agnostic" models incorporating data from multiple platforms will be necessary to advance precision transplant medicine. This will allow for the use of previously collected data and existing registries to be applied to current problems using a "Big Data" framework.

Conclusion

Precision transplant medicine offers the opportunity to advance end stage organ care and expand the number of people for whom transplant would be a treatment. Continued work to integrate biologic data from multiple modalities is being pursued to maximize our ability to predict graft outcomes and monitor allografts. Ultimately, precision diagnostics will allow clinicians to more accurately care for the complex ecosystem that is the transplant patient.

References

[1] Waiting list candidates by organ type [Internet]. UNOS; 2018. [Updated 2018 May 20; Cited 2018 May 20]. Available from: https://unos.org/data/transplant-trends/waiting-list-candidates-by-organ-type/.

[2] Haas M, Loupy A, Lefaucheur C, Roufosse C, Glotz D, Seron D, et al. The Banff 2017 kidney meeting report: revised diagnostic criteria for chronic active T cell-mediated rejection, antibody-mediated rejection, and prospects for integrative endpoints for next-generation clinical trials. Am J Transplant 2018;18(2):293–307.

[3] Halloran PF, Famulski KS, Reeve J. Molecular assessment of disease states in kidney transplant biopsy samples. Nat Rev Nephrol 2016;12(9):534–48.

[4] Reeve J, Sellarés J, Mengel M, Sis B, Skene A, Hidalgo L, et al. Molecular diagnosis of T cell-mediated rejection in human kidney transplant biopsies. Am J Transplant 2013;13(3):645–55.

[5] Sarwal M, Chua M-S, Kambham N, Hsieh S-C, Satterwhite T, Masek M, et al. Molecular heterogeneity in acute renal allograft rejection identified by DNA microarray profiling. N Engl J Med 2003;349(2):125–38.

[6] Halloran PF, Merino Lopez M, Barreto Pereira A. Identifying subphenotypes of antibody-mediated rejection in kidney transplants. Am J Transplant 2016 Jan 6;16(3):908–20.

[7] Halloran PF, Reeve J, Akalin E, Aubert O, Bohmig GA, Brennan D, et al. Real time central assessment of kidney transplant indication biopsies by microarrays: the INTERCOMEX study. Am J Transplant 2017 May 30;17(11):2851–62.

[8] Morgan TA, Chandran S, Burger IM, Zhang CA, Goldstein RB. Complications of ultrasound-guided renal transplant biopsies. Am J Transplant 2016;16(4):1298–305.

[9] Burnham P, Khush K. De Vlaminck annals of the American I. myriad applications of circulating cell-free DNA in precision organ transplant monitoring. Ann Am Thorac Soc 2017;15(Suppl. 3):S237–41.

[10] Macher HC, Suárez-Artacho G, Guerrero JM, Gómez-Bravo MA, Álvarez-Gómez S, Bernal-Bellido C, et al. Monitoring of transplanted liver health by quantification of organ-specific genomic marker in circulating DNA from receptor. Bonino F, editor, PLoS One 2014;9(12):e113987-18.

[11] Sigdel TK, Vitalone MJ, Tran TQ, Dai H, Hsieh S-C, Salvatierra O, et al. A rapid noninvasive assay for the detection of renal transplant injury. Transplantation 2013;96(1):97–101.

[12] Beck J, Oellerich M, Schulz U, Schauerte V, Reinhard L, Fuchs U, et al. Donor-derived cell-free DNA is a novel universal biomarker for allograft rejection in solid organ transplantation. Transplant Proc 2015;47(8):2400–3.

[13] De Vlaminck I, Valantine HA, Snyder TM, Strehl C, Cohen G, Luikart H, et al. Circulating cell-free DNA enables noninvasive diagnosis of heart transplant rejection. Sci Transl Med 2014;6(241):241ra77–7.

[14] De Vlaminck I, Martin L, Kertesz M, Patel K, Kowarsky M, Strehl C, et al. Noninvasive monitoring of infection and rejection after lung transplantation. Proc Natl Acad Sci USA 2015;112(43):13336–41.

[15] Kurian SM, Williams AN, Gelbart T, Campbell D, Mondala TS, Head SR, et al. Molecular classifiers for acute kidney transplant rejection in peripheral blood by whole genome gene expression profiling. Am J Transplant 2014;14(5):1164–72.

[16] Roedder S, Sigdel T, Salomonis N, Hsieh S, Dai H, Bestard O, et al. The kSORT assay to detect renal transplant patients at high risk for acute rejection: results of the multicenter AART study. Remuzzi G, editor, PLoS Med 2014;11(11):e1001759-15.

[17] Pham MX, Teuteberg JJ, Kfoury AG, Starling RC, Deng MC, Cappola TP, et al. Gene-expression profiling for rejection surveillance after cardiac transplantation. N Engl J Med 2010;362(20):1890–900.

[18] Suthanthiran M, Schwartz JE, Ding R, Abecassis M, Dadhania D, Samstein B, et al. Urinary-cell mRNA profile and acute cellular rejection in kidney allografts. N Engl J Med 2013;369(1):20–31.

[19] Keslar KS, Lin M, Zmijewska AA, Sigdel TK, Tran TQ, Ma L, et al. Multicenter evaluation of a standardized protocol for noninvasive gene expression profiling. Am J Transplant 2013;13(7):1891–7.

[20] Lorenzen JM, Volkmann I, Fiedler J, Schmidt M, Scheffner I, Haller H, et al. Urinary miR-210 as a mediator of acute T-cell mediated rejection in renal allograft recipients. Am J Transplant 2011;11(10):2221–7.

[21] Matz M, Lorkowski C, Fabritius K, Durek P, Wu K, Rudolph B, et al. Free microRNA levels in plasma distinguish T-cell mediated rejection from stable graft function after kidney transplantation. Transpl Immunol 2016;39(C):52–9.

[22] Sigdel TK, Gao Y, He J, Wang A, Nicora CD, Fillmore TL, et al. Mining the human urine proteome for monitoring renal transplant injury. Kidney Int 2016;89(6):1244–52.

[23] Suhre K, Schwartz JE, Sharma VK, Chen Q, Lee JR, Muthukumar T, et al. Urine metabolite profiles predictive of human kidney allograft status. J Am Soc Nephrol 2016;27(2):626–36.

[24] Toby TK, Abecassis M, Kim K, Thomas PM, Fellers RT, LeDuc RD, et al. Proteoforms in peripheral blood mononuclear cells as novel rejection biomarkers in liver transplant recipients. Am J Transplant 2017;17(9):2458–67.

[25] Oh PL, Martínez I, Sun Y, Walter J, Peterson DA, Mercer DF. Characterization of the Ileal microbiota in rejecting and nonrejecting recipients of small bowel transplants. Am J Transplant 2011;12(3):753–62.

[26] De Vriese AS, Sethi S, Nath KA, Glassock RJ, Fervenza FC. Differentiating primary, genetic, and secondary FSGS in adults: a clinicopathologic approach. J Am Soc Nephrol 2018;29(3):759–74.

[27] Franco Palacios CR, Lieske JC, Wadei HM, Rule AD, Fervenza FC, Voskoboev N, et al. Urine but not serum soluble Urokinase receptor (suPAR) may identify cases of recurrent FSGS in kidney transplant candidates. Transplantation 2013;96(4):394–9.

[28] Delville M, Sigdel TK, Wei C, Li J, Hsieh S-C, Fornoni A, et al. A circulating antibody panel for pretransplant prediction of FSGS recurrence after kidney transplantation. Sci Transl Med 2014;6(256):256ra136–6.

[29] Berthoux F, Suzuki H, Mohey H, Maillard N, Mariat C, Novak J, et al. Prognostic value of serum biomarkers of autoimmunity for recurrence of IgA nephropathy after kidney transplantation. J Am Soc Nephrol 2017;28(6):1943–50.

[30] Faisal N, Renner EL. Recurrence of autoimmune liver diseases after liver transplantation. World J Hepatol 2015;7(29):2896–905.

[31] Mazzaferro V, Regalia E, Andreola S, Pulvirenti A, Bozetti F, Monalto F, et al. Liver transplantation for the treatment of small hepatocellular carcinomas in patients with cirrhosis. N Engl J Med 1996;334(11):693–700.

[32] Xu X, Lu D, Ling Q, Wei X, Wu J, Zhou L, et al. Liver transplantation for hepatocellular carcinoma beyond the Milan criteria. Gut 2016;65(6):1035–41.

[33] Mehta N, Heimbach J, Harnois DM, Sapisochin G, Dodge JL, Lee D, et al. Validation of a risk estimation of tumor recurrence after transplant (RETREAT) score for hepatocellular carcinoma recurrence after liver transplant. JAMA Oncol 2017;3(4):493–8.

[34] Sasaki K, Firi D, Hashimoto K, Fujiki M, Diago-Uso T, Quintini C, et al. Development and validation of the HALT-HCC score to predict mortality in liver transplant recipients with hepatocellular carcinoma: a retrospective cohort analysis. Lancet Gastroenterol Hepatol 2017;2(8):595–603.

[35] Genovese G, Friedman DJ, Ross MD, Lecordier L, Uzureau P, Freedman BI, et al. Association of Trypanolytic ApoL1 variants with kidney disease in African Americans. Science 2010;329(5993):841–5.

[36] Julian BA, Gaston RS, Brown WM, Reeves-Daniel AM, Israni AK, Schladt DP, et al. Effect of replacing race with Apolipoprotein L1 genotype in calculation of kidney donor risk index. Am J Transplant 2017;17(6):1540–8.

[37] Locke JE, Sawinski D, Reed RD, Shelton B, MacLennan PA, Kumar V, et al. Apolipoprotein L1 and chronic kidney disease risk in young potential living kidney donors. Ann Surg 2018;267(6):1161–8.

[38] Doshi MD, Ortigosa-Goggins M, Garg AX, Li L, Poggio ED, Winkler CA, et al. APOL1 Genotype and renal function of black living donors. J Am Soc Nephrol 2018;29(4):1309–16.

[39] Held PJ, Kahan BD, Hunsicker LG, Liska D, Wolfe RA, Port FK, et al. The impact of HLA mismatches on the survival of first cadaveric kidney transplants. N Engl J Med 1994;331(12):765–70.

[40] Wiebe C, Pochinco D, Blydt-Hansen TD, Ho J, Birk PE, Karpinski M, et al. Class II HLA epitope matching-a strategy to minimize De NovoDonor-specific antibody development and improve outcomes. Am J Transplant 2013;13(12):3114–22.

[41] Kosmoliaptsis V, Mallon DH, Chen Y, Bolton EM, Bradley JA, Taylor CJ. Alloantibody responses after renal transplant failure can be better predicted by donor-recipient HLA amino acid sequence and physicochemical disparities than conventional HLA matching. Am J Transplant 2016;16(7):2139–47.

[42] Wiebe C, Nickerson P. Strategic use of epitope matching to improve outcomes. Transplantation 2016;100(10):2048–52.

[43] Terasaki PI, Ozawa M, Castro R. Four-year follow-up of a prospective trial of HLA and MICA antibodies on kidney graft survival. Am J Transplant 2007;7(2):408–15. Wiley/Blackwell (10.1111).

[44] Wiebe C, Gibson IW, Blydt-Hansen TD, Karpinski M, Ho J, Storsley LJ, et al. Evolution and clinical pathologic correlations of de novo donor-specific HLA antibody post kidney transplant. Am J Transplant 2012 Mar 19;12(5):1157–67.

[45] Geneugelijk K, Niemann M, Drylewicz J, van Zuilen AD, Joosten I, Allebes WA, et al. PIRCHE-II is related to graft failure after kidney transplantation. Front Immunol 2018;9:23–9.

[46] O'Connell PJ, Zhang W, Menon MC, Yi Z, Schröppel B, Gallon L, et al. Biopsy transcriptome expression profiling to identify kidney transplants at risk of chronic injury: a multicentre, prospective study. Lancet 2016;388(10048):983–93.

[47] Reeve J, Böhmig GA, Eskandary F, Einecke G, Lefaucheur C, Loupy A, et al. Assessing rejection-related disease in kidney transplant biopsies based on archetypal analysis of molecular phenotypes. JCI Insight 2017;2(12):e94197.

[48] Loupy A, Lefaucheur C, Vernerey D, Chang J, Hidalgo LG, Beuscart T, et al. Molecular microscope strategy to improve risk stratification in early antibody-mediated kidney allograft rejection. J Am Soc Nephrol 2014;25(10):2267–77.

[49] Sharma S, Thomas PG. The two faces of heterologous immunity: protection or immunopathology. J Leukoc Biol 2013;95(3):405–16.

[50] Lefaucheur C, Loupy A, Hill GS, Andrade J, Nochy D, Antoine C, et al. Preexisting donor-specific HLA antibodies predict outcome in kidney transplantation. J Am Soc Nephrol 2010;21(8):1398–406.

[51] Kim M, Townsend KR, Wood IG, Boukedes S, Guleria I, Gabardi S, et al. Impact of pretransplant anti-HLA antibodies on outcomes in lung transplant candidates. Am J Respir Crit Care Med 2014;189(10):1234–9.

[52] Tran A, Fixler D, Huang R, Meza T, Lacelle C, Das BB. Donor-specific HLA alloantibodies: impact on cardiac allograft vasculopathy, rejection, and survival after pediatric heart transplantation. J Heart Lung Transplant 2016;35(1):87–91.

[53] Adebiyi OO, Gralla J, Klem P, Freed B, Davis S, Wiseman AC, et al. Clinical significance of pretransplant donor-specific antibodies in the setting of negative cell-based flow cytometry crossmatching in kidney transplant recipients. Am J Transplant, 2nd ed. 2016;16(12):3458–67.

[54] Modena BD, Milam R, Harrison F, Cheeseman JA, Abecassis MM, Friedewald JJ, et al. Changes in urinary microbiome populations correlate in kidney transplants with interstitial fibrosis and tubular atrophy documented in early surveillance biopsies. Am J Transplant 2016;17(3):712–23.

[55] Hsin MK, Zamel R, Cypel M, Wishart D, Han B, Keshavjee S, et al. Metabolic profile of ex vivo lung perfusate yields biomarkers for lung transplant outcomes. Ann Surg 2018;267(1):196–7.

[56] Greenland JR, Jewell NP, Gottschall M, Trivedi NN, Kukreja J, Hays SR, et al. Bronchoalveolar lavage cell immunophenotyping facilitates diagnosis of lung allograft rejection. Am J Transplant 2014;14(4):831–40.

[57] Newell KA, Asare A, Sanz I, Wei C, Rosenberg A, Gao Z, et al. Longitudinal studies of a B cell-derived signature of tolerance in renal transplant recipients. Am J Transplant 2015;15(11):2908–20.

[58] Crespo E, Roedder S, Sigdel T, Hsieh S-C, Luque S, Cruzado JM, et al. Molecular and functional noninvasive immune monitoring in the ESCAPE study for prediction of subclinical renal allograft rejection. Transplantation 2017;101(6):1400–9.

[59] Peyster EG, Madabhushi A, Margulies KB. Advanced Morphologic Analysis for Diagnosing Allograft Rejection: The Case of Cardiac Transplant Rejection. Transplantation 2018;102(8):1230–9.

[60] Bakir M, Jackson NJ, Han SX, Bui A, Chang E, Liem DA, et al. Clinical phenomapping and outcomes after heart transplantation. J Heart Lung Transplant 2018;37(8):1–11.

Chapter 21

Molecular prognosis in liver cirrhosis

Sai Krishna Athuluri-Divakar, Yujin Hoshida
Liver Tumor Translational Research Program, Simmons Comprehensive Cancer Center, Division of Digestive and Liver Diseases, Department of Internal Medicine, University of Texas Southwestern Medical Center, Dallas, TX, United States

Introduction

Liver cirrhosis is the final consequence of chronic inflammation and progressive fibrosis in the liver mainly affected with hepatitis B virus (HBV), hepatitis C virus (HCV), excess alcohol intake, or non-alcoholic fatty liver disease (NAFLD) [1,2]. The prevalence of cirrhosis keeps rising, and its mortality accounts for 2% of global deaths, an increase of 47% between 1990 and 2013 [3]. Cirrhosis is the major predisposing factor of liver cancer, with hepatocellular carcinoma (HCC) as the major histological type, which is second leading cause of cancer death worldwide [4]. Thus, close clinical monitoring of the patients is critical to manage development of the lethal complication. However, prognosis of cirrhosis varies substantially varies across patients due to severity of cirrhosis as well as currently unknown factors, making determination of prognosis for each individual patient is still challenging. The high mortality and global prevalence of cirrhosis have highlighted the necessity to further refine capacity to determine prognosis. Previous studies have attempted to develop prognostic indicators based on clinical variables as well as molecular biomarkers, some of which may be adopted in actual clinical management of the patients to assist physicians' clinical decision making and optimize allocation of limited medical resources once successfully validated. In this chapter, we overview the natural history of cirrhosis in the context of prognosis, clinical and molecular prognostic predictors for cirrhotic subjects in literature, and discuss potential applications and challenges in the clinical development.

Natural history and pathogenesis of cirrhosis

It is clinically well known that cirrhosis develops from gradually progressing liver fibrosis caused by chronic exposure to the etiological and environmental

Genomic and Precision Medicine. https://doi.org/10.1016/B978-0-12-801496-7.00021-6

factors. In the early asymptomatic stage of compensated cirrhosis, portal venous pressure measured by hepatic venous pressure gradient (HVPG) is generally below 10 mmHg, and the patients are free from esophageal and gastric varices [5]. As the liver fibrosis progresses, increased portal venous pressure and impaired hepatic protein synthetic function lead to development of ascites, portal hypertension-induced variceal hemorrhage, hepatic encephalopathy and/or jaundice. The occurrence of any of these complications indicates transition from compensated to decompensated phase of the disease [6–8]. Estimated rate of progression from a compensated to a decompensated stage is approximately 5–7% annually, and median survival time sharply decreases from 12 years in compensated disease to approximately 2 years in decompensated cirrhosis [8]. Increase of portal venous pressure measured as HVPG >16–20 mmHg often leads to development of severe complications of cirrhosis such as refractory ascites, bacterial infection, recurrent variceal hemorrhage, hepatorenal syndrome, and death. A clinical classification stages cirrhosis severity from stage 1, which includes cirrhotic patients with no ascites, no esophageal varices, and a low mortality, to stage 4, which is characterized by gastrointestinal variceal bleeding and a high mortality of over 50% at 1 year of follow-up [8]. Cirrhosis is also the major risk factor for developing HCC. The risk of developing HCC in cirrhosis depends largely on the underlying condition. Five-year cumulative HCC risk is 17–30% in HCV-related cirrhosis, 21% in hemochromatosis, and 8–12% in alcoholic cirrhosis. In contrast, only 4% of biliary cirrhosis patients develop [9–11]. Importantly, HCC can also occur on the background of non-cirrhotic liver disease, especially in the context of chronic HBV infection and non-cirrhotic NAFLD [12]. Prediction of HCC risk in liver disease remains an ongoing challenge that requires improved prediction of HCC risk across the multiple liver disease etiologies.

Cirrhosis is the manifestation of terminal-stage fibrosis in chronically inflamed liver as characterized histologically by the formation of parenchymal nodules separated by fibrotic septa and associated with major distortion in vascular architecture as the consequence of continuous liver injury and regeneration [13,14]. Fibrogenesis is a ubiquitous pathological process involved in wound healing, resulting from cellular and molecular responses triggered by tissue injury, ultimately leading to parenchymal scarring and organ dysfunction [15]. Fibrogenesis can affect virtually any solid organ including heart, liver, kidney, pancreas, and lung. Fibrosis stage is linked with step-wise increase of annual HCC incidence in HCV-infected patients [16]. Architectural vascular alterations in cirrhotic liver include angiogenesis, vascular occlusion leading to parenchymal extinction, and formation of intrahepatic shunts. Increased resistance to portal blood flow and splanchnic vasodilatation mediated through increased NO and reduced response to vasoconstrictors lead to portal venous hypertension and resulting in the complications such as ascites and variceal bleeding.

Chronic damage to hepatocytes or biliary epithelial cells leads to release of inflammatory and fibrogenic mediators such as reactive oxygen species, cell death signals, ligands of cellular signaling, and nucleotides [14]. A complex series of events such as intracellular inflammasome activation, induced signaling via nuclear receptors such as farsenoid-X-receptor and peroxisome proliferator-activated receptors, and other transcriptional activators contribute to activation of hepatic stellate cell, the major driver of fibrogenesis [2]. Autophagy linked to endoplasmic stress and the unfolded protein response plays a role in providing energy for the activation of hepatic stellate cells [17,18]. Notably, composition of dietary fat and altered microbiome can increase fibrogenic potential in animal models, possibly mediated by pathogen-associated molecular signaling such as activation of toll-like receptors [19]. Perpetuated activation of stellate cells is supported by various cell types, including injured hepatocytes, hepatic progenitor cells, Kupffer cells, sinusoidal endothelial cells, platelets, and infiltrating immune cells through a wide variety of mediators [2,14].

Identifying subjects at higher risk of progressive fibrosis toward HCC development has been an important unmet need given the high prevalence of cirrhosis affecting more than 630,000 adults in the United States alone according to recent population-based estimations [20]. The vast size of the patient population hampers application of the practice guideline-recommended regular monitoring and screening of the clinically at-risk patients with the currently available limited medical resources [1]. For example, only less than 20% of HCC patients are diagnosed at early-stage amenable to curative therapy through the regular biannual screening even in developed countries [21–23]. Thus, clinically applicable prognostic indicators for cirrhosis are urgently needed to enable effective clinical management of the patients [24]. Indeed, individual risk-based personalized HCC screening is cost-effective as suggested by a comprehensive model-based cost-effectiveness analysis [25].

Clinical prognostic indicators in cirrhosis

Non-invasive and invasive clinical prognostic indicators and scores have been proposed to date in various clinical settings (Table 1). Severity of cirrhosis is clinically presented as impaired healthy liver function, which can be estimated by using available clinical symptoms and laboratory data to guide strategies of follow up and indication of medical interventions such as treatment for gastroesophageal varices and liver transplantation for advanced cirrhosis. One of the earlier prognostic score is the Child-Turcotte-Pugh (CTP) score, adopted in the United States in 1998 for donor liver allocation, which was later replaced by the model for end-stage liver disease (MELD) in 2002 based on more objective variables [26,27]. The MELD score, consisting of bilirubin, creatinine, and international normalized ratio of prothrombin time (INR), was initially developed as a prognostic tool in cirrhotic patients undergoing a transjugular intrahepatic portosystemic shunt (TIPS) [28]. The score has been subsequently adopted by liver transplant programs across the world for more accurate prognostication compared to CTP score [29,30]. MELD score has been evaluated

TABLE 1 Clinical prognostic indicators in cirrhosis

Risk indicator	Clinical outcome	Major etiology	Major race/ ethnicity
LSM-HCC score	HCC	HBV	Asian
REACH-B	HCC	HBV	Asian
CU-HCC	HCC	HBV	Asian
PAGE-B	HCC	HBV	White
FIB-4	HCC	HBV	Asian
GAG-HCC	HCC	HBV	Asian
REVEAL-HCV	HCC	HCV	Asian
ADRESS-HCC	HCC	HCV, alcohol, NAFLD	White, Hispanic
FILI score	Fibrosis after lifestyle intervention	NAFLD	n.a.
ELF score	Fibrosis (F3–4)	HCV, HBV	n.a.

HCC, hepatocellular carcinoma; REACH-B, Risk estimation for hepatocellular carcinoma in chronic hepatitis B; LSM, liver stiffness measurement; HBV, hepatitis B virus; HCV, hepatitis C virus; REVEAL-HCV, Risk Evaluation of Viral Load Elevation and Associated Liver Disease/Cancer in HCV; ADRESS-HCC, age, diabetes, race, etiology of cirrhosis, sex and severity of liver dysfunction-HCC; NAFLD, non-alcoholic steatohepatitis; FILI, fibrosis improvement after lifestyle interventions; ELF, enhanced liver fibrosis.

for other potential applications such as prior assessment for non-hepatic surgery, variceal bleeding, hepatorenal syndrome, and mortality from severe alcoholic hepatitis [31–34].

Histological assessment of liver tissue has been used as the gold standard of disease staging and often for prognostication in a variety of liver diseases. Several histological morphology-based classification systems have been developed and evaluated using thickness of fibrotic septa, number and size of cirrhotic nodules, type/location of hepatic necrosis and infiltrating cells, and correlated with clinical prognosis as well as surrogate measures such as HVPG [35–38]. Fibrotic collagen content can be quantified by using digital image analysis of collagen staining, which has been correlated with HVPG and clinical outcomes such as fibrosis progression and emergence of decompensation, mostly in the setting of liver transplantation in HCV-infected patients [39–41]. Automated assessment of quantity and architectural features of collagen band can also be performed on unstained histological slides [42]. Collagen proportionate area (CPA) may perform better than traditional histological measures to predict risk of progressing into decompensated stage [43]. Liver biopsy-based histological assessment supplemented with other measures such as HVPG provide deterministic evidence of cirrhosis, although the histological staging can be affected by sampling variability in biopsy. In addition, these methods are relatively invasive and may not be indicated for patients especially with more advanced cirrhosis with impaired blood coagulation. Alternatively, many clinical variable-based scoring systems have also been proposed for prognostic prediction in cirrhotic patients with various etiologies (Table 1). Liver stiffness measurement by transient elastography or MR-elastography are non-invasive, imaging-based techniques, which were developed as diagnostic tools to assess liver fibrosis severity measures such as HVPG [44,45]. Besides prediction of cirrhosis progression and mortality, HCC development is another important clinical endpoint to predict. Numerous clinical HCC scores have been developed especially in HBV- and HCV-related liver diseases to assess HCC risk (Table 1). However, none of them has been adopted into clinical practice, and no universal risk score has been established across the liver disease etiologies.

Molecular prognostic indicators in cirrhosis

The clinical variable-based prognostic indicators enable stratification of patients into subgroups with either severer or milder fibrosis/cirrhosis. However, even within the stratified subgroup of patients such as clinically asymptomatic early-stage cirrhosis, individual prognosis greatly varies across the patients. This is also still a sizable patient population to regularly follow up with biannual HCC screening by abdominal ultrasound and serum alpha-fetoprotein measurement [23]. Performance of the clinical variable-based prognostic predictors is limited within the clinical prognostic subgroup because the population is homogeneous based on the clinical variables used even with the use of sophisticated machine-learning algorithms [46,47]. Genome-wide molecular profiling is an approach to potentially overcome the issue by allowing exploration from far larger pool of molecular variables for their prognostic association.

Genome-wide assessment of liver tissue transcriptome has been widely used for exploration of molecular prognostic indicators (Table 2). A 186- or 32-gene expression signature, Prognostic Liver Signature (PLS), was derived from non-tumor liver tissues from cirrhotic subjects with or without HCC, and has been validated for prediction of HCC risk, liver cirrhosis progression, and overall survival in patients affected with HBV, HCV (with or without active infection), alcohol, and NAFLD [48–52]. PLS was present in the liver of rodent models of cirrhosis-driven HCC, and the poor prognosis pattern of the signature was reversed by candidate HCC chemoprevention agents such as an epidermal growth factor (EGF) receptor inhibitor, erlotinib (being tested in a clinical trial: NCT02273362), and lysophosphatidic acid (LPA) pathway inhibitors [51,53]. These experimental data support the use of PLS not only for prognostic prediction, but also as a companion biomarker for HCC chemoprevention therapies. Interestingly, one of the PLS member gene, *IGF1*, encoding insulin-like growth factor 1 (IGF-1) has been associated with hepatocellular dysfunction due to

TABLE 2 Molecular prognostic indicators in cirrhosis

Risk indicator	Clinical outcome	Major etiology	Major race/ ethnicity
186/32-gene Prognostic Liver Signature (PLS)	HCC, cirrhosis progression, death	HCV, HBV, alcohol, NAFLD	Asian, white
IGF 1 (serum protein)	HCC	HCV	White
HSC signature (liver tissue transcriptome)	HCC, HCC recurrence	HCV, HBV	White/Asian
Activated HSC signature (liver tissue transcriptome)	HCC recurrence	HBV	Asian
HIR signature (liver tissue transcriptome)	Late HCC recurrence	HBV	Asian
EGF 61*G (SNP, rs4444903)	HCC	HCV	White, black
TLL1 (SNP, rs17047200)	HCC	HCV after SVR	Asian
PNPLA3 444*G (SNP, rs738409)	HCC	Alcohol, HCV	White

HCC, hepatocellular carcinoma; HBV, hepatitis B virus; HCV, hepatitis C virus; IGF 1, insulin-like growth factor 1; SVR, sustained virologic response; NAFLD, non-alcoholic fatty liver disease.

impaired protein synthetic function of hepatocytes and decreased expression of growth hormone receptors [54]. Indeed, serum levels of IGF-1 protein was associated with liver failure and risk of HCC [55,56]. Similarly, other liver tissue-derived gene-expression signatures have been associated with multicentric HCC development and late recurrence after curative HCC treatment (>1–2 years after the therapies) attributable to de novo HCC development [57–59]. A gene signature of hepatic injury and regeneration was associated with late HCC recurrence and hepatic stellate cell gene signatures were also reported for their association with HCC recurrence and survival [58,60].

Several germline single nucleotide polymorphisms (SNPs) were reported as potential markers of increased risk of developing HCC and other cirrhosis complications (Table 2). A 7-gene SNP panel (so-called cirrhosis risk score) was associated with development of cirrhosis in HCV-infected individuals and fibrosis progression after liver transplantation for HCV-related cirrhosis [61,62]. *EGF* 61*G allele was associated with HCC risk in Eastern and Western patients with fibrotic liver diseases mostly caused by viral hepatitis [63–65]. The risk allele in transplant donors was associated with accelerated fibrosis progression in the graft liver [66]. *EGF* over-expression in hepatic stellate cells and macrophages drives hepatocarcinogenesis in rodents [53,67]. A SNP in *IFNL3* was discovered as a predictor of spontaneous HCV clearance and response to antiviral therapies [68–71]. The variant was associated also with HCC development in patients with HCV genotype 1 infection and treated with interferon-based therapies and recurrence after curative HCC treatment, assumedly by modulating hepatic inflammation and fibrogenesis [72–75]. A genome-wide association study (GWAS) identified an HCC susceptibility-associated SNP in *MICA*, encoding a major histocompatibility complex-related protein that activates anti-tumor immunity through natural killer cells and $CD8^+$ T cells [76–79]. Another GWAS of Japanese hepatitis C patients found an HCC risk-associated SNP in *DEPDC5* [80]. The variant was also associated with fibrosis progression in European liver disease patients [81]. A retrospective GWAS identified a SNP in *TLL1* as a predictor of HCC in patients previously infected with HCV but successfully treated by antiviral therapies [82,83]. In HBV-infected patients, an intronic SNP in *KIF1B* was associated with HCC risk in multiple independent Chinese patient cohorts, although it was not replicated in patients in other regions [84–86]. Another GWAS involving >2000 chronic HBV carries identified SNPs in *STAT4* and *HLA-DQB1/HLA-DBA2* as risk variants [87]. A genomic DNA copy number variation (CNV) analysis of Chinese HBV carries identified an HCC risk associated duplication in chromosome 15q13.3, which was correlated with hepatic *SNORA18L5* expression, suppressing p53 function [88].

Genetic variants specifically relevant to metabolic liver disease etiologies, i.e., alcohol abuse and NAFLD, have also been explored and reported. A SNP in *PNPLA3* was identified as a risk allele associated with NAFLD [89]. Follow-up studies found its association with fibrosis progression in patients with HCV infection, alcoholic liver disease, and NAFLD [90–92]. Meta-analyses also confirmed association of the SNP with HCC risk in patients with alcoholic liver disease and NAFLD as well as HBV/HCV infection [93–97]. A SNP in *TM6SF2* was identified for its association with presence of NAFLD and alcoholic cirrhosis as well as hepatic steatosis and dyslipidemia in viral hepatitis [98–100]. It was also significantly associated with HCC in alcohol-related cirrhosis [95]. *MBOAT7* SNP was reported as a risk locus for alcoholic cirrhosis and NAFLD severity [99,101]. The SNP was independently associated with NAFLD-related HCC even in patients without advanced liver fibrosis, whereas the *PNPLA3* SNP was associated with HCC in patients with severer fibrosis [102]. A SNP in *DYSF* gene was found to be associated with NASH-related HCC [103]. A recent GWAS identified a splicing variant in *HSD17B13*, which was protective against disease progression in alcoholic and non-alcoholic liver diseases, and possibly mitigates hepatic injury associated with the *PNPLA3* risk allele [104]. A small retrospective study suggested that its protein abundance in peritumoral liver may be negatively associated with HCC recurrence after curative treatment [105].

Clinical utility of prognostic prediction in cirrhosis

Molecular prognostic biomarkers are expected to predict risk of major clinical endpoints such as fibrosis progression, transition to decompensated state, HCC development, liver failure that requires transplantation, and liver-related mortality, and complement clinical variable-based prognostication. Furthermore, molecular prognostic indicators may provide clues to new molecular-targeted treatment and/or prevention strategy guided by the biomarker. The value of molecular prognostic biomarkers especially in the setting of chemoprevention cannot be overemphasized, besides its utility to refine risk-based patient follow-up. Chemoprevention clinical trials have been deemed highly resource-demanding in terms of sample size and follow-up period despite the extremely low success rates [106–108]. Molecular risk biomarker-based enrichment of high-risk patients will drastically lower the bar to design and conduct chemoprevention trials by substantially reducing required sample size and the duration of observation [24]. Technical difficulty and financial requirement are the major factors that hamper clinical translation of molecular prognostic biomarker candidates [109]. Acquisition of liver tissues will be critical to facilitate development of reliable biomarker assays. Resources to support biomarker validation is another indispensable component. Computational resources for in silico biomarker validation will also play a significant role in the process [110–112]. Processes of regulatory approval and reimbursement are additional factors that determine successful clinical deployment of molecular prognostic biomarker assays [109].

Conclusions

Ever-evolving omics assay technologies have significantly facilitated the process of discovering candidate prognostic biomarkers in inflammation/fibrosis-driven chronic liver diseases. Studies have suggested their promising performance to eventually alter standard of clinical care by informing physicians about accurate estimate of future risk of developing lethal complications such as liver failure and HCC development. Such biomarkers will also guide identification and development of new therapeutic and/or preventive approaches that could lead to realization of "precision medicine" approaches in chronic liver diseases, and substantially improve the dismal prognosis of the deadly diseases [113].

References

[1] Fujiwara N, Friedman SL, Goossens N, Hoshida Y. Risk factors and prevention of hepatocellular carcinoma in the era of precision medicine. J Hepatol 2018;68(3):526–49. https://doi.org/10.1016/j.jhep.2017.09.016.

[2] Higashi T, Friedman SL, Hoshida Y. Hepatic stellate cells as key target in liver fibrosis. Adv Drug Deliv Rev 2017;121:27–42. https://doi.org/10.1016/j.addr.2017.05.007.

[3] GBD 2013 Mortality and Causes of Death Collaborators. Global, regional, and national age-sex specific all-cause and cause-specific mortality for 240 causes of death, 1990-2013: a systematic analysis for the Global Burden of Disease Study 2013. Lancet 2015;385(9963):117–71. https://doi.org/10.1016/S0140-6736(14)61682-2.

[4] Ferlay J, Soerjomataram I, Dikshit R, Eser S, Mathers C, Rebelo M, et al. Cancer incidence and mortality worldwide: sources, methods and major patterns in GLOBOCAN 2012. Int J Cancer 2015;136(5):E359–86. https://doi.org/10.1002/ijc.29210.

[5] Bosch J, Iwakiri Y. The portal hypertension syndrome: etiology, classification, relevance, and animal models. Hepatol Int 2018;12(Suppl 1):1–10. https://doi.org/10.1007/s12072-017-9827-9.

[6] Gines P, Quintero E, Arroyo V, Teres J, Bruguera M, Rimola A, et al. Compensated cirrhosis: natural history and prognostic factors. Hepatology 1987;7(1):122–8.

[7] Saunders JB, Walters JR, Davies AP, Paton A. A 20-year prospective study of cirrhosis. Br Med J (Clin Res Ed) 1981;282(6260):263–6.

[8] D'Amico G, Garcia-Tsao G, Pagliaro L. Natural history and prognostic indicators of survival in cirrhosis: a systematic review of 118 studies. J Hepatol 2006;44(1):217–31. https://doi.org/10.1016/j.jhep.2005.10.013.

[9] Mancebo A, González–Diéguez ML, Cadahía V, Varela M, Pérez R, Navascués CA, et al. Annual incidence of hepatocellular carcinoma among patients with alcoholic cirrhosis and identification of risk groups. Clin Gastroenterol Hepatol 2013;11(1):95–101. https://doi.org/10.1016/j.cgh.2012.09.007.

[10] Fattovich G, Stroffolini T, Zagni I, Donato F. Hepatocellular carcinoma in cirrhosis: incidence and risk factors. Gastroenterology 2004;127(5 Suppl 1):S35–50.

[11] El-Serag HB, Kanwal F. Epidemiology of hepatocellular carcinoma in the United States: where are we? Where do we go? Hepatology 2014;60(5):1767–75. https://doi.org/10.1002/hep.27222.

[12] Michelotti GA, Machado MV, Diehl AM. NAFLD, NASH and liver cancer. Nat Rev Gastroenterol Hepatol 2013;10(11):656–65. https://doi.org/10.1038/nrgastro.2013.183.

[13] Tsochatzis EA, Bosch J, Burroughs AK. Liver cirrhosis. Lancet 2014;383(9930):1749–61. https://doi.org/10.1016/S0140-6736(14)60121-5.

[14] Lee YA, Wallace MC, Friedman SL. Pathobiology of liver fibrosis: a translational success story. Gut 2015;64(5):830–41. https://doi.org/10.1136/gutjnl-2014-306842.

[15] Rockey DC, Bell PD, Hill JA. Fibrosis—a common pathway to organ injury and failure. N Engl J Med 2015;372(12):1138–49. https://doi.org/10.1056/NEJMra1300575.

[16] Yoshida H, Shiratori Y, Moriyama M, Arakawa Y, Ide T, Sata M, et al. Interferon therapy reduces the risk for hepatocellular carcinoma: national surveillance program of cirrhotic and noncirrhotic patients with chronic hepatitis C in Japan. Ann Intern Med 1999;131(3):174–81.

[17] Hernandez-Gea V, Hilscher M, Rozenfeld R, Lim MP, Nieto N, Werner S, et al. Endoplasmic reticulum stress induces fibrogenic activity in hepatic stellate cells through autophagy. J Hepatol 2013;59(1):98–104. https://doi.org/10.1016/j.jhep.2013.02.016.

[18] Hernandez-Gea V, Ghiassi-Nejad Z, Rozenfeld R, Gordon R, Fiel MI, Yue Z, et al. Autophagy releases lipid that promotes fibrogenesis by activated hepatic stellate cells in mice and in human tissues. Gastroenterology 2012;142(4):938–46. https://doi.org/10.1053/j.gastro.2011.12.044.

[19] De Minicis S, Rychlicki C, Agostinelli L, Saccomanno S, Candelaresi C, Trozzi L, et al. Dysbiosis contributes to fibrogenesis in the course of chronic liver injury in mice. Hepatology 2014;59(5):1738–49. https://doi.org/10.1002/hep.26695.

[20] Scaglione S, Kliethermes S, Cao G, Shoham D, Durazo R, Luke A, et al. The epidemiology of cirrhosis in the United States: a population-based study. J Clin Gastroenterol 2015;49:690–6. https://doi.org/10.1097/mcg.0000000000000208.

[21] Davila JA, Henderson L, Kramer JR, Kanwal F, Richardson PA, Duan Z, et al. Utilization of surveillance for hepatocellular carcinoma among hepatitis C virus-infected veterans in the United States. Ann Intern Med 2011;154(2):85–93. https://doi.org/10.7326/0003-4819-154-2-201101180-00006.

[22] El-Serag HB, Kramer JR, Chen GJ, Duan Z, Richardson PA, Davila JA. Effectiveness of AFP and ultrasound tests on hepatocellular carcinoma mortality in HCV-infected patients in the USA. Gut 2011;60(7):992–7. https://doi.org/10.1136/gut.2010.230508.

[23] Davila JA, Morgan RO, Richardson PA, Du XL, McGlynn KA, El-Serag HB. Use of surveillance for hepatocellular carcinoma among patients with cirrhosis in the United States. Hepatology 2010;52(1):132–41. https://doi.org/10.1002/hep.23615.

[24] Hoshida Y, Fuchs BC, Bardeesy N, Baumert TF, Chung RT. Pathogenesis and prevention of hepatitis C virus-induced hepatocellular carcinoma. J Hepatol 2014;61(1):S79–90. https://doi.org/10.1016/j.jhep.2014.07.010.

[25] Goossens N, Singal AG, King LY, Andersson KL, Fuchs BC, Besa C, et al. Cost-effectiveness of risk score-stratified hepatocellular carcinoma screening in patients with cirrhosis. Clin Transl Gastroenterol 2017;8(6):e101. https://doi.org/10.1038/ctg.2017.26.

[26] Smith JM, Biggins SW, Haselby DG, Kim WR, Wedd J, Lamb K, et al. Kidney, pancreas and liver allocation and distribution in the United States. Am J Transplant 2012;12(12):3191–212. https://doi.org/10.1111/j.1600-6143.2012.04259.x.

[27] Freeman Jr. RB, Wiesner RH, Harper A, McDiarmid SV, Lake J, Edwards E, et al. The new liver allocation system: moving toward evidence-based transplantation policy. Liver Transpl 2002;8(9):851–8. https://doi.org/10.1053/jlts.2002.35927.

[28] Malinchoc M, Kamath PS, Gordon FD, Peine CJ, Rank J, ter Borg PC. A model to predict poor survival in patients undergoing transjugular intrahepatic portosystemic shunts. Hepatology 2000;31(4):864–71. https://doi.org/10.1053/he.2000.5852.

[29] Said A, Williams J, Holden J, Remington P, Gangnon R, Musat A, et al. Model for end stage liver disease score predicts mortality across a broad spectrum of liver disease. J Hepatol 2004;40(6):897–903. https://doi.org/10.1016/j.jhep.2004.02.010.

[30] Wiesner R, Edwards E, Freeman R, Harper A, Kim R, Kamath P, et al. Model for end-stage liver disease (MELD) and allocation of donor livers. Gastroenterology 2003;124(1):91–6. https://doi.org/10.1053/gast.2003.50016.

[31] Teh SH, Nagorney DM, Stevens SR, Offord KP, Therneau TM, Plevak DJ, et al. Risk factors for mortality after surgery in patients with cirrhosis. Gastroenterology 2007;132(4):1261–9. https://doi.org/10.1053/j.gastro.2007.01.040.

[32] Reverter E, Tandon P, Augustin S, Turon F, Casu S, Bastiampillai R, et al. A MELD-based model to determine risk of mortality among patients with acute variceal bleeding. Gastroenterology 2014;146(2):412–3. https://doi.org/10.1053/j.gastro.2013.10.018.

[33] Alessandria C, Ozdogan O, Guevara M, Restuccia T, Jiménez W, Arroyo V, et al. MELD score and clinical type predict prognosis in hepatorenal syndrome: relevance to liver transplantation. Hepatology 2005;41(6):1282–9. https://doi.org/10.1002/hep.20687.

[34] Dunn W, Jamil LH, Brown LS, Wiesner RH, Kim WR, Menon KV, et al. MELD accurately predicts mortality in patients with alcoholic hepatitis. Hepatology 2005;41(2):353–8. https://doi.org/10.1002/hep.20503.

[35] Sethasine S, Jain D, Groszmann RJ, Garcia-Tsao G. Quantitative histological-hemodynamic correlations in cirrhosis. Hepatology 2012;55(4):1146–53. https://doi.org/10.1002/hep.24805.

[36] Rastogi A, Maiwall R, Bihari C, Ahuja A, Kumar A, Singh T, et al. Cirrhosis histology and Laennec staging system correlate with high portal pressure. Histopathology 2013;62(5):731–41. https://doi.org/10.1111/his.12070.

[37] Kim MY, Cho MY, Baik SK, Park HJ, Jeon HK, Im CK, et al. Histological subclassification of cirrhosis using the Laennec fibrosis scoring system correlates with clinical stage and grade of portal hypertension. J Hepatol 2011;55(5):1004–9. https://doi.org/10.1016/j.jhep.2011.02.012.

[38] Kim SU, Oh HJ, Wanless IR, Lee S, Han KH, Park YN. The Laennec staging system for histological sub-classification of cirrhosis is useful for stratification of prognosis in patients with liver cirrhosis. J Hepatol 2012;57(3):556–63. https://doi.org/10.1016/j.jhep.2012.04.029.

[39] Calvaruso V, Burroughs AK, Standish R, Manousou P, Grillo F, Leandro G, et al. Computer-assisted image analysis of liver collagen: relationship to Ishak scoring and hepatic venous pressure gradient. Hepatology 2009;49(4):1236–44. https://doi.org/10.1002/hep.22745.

[40] Manousou P, Burroughs AK, Tsochatzis E, Isgro G, Hall A, Green A, et al. Digital image analysis of collagen assessment of progression of fibrosis in recurrent HCV after liver transplantation. J Hepatol 2013;58(5):962–8. https://doi.org/10.1016/j.jhep.2012.12.016.

[41] Calvaruso V, Dhillon AP, Tsochatzis E, Manousou P, Grillo F, Germani G, et al. Liver collagen proportionate area predicts decompensation in patients with recurrent hepatitis C virus cirrhosis after liver transplantation. J Gastroenterol Hepatol 2012;27(7):1227–32. https://doi.org/10.1111/j.1440-1746.2012.07136.x.

[42] Xu S, Wang Y, Tai DC, Wang S, Cheng CL, Peng Q, et al. qFibrosis: a fully-quantitative innovative method incorporating histological features to facilitate accurate fibrosis scoring in animal model and chronic hepatitis B patients. J Hepatol 2014;61(2):260–9.

[43] Tsochatzis E, Bruno S, Isgro G, Hall A, Theocharidou E, Manousou P, et al. Collagen proportionate area is superior to other histological methods for sub-classifying cirrhosis and determining prognosis. J Hepatol 2014;60(5):948–54. https://doi.org/10.1016/j.jhep.2013.12.023.

[44] Robic MA, Procopet B, Metivier S, Peron JM, Selves J, Vinel JP, et al. Liver stiffness accurately predicts portal hypertension related complications in patients with chronic liver disease: a prospective study. J Hepatol 2011;55(5):1017–24. https://doi.org/10.1016/j.jhep.2011.01.051.

[45] European Association for Study of Liver; Asociacion Latinoamericana para el Estudio del Higado. EASL-ALEH clinical practice guidelines: non-invasive tests for evaluation of liver disease severity and prognosis. J Hepatol 2015;63:237–64. https://doi.org/10.1016/j.jhep.2015.04.006.

[46] El-Serag HB, Kanwal F, Davila JA, Kramer J, Richardson P. A new laboratory-based algorithm to predict development of hepatocellular carcinoma in patients with hepatitis C and cirrhosis. Gastroenterology 2014;146(5):1249–1255.e1. https://doi.org/10.1053/j.gastro.2014.01.045.

[47] Singal AG, Mukherjee A, Joseph Elmunzer B, Higgins PDR, Lok AS, Zhu J, et al. Machine learning algorithms outperform conventional regression models in predicting development of hepatocellular carcinoma. Am J Gastroenterol 2013;108(11):1723–30. https://doi.org/10.1038/ajg.2013.332.

[48] Hoshida Y, Villanueva A, Sangiovanni A, Sole M, Hur C, Andersson KL, et al. Prognostic gene expression signature for patients with hepatitis C-related early-stage cirrhosis. Gastroenterology 2013;144(5):1024–30. https://doi.org/10.1053/j.gastro.2013.01.021.

[49] King LY, Canasto-Chibuque C, Johnson KB, Yip S, Chen X, Kojima K, et al. A genomic and clinical prognostic index for hepatitis C-related early-stage cirrhosis that predicts clinical deterioration. Gut 2015;64:1296–302. https://doi.org/10.1136/gutjnl-2014-307862.

[50] Hoshida Y, Villanueva A, Kobayashi M, Peix J, Chiang DY, Camargo A, et al. Gene expression in fixed tissues and outcome in hepatocellular carcinoma. N Engl J Med 2008;359(19):1995–2004. https://doi.org/10.1056/NEJMoa0804525.

[51] Nakagawa S, Wei L, Song WM, Higashi T, Ghoshal S, Kim RS, et al. Molecular liver cancer prevention in cirrhosis by organ transcriptome analysis and lysophosphatidic acid pathway inhibition. Cancer Cell 2016;30(6):879–90. https://doi.org/10.1016/j.ccell.2016.11.004.

[52] Ono A, Goossens N, Finn RS, Schmidt WN, Thung SN, Im GY, et al. Persisting risk of hepatocellular carcinoma after hepatitis C virus cure monitored by a liver transcriptome signature. Hepatology 2017;66(4):1344–6. https://doi.org/10.1002/hep.29203.

[53] Fuchs BC, Hoshida Y, Fujii T, Wei L, Yamada S, Lauwers GY, et al. Epidermal growth factor receptor inhibition attenuates liver fibrosis and development of hepatocellular carcinoma. Hepatology 2014;59(4):1577–90. https://doi.org/10.1002/hep.26898.

[54] Assy N, Pruzansky Y, Gaitini D, Orr ZS, Ze H, Baruch Y. Growth hormone-stimulated IGF-1 generation in cirrhosis reflects hepatocellular dysfunction. J Hepatol 2008;49(1):34–42.
[55] Khoshnood A, Toosi MN, Faravash MJ, Esteghamati A, Froutan H, Ghofrani H, et al. A survey of correlation between insulin-like growth factor-I (IGF-I) levels and severity of liver cirrhosis. Hepat Mon 2013;13(2):e6181.
[56] Mazziotti G, Sorvillo F, Morisco F, Carbone A, Rotondi M, Stornaiuolo G, et al. Serum insulin-like growth factor I evaluation as a useful tool for predicting the risk of developing hepatocellular carcinoma in patients with hepatitis C virus-related cirrhosis. Cancer 2002;95(12):2539–45.
[57] Okamoto M, Utsunomiya T, Wakiyama S, Hashimoto M, Fukuzawa K, Ezaki T, et al. Specific gene-expression profiles of noncancerous liver tissue predict the risk for multicentric occurrence of hepatocellular carcinoma in hepatitis C virus–positive patients. Ann Surg Oncol 2006;13(7):947–54. https://doi.org/10.1245/aso.2006.07.018.
[58] Kim JH, Sohn BH, Lee HS, Kim SB, Yoo JE, Park YY, et al. Genomic predictors for recurrence patterns of hepatocellular carcinoma: model derivation and validation. PLoS Med 2014;11(12):e1001770 https://doi.org/10.1371/journal.pmed.1001770.
[59] Utsunomiya T, Shimada M, Imura S, Morine Y, Ikemoto T, Mori M. Molecular signatures of noncancerous liver tissue can predict the risk for late recurrence of hepatocellular carcinoma. J Gastroenterol 2010;45(2):146–52. https://doi.org/10.1007/s00535-009-0164-1.
[60] Ji J, Eggert T, Budhu A, Forgues M, Takai A, Dang H, et al. Hepatic stellate cell and monocyte interaction contributes to poor prognosis in hepatocellular carcinoma. Hepatology 2015;62:481–95. https://doi.org/10.1002/hep.27822.
[61] Huang H, Shiffman ML, Friedman S, Venkatesh R, Bzowej N, Abar OT, et al. A 7 gene signature identifies the risk of developing cirrhosis in patients with chronic hepatitis C. Hepatology 2007;46(2):297–306. https://doi.org/10.1002/hep.21695.
[62] do ON, Eurich D, Schmitz P, Schmeding M, Heidenhain C, Bahra M, et al. A 7-gene signature of the recipient predicts the progression of fibrosis after liver transplantation for hepatitis C virus infection. Liver Transpl 2012;18(3):298–304. https://doi.org/10.1002/lt.22475.
[63] Tanabe KK, Lemoine A, Finkelstein DM, Kawasaki H, Fujii T, Chung RT, et al. Epidermal growth factor gene functional polymorphism and the risk of hepatocellular carcinoma in patients with cirrhosis. JAMA 2008;299(1):53–60. https://doi.org/10.1001/jama.2007.65.
[64] Abu Dayyeh BK, Yang M, Fuchs BC, Karl DL, Yamada S, Sninsky JJ, et al. A functional polymorphism in the epidermal growth factor gene is associated with risk for hepatocellular carcinoma. Gastroenterology 2011;141(1):141–9. https://doi.org/10.1053/j.gastro.2011.03.045.
[65] Jiang G, Yu K, Shao L, Yu X, Hu C, Qian P, et al. Association between epidermal growth factor gene +61A/G polymorphism and the risk of hepatocellular carcinoma: a meta-analysis based on 16 studies. BMC Cancer 2015;15:314 https://doi.org/10.1186/s12885-015-1318-6.
[66] Mueller JL, King LY, Johnson KB, Gao T, Nephew LD, Kothari D, et al. Impact of EGF, IL28B, and PNPLA3 polymorphisms on the outcome of allograft hepatitis C: a multicenter study. Clin Transplant 2016;30(4):452–60. https://doi.org/10.1111/ctr.12710.
[67] Lanaya H, Natarajan A, Komposch K, Li L, Amberg N, Chen L, et al. EGFR has a tumour-promoting role in liver macrophages during hepatocellular carcinoma formation. Nat Cell Biol 2014;16(10):972–7. https://doi.org/10.1038/ncb3031.
[68] Thomas DL, Thio CL, Martin MP, Qi Y, Ge D, O'Huigin C, et al. Genetic variation in IL28B and spontaneous clearance of hepatitis C virus. Nature 2009;461(7265):798–801. https://doi.org/10.1038/nature08463.
[69] Suppiah V, Moldovan M, Ahlenstiel G, Berg T, Weltman M, Abate ML, et al. IL28B is associated with response to chronic hepatitis C interferon-alpha and ribavirin therapy. Nat Genet 2009;41(10):1100–4. https://doi.org/10.1038/ng.447.

[70] Tanaka Y, Nishida N, Sugiyama M, Kurosaki M, Matsuura K, Sakamoto N, et al. Genome-wide association of IL28B with response to pegylated interferon-alpha and ribavirin therapy for chronic hepatitis C. Nat Genet 2009;41(10):1105–9. https://doi.org/10.1038/ng.449.

[71] Ge D, Fellay J, Thompson AJ, Simon JS, Shianna KV, Urban TJ, et al. Genetic variation in IL28B predicts hepatitis C treatment-induced viral clearance. Nature 2009;461(7262):399–401. https://doi.org/10.1038/nature08309.

[72] Asahina Y, Tsuchiya K, Nishimura T, Muraoka M, Suzuki Y, Tamaki N, et al. Genetic variation near interleukin 28B and the risk of hepatocellular carcinoma in patients with chronic hepatitis C. J Gastroenterol 2014;49(7):1152–62. https://doi.org/10.1007/s00535-013-0858-2.

[73] Chang KC, Tseng PL, Wu YY, Hung HC, Huang CM, Lu SN, et al. A polymorphism in interferon L3 is an independent risk factor for development of hepatocellular carcinoma after treatment of hepatitis C virus infection. Clin Gastroenterol Hepatol 2015;13(5):1017–24. https://doi.org/10.1016/j.cgh.2014.10.035.

[74] Hodo Y, Honda M, Tanaka A, Nomura Y, Arai K, Yamashita T, et al. Association of interleukin-28B genotype and hepatocellular carcinoma recurrence in patients with chronic hepatitis C. Clin Cancer Res 2013;19(7):1827–37. https://doi.org/10.1158/1078-0432.CCR-12-1641.

[75] Eslam M, McLeod D, Kelaeng KS, Mangia A, Berg T, Thabet K, et al. IFN-lambda3, not IFN-lambda4, likely mediates IFNL3-IFNL4 haplotype-dependent hepatic inflammation and fibrosis. Nat Genet 2017;49(5):795–800. https://doi.org/10.1038/ng.3836.

[76] Kumar V, Kato N, Urabe Y, Takahashi A, Muroyama R, Hosono N, et al. Genome-wide association study identifies a susceptibility locus for HCV-induced hepatocellular carcinoma. Nat Genet 2011;43(5):455–8. https://doi.org/10.1038/ng.809.

[77] Hai H, Tamori A, Thuy LTT, Yoshida K, Hagihara A, Kawamura E, et al. Polymorphisms in MICA, but not in DEPDC5, HCP5 or PNPLA3, are associated with chronic hepatitis C-related hepatocellular carcinoma. Sci Rep 2017;7(1):11912 https://doi.org/10.1038/s41598-017-10363-5.

[78] Huang CF, Huang CY, Yeh ML, Wang SC, Chen KY, Ko YM, et al. Genetics variants and serum levels of MHC class I chain-related a in predicting hepatocellular carcinoma development in chronic hepatitis C patients post antiviral treatment. EBioMedicine 2017;15:81–9. https://doi.org/10.1016/j.ebiom.2016.11.031.

[79] Lange CM, Bibert S, Dufour JF, Cellerai C, Cerny A, Heim MH, et al. Comparative genetic analyses point to HCP5 as susceptibility locus for HCV-associated hepatocellular carcinoma. J Hepatol 2013;59(3):504–9. https://doi.org/10.1016/j.jhep.2013.04.032.

[80] Miki D, Ochi H, Hayes CN, Abe H, Yoshima T, Aikata H, et al. Variation in the DEPDC5 locus is associated with progression to hepatocellular carcinoma in chronic hepatitis C virus carriers. Nat Genet 2011;43(8):797–800. https://doi.org/10.1038/ng.876.

[81] Burza MA, Motta BM, Mancina RM, Pingitore P, Pirazzi C, Lepore SM, et al. DEPDC5 variants increase fibrosis progression in Europeans with chronic hepatitis C virus infection. Hepatology 2016;63(2):418–27. https://doi.org/10.1002/hep.28322.

[82] Matsuura K, Sawai H, Ikeo K, Ogawa S, Iio E, Isogawa M, et al. Genome-wide association study identifies TLL1 variant associated with development of hepatocellular carcinoma after eradication of hepatitis C virus infection. Gastroenterology 2017;152(6):1383–94. https://doi.org/10.1053/j.gastro.2017.01.041.

[83] Seko Y, Yamaguchi K, Mizuno N, Okuda K, Takemura M, Taketani H, et al. Combination of PNPLA3 and TLL1 polymorphism can predict advanced fibrosis in Japanese patients with nonalcoholic fatty liver disease. J Gastroenterol 2018;53(3):438–48. https://doi.org/10.1007/s00535-017-1372-8.

[84] Zhang H, Zhai Y, Hu Z, Wu C, Qian J, Jia W, et al. Genome-wide association study identifies 1p36.22 as a new susceptibility locus for hepatocellular carcinoma in chronic hepatitis B virus carriers. Nat Genet 2010;42(9):755–8. https://doi.org/10.1038/ng.638.

[85] Sawai H, Nishida N, Mbarek H, Matsuda K, Mawatari Y, Yamaoka M, et al. No association for Chinese HBV-related hepatocellular carcinoma susceptibility SNP in other East Asian populations. BMC Med Genet 2012;13:47 https://doi.org/10.1186/1471-2350-13-47.

[86] Sopipong W, Tangkijvanich P, Payungporn S, Posuwan N, Poovorawan Y. The KIF1B (rs17401966) single nucleotide polymorphism is not associated with the development of HBV-related hepatocellular carcinoma in Thai patients. Asian Pac J Cancer Prev 2013; 14(5):2865–9.

[87] Jiang DK, Sun J, Cao G, Liu Y, Lin D, Gao YZ, et al. Genetic variants in STAT4 and HLA-DQ genes confer risk of hepatitis B virus-related hepatocellular carcinoma. Nat Genet 2013; 45(1):72–5. https://doi.org/10.1038/ng.2483.

[88] Cao P, Yang A, Wang R, Xia X, Zhai Y, Li Y, et al. Germline duplication of SNORA18L5 increases risk for HBV-related hepatocellular carcinoma by altering localization of ribosomal proteins and decreasing levels of p53. Gastroenterology 2018;155:542–56. https://doi.org/10.1053/j.gastro.2018.04.020.

[89] Romeo S, Kozlitina J, Xing C, Pertsemlidis A, Cox D, Pennacchio LA, et al. Genetic variation in PNPLA3 confers susceptibility to nonalcoholic fatty liver disease. Nat Genet 2008;40(12):1461–5. https://doi.org/10.1038/ng.257.

[90] Valenti L, Rumi M, Galmozzi E, Aghemo A, Del Menico B, De Nicola S, et al. Patatin-like phospholipase domain-containing 3 I148M polymorphism, steatosis, and liver damage in chronic hepatitis C. Hepatology 2011;53(3):791–9. https://doi.org/10.1002/hep.24123.

[91] Burza MA, Molinaro A, Attilia ML, Rotondo C, Attilia F, Ceccanti M, et al. PNPLA3 I148M (rs738409) genetic variant and age at onset of at-risk alcohol consumption are independent risk factors for alcoholic cirrhosis. Liver Int 2014;34(4):514–20. https://doi.org/10.1111/liv.12310.

[92] Rotman Y, Koh C, Zmuda JM, Kleiner DE, Liang TJ, Nash CRN. The association of genetic variability in patatin-like phospholipase domain-containing protein 3 (PNPLA3) with histological severity of nonalcoholic fatty liver disease. Hepatology 2010;52(3):894–903. https://doi.org/10.1002/hep.23759.

[93] Singal AG, Manjunath H, Yopp AC, Beg MS, Marrero JA, Gopal P, et al. The effect of PNPLA3 on fibrosis progression and development of hepatocellular carcinoma: a meta-analysis. Am J Gastroenterol 2014;109(3):325–34. https://doi.org/10.1038/ajg.2013.476.

[94] Trepo E, Nahon P, Bontempi G, Valenti L, Falleti E, Nischalke HD, et al. Association between the PNPLA3 (rs738409 C>G) variant and hepatocellular carcinoma: evidence from a meta-analysis of individual participant data. Hepatology 2014;59(6):2170–7. https://doi.org/10.1002/hep.26767.

[95] Stickel F, Buch S, Nischalke HD, Weiss KH, Gotthardt D, Fischer J, et al. Genetic variants in PNPLA3 and TM6SF2 predispose to the development of hepatocellular carcinoma in individuals with alcohol-related cirrhosis. Am J Gastroenterol 2018;113:1475–83. https://doi.org/10.1038/s41395-018-0041-8.

[96] Brouwer WP, van der Meer AJ, Boonstra A, Pas SD, de Knegt RJ, de Man RA, et al. The impact of PNPLA3 (rs738409 C>G) polymorphisms on liver histology and long-term clinical outcome in chronic hepatitis B patients. Liver Int 2015;35(2):438–47. https://doi.org/10.1111/liv.12695.

[97] Ali M, Yopp A, Gopal P, Beg MS, Zhu H, Lee W, et al. A variant in PNPLA3 associated with fibrosis progression but not hepatocellular carcinoma in patients with hepatitis C virus infection. Clin Gastroenterol Hepatol 2016;14(2):295–300. https://doi.org/10.1016/j.cgh.2015.08.018.

[98] Kozlitina J, Smagris E, Stender S, Nordestgaard BG, Zhou HH, Tybjaerg-Hansen A, et al. Exome-wide association study identifies a TM6SF2 variant that confers susceptibility to nonalcoholic fatty liver disease. Nat Genet 2014;46(4):352–6. https://doi.org/10.1038/ng.2901.

[99] Buch S, Stickel F, Trepo E, Way M, Herrmann A, Nischalke HD, et al. A genome-wide association study confirms PNPLA3 and identifies TM6SF2 and MBOAT7 as risk loci for alcohol-related cirrhosis. Nat Genet 2015;47(12):1443–8. https://doi.org/10.1038/ng.3417.

[100] Eslam M, Mangia A, Berg T, Chan HL, Irving WL, Dore GJ, et al. Diverse impacts of the rs58542926 E167K variant in TM6SF2 on viral and metabolic liver disease phenotypes. Hepatology 2016;64(1):34–46. https://doi.org/10.1002/hep.28475.

[101] Mancina RM, Dongiovanni P, Petta S, Pingitore P, Meroni M, Rametta R, et al. The MBOAT7-TMC4 variant rs641738 increases risk of nonalcoholic fatty liver disease in individuals of European descent. Gastroenterology 2016;150(5):1219–6. https://doi.org/10.1053/j.gastro.2016.01.032.

[102] Donati B, Dongiovanni P, Romeo S, Meroni M, McCain M, Miele L, et al. MBOAT7 rs641738 variant and hepatocellular carcinoma in non-cirrhotic individuals. Sci Rep 2017;7(1):4492. https://doi.org/10.1038/s41598-017-04991-0.

[103] Kawaguchi T, Shima T, Mizuno M, Mitsumoto Y, Umemura A, Kanbara Y, et al. Risk estimation model for nonalcoholic fatty liver disease in the Japanese using multiple genetic markers. PLoS One 2018;13(1):e0185490 https://doi.org/10.1371/journal.pone.0185490.

[104] Abul-Husn NS, Cheng X, Li AH, Xin Y, Schurmann C, Stevis P, et al. A protein-truncating HSD17B13 variant and protection from chronic liver disease. N Engl J Med 2018;378(12):1096–106. https://doi.org/10.1056/NEJMoa1712191.

[105] Chen J, Zhuo JY, Yang F, Liu ZK, Zhou L, Xie HY, et al. 17-beta-hydroxysteroid dehydrogenase 13 inhibits the progression and recurrence of hepatocellular carcinoma. Hepatobiliary Pancreat Dis Int 2018;17:220–6. https://doi.org/10.1016/j.hbpd.2018.04.006.

[106] Hoshida Y, Fuchs BC, Tanabe KK. Prevention of hepatocellular carcinoma: potential targets, experimental models, and clinical challenges. Curr Cancer Drug Targets 2012;12(9):1129–59.

[107] Lippman SM, Klein EA, Goodman PJ, Lucia MS, Thompson IM, Ford LG, et al. Effect of selenium and vitamin E on risk of prostate cancer and other cancers: the selenium and vitamin E cancer prevention trial (SELECT). JAMA 2009;301(1):39–51. https://doi.org/10.1001/jama.2008.864.

[108] Di Bisceglie AM, Shiffman ML, Everson GT, Lindsay KL, Everhart JE, Wright EC, et al. Prolonged therapy of advanced chronic hepatitis C with low-dose peginterferon. N Engl J Med 2008;359(23):2429–41. https://doi.org/10.1056/NEJMoa0707615.

[109] Goossens N, Nakagawa S, Sun X, Hoshida Y. Cancer biomarker discovery and validation. Transl Cancer Res 2015;4(3):256–69. https://doi.org/10.3978/j.issn.2218-676X.2015.06.04.

[110] Chen X, Sun X, Hoshida Y. Survival analysis tools in genomics research. Hum Genomics 2014;8(1):21. https://doi.org/10.1186/s40246-014-0021-z.

[111] Kim RS, Goossens N, Hoshida Y. Use of big data in drug development for precision medicine. Expert Rev Precis Med Drug Dev 2016;1(3):245–53. https://doi.org/10.1080/23808993.2016.1174062.

[112] Wooden B, Goossens N, Hoshida Y, Friedman SL. Using big data to discover diagnostics and therapeutics for gastrointestinal and liver diseases. Gastroenterology 2017;152(1):53–3. https://doi.org/10.1053/j.gastro.2016.09.065.

[113] Collins FS, Varmus H. A new initiative on precision medicine. N Engl J Med 2015;372(9):793–5. https://doi.org/10.1056/NEJMp1500523.

Chapter 22

Biomarkers and osteoarthritis☆

Virginia Byers Kraus[a,b], **Ming-Feng Hsueh**[b]
[a]Department of Medicine, Duke University School of Medicine, Durham, NC 27701, United States, [b]Duke Molecular Physiology Institute, Duke University School of Medicine, Durham, NC 27701, United States

Personalized medicine—choice *versus* cost

In formulating an individual care plan, traditionally one considers a patient's signs and symptoms, family history, social circumstances, environment, risk factors and behaviors. Using a personalized medicine approach, one seeks to optimize and tailor the care of the individual patient through reliance on traditional medical information complemented by some form of companion diagnostic to more fully characterize the condition and disease or health status of the individual patient. For treatment of osteoarthritis (OA), a musculoskeletal disease, the companion diagnostic can take the form of an *in vitro* or an *in vivo* test. *In vitro diagnostics* could be based in genetics, genomics, proteomics, metabolomics or any other molecular profiling mechanism [1]. *In vivo diagnostics* for OA take the form of specialized imaging methodologies such as magnetic resonance imaging (MRI), including sodium MRI [2] and T2 mapping, T1rho mapping, and delayed gadolinium-enhanced magnetic resonance imaging of cartilage (dGEMRIC) [3], positron emission tomography and scintigraphy [4], extremity cone-beam computed tomography [5], contrast-enhanced musculoskeletal ultrasonography [6], and analysis of X-ray bone trabecular texture by fractal signature analysis [7], to name a few.

Although diagnostic advances make aspects of personalized medicine a reality, invariably, economic considerations play an increasing role in shaping health care choices. Moreover, identification of those who cannot safely stay on 'standard care' does not necessarily imply that there will be third-party reimbursement for individualized non-standard care. The current practice of evidence-based medicine is guided by cost-effectiveness research. The cost-effectiveness of a therapeutic or preventive intervention is the ratio of the cost of the intervention to a relevant measure of its effect. To reach some theoretical threshold of cost-effectiveness, it would generally be necessary for an intervention

☆VBK is supported by 5P30AG028716 and AR50245 from the NIA/NIH and NIAMS/NIH.

Genomic and Precision Medicine. https://doi.org/10.1016/B978-0-12-801496-7.00022-8

with only incremental benefit to cost very little, or for a costly intervention to improve the health of many or result in a dramatic health benefit for a few. Cost-effectiveness analyses are designed to generally address which "one size" fits all and provides benefit for the most individuals at the least cost.

So long as the focus continues to be on late stage manifestations of OA, these two approaches to medical therapy, personalization *versus* cost-effectiveness can be at odds. This is because late stage OA, diagnosed by X-ray, is a prevalent, recalcitrant, chronic disease process associated with a high burden of mobility disability (defined as needing help walking or climbing stairs) that for knee OA is greater than that attributable to any other medical condition in people aged 65 years and older [7a, 7b] and for knee OA, excess mortality *(https://www.oarsi.org/research/oa-serious-disease)*. Personalized medicine targeting end-stage subsets of OA likely would entail a proliferation of expensive salvage procedures for each joint type and site in the body. In contrast, a focus on development of pharmacologic agents targeting specific etiologic pathways of early molecular and pre-radiographic disease would likely constitute a more cost-effective strategy and have the added prospective cost and health benefits accruing from the prevention of disability.

Osteoarthritis—the magnitude of the problem

OA is the most common joint disorder in the US and worldwide; worldwide estimates are that 9.6% of men and 18.0% of women over 60 years old have symptomatic OA (http://www.who.int/chp/topics/rheumatic/en/index.html). OA increases with age, and is estimated to affect 40% of individuals over 70 years of age [8, 9]. In the United States, doctor-diagnosed arthritis prevalence was 54.4 million (22.7%) in adults during 2013–2015; 23.7 million (43.5% of those with arthritis) had arthritis-attributable activity limitations (approximately 20% increase since 2002) [10]. The World Health Organization estimates that 80% of those with OA will have limitations in movement, and that 25% cannot perform the major daily activities of their lives (http://www.who.int/chp/topics/rheumatic/en/index.html). In 2013, the national total medical expenditures and earnings losses attributable to arthritis in the US were estimated to be as high as $303.5 billion [11]. Thus, the potential cost implications of personalized medicine for such a prevalent and disabling condition as OA are quite great. For aspects of personalized medicine to be adopted as mainstream medical practice, there will need to be a reconciliation of these two important approaches and aspects of medical care—namely a reconciliation of the focus on the needs of the individual patient and the often conflicting focus on cost-effectiveness.

Where personalized medicine and cost-effective therapy are likely to synergize is in the early identification and prevention of a chronic disease process. The cost of identifying and treating the disease in its early stages would likely be a fraction of the cost of treating the established late stages, and have the added benefit of greater likelihood of disease modification and prevention of disability. Further cost-savings could likely be realized with a strategy for identifying and screening individuals more susceptible to OA. Thus, development of an early identification strategy is of fundamental importance as it would promote cost-effective prevention and early-stage treatment strategies. The cost-effective means of identifying early OA and monitoring treatment of the molecular and pre-radiographic stages of the disease clearly lies in biomarker discovery and validation. So, in this article, we consider the application of biomarkers in a personalized medicine strategy for OA based on the targeting of pre-radiographic OA and the early molecular stage of OA.

Paradigms for studying early osteoarthritis events

OA follows a consistent pattern of pathological progression [12] to common end-stage radiographic manifestations. Radiographic OA prevalence increases in conjunction with estrogen deficiency and aging [13]. Based on current knowledge, susceptibility to OA involves multiple pathways including signaling cascades that regulate skeletal morphogenesis, extracellular matrix proteins and chondrocyte phenotype, apoptosis pathways as well as inflammatory molecules related to senescence phenotypic changes in joint tissues with cytokine production, prostaglandin and arachidonic acid metabolism and proteases [9, 14–16]. To date, an increasing number of loci demonstrate compelling association with OA across a broad range of ethnic groups [16a]: the 7q22 locus, the *GDF5* gene encoding Growth/differentiation factor 5 that has a role in joint development, and the *DIO2* gene encoding Type II iodothyronine deiodinase responsible for conversion of thyroid hormone to its active form [17]. The identification of OA loci has confirmed the joint-specific effects, with loci often contributing to disease risk at a particular skeletal site; OA loci show gender differences in molecular mechanisms underlying OA susceptibility [18]. There is also an interaction between hereditary and environmental risk factors for OA [19] suggesting that genetic profiling early in life could contribute to preventive strategies to minimize risk of OA. Synovial fluid analyses suggest that OA involves inflammasome activation and innate immunity [20].

OA may also occur as a sequela of major joint insult such as infection or joint injury [21, 22]. Unlike idiopathic OA (traditionally called primary OA), post-traumatic OA has a known time of onset, making it a much more tractable type of OA for monitoring and treating events of the early stages of the disease process. Acute trauma to the anterior cruciate ligament (ACL) or meniscus has been demonstrated to be a major risk factor for the development of knee OA, with a 50% chance of a patient developing symptomatic OA 10–20 years post-injury [21]. Using sensitive MRI techniques (T2 mapping and T1 rho), signs of early OA are discernible as early as 6 months after ACL injury [23]. Post-traumatic OA is believed to account for up to 5.6 million cases per year, or 12% of the total US cases of symptomatic OA, with an estimated cost in 2006 of $3.06 billion per year or 0.15% of total US direct health care costs [24, 25].

The predisposition to post-traumatic OA is also likely influenced by genetic determinants suggested by the increased prevalence of post-traumatic OA after meniscectomy in individuals with hand OA [19, 26], a form of OA with a strong genetic influence [27].

Although there are some additional factors that predict risk of progression for established disease, such as subchondral bone trabecular texture and knee malalignment [7], age, generalized OA, and vitamin D deficiency [28], to name a few, to date only a few factors, such as joint injury, obesity, minor radiographic degenerative features [28, 29] and synovial inflammation [30], have been identified as predictors of incident disease. There are no biochemical biomarkers yet qualified as indicators of the pre-radiographic molecular stage of the disease although there are a few promising biomarkers under study. A prior candidate biomarker study in the Chingford cohort yielded two serum biomarkers, COMP (higher) and aggrecan (lower), that predicted in advance, by as much as 10 years, incident radiographic knee OA [31]. Plasma chemokine ligand 3 (CCL3) is reported to detect pre-radiographic OA and correlate with the severity of knee damage in OA [32]. An additional recent study identifies CCL2 as a promising marker for predicting subsequent radiographic OA [33]. These results provide hope that in future it may be possible to identify the early phases of the disease with molecular profiling provided by *in vitro* and *in vivo* biomarkers.

Identifying the molecular stage of osteoarthritis

Based on studies of established OA, biomarkers with particular utility for monitoring OA onset and disease process are macromolecules originating from joint structures whose levels in serum, urine, and synovial fluid reflect processes taking place locally in the joint (see comprehensive review by van Spil) [34]. Direct evidence for a molecular stage of OA comes from only a few studies to date. In addition to the aforementioned studies, one study demonstrated that four serum proteins (matrix metalloproteinase (MMP)-7, interleukin (IL)-15, plasminogen activator inhibitor (PAI)-1 and soluble vascular adhesion protein (sVAP)-1) were already different in samples obtained ten years before the appearance of incident radiographic knee or hand OA [35]; of note, IL-15, MMP-7 and VAP-1 increased and PAI-I decreased in incident OA cases relative to controls. Another study demonstrated that systemic concentrations of joint tissue components, COMP and Hyaluronan (HA), predicted, up to 7 years later, the occurrence of incident knee joint space narrowing (by both COMP and HA) and osteophyte (by COMP) [36]. Based on recent evidence for differences in molecular constituents in different cartilages throughout the body [37, 38], it may even be possible to predict and monitor early OA in different joints using joint-specific systemic biomarkers. Potential proof of this concept comes from analyses of 450 participants from the Johnston County Osteoarthritis Project wherein serum COMP was been shown to be associated with radiographic knee OA severity, whereas deamidated serum COMP, abundant in adult

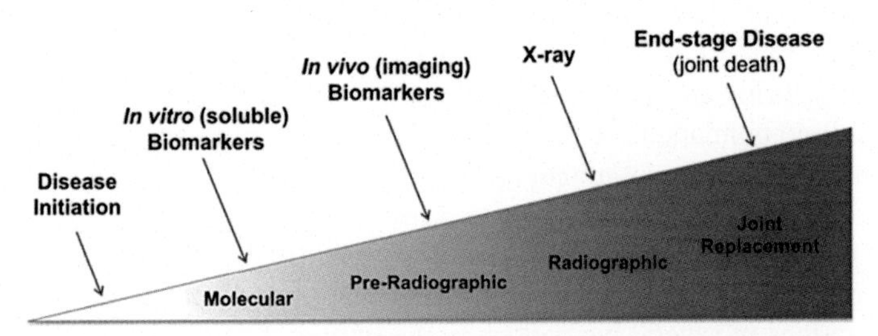

FIG. 1 The natural history of osteoarthritis (OA)—a new paradigm. The triangle denotes the progression of disease from a clinically undetectable but serologically detectable molecular stage to the clinically detectable pre-radiographic and radiographic stages. The pre-radiographic stage refers to sensitive detection of disease features by magnetic resonance imaging or ultrasound or other sensitive imaging modality. The radiographic stage refers to the detection of disease by the traditional method of X-ray imaging. The uphill slant of the triangle depicts the increasing difficulty, as disease progresses, to repair or modify the accrued structural damage. This graphic portrays the increased likelihood of successfully instituting cost-effective personalized medicine approaches with the earliest possible diagnosis and intervention.

hip cartilage, was associated with radiographic hip OA severity [39]. Although these biomarkers could be considered prognostic of incident radiographic OA, alternatively they could be considered diagnostic of the molecular stages of the disease. Other studies using MRI have confirmed the existence of a symptomatic pre-radiographic stage of the disease [40]. Taken together, these studies demonstrate that degenerative changes in the articular cartilage occur long before radiological changes are observed. They provide data to support a new concept of the natural history of the OA disease process (Fig. 1 based on [41, 42]).

Joint injury as a paradigm for early osteoarthritis identification and treatment

As mentioned above, the joint injury paradigm provides a potential gateway to understanding the early stages of OA. Unlike primary idiopathic OA, OA arising from joint injury has a known date of onset, so the early events can be unequivocally dated and monitored. Biomarkers after joint injury can be monitored in the serum, urine, and, most proximally to the joint, through the joint (synovial) fluid. Longitudinal studies of the aftermath of severe joint injury have incontrovertibly established the existence of a prolonged molecular phase of the disease characterized by biomarker abnormalities [43–50]. After knee injury, cartilage degradation is favored over repair, with increased collagen cleavage [51–54]. Within the first month after joint injury in humans, elevations have been documented in synovial fluid concentrations of cartilage proteoglycan fragments [43, 53], metalloproteinases (MMP-3/stromelysin-1) [45, 48], and collagen fragments [53, 55], with documented elevation of collagen fragments for up to 7 years after injury [54]. Aggrecanase cleavage of the interglobular domain (IGD) of aggrecan at the 392Glu-393Ala bond that releases N-terminal 393ARGS fragments has been shown to be one of the early key events in arthritis and joint injuries [49]. Our recent work has suggested that synovial fluid biomarkers on the day of ACL reconstructive surgery, particularly high synovial fluid glycosaminoglycan, predicts cartilage degeneration by MRI starting at 6 months and continuing to 3 years after surgery [56]. The challenge will be to extend these discoveries from synovial fluid to the serum. In this regard, a recent study showed a significant correlation of serum and synovial fluid concentrations of several biomarkers in the setting of acute human knee injury [53], including serum CTX-I, NTX, osteocalcin, and MMP-3. Five years after joint injury, concentrations of aggrecan ARGS fragment in synovial fluid and serum were also shown to be correlated [50]. Serum and synovial fluid type II collagen epitope (C2C) were significantly correlated in knee injury patients 0 days and 7 years after injury [54]. These results suggest that it may be possible to monitor joint health and risk of early OA development with a select group of molecular signatures and systemic biomarkers. These results also suggest that biomarkers may facilitate rapid, rational and aggressive treatment responses in the individuals at high risk for progressive and severe disease.

We noted that the pattern of biomarker alterations observed after joint injury matches the pattern of cartilage components released from cartilage stimulated *in vitro* with pro-inflammatory cytokines [53] (Fig. 2). These early events are characterized by a loss of proteoglycan followed by loss of collagen, considered a critical and possibly irreversible step in a course of inexorable joint degeneration and functional decline. This is supported by a quantitative proteomic study showing sequential degradation of aggrecan followed by collagens released from human knee cartilage stimulated *in vitro* with injurious compression and cytokines [57]. This is also supported by the fact that elevations of cartilage components in the serum can persist over decades after joint injury [44–48]. The sustained increased release of cartilage macromolecular fragments after joint trauma is thought to be responsible for the frequent development, as high as 50%, of post-traumatic OA in patients with joint injuries [21]. These biomarker observations provide great hope that disease-modifying therapies are within reach since there are already many known pharmacologic agents with chondroprotective effects in the cartilage explant model and acute injury animal models of OA [53, 57, 58].

It is also likely that early OA biomarkers identified through the study of joint injury will inform strategies for identifying early spontaneous idiopathic OA. This is illustrated by studies evaluating guinea pig knee OA comparing and contrasting histology and biomarkers in joint fluid from spontaneous OA *versus* injury-induced (ACL transection) cartilage damage [59, 60]. Animals with ACL injury at 5.5 months of age had comparable cartilage damage, synovial fluid stromal cell-derived-factor (SDF-1) and matrix metalloproteinase-13

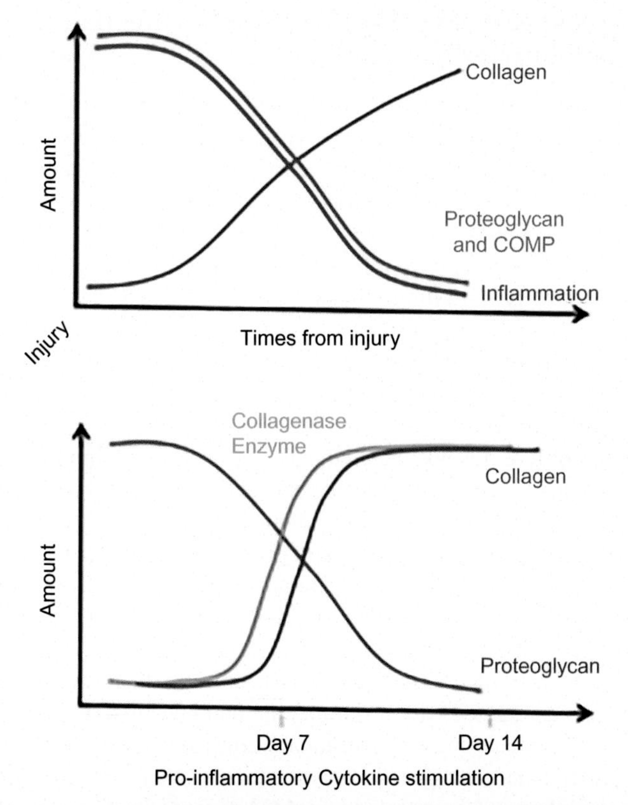

FIG. 2 The pattern of synovial fluid biomarkers after severe joint injury. During the first 4–6 weeks after a severe sports injury, the inflamed joint contains molecular evidence of the breakdown and turnover of the cartilage. The same pattern of cartilage breakdown and turnover can be achieved by adding proinflammatory cytokines to a cartilage specimen *in vitro*. There is an initial phase of loss of proteoglycan fragments followed by a phase of loss of collagen fragments, considered a probable irreversible event. (Graphic kindly provided by Dr. Jonathan Catterall and based on data reported in [53]).

(MMP-13) but more synovitis than animals with spontaneous OA at 12 months of age. By 12 months of age (9 months after injury) there was no qualitative difference in the biomarkers compared with animals with spontaneous OA at 12 months of age, just a quantitative difference (higher levels post-injury than in spontaneous OA) including higher synovial fluid SDF-1, collagen fragment C2C, proteoglycan fragments, MMP-13, and interleukin-1beta and lower lubricin. In another study comparing lipid peroxidation and antioxidant biomarkers in joint fluid from primary end-stage knee OA and post-injury cartilage damage, a number of biomarkers (thiobarbituric acid reactive substances, glutathione, glutathione peroxidase, and superoxide dismutase) were qualitatively and quantitatively similar in the two conditions [61]. This suggests that the two processes are, in some respects, qualitatively similar and findings in one could inform diagnostics and treatment strategies for the other.

Advantages of early arthritis identification illustrated by rheumatoid arthritis

Interestingly, a precedent now exists in rheumatoid arthritis (RA) for earlier diagnosis. A revision in 2010 of the criteria (formerly from 1987) for an RA diagnosis [62] were motivated by a desire to increase the likelihood of earlier diagnosis so that prompt disease-modifying treatment could be initiated [63] because it is now well recognized that delays in initiating therapy have a deleterious effect on long-term outcomes [64]. The presence of RA joint erosions on radiographs was omitted from the criteria because of the new emphasis on early diagnosis. This shift of focus from diagnosing and alleviating the consequences of established joint damage to early diagnosis and intervention presents a compelling paradigm that should inform the approach to the diagnosis and treatment of OA.

In the case of RA, it is postulated that the early period of immune dysregulation is potentially reversible and that the subsequent development of pannus and erosive joint damage represents a stage of auto-immune dysfunction that is less amenable to immunomodulatory therapy [64]. In RA, the persistence of synovitis and cumulative exposure to inflammation correlate with disease progression (summarized in [64]). We identified inflammasome activation and innate immunity as an indicator of severity and risk of progression of OA [20]. The systemic and local burden of pathogenic molecules, such as lipopolysaccharide (LPS), are also associated with inflammation and knee OA severity *via* the activation of macrophage-mediated inflammation [65]. We and others [66, 67] find that synovial macrophages are key effectors in OA and both associated with production of inflammatory cytokines and OA progression [68, 69]. These findings suggest that in OA, as in RA, cumulative exposure to inflammation may be a key factor in the severity and progression of OA and represent a key target for intervention.

In addition to genetic predisposition, RA as hypothesized for OA, also has an asymptomatic stage characterized by serological abnormalities; in RA this is characterized by autoantibody production and high C-reactive protein prior to onset of clinical manifestations of RA [70]. The appropriate pre-clinical markers have been posited to be potential triggers for pre-emptive treatments of 'at risk' individuals [64]. In early RA, high risk for progression has been associated with high baseline serum COMP, Glc-Gal-PyD, CTX-I and CTX-II, MMP-3, first year time-averaged Erythrocyte Sedimentation Rate (ESR), the serum osteoprotegerin:RANKL ratio, positive rheumatoid factor, anti-cyclic citrullinated peptide antibody and the HLA-DRB1 shared allele (reviewed in [71]).

Efforts are ongoing to identify predictors of response to particular therapies [72] to enable prompt treatment upon the first unequivocal clinical signs of disease. In the case of RA, this approach is anticipated to minimize joint destruction through early intervention, minimize adverse events through minimizing exposure of patients to ineffective treatments, and ensure more appropriate use of health care funds. For instance, low serum COMP, a measure of joint tissue degradation, predicts a more rapid and sustained clinical response to treatment in RA with an anti-TNF agent compared with high baseline serum COMP [73, 74] suggesting that serum COMP may reflect some irreversible aspects of joint deterioration in RA. In another RA study, serum MRP8/14 protein complexes, were consistently higher in responders to treatment with biologics (adalimumab, infliximab, and rituximab), independent of the specific mechanism of action [75]. Another study investigated circulating C-reactive protein (CRP) as a potential marker of early response to infliximab plus methotrexate (MTX) in 2 cohorts with RA [76]. Although CRP declined markedly at Week 2, patients who were EULAR non-responders at Week 14 showed a CRP rebound at Weeks 6 and 14. Efforts have also been made to predict the most appropriate biological therapy based on serum cytokine/chemokine/soluble receptor profiles in RA patients [77]. Based on analyses of thirty-one cytokines/chemokines/soluble receptors, baseline serum sgp130, IL-6, IL-8, Eotaxin, IP-10, VEGF, GM-CSF, sTNFR-I and -II predicted a better response to tocilizumab than etanercept over 16 weeks in 138 RA patients. Although less reliable than those for tocilizumab, a few possible biomarkers predicted better response to etanercept therapy were also identified. In RA, promising genetic and genomics markers predicting response to biological treatments have also been identified. Several RA susceptibility genes (HLA-DRB1, PTPRC, PTPN2, AFF3 and CD226), and gene loci (PDE3A–SLCO1C1, NUBPL, CNTN5, VAV1, SPRED2, PDZD2, EYA4, CENTD1, MAFB2, QK1, LASS6, IFNK, CST5, GBP6, PON1, LMO4 and MOBKL2B) are associated with response to anti-TNF therapy [78]. A recent large study ($n = 1239$ RA patients and 1229 healthy controls) showed that RA patients with the IFNGrs2069705C allele responded significantly better to anti-TNF drugs than patients carrying the wild-type allele ($P = 0.0075$) [79]. As a final example, various transforming growth factor β1 single nucleotide polymorphisms (SNPs) are associated with a good response to rituximab treatment [80].

Outlook for the future

A combination of biomarkers and sensitive imaging modalities improves the ability to diagnose and prognose established OA [1, 81], so it is also likely that combinatorial markers will provide the best discriminatory power for pre-radiographic disease. As noted by a working group on OA prevention and risk reduction [82], it will be necessary to discern when a specific value of the biomarker(s) truly represents a "disease-free" state, and to establish an understanding of the rapidity of the biomarker change (if treated as a continuous variable) or conversion (if treated as a discrete variable) in relation to the development of disease. Ideally, the biomarker will have been previously validated against a clinically relevant endpoint for its use as a surrogate measure [83].

Although the majority (95%) of drugs used are for chronic or complex diseases, there are currently no FDA- or EMA-approved pharmacological interventions to modify the OA disease process. This is in large part due to the dearth of sensitive *in vitro* or *in vivo* biomarkers suitable for monitoring efficacy of interventions. However, there are many OA-related biomarkers in various stages of qualification and many ways in which biomarkers may contribute to the development of drugs for OA [42]. An Osteoarthritis Research Society International (OARSI) working group on OA biomarkers has suggested that the biomarkers likely to have the earliest beneficial impact on clinical trials include those that would identify subjects likely to either respond and/or progress within a short time frame, and those that would provide early feedback for preclinical decision-making that a drug was having the desired biochemical effect [42]. However, the identification of novel early OA biomarkers remains challenging, even though the research in OA biomarkers has been steadily progressing in the recent years [84]. The Osteoarthritis Research Society International (OARSI) has sponsored efforts to facilitate and improve clinical trial methodology in OA through fostering an understanding of the qualification of biomarkers as drug development tools and the various contexts for which OA biomarkers may be used in clinical trials [42, 85] which has led to the initial qualification of imaging [86] and biochemical markers [87, 88] in OA for predicting clinically relevant progression (the combination of radiographic and symptomatic worsening). This is a major step forward for the future development and the clinical use of OA biomarkers.

The ultimate in personalized medicine is to make new organs from a patient's own cells. In contrast to the current dearth of pharmacological therapy to impact joint degeneration, an FDA-approved autologous tissue repair procedure for OA [http://www.fda.gov/BiologicsBloodVaccines/CellularGeneTherapyProducts/ApprovedProducts/ucm134025.htm] was introduced in 1997. Engineered tissue offers great promise for cartilage regeneration. Allogeneic chondrocytes derived from human juvenile articular cartilage offer an alternative and may represent an improved source of cells for cartilage repair [89]. This rapidly-evolving area appears to be forming the basis of a growing armamentarium of potential OA therapeutics. However, large random controlled and long-term follow-up studies are needed to confirm the safety and effectiveness of these regenerative therapies [90–92]. It is expected that both *in vivo* and *in vitro* biomarkers will be central to the monitoring of these personalized medicine approaches to the treatment of OA.

In summary (Table 1), biomarkers are central to personalized medicine as they establish the foundation for identifying early disease stages and the molecular pathophysiology to inform individualized therapeutic recommendations and decision-making. Early disease identification with biomarkers has great potential for stimulating cost-effective personalized medicine approaches for the treatment of OA with the hope of preventing the long-term disabling consequences. Perhaps one day, a "joint health panel" of blood tests will become part of an

TABLE 1 Summary of role of biomarkers for early Osteoarthritis (OA) identification

The most cost-effective application of personalized medicine in the chronic disease of OA is likely to be related to the identification, monitoring and treatment of early OA, *i.e.* in the pre-radiographic and molecular stages.
Rapid advances in the genetic and epigenetic knowledge of molecular pathways involved in OA susceptibility will inform and facilitate the development of biomarkers for early disease detection and monitoring.
Biomarkers indicative of joint tissue metabolism in the first several years after severe acute joint injury could provide indicators of biomarkers that could detect the early molecular stages of spontaneous OA.
A combinatorial biomarker will likely provide greatest discriminatory power for early OA detection.
The list of biomarkers for OA needs to expand beyond established radiographic OA to encompass pre-radiographic OA and molecular stages of OA, incidence and progression.

annual health screening, particularly for high-risk individuals with known genetic susceptibility and/or joint trauma history. As for other 'silent' diseases such as hypertension, osteoporosis and early heart disease [41], we may establish a means of detecting the disease prior to the stroke, fracture, or heart attack equivalent, namely, prior to the appearance of joint damage on an X-ray or MRI. We might thereby institute healthy lifestyle modifications, pharmacologic and/or biologic agents to slow or halt the disease process prior to loss of organ function.

References

[1] Garnero P. Use of biochemical markers to study and follow patients with osteoarthritis. Curr Rheumatol Rep 2006;8(1):37–44.

[2] Madelin G, Lee J-S, Inati S, Jerschow A, Regatte R. Sodium inversion recovery MRI of the knee joint in Vivo at 7T. J Magn Reson 2010; https://doi.org/10.1016/j.jmr.2010.08.003.

[3] Burstein D, Gray M, Mosher T, Dardzinski B. Measures of molecular composition and structure in osteoarthritis. Radiol Clin North Am 2009;47(4):675–86.

[4] Omoumi P, Mercier GA, Lecouvet F, Simoni P, Vande Berg BC. CT arthrography, MR arthrography, PET, and scintigraphy in osteoarthritis. Radiol Clin North Am 2009;47(4):595–615.

[5] Thawait GK, Demehri S, Almuhit A, Zbijweski W, Yorkston J, Del Grande F, Zikria B, Carrino JA, Siewerdsen JH. Extremity cone-beam CT for evaluation of medial tibiofemoral osteoarthritis: Initial experience in imaging of the weight-bearing and non-weight-bearing knee. Eur J Radiol 2015;84(12):2564–70.

[6] Song IH, Althoff CE, Hermann KG, Scheel AK, Knetsch T, Schoenharting M, et al. Knee osteoarthritis. Efficacy of a new method of contrast-enhanced musculoskeletal ultrasonography in detection of synovitis in patients with knee osteoarthritis in comparison with magnetic resonance imaging. Ann Rheum Dis 2008;67(1):19–25.

[7] Kraus VB, Collins JE, Charles HC, Pieper CF, Whitley L, Losina E, et al. Predictive validity of radiographic trabecular bone texture in knee osteoarthritis: the osteoarthritis research society international/foundation for the national institutes of health osteoarthritis biomarkers consortium. Arthritis Rheumatol 2018;70(1):80–7.
[7a] Centers for Disease Control and Prevention. Prevalence of disabilities and associated health conditions among adults – United States, 1999. MMWR Morb Mortal Wkly Rep 2001;50(7):120–5.
[7b] Guccione AA, Felson DT, Anderson JJ, Anthony JM, Zhang Y, Wilson PW, et al. The effects of specific medical conditions on the functional limitations of elders in the Framingham Study. Am J Public Health 1994;84(3):351–8.
[8] Dieppe PA, Lohmander LS. Pathogenesis and management of pain in osteoarthritis. Lancet 2005;365(9463):965–73.
[9] Valdes A. Molecular pathogenesis and genetics of osteoarthritis: implications for personalized medicine. Pers Med 2010;7(1):49–63.
[10] Barbour KE, Helmick CG, Boring M, Brady TJ. Vital signs: prevalence of doctor-diagnosed arthritis and arthritis-attributable activity limitation—United States, 2013–2015. MMWR Morb Mortal Wkly Rep 2017;66(9):246–53.
[11] Murphy LB, Cisternas MG, Pasta DJ, Helmick CG, Yelin EH. Medical expenditures and earnings losses among US adults with arthritis in 2013. Arthritis Care Res (Hoboken) 2018;70(6):869–76.
[12] Goldring MB, Goldring SR. Articular cartilage and subchondral bone in the pathogenesis of osteoarthritis. Ann N Y Acad Sci 2010;1192(1):230–7.
[13] Herrero-Beaumont G, Roman-Blas JA, Castaneda S, Jimenez SA. Primary osteoarthritis no longer primary: three subsets with distinct etiological, clinical, and therapeutic characteristics. Semin Arthritis Rheum 2009;39(2):71–80.
[14] Bos SD, Slagboom PE, Meulenbelt I. New insights into osteoarthritis: early developmental features of an ageing-related disease. Curr Opin Rheumatol 2008;20(5):553–9.
[15] Valdes AM, Spector TD. The clinical relevance of genetic susceptibility to osteoarthritis. Best Pract Res Clin Rheumatol 2010;24(1):3–14.
[16] Jeon OH, Kim C, Laberge RM, Demaria M, Rathod S, Vasserot AP, et al. Local clearance of senescent cells attenuates the development of post-traumatic osteoarthritis and creates a pro-regenerative environment. Nat Med 2017;23(6):775–81.
[16a] Tachmazidou I, Hatzikotoulas K, Southam L, Esparza-Gordillo J, Haberland V, Zheng J, et al. Identification of new therapeutic targets for osteoarthritis through genome-wide analyses of UK Biobank data. Nat Genet 2019;51(2):230–6.
[17] Loughlin J. Osteoarthritis year 2010 in review: genetics. Osteoarthr Cartil 2011;19(4):342–5.
[18] Reynard LN. Analysis of genetics and DNA methylation in osteoarthritis: what have we learnt about the disease? Semin Cell Dev Biol 2017;62:57–66.
[19] Englund M, Paradowski PT, Lohmander LS. Association of radiographic hand osteoarthritis with radiographic knee osteoarthritis after meniscectomy. Arthritis Rheum 2004;50(2):469–75.
[20] Denoble AE, Huffman KM, Stabler TV, Kelly SJ, Hershfield MS, McDaniel GE, et al. Uric acid is a danger signal of increasing risk for osteoarthritis through inflammasome activation. Proc Natl Acad Sci U S A 2011;108(5):2088–93.
[21] Lohmander LS, Englund PM, Dahl LL, Roos EM. The long-term consequence of anterior cruciate ligament and meniscus injuries: osteoarthritis. Am J Sports Med 2007;35(10):1756–69.
[22] Roos EM. Joint injury causes knee osteoarthritis in young adults. Curr Opin Rheumatol 2005;17(2):195–200.

[23] Amano K, Huebner J, Tanaka M, Stabler T, Loback I, McCulloch C, et al. Synovial fluid profile at the time of anterior cruciate ligament reconstruction and its association with cartilage matrix composition 3 years after surgery. Am J Sports Med 2018;46(4):890–9.

[24] Brown TD, Johnston RC, Saltzman CL, Marsh JL, Buckwalter JA. Posttraumatic osteoarthritis: a first estimate of incidence, prevalence, and burden of disease. J Orthop Trauma 2006;20(10):739–44.

[25] Buckwalter JA, Saltzman C, Brown T. The impact of osteoarthritis: implications for research. Clin Orthop Relat Res 2004;427S(Oct):S6–15.

[26] Doherty M, Watt I, Dieppe P. Influence of primary generalised osteoarthritis on development of secondary osteoarthritis. Lancet 1983;2(8340):8–11.

[27] Michou L. Genetics of digital osteoarthritis. Joint Bone Spine; 2010.

[28] Cheung PP, Gossec L, Dougados M. What are the best markers for disease progression in osteoarthritis (OA)? Best Pract Res Clin Rheumatol 2010;24(1):81–92.

[29] Kerkhof HJM, Bierma-Zeinstra SMA, Arden NK, Metrustry S, Castano-Betancourt M, Hart DJ, et al. Prediction model for knee osteoarthritis incidence, including clinical, genetic and biochemical risk factors. Ann Rheum Dis 2014;73(12):2116–21.

[30] Atukorala I, Kwoh CK, Guermazi A, Roemer FW, Boudreau RM, Hannon MJ, et al. Synovitis in knee osteoarthritis: a precursor of disease? Ann Rheum Dis 2016;75(2):390–5.

[31] Blumenfeld O, Williams FM, Hart DJ, Spector TD, Arden N, Livshits G. Association between cartilage and bone biomarkers and incidence of radiographic knee osteoarthritis (RKOA) in UK females: a prospective study. Osteoarthr Cartil 2013;21(7):923–9.

[32] Zhao XY, Yang ZB, Zhang ZJ, Zhang ZQ, Kang Y, Huang GX, et al. CCL3 serves as a potential plasma biomarker in knee degeneration (osteoarthritis). Osteoarthr Cartil 2015;23(8):1405–11.

[33] Longobardi L, Jordan JM, Shi XA, Renner JB, Schwartz TA, Nelson AE, et al. Associations between the chemokine biomarker CCL2 and knee osteoarthritis outcomes: the Johnston County osteoarthritis project. Osteoarthr Cartil 2018;.

[34] van Spil WE, Degroot J, Lems WF, Oostveen JC, Lafeber FP. Serum and urinary biochemical markers for knee and hip-osteoarthritis: a systematic review applying the consensus BIPED criteria. Osteoarthr Cartil 2010;18(5):605–12.

[35] Ling SM, Patel DD, Garnero P, Zhan M, Vaduganathan M, Muller D, et al. Serum protein signatures detect early radiographic osteoarthritis. Osteoarthr Cartil 2009;17(1):43–8.

[36] Golightly Y, Marshall S, Kraus V, Renner J, Villaveces A, Casteel C, et al. Serum cartilage oligomeric matrix protein, hyaluronan, high-sensitivity C-reactive protein, and keratan sulfate as predictors of incident radiographic knee osteoarthritis: differences by chronic knee symptoms. Osteoarthr Cartil 2010;18:S62–3. Suppl. 2.

[37] Onnerfjord P, Khabut A, Reinholt FP, Svensson O, Heinegard D. Quantitative proteomic analysis of eight cartilaginous tissues reveals characteristic differences as well as similarities between subgroups. J Biol Chem 2012;287(23):18913–24.

[38] Hsueh MF, Kraus VB, Onnerfjord P. Cartilage matrix remodelling differs by disease state and joint type. Eur Cell Mater 2017;34:70–82.

[39] Catterall JB, Hsueh MF, Stabler TV, McCudden CR, Bolognesi M, Zura R, et al. Protein modification by deamidation indicates variations in joint extracellular matrix turnover. J Biol Chem 2012;287(7):4640–51.

[40] Cibere J, Zhang H, Garnero P, Poole AR, Lobanok T, Saxne T, et al. Association of biomarkers with pre-radiographically defined and radiographically defined knee osteoarthritis in a population-based study. Arthritis Rheum 2009;60(5):1372–80.

[41] Kraus V. Waiting for action on the osteoarthritis front. Curr Drug Targets 2010;11(4):1–2.

[42] Kraus VB, Burnett B, Coindreau J, Cottrell S, Eyre D, Gendreau M, et al. Application of biomarkers in the development of drugs intended for the treatment of osteoarthritis. Osteoarthr Cartil 2011;19(5):515–42.

[43] Lohmander LS, Dahlberg L, Ryd L, Heinegard D. Increased levels of proteoglycan fragments in knee joint fluid after injury. Arthritis Rheum 1989;32(11):1434–42.

[44] Lohmander LS. The release of aggrecan fragments into synovial fluid after joint injury and in osteoarthritis. J Rheumatol Suppl 1995;43:75–7.

[45] Lohmander LS, Hoerrner LA, Dahlberg L, Roos H, Bjornsson S, Lark MW. Stromelysin, tissue inhibitor of metalloproteinases and proteoglycan fragments in human knee joint fluid after injury. J Rheumatol 1993;20(8):1362–8.

[46] Lohmander LS, Hoerrner LA, Lark MW. Metalloproteinases, tissue inhibitor, and proteoglycan fragments in knee synovial fluid in human osteoarthritis. Arthritis Rheum 1993;36(2):181–9.

[47] Lohmander LS, Neame PJ, Sandy JD. The structure of aggrecan fragments in human synovial fluid. Evidence that aggrecanase mediates cartilage degradation in inflammatory joint disease, joint injury, and osteoarthritis. Arthritis Rheum 1993;36(9):1214–22.

[48] Lohmander LS, Roos H, Dahlberg L, Hoerrner LA, Lark MW. Temporal patterns of stromelysin-1, tissue inhibitor, and proteoglycan fragments in human knee joint fluid after injury to the cruciate ligament or meniscus. J Orthop Res 1994;12(1):21–8.

[49] Larsson S, Lohmander LS, Struglics A. Synovial fluid level of aggrecan ARGS fragments is a more sensitive marker of joint disease than glycosaminoglycan or aggrecan levels: a cross-sectional study. Arthritis Res Ther 2009;11(3):R92.

[50] Struglics A, Larsson S, Kumahashi N, Frobell R, Lohmander LS. Changes in cytokines and aggrecan ARGS neoepitope in synovial fluid and serum and in C-terminal crosslinking telopeptide of Type II collagen and N-terminal crosslinking telopeptide of Type I collagen in urine over five years after anterior cruciate ligament rupture: an exploratory analysis in the knee anterior cruciate ligament, nonsurgical versus surgical treatment trial. Arthritis Rheumatol 2015;67(7):1816–25.

[51] Lohmander LS, Atley LM, Pietka TA, Eyre DR. The release of crosslinked peptides from type II collagen into human synovial fluid is increased soon after joint injury and in osteoarthritis. Arthritis Rheum 2003;48(11):3130–9.

[52] Aurich M, Squires GR, Reiner A, Mollenhauer JA, Kuettner KE, Poole AR, et al. Differential matrix degradation and turnover in early cartilage lesions of human knee and ankle joints. Arthritis Rheum 2005;52(1):112–9.

[53] Catterall JB, Stabler TV, Flannery CR, Kraus VB. Changes in serum and synovial fluid biomarkers after acute injury (NCT00332254). Arthritis Res Ther 2010;12(6):R229.

[54] Kumahashi N, Sward P, Larsson S, Lohmander LS, Frobell R, Struglics A. Type II collagen C2C epitope in human synovial fluid and serum after knee injury—associations with molecular and structural markers of injury. Osteoarthr Cartil 2015;23(9):1506–12.

[55] Lattermann C, Jacobs CA, Proffitt Bunnell M, Huston LJ, Gammon LG, Johnson DL, et al. A multicenter study of early anti-inflammatory treatment in patients with acute anterior cruciate ligament tear. Am J Sports Med 2017;45(2):325–33.

[56] Amano K, Pedoia V, Su F, Souza RB, Li X, Ma CB. Persistent biomechanical alterations after ACL reconstruction are associated with early cartilage matrix changes detected by quantitative MR. Orthop J Sports Med 2016;4(4):2325967116644421.

[57] Wang Y, Li Y, Khabut A, Chubinskaya S, Grodzinsky AJ, Önnerfjord P. Quantitative proteomics analysis of cartilage response to mechanical injury and cytokine treatment. Matrix Biol 2016;63:11–22.

[58] Lotz MK, Kraus VB. New developments in osteoarthritis. Posttraumatic osteoarthritis: pathogenesis and pharmacological treatment options. Arthritis Res Ther 2010;12(3):211.

[59] Wei L, Fleming BC, Sun X, Teeple E, Wu W, Jay GD, et al. Comparison of differential biomarkers of osteoarthritis with and without posttraumatic injury in the Hartley guinea pig model. J Orthop Res 2010;28(7):900–6.

[60] Thomas NP, Wu WJ, Fleming BC, Wei F, Chen Q, Wei L. Synovial inflammation plays a greater role in post-traumatic osteoarthritis compared to idiopathic osteoarthritis in the Hartley guinea pig knee. BMC Musculoskelet Disord 2017;18(1):556.

[61] Sutipornpalangkul W, Morales NP, Charoencholvanich K, Harnroongroj T. Lipid peroxidation, glutathione, vitamin E, and antioxidant enzymes in synovial fluid from patients with osteoarthritis. Int J Rheum Dis 2009;12(4):324–8.

[62] Bykerk VP, Massarotti EM. The new ACR/EULAR classification criteria for RA: how are the new criteria performing in the clinic? Rheumatology (Oxford) 2012;51(Suppl 6):vi10–5.

[63] Aletaha D, Neogi T, Silman AJ, Funovits J, Felson DT, Bingham 3rd CO, et al. 2010 rheumatoid arthritis classification criteria: an American College of Rheumatology/European League Against Rheumatism collaborative initiative. Ann Rheum Dis 2010;69(9):1580–8.

[64] Dale J, Porter D. Pharmacotherapy: concepts of pathogenesis and emerging treatments. Optimising the strategy of care in early rheumatoid arthritis. Best Pract Res Clin Rheumatol 2010;24(4):443–55.

[65] Huang ZY, Stabler T, Pei FX, Kraus VB. Both systemic and local lipopolysaccharide (LPS) burden are associated with knee OA severity and inflammation. Osteoarthr Cartil 2016;24(10):1769–75.

[66] Manferdini C, Paolella F, Gabusi E, Silvestri Y, Gambari L, Cattini L, et al. From osteoarthritic synovium to synovial-derived cells characterization: synovial macrophages are key effector cells. Arthritis Res Ther 2015;18:83.

[67] Kraus VB, McDaniel G, Huebner JL, Stabler TV, Pieper CF, Shipes SW, et al. Direct in vivo evidence of activated macrophages in human osteoarthritis. Osteoarthr Cartil 2016;24(9):1613–21.

[68] Daghestani HN, Pieper CF, Kraus VB. Soluble macrophage biomarkers indicate inflammatory phenotypes in patients with knee osteoarthritis. Arthritis Rheumatol 2015;67(4):956–65.

[69] Hsueh MF, Lu Y, Wellman SS, Bolognesi MP, Kraus VB. Functional folate receptor cell-associated inflammatory cytokines predict the progression of knee osteoarthritis. Osteoarthr Cartil 2018;26:S121–2.

[70] Deane KD, Norris JM, Holers VM. Preclinical rheumatoid arthritis: identification, evaluation, and future directions for investigation. Rheum Dis Clin North Am 2010;36(2):213–41.

[71] Landewe R. Predictive markers in rapidly progressing rheumatoid arthritis. J Rheumatol Suppl 2007;80:8–15.

[72] Marotte H, Miossec P. Biomarkers for prediction of TNFalpha blockers response in rheumatoid arthritis. Joint Bone Spine 2010;77(4):297–305.

[73] Morozzi G, Fabbroni M, Bellisai F, Cucini S, Simpatico A, Galeazzi M. Low serum level of COMP, a cartilage turnover marker, predicts rapid and high ACR70 response to adalimumab therapy in rheumatoid arthritis. Clin Rheumatol 2007;26(8):1335–8.

[74] Morozzi G, Fabbroni M, Bellisai F, Pucci G, Galeazzi M. Cartilage oligomeric matrix protein level in rheumatic diseases: potential use as a marker for measuring articular cartilage damage and/or the therapeutic efficacy of treatments. Ann N Y Acad Sci 2007;1108:398–407.

[75] Choi IY, Gerlag DM, Herenius MJ, Thurlings RM, Wijbrandts CA, Foell D, et al. MRP8/14 serum levels as a strong predictor of response to biological treatments in patients with rheumatoid arthritis. Ann Rheum Dis 2015;74(3):499–505.

[76] Westhovens R, van Vollenhoven RF, Boumpas DT, Brzosko M, Svensson K, Bjorneboe O, et al. The early clinical course of infliximab treatment in rheumatoid arthritis: results from the REMARK observational study. Clin Exp Rheumatol 2014;32(3):315–23.

[77] Uno K, Yoshizaki K, Iwahashi M, Yamana J, Yamana S, Tanigawa M, et al. Pretreatment prediction of individual rheumatoid arthritis patients' response to anti-cytokine therapy using serum cytokine/chemokine/soluble receptor biomarkers. PLoS One 2015;10(7):e0132055.

[78] Goulielmos GN, Zervou MI, Myrthianou E, Burska A, Niewold TB, Ponchel F. Genetic data: the new challenge of personalized medicine, insights for rheumatoid arthritis patients. Gene 2016;583(2):90–101.

[79] Canet LM, Caliz R, Lupianez CB, Canhao H, Martinez M, Escudero A, et al. Genetic variants within immune-modulating genes influence the risk of developing rheumatoid arthritis and anti-TNF drug response: a two-stage case-control study. Pharmacogenet Genomics 2015;25(9):432–43.

[80] Daien CI, Fabre S, Rittore C, Soler S, Daien V, Tejedor G, et al. TGF beta1 polymorphisms are candidate predictors of the clinical response to rituximab in rheumatoid arthritis. Joint Bone Spine 2012;79(5):471–5.

[81] Dam EB, Loog M, Christiansen C, Byrjalsen I, Folkesson J, Nielsen M, et al. Identification of progressors in osteoarthritis by combining biochemical and MRI-based markers. Arthritis Res Ther 2009;11(4):R115.

[82] Jordan JM, Sowers MF, Messier SP, Bradley J, Arangio G, Katz JN, et al. Methodologic issues in clinical trials for prevention or risk reduction in osteoarthritis. Osteoarthr Cartil 2011;19(5):500–8.

[83] Bauer DC, Hunter DJ, Abramson SB, Attur M, Corr M, Felson D, et al. Classification of osteoarthritis biomarkers: a proposed approach. Osteoarthr Cartil 2006;14(8):723–7.

[84] Mobasheri A, Bay-Jensen AC, van Spil WE, Larkin J, Levesque MC. Osteoarthritis year in review 2016: biomarkers (biochemical markers). Osteoarthr Cartil 2017;25(2):199–208.

[85] Kraus VB, Blanco FJ, Englund M, Henrotin Y, Lohmander LS, Losina E, et al. OARSI clinical trials recommendations: soluble biomarker assessments in clinical trials in osteoarthritis. Osteoarthr Cartil 2015;23(5):686–97.

[86] Collins JE, Losina E, Nevitt MC, Roemer FW, Guermazi A, Lynch JA, et al. Semiquantitative imaging biomarkers of knee osteoarthritis progression: data from the foundation for the National Institutes of Health Osteoarthritis Biomarkers Consortium. Arthritis Rheumatol (Hoboken, NJ) 2016;68(10):2422–31.

[87] Kraus VB, Hargrove DE, Hunter DJ, Renner JB, Jordan JM. Establishment of reference intervals for osteoarthritis-related soluble biomarkers: the FNIH/OARSI OA biomarkers consortium. Ann Rheum Dis 2017;76(1):179–85.

[88] Kraus VB, Collins JE, Hargrove D, Losina E, Nevitt M, Katz JN, et al. Predictive validity of biochemical biomarkers in knee osteoarthritis: data from the FNIH OA biomarkers consortium. Ann Rheum Dis 2017;76(1):186–95.

[89] HDt A, Martin JA, Amendola RL, Milliman C, Mauch KA, Katwal AB, et al. The potential of human allogeneic juvenile chondrocytes for restoration of articular cartilage. Am J Sports Med 2010;38(7):1324–33.

[90] Nixon AJ, Goodrich LR, Scimeca MS, Witte TH, Schnabel LV, Watts AE, et al. Gene therapy in musculoskeletal repair. Ann N Y Acad Sci 2007;1117:310–27.

[91] Evans CH, Ghivizzani SC, Herndon JH, Robbins PD. Gene therapy for the treatment of musculoskeletal diseases. J Am Acad Orthop Surg 2005;13(4):230–42.

[92] Zhang W, Ouyang H, Dass CR, Xu J. Current research on pharmacologic and regenerative therapies for osteoarthritis. Bone Res 2016;4:15040.

Chapter 23

The microbiota and infectious diseases

Neeraj K. Surana
Departments of Pediatrics, Molecular Genetics and Microbiology, and Immunology, Duke University, Durham, NC, United States

Since the beginning of humankind, scholars have been investigating the underpinnings of disease with an almost singular focus on the human side of the equation. Microbes were not recognized as an important cause of disease until the inception of the "germ theory" in the late nineteenth century. During the first century of medical microbiology, research largely centered on the role of microbes as pathogens. Only recently has there been a resurgence of interest in understanding how commensal organisms—the bacteria, viruses, fungi, and Archaea that make up the *microbiota*—impact human physiology. The idea that these microorganisms are vital to the well-being of humans has challenged our traditional notions of "self." Indeed, a human being can most accurately be described as a *holobiont*: a complex assemblage of human cells and microorganisms interacting in an elaborate *pas de deux* that drives normal physiologic processes.

Aimed at a better understanding of this relationship, myriad studies during the past decade have begun to catalogue the microbiota at various body sites and in a multitude of disease conditions [1–5]. Diseases in virtually every organ system have been associated with changes in the microbiota. Indeed, the microbiota has been linked to intestinal disorders, disturbances in metabolic function, autoimmune diseases, and psychiatric conditions and has been shown to influence susceptibility to infection and the efficacy of pharmaceutical therapies. Knowledge of the specific mechanism(s) underlying most of these microbe–disease associations is lacking; it remains unclear whether the disease-associated alterations in the microbiota represent mere biomarkers of disease, a causal relationship, or a combination of the two. Although cause-and-effect relationships are still being elucidated for many diseases, it is clear that humans coexist in an intricate relationship with commensal organisms. Although this host–commensal relationship impacts numerous diseases across all organ systems, this chapter will more specifically focus on some exemplar data as it relates to infectious diseases.

Genomic and Precision Medicine. https://doi.org/10.1016/B978-0-12-801496-7.00023-X

Overview of the microbiota

Initial estimates from the 1970s suggested that there were ~10-fold more bacteria in the body that there were human cells [6,7]. However, a more recent estimate has suggested that there are only ~1.3 times more bacteria in the body that there are human cells [8]. Of note, this more recent study does not include the number of viruses (~10-fold more abundant than other microbes), fungi, or Archaea. Including these additional classes of microbes likely makes correct the idea that microbes outnumber human cells by a 10:1 ratio. With regards to genetic potential, the human contribution is even smaller: in contrast to the ~20,000 genes in the human genome, there are estimated to be >2,000,000 genes in the microbiota—a 100:1 ratio in favor of microbes [1,9].

In terms of overall diversity, more than 10,000 different bacterial species are present in the collective human microbiota; the intestines alone contain more than 1000 species [9]. At any given time, the body of any given individual harbors 500–1000 bacterial species [10], with 100–200 bacterial species in the gut alone [9]. If one considers different strains of the same bacterial species, which may be functionally different from one another, the diversity of the microbiota is probably at least a magnitude greater [11].

A landmark study that cataloged the microbiome in multiple anatomic sites in healthy adults found that the composition of the microbiota differs by body site, there is tremendous inter-individual variation, and the microbial gene content is relatively conserved irrespective of the body site or individual [12]. In fact, the effect of the anatomic site on microbial composition is far greater than the effect of heterogeneity between individuals [i.e., all samples from a given body region (e.g., skin) were more similar to each other than they were to samples from a different body region (e.g., stool), even in the same individual] [12]. Taken together, these findings highlight the remarkable personalization of the human microbiome. While the human genome is typically >99.5% identical in different people [13], the microbiotas of two individuals may not overlap at all. Although the "precision medicine" approach currently focuses on teasing out how differences in the human genome relate to different clinical end points, the human microbiome clearly represents a critical component for consideration.

Determinants of the microbiota

Identification of the factors that influence the microbiota's composition is critical to an understanding of what leads to and controls intra- and inter-individual variation. Some studies have demonstrated that host genetics have a small but statistically significant effect on the microbiota's composition [14–16], but there is some controversy over this finding [17,18]. Moreover, the microbiota changes with age, with pronounced changes apparent particularly in early infancy and in the elderly [19,20]. In addition to these factors which are difficult to control, an individual's specific microbial configuration is quickly altered in response to subtle changes in the microenvironments in which the bacteria reside. On a day-to-day basis, these changes usually reflect alterations in the relative abundance of the various microbes. However, some exposures have a greater effect on the microbiota and can shift the microbial population to a new equilibrium via the loss of specific species and/or the acquisition of others; this new microbial equilibrium can be associated with either health or a disease state. Diet is perhaps one of the largest drivers of the intestinal microbiota as it provides nutrients needed not only by our own cells but also by the microbes living in the alimentary tract [21,22]. Other factors that can alter the microbiota in a dynamic manner include lifestyle choices (e.g., choice of household members, presence of pets, living in a rural or urban setting, country of residence) and circadian rhythms [1,23]. Furthermore, virtually all drugs have the capacity to change the microbiota by altering the chemical landscape in which the microorganisms live (e.g., statins, bile acid sequestrants), modulating the host's ability to recognize and react to microbes (e.g., immunosuppressants), and/or directly interfering with the microbiota's constituents (e.g., antibiotics) [24–27]. While it is clear that the microbiota can be altered through these various mechanisms, it is not yet clear whether these changes are biologically significant.

The role of the microbiota in modulating infectious diseases

The increased susceptibility of antibiotic-treated mice to infection with a wide range of enteric pathogens was initially observed in the 1950s and led soon thereafter to the concept of *colonization resistance*, which holds that the normal intestinal microbiota plays a critical role in preventing colonization—and therefore disease production—by invading pathogens [28,29]. Seminal work in the 1970s demonstrated that this protection is largely reliant on anaerobic organisms [29,30], and the subsequent half-century has been spent trying to identify the specific microbes involved. Although much of the work relating the microbiota to infection has focused on enteric pathogens, the intestinal microbiota has also been clearly linked to bacterial pneumonia in mouse models, and changes in the microbial composition of the gut have been causally related to changes in the severity of disease [31,32]. Although this gut–lung axis clearly exists in animals, its relevance in humans is still unclear. Several groups are beginning to study the human lung microbiome in the context of pneumonia and tuberculosis. Moreover, the relationships between the microbiota and both systemic infections (e.g., HIV infection, sepsis) and the response to vaccination are starting to be explored.

Enteric infections

Clostridium difficile infection (CDI) represents a growing worldwide epidemic and is the leading cause of antibiotic-associated diarrhea [33]. Roughly 15–30% of patients who are successfully treated for CDI end up with recurrent disease [34,35]. The strong association between antibiotic exposure and CDI initially raised the idea that the microbiota is inextricably linked to acquisition of disease, presumably because of the loss of colonization resistance [36]. Consistent with the epidemiologic data, characterization of the fecal microbiota of patients with CDI revealed that it is a markedly less diverse, dysbiotic community [37]. Fecal microbiota transplantation (FMT)—the "transplantation" of stool from a healthy individual into patients with disease—was successfully used in the 1950s to treat four patients with severe CDI and has recently been demonstrated in numerous studies to be an effective therapy for recurrent CDI, with clinical cure in $\geq$85% of patients [38–40]. Thus, FMT for recurrent CDI has become the "poster child" for the idea that microbiome-based therapies may transform the management of many diseases previously been considered to be refractory to medical therapy. Although FMT is agnostic as to the underlying mechanism of protection, work is ongoing to identify specific microbes and host pathways that can protect against CDI. Studying mice with differential susceptibilities to CDI due to antibiotic-induced changes in their microbiota, investigators identified a cocktail of four bacteria (*Clostridium scindens*, *Barnesiella intestihominis*, *Pseudoflavonifractor capillosus*, and *Blautia hansenii*) that conferred protection against CDI in a mouse model [41]. Intriguingly, treatment of mice with just *C. scindens* offered significant, though not complete, protection in a bile acid–dependent manner [41]. Clinical data from patients who underwent HSCT also associated *C. scindens* with protection from CDI, an observation that suggests the possibility of translating these findings from mice to humans [41]. This study provides an example of the identification of relevant bacterial factors through examination of microbial differences in populations that differ in disease risk.

In contrast to CDI where a loss of microbial diversity typically precedes infection, microbiome-related changes associated with *Vibrio cholerae* infection include a striking loss of diversity (largely due to *V. cholerae*'s becoming the dominant member of the microbiota) and an altered composition that rapidly follows the onset of disease [42]. These changes, which occur in a reproducible and stereotypical manner, are reversible with treatment of the disease [42]. This recovery phase involves a microbial succession that is similar to the assembly and maturation of the microbiota of healthy infants [43]. In addition to *V. cholerae*, streptococcal and fusobacterial species bloom during the early phases of diarrhea, and the relative abundances of *Bacteroides*, *Prevotella*, *Ruminococcus/Blautia*, and *Faecalibacterium* species increase during the resolution phase and mark the return to a healthy adult microbiota. Analysis of these microbial changes occurring in patients with cholera and in healthy children led to the selection of 14 bacteria that were transplanted into gnotobiotic mice, which were then challenged with *V. cholerae*. Bioinformatic analysis of specific taxa changing during cholera determined that *Ruminococcus obeum* restrained *V. cholerae* growth [42]. Subsequently, this relationship was experimentally confirmed, and the *R. obeum* quorum-sensing molecule AI-2 (autoinducer 2) was found to be responsible for restricting *V. cholerae* colonization via an unclear mechanism [42].

These studies highlight the potential for use of microbiome-based therapies to prevent and/or treat infectious diseases. Moreover, they suggest that temporal analysis of longitudinal microbiome data may be an effective strategy for identifying microbes with causal relationships to disease.

HIV infection

The augmentation of HIV pathogenesis by some viral, bacterial, and parasitic co-infections suggests that a patient's underlying microbial environment can influence the severity of HIV disease [44]. Moreover, it has been hypothesized that the intestinal immune system plays a significant role in regulating HIV-induced immune activation; this seems particularly likely since the intestines are an early site for viral replication and exhibit immune defects before peripheral CD4+ T cell counts decrease [45]. Several studies have examined the intestinal microbiotas of HIV-infected individuals. Initial studies performed in nonhuman primates infected with simian immunodeficiency virus found no alteration in the bacterial components of the fecal microbiota; however, there were profound changes in the enteric virome [46]. In contrast, many recent studies exploring this issue in human patients have identified substantial differences in the HIV-associated fecal microbiota that correlate with systemic markers of inflammation [47–49]. Curiously, these microbial changes do not necessarily normalize with antiretroviral therapy, a finding that suggests the microbiota may have some "memory" of the previously high HIV loads and/or that HIV infection helps reset the "normal" microbiota. This memory-like capacity of the microbiota has been demonstrated in animal models in the context of other infections and in response to dieting [21,50–52].

Given that the majority of new HIV transmission events follow heterosexual intercourse, there has been significant interest in examining the relationship between the vaginal microbiota and HIV acquisition. A longitudinal study of South African adolescent girls who underwent high-frequency testing for incident HIV infection facilitated the identification of bacteria that were associated with reduced risk of HIV acquisition (*Lactobacillus* species other than *L. iners*) or with enhanced risk (*Prevotella melaninogenica*, *Prevotella bivia*, *Veillonella montpellierensis*, *Mycoplasma*, and *Sneathia sanguinegens*) [53]. In mice inoculated intravaginally with *Lactobacillus crispatus* or *P. bivia*, the latter organism induced a greater number of activated CD4+ T cells in the female genital tract, suggesting that the increased risk of HIV acquisition associated with *P. bivia* may be secondary to the increased presence of target cells [53]. In a separate study, the composition of the vaginal microbiota was shown to modulate the antiviral efficacy of a tenofovir gel microbicide. Although tenofovir reduced HIV acquisition by 61% in women who had a *Lactobacillus*-dominant vaginal microbiota, it reduced HIV acquisition by only 18% in women whose vaginal microbiota comprised primarily *Gardnerella vaginalis* and other anaerobes [54]. This difference in efficacy was due to the ability of *G. vaginalis* to metabolize tenofovir faster than the target cells can take up the drug and convert it into its active form, tenofovir diphosphate. These findings, the fundamental basis of which have been independently replicated [55], illustrate how microbial ecology can be an important consideration in choosing effective treatment regimens.

Response to vaccination

Second only to the provision of clean water, vaccination has been the most effective public health intervention in the prevention of serious infectious diseases [56]. Its effects are mediated by antigen-specific antibodies and, in some cases, effector T cell responses. Although vaccines are clearly effective on a population scale, the magnitude of the immune response to vaccines can vary among individuals by 10-fold to 100-fold [57–59]. Although many factors (e.g., genetics, maternal antibody levels, prior antigen exposures) can affect vaccine immunogenicity, the microbiota is now recognized as another important factor [60–63]. Analysis of the fecal microbiotas of 48 Bangladeshi children identified specific taxa that exhibited positive associations (e.g., *Actinomyces*, *Rothia*, and *Bifidobacterium* species) and negative associations (e.g., *Acinetobacter*, *Prevotella*, and staphylococcal species) with responses to vaccines against polio, tuberculosis (bacille Calmette-Guérin), tetanus, and hepatitis B [64]. A study of infants from Ghana revealed an inverse relationship between the fecal abundance of Bacteroidetes and a response to the rotavirus vaccine [65]. Moreover, the nasal microbiota has been implicated as a factor that contributes to the IgA response to live, attenuated influenza vaccines [66]. These correlations based on clinical data have been partially confirmed in animal studies. The best example is the demonstration that the responses to non-adjuvanted viral subunit vaccines (inactivated influenza and polio vaccines) are reliant on the microbiota, whereas the responses to live or adjuvanted vaccines (live attenuated yellow fever, Tdap/alum, an HIV envelope protein/alum vaccine) are not [63,67]. An interesting note is that the antibody response to inactivated influenza vaccine may be dependent on recognition of the microbiota by Toll-like receptor 5, presumably via flagellin-expressing microbes [63], though this finding may result from an experimental artifact [67]. These data suggest that the microbiota may serve as an adjuvant for certain vaccine types. Confirmation of these findings in clinical settings may suggest ways to improve vaccine efficacy in the future.

Translating microbiome science to the clinic

The numerous microbiome–disease associations identified thus far have generated a great deal of hope that understanding the relevant microbe–host interactions will open the door to unlimited therapeutic applications. Microbiome-based therapies offer several potential benefits. Patients often view such treatment as more "natural" than conventional drug therapy and are therefore more likely to comply with it. Biologically, microbiome-based therapies are more likely to address one of the root causes of disease (microbial dysbiosis) rather than simply affecting the downstream sequelae. Finally, a given microbiome-based therapy may serve as a "polypill" that is effective against several different diseases stemming from similar microbial changes. Despite tremendous interest in therapeutically exploiting the microbiome, there have thus far been few clinical successes along these lines.

The most successful therapeutic application of microbiome science has been the use of FMT, particularly for CDI. As mentioned earlier, FMT involves "transplanting" stool from a healthy individual to a diseased patient, with the idea that the "healthy" microbiota will correct whatever derangement may exist in the ill patient and therefore will alleviate symptoms. Fundamentally, this notion is agnostic as to the specific microbial dysbiosis and holds that any healthy microbiota will be curative. The idea of FMT dates back to at least the fourth century, when traditional Chinese doctors used a "yellow soup" (fresh human fecal suspension) to successfully treat food poisoning and severe diarrhea [68]. The continued use of FMT through the centuries for the treatment of diarrheal illnesses in both humans and animals, along with the growing appreciation in recent years of the importance of the microbiota, laid the groundwork for using FMT to treat CDI. Since the first major prospective trial assessing FMT for recurrent CDI in 2013 [40], most of the numerous studies of FMT for CDI have demonstrated remarkable efficacy, with an average clinical cure rate of ≥85% [69,70]. The donor stool can be fresh or frozen (use of the latter allows biobanking of samples from a limited number of pre-screened donors) and can be administered via nasogastric tube, nasoduodenal tube, colonoscopy, enema, or oral capsules; the cure rate is slightly higher with lower-gastrointestinal administration than with upper-gastrointestinal treatment [70]. The optimal screening, preparation, and concentration of infused donor stool have not yet been determined. The most common adverse effects of FMT include altered gastrointestinal motility (with constipation or diarrhea), abdominal cramps, and bloating, all of which are generally transient and resolve within 48 h [69,70]. At least 80 immunosuppressed patients have undergone FMT with no serious adverse events noted during 3 months of follow-up [70].

The successful use and the favorable short-term safety profile of FMT for CDI have led to its expanded application for other indications. At the end of 2018, 195 active trials (listed at ClinicalTrials.gov) were investigating the efficacy of FMT for a range of diseases, including CDI, IBD (ulcerative colitis and Crohn's disease), obesity, eradication of multidrug-resistant organisms, anxiety and depression, cirrhosis, and type 2 diabetes. The few published studies regarding indications other than CDI have generally included small sample sizes and have offered mixed results [71–73]. In contrast to the successes in CDI, the results have been more varied for patients with IBD [74], which is perhaps the second-best-studied indication. It is not clear whether these discrepancies are due to heterogeneity in recipients (e.g., in terms of underlying disease mechanisms or endogenous microbiotas), the donor material, and/or the logistical details of FMT administration (e.g., route, frequency, dose).

Although FMT offers an important proof of concept that microbiome-based therapies can be effective, treatment is difficult to standardize across large populations because of variability among stool donors and among the endogenous microbiotas of recipients. In addition, FMT is fraught with safety concerns, and its mechanism(s) of action are unclear. FMT likely represents the first generation of microbiome-based therapies; subsequent generations will include the use of more refined bacterial cocktails, single strains of bacteria, or bacterial metabolites as the therapeutic intervention. The field of probiotics has a complicated history: many different strains have been tested against a multitude of diseases. Several meta-analyses have combined results across bacterial strains and/or disease indications and have generally concluded that the data are not yet convincing enough to support the use of the tested regimens [74–77]. It should be noted that the tested organisms have been chosen mainly on the basis of their presumed safety profile rather than in light of a plausible biological link to disease. The hope is that more focused, mechanistic microbiome studies will identify specific commensal organisms—and their underlying mechanisms of action—that are involved in disease pathogenesis and that will serve as the basis for the next wave of rationally chosen probiotics, a few of which are currently in clinical trials. The main hurdle in this endeavor has been identifying specific microbes that are causally related to protection from disease [78,79].

Parting thoughts

The medical view of microbes has changed radically, moving from the early-twentieth-century notion that we are engaged in a constant struggle with microbes—an "us-versus-them" mentality that focused on the necessity of eradicating bacteria—to the more recent understanding that we live in a carefully negotiated state of détente with our commensal organisms. Instead of holding a simple view of bacteria as enemies to be eliminated with antibiotics, scientists are increasingly recognizing the critical role these organisms play in maintaining human health. Loss of these host–microbe interactions in the increasingly sterile environment typical of Western civilization may have predisposed to the increased incidence of autoimmune and inflammatory diseases. The field of microbiome research has made great strides over the past decade in cataloguing the normal microbiota and is now on the cusp of being able to identify clinically actionable microbe–host relationships.

The recent explosion of "–omics" technologies (e.g., metagenomics, metatranscriptomics, metabolomics) has enabled the generation of vast amounts of data, but it is not yet clear how best to integrate datasets in order to gain useful insights into host–microbe relationships. The use of FMT has demonstrated that modulation of an individual's microbiota can effectively treat certain diseases; however, models with which to predict specifically how a microbiota will change after modulation—and what potentially untoward effects these changes might have—are still lacking. Implicit in this limitation is our ignorance about what microbial configuration is optimal and how a given microbiota should be rationally altered to obtain an ideal outcome.

Despite initial hyperbolic hype and a few false starts, microbiome research now stands at the precipice of an ability to treat the fundamental basis of many diseases. As the field continues to mature, it will need to move beyond correlations and address causation. The identification of causal microbes and their mechanisms of action will create a "microbial toolbox" from which relevant bioactive strains can be chosen on a per-patient basis to correct specific underlying microbial dysbiosis. In the near future, our knowledge base regarding the microbiome and its relationship to health and disease will be robust enough that it will be critical to incorporate information about the patient's microbiome when making treatment decisions.

References

[1] Gilbert JA, Blaser MJ, Caporaso JG, Jansson JK, Lynch SV, Knight R. Current understanding of the human microbiome. Nat Med 2018;24(4):392–400.

[2] Gilbert JA, Quinn RA, Debelius J, Xu ZZ, Morton J, Garg N, et al. Microbiome-wide association studies link dynamic microbial consortia to disease. Nature 2016;535(7610):94–103.

[3] Honda K, Littman DR. The microbiome in infectious disease and inflammation. Annu Rev Immunol 2012;30:759–95.

[4] Lynch SV, Pedersen O. The human intestinal microbiome in health and disease. N Engl J Med 2016;375(24):2369–79.

[5] Surana NK, Kasper DL. Deciphering the tete-a-tete between the microbiota and the immune system. J Clin Invest 2014;124(10):4197–203.

[6] Dobzhansky T. Genetics of the evolutionary process. New York: Columbia University Press; 1971.

[7] Luckey TD. Introduction to intestinal microecology. Am J Clin Nutr 1972;25(12):1292–4.

[8] Sender R, Fuchs S, Milo R. Are we really vastly outnumbered? revisiting the ratio of bacterial to host cells in humans. Cell 2016;164(3):337–40.

[9] Qin J, Li R, Raes J, Arumugam M, Burgdorf KS, Manichanh C, et al. A human gut microbial gene catalogue established by metagenomic sequencing. Nature 2010;464(7285):59–65.

[10] Turnbaugh PJ, Ley RE, Hamady M, Fraser-Liggett CM, Knight R, Gordon JI. The human microbiome project. Nature 2007;449(7164):804–10.

[11] Locey KJ, Lennon JT. Scaling laws predict global microbial diversity. Proc Natl Acad Sci U S A 2016;113(21):5970–5.

[12] Human Microbiome Project C. Structure, function and diversity of the healthy human microbiome. Nature 2012;486(7402):207–14.

[13] Genomes Project C, Auton A, Brooks LD, Durbin RM, Garrison EP, Kang HM, et al. A global reference for human genetic variation. Nature 2015;526(7571):68–74.

[14] Goodrich JK, Waters JL, Poole AC, Sutter JL, Koren O, Blekhman R, et al. Human genetics shape the gut microbiome. Cell 2014;159(4):789–99.

[15] Silverman M, Kua L, Tanca A, Pala M, Palomba A, Tanes C, et al. Protective major histocompatibility complex allele prevents type 1 diabetes by shaping the intestinal microbiota early in ontogeny. Proc Natl Acad Sci U S A 2017;114(36):9671–6.

[16] Kubinak JL, Stephens WZ, Soto R, Petersen C, Chiaro T, Gogokhia L, et al. MHC variation sculpts individualized microbial communities that control susceptibility to enteric infection. Nat Commun 2015;6:8642.

[17] Rothschild D, Weissbrod O, Barkan E, Kurilshikov A, Korem T, Zeevi D, et al. Environment dominates over host genetics in shaping human gut microbiota. Nature 2018;555(7695):210–5.

[18] Turnbaugh PJ, Hamady M, Yatsunenko T, Cantarel BL, Duncan A, Ley RE, et al. A core gut microbiome in obese and lean twins. Nature 2009;457(7228):480–4.

[19] Yatsunenko T, Rey FE, Manary MJ, Trehan I, Dominguez-Bello MG, Contreras M, et al. Human gut microbiome viewed across age and geography. Nature 2012;486(7402):222–7.

[20] O'Toole PW, Jeffery IB. Gut microbiota and aging. Science 2015;350(6265):1214–5.

[21] David LA, Maurice CF, Carmody RN, Gootenberg DB, Button JE, Wolfe BE, et al. Diet rapidly and reproducibly alters the human gut microbiome. Nature 2014;505(7484):559–63.

[22] Turnbaugh PJ, Ridaura VK, Faith JJ, Rey FE, Knight R, Gordon JI. The effect of diet on the human gut microbiome: a metagenomic analysis in humanized gnotobiotic mice. Sci Transl Med 2009;1(6):6ra14.

[23] Thaiss CA, Levy M, Korem T, Dohnalova L, Shapiro H, Jaitin DA, et al. Microbiota diurnal rhythmicity programs host transcriptome oscillations. Cell 2016;167(6):1495–510. [e12].

[24] Dethlefsen L, Huse S, Sogin ML, Relman DA. The pervasive effects of an antibiotic on the human gut microbiota, as revealed by deep 16S rRNA sequencing. PLoS Biol 2008;6(11):e280.

[25] Dethlefsen L, Relman DA. Incomplete recovery and individualized responses of the human distal gut microbiota to repeated antibiotic perturbation. Proc Natl Acad Sci U S A 2011;108(Suppl. 1):4554–61.

[26] Forslund K, Hildebrand F, Nielsen T, Falony G, Le Chatelier E, Sunagawa S, et al. Disentangling type 2 diabetes and metformin treatment signatures in the human gut microbiota. Nature 2015;528(7581):262–6.

[27] Maier L, Pruteanu M, Kuhn M, Zeller G, Telzerow A, Anderson EE, et al. Extensive impact of non-antibiotic drugs on human gut bacteria. Nature 2018;555(7698):623–8.

[28] Bohnhoff M, Drake BL, Miller CP. Effect of streptomycin on susceptibility of intestinal tract to experimental Salmonella infection. Proc Soc Exp Biol Med 1954;86(1):132–7.

[29] van der Waaij D, Berghuis-de Vries JM, Lekkerkerk L-v. Colonization resistance of the digestive tract in conventional and antibiotic-treated mice. J Hyg (Lond) 1971;69(3):405–11.

[30] van der Waaij D, Berghuis JM, Lekkerkerk JE. Colonization resistance of the digestive tract of mice during systemic antibiotic treatment. J Hyg (Lond) 1972;70(4):605–10.

[31] Gauguet S, D'Ortona S, Ahnger-Pier K, Duan B, Surana NK, Lu R, et al. Intestinal microbiota of mice influences resistance to Staphylococcus aureus pneumonia. Infect Immun 2015;83(10):4003–14.

[32] McAleer JP, Kolls JK. Contributions of the intestinal microbiome in lung immunity. Eur J Immunol 2018;48(1):39–49.

[33] McDonald LC, Gerding DN, Johnson S, Bakken JS, Carroll KC, Coffin SE, et al. Clinical practice guidelines for clostridium difficile infection in adults and children: 2017 update by the infectious diseases society of america (IDSA) and society for healthcare epidemiology of america (SHEA). Clin Infect Dis 2018;66(7):e1–48.

[34] Cornely OA, Crook DW, Esposito R, Poirier A, Somero MS, Weiss K, et al. Fidaxomicin versus vancomycin for infection with Clostridium difficile in Europe, Canada, and the USA: a double-blind, non-inferiority, randomised controlled trial. Lancet Infect Dis 2012;12(4):281–9.

[35] Louie TJ, Miller MA, Mullane KM, Weiss K, Lentnek A, Golan Y, et al. Fidaxomicin versus vancomycin for clostridium difficile infection. N Engl J Med 2011;364(5):422–31.

[36] Britton RA, Young VB. Role of the intestinal microbiota in resistance to colonization by Clostridium difficile. Gastroenterology 2014;146(6):1547–53.

[37] Chang JY, Antonopoulos DA, Kalra A, Tonelli A, Khalife WT, Schmidt TM, et al. Decreased diversity of the fecal microbiome in recurrent Clostridium difficile-associated diarrhea. J Infect Dis 2008;197(3):435–8.

[38] Eiseman B, Silen W, Bascom GS, Kauvar AJ. Fecal enema as an adjunct in the treatment of pseudomembranous enterocolitis. Surgery 1958;44(5):854–9.

[39] Cammarota G, Ianiro G, Gasbarrini A. Fecal microbiota transplantation for the treatment of clostridium difficile infection: a systematic review. J Clin Gastroenterol 2014;48(8):693–702.

[40] van Nood E, Vrieze A, Nieuwdorp M, Fuentes S, Zoetendal EG, de Vos WM, et al. Duodenal infusion of donor feces for recurrent Clostridium difficile. N Engl J Med 2013;368(5):407–15.

[41] Buffie CG, Bucci V, Stein RR, McKenney PT, Ling L, Gobourne A, et al. Precision microbiome reconstitution restores bile acid mediated resistance to clostridium difficile. Nature 2015;517(7533):205–8.

[42] Hsiao A, Ahmed AM, Subramanian S, Griffin NW, Drewry LL, Petri Jr. WA, et al. Members of the human gut microbiota involved in recovery from vibrio cholerae infection. Nature 2014;515(7527):423–6.

[43] Subramanian S, Huq S, Yatsunenko T, Haque R, Mahfuz M, Alam MA, et al. Persistent gut microbiota immaturity in malnourished Bangladeshi children. Nature 2014;510(7505):417–21.

[44] Moir S, Chun TW, Fauci AS. Pathogenic mechanisms of HIV disease. Annu Rev Pathol 2011;6:223–48.

[45] Brenchley JM, Price DA, Schacker TW, Asher TE, Silvestri G, Rao S, et al. Microbial translocation is a cause of systemic immune activation in chronic HIV infection. Nat Med 2006;12(12):1365–71.

[46] Handley SA, Thackray LB, Zhao G, Presti R, Miller AD, Droit L, et al. Pathogenic simian immunodeficiency virus infection is associated with expansion of the enteric virome. Cell 2012;151(2):253–66.

[47] Lozupone CA, Li M, Campbell TB, Flores SC, Linderman D, Gebert MJ, et al. Alterations in the gut microbiota associated with HIV-1 infection. Cell Host Microbe 2013;14(3):329–39.

[48] McHardy IH, Li X, Tong M, Ruegger P, Jacobs J, Borneman J, et al. HIV Infection is associated with compositional and functional shifts in the rectal mucosal microbiota. Microbiome 2013;1(1):26.

[49] Vujkovic-Cvijin I, Dunham RM, Iwai S, Maher MC, Albright RG, Broadhurst MJ, et al. Dysbiosis of the gut microbiota is associated with hiv disease progression and tryptophan catabolism. Sci Transl Med 2013;5(193):193ra91.

[50] Fonseca DM, Hand TW, Han SJ, Gerner MY, Glatman Zaretsky A, Byrd AL, et al. Microbiota-dependent sequelae of acute infection compromise tissue-specific immunity. Cell 2015;163(2):354–66.

[51] Hand TW, Dos Santos LM, Bouladoux N, Molloy MJ, Pagan AJ, Pepper M, et al. Acute gastrointestinal infection induces long-lived microbiota-specific T cell responses. Science 2012;337(6101):1553–6.

[52] Kamdar K, Khakpour S, Chen J, Leone V, Brulc J, Mangatu T, et al. Genetic and metabolic signals during acute enteric bacterial infection alter the microbiota and drive progression to chronic inflammatory disease. Cell Host Microbe 2016;19(1):21–31.

[53] Gosmann C, Anahtar MN, Handley SA, Farcasanu M, Abu-Ali G, Bowman BA, et al. Lactobacillus-deficient cervicovaginal bacterial communities are associated with increased HIV acquisition in young south african women. Immunity 2017;46(1):29–37.

[54] Klatt NR, Cheu R, Birse K, Zevin AS, Perner M, Noel-Romas L, et al. Vaginal bacteria modify HIV tenofovir microbicide efficacy in African women. Science 2017;356(6341):938–45.
[55] Taneva E, Sinclair S, Mesquita PM, Weinrick B, Cameron SA, Cheshenko N, et al. Vaginal microbiome modulates topical antiretroviral drug pharmacokinetics. JCI Insight 2018;3(13).
[56] Centers for Disease Control and Prevention (CDC). Ten great public health achievements—United States, 1900–1999. MMWR Morb Mortal Wkly Rep 1999;48(12):241–3.
[57] Junqueira AL, Tavares VR, Martins RM, Frauzino KV, da Costa e Silva AM, Minamisava R, et al. Safety and immunogenicity of hepatitis B vaccine administered into ventrogluteal vs. anterolateral thigh sites in infants: a randomised controlled trial. Int J Nurs Stud 2010;47(9):1074–9.
[58] Nakaya HI, Hagan T, Duraisingham SS, Lee EK, Kwissa M, Rouphael N, et al. Systems analysis of immunity to influenza vaccination across multiple years and in diverse populations reveals shared molecular signatures. Immunity 2015;43(6):1186–98.
[59] Querec TD, Akondy RS, Lee EK, Cao W, Nakaya HI, Teuwen D, et al. Systems biology approach predicts immunogenicity of the yellow fever vaccine in humans. Nat Immunol 2009;10(1):116–25.
[60] Mentzer AJ, O'Connor D, Pollard AJ, Hill AV. Searching for the human genetic factors standing in the way of universally effective vaccines. Philos Trans R Soc Lond B Biol Sci 2015;370(1671).
[61] Munoz FM, Van Damme P, Dinleyici E, Clarke E, Kampmann B, Heath PT, et al. The fourth international neonatal and maternal immunization symposium (INMIS 2017): toward integrating maternal and infant immunization programs. mSphere 2018;3(6).
[62] Nguyen QN, Himes JE, Martinez DR, Permar SR. The impact of the gut microbiota on humoral immunity to pathogens and vaccination in early infancy. PLoS Pathog 2016;12(12):e1005997.
[63] Oh JZ, Ravindran R, Chassaing B, Carvalho FA, Maddur MS, Bower M, et al. TLR5-mediated sensing of gut microbiota is necessary for antibody responses to seasonal influenza vaccination. Immunity 2014;41(3):478–92.
[64] Huda MN, Lewis Z, Kalanetra KM, Rashid M, Ahmad SM, Raqib R, et al. Stool microbiota and vaccine responses of infants. Pediatrics 2014;134(2):e362–72.
[65] Harris VC, Armah G, Fuentes S, Korpela KE, Parashar U, Victor JC, et al. Significant correlation between the infant gut microbiome and rotavirus vaccine response in rural Ghana. J Infect Dis 2017;215(1):34–41.
[66] Salk HM, Simon WL, Lambert ND, Kennedy RB, Grill DE, Kabat BF, et al. Taxa of the nasal microbiome are associated with influenza-specific IgA response to live attenuated influenza vaccine. PLoS One 2016;11(9):e0162803.
[67] Lynn MA, Tumes DJ, Choo JM, Sribnaia A, Blake SJ, Leong LEX, et al. Early-life antibiotic-driven dysbiosis leads to dysregulated vaccine immune responses in mice. Cell Host Microbe 2018;23(5):653–5.
[68] Young VB. Therapeutic manipulation of the microbiota: past, present, and considerations for the future. Clin Microbiol Infect 2016;22(11):905–9.
[69] Ianiro G, Maida M, Burisch J, Simonelli C, Hold G, Ventimiglia M, et al. Efficacy of different faecal microbiota transplantation protocols for Clostridium difficile infection: a systematic review and meta-analysis. United European Gastroenterol J 2018;6(8):1232–44.
[70] Quraishi MN, Widlak M, Bhala N, Moore D, Price M, Sharma N, et al. Systematic review with meta-analysis: the efficacy of faecal microbiota transplantation for the treatment of recurrent and refractory Clostridium difficile infection. Aliment Pharmacol Ther 2017;46(5):479–93.

[71] Halkjaer SI, Christensen AH, Lo BZS, Browne PD, Gunther S, Hansen LH, et al. Faecal microbiota transplantation alters gut microbiota in patients with irritable bowel syndrome: results from a randomised, double-blind placebo-controlled study. Gut 2018;67(12):2107–15.
[72] Kang DW, Adams JB, Gregory AC, Borody T, Chittick L, Fasano A, et al. Microbiota transfer therapy alters gut ecosystem and improves gastrointestinal and autism symptoms: an open-label study. Microbiome 2017;5(1):10.
[73] Singh R, de Groot PF, Geerlings SE, Hodiamont CJ, Belzer C, Berge I, et al. Fecal microbiota transplantation against intestinal colonization by extended spectrum beta-lactamase producing Enterobacteriaceae: a proof of principle study. BMC Res Notes 2018;11(1):190.
[74] Imdad A, Nicholson MR, Tanner-Smith EE, Zackular JP, Gomez-Duarte OG, Beaulieu DB, et al. Fecal transplantation for treatment of inflammatory bowel disease. Cochrane Database Syst Rev 2018;(11) CD012774.
[75] Dalal R, McGee RG, Riordan SM, Webster AC. Probiotics for people with hepatic encephalopathy. Cochrane Database Syst Rev 2017;(2) CD008716.
[76] Grev J, Berg M, Soll R. Maternal probiotic supplementation for prevention of morbidity and mortality in preterm infants. Cochrane Database Syst Rev 2018;(12) CD012519.
[77] Schwenger EM, Tejani AM, Loewen PS. Probiotics for preventing urinary tract infections in adults and children. Cochrane Database Syst Rev 2015;(12) CD008772.
[78] Fischbach MA. Microbiome: Focus on Causation and Mechanism. Cell 2018;174(4):785–90.
[79] Surana NK, Kasper DL. Moving beyond microbiome-wide associations to causal microbe identification. Nature 2017;552(7684):244–7.

Index

Note: Page numbers followed by *f* indicate figures, *t* indicate tables, and *b* indicate boxes.

D

E

F

I

K

L

M

N

R

S

Printed in the United States
By Bookmasters